Polymers from Renewable Resources

Polymers from Renewable Resources

Editors

Valentina Siracusa
Nadia Lotti
Michelina Soccio
Alexey L. Iordanskii

Basel • Beijing • Wuhan • Barcelona • Belgrade • Novi Sad • Cluj • Manchester

Editors

Valentina Siracusa
Department of Chemical Science
University of Catania
Catania, Italy

Nadia Lotti
Department of Civil, Chemical, Environmental and Materials Engineering
University of Bologna
Bologna, Italy

Michelina Soccio
Department of Civil, Chemical, Environmental and Materials Engineering
University of Bologna
Bologna, Italy

Alexey L. Iordanskii
N.N. Semenov Federal Research Center for Chemical Physics
Russian Academy of Sciences
Moscow, Russia

Editorial Office
MDPI
St. Alban-Anlage 66
4052 Basel, Switzerland

This is a reprint of articles from the Topic published online in the open access journals *Polymers* (ISSN 2073-4360), *Sustainability* (ISSN 2071-1050), and *Polysaccharides* (ISSN 2673-4176) (available at: https://www.mdpi.com/topics/polymers_renewable).

For citation purposes, cite each article independently as indicated on the article page online and as indicated below:

Lastname, A.A.; Lastname, B.B. Article Title. *Journal Name* **Year**, *Volume Number*, Page Range.

ISBN 978-3-0365-8578-9 (Hbk)
ISBN 978-3-0365-8579-6 (PDF)
doi.org/10.3390/books978-3-0365-8579-6

© 2023 by the authors. Articles in this book are Open Access and distributed under the Creative Commons Attribution (CC BY) license. The book as a whole is distributed by MDPI under the terms and conditions of the Creative Commons Attribution-NonCommercial-NoDerivs (CC BY-NC-ND) license.

Contents

About the Editors vii

Valentina Siracusa, Nadia Lotti, Michelina Soccio and Alexey L. Iordanskii
"Polymers from Renewable Resources": Key Findings from This Topic Special Issue
Reprinted from: *Polymers* **2023**, *15*, 3300, doi:10.3390/polym15153300 1

Mariana Ichim, Ioan Filip, Lucia Stelea, Gabriela Lisa and Emil Ioan Muresan
Recycling of Nonwoven Waste Resulting from the Manufacturing Process of Hemp Fiber-Reinforced Recycled Polypropylene Composites for Upholstered Furniture Products
Reprinted from: *Sustainability* **2023**, *15*, 3635, doi:10.3390/su15043635 7

Miriane Maria de Sousa, Vânia Miria C. Clemente, Rosilene Maria de S. Santos, Mariane Oliveira, José Osvaldo Ramos Silva, Laís Fernanda Batista, et al.
Development and Characterization of Sustainable Antimicrobial Films Incorporated with Natamycin and Cellulose Nanocrystals for Cheese Preservation
Reprinted from: *Polysaccharides* **2023**, *4*, 4, doi:10.3390/polysaccharides4010004 27

Beatriz Marjorie Marim, Janaina Mantovan, Gina Alejandra Gil-Giraldo, Jéssica Fernanda Pereira, Bruno Matheus Simões, Fabio Yamashita and Suzana Mali
Reactive Extrusion-Assisted Process to Obtain Starch Hydrogels through Reaction with Organic Acids
Reprinted from: *Polysaccharides* **2022**, *3*, 46, doi:10.3390/polysaccharides3040046 39

Stefania Marano, Emiliano Laudadio, Cristina Minnelli and Pierluigi Stipa
Tailoring the Barrier Properties of PLA: A State-of-the-Art Review for Food Packaging Applications
Reprinted from: *Polymers* **2022**, *14*, 1626, doi:10.3390/polym14081626 51

Juliana V. C. Azevedo, Esther Ramakers-van Dorp, Roman Grimmig, Berenika Hausnerova and Bernhard Möginger
Process-Induced Morphology of Poly(Butylene Adipate Terephthalate)/Poly(Lactic Acid) Blown Extrusion Films Modified with Chain-Extending Cross-Linkers
Reprinted from: *Polymers* **2022**, *14*, 1939, doi:10.3390/polym14101939 95

Mitul Kumar Patel, Marta Zaccone, Laurens De Brauwer, Rakesh Nair, Marco Monti, Vanesa Martinez-Nogues, et al.
Improvement of Poly(lactic acid)-Poly(hydroxy butyrate) Blend Properties for Use in Food Packaging: Processing, Structure Relationships
Reprinted from: *Polymers* **2022**, *14*, 5104, doi:10.3390/polym14235104 107

Svetlana Rogovina, Lubov Zhorina, Anastasia Yakhina, Alexey Shapagin, Alexey Iordanskii and Alexander Berlin
Hydrolysis, Biodegradation and Ion Sorption in Binary Biocomposites of Chitosan with Polyesters: Polylactide and Poly(3-Hydroxybutyrate)
Reprinted from: *Polymers* **2023**, *15*, 645, doi:10.3390/polym15030645 123

Tej Singh, Amar Patnaik, Lalit Ranakoti, Gábor Dogossy and László Lendvai
Thermal and Sliding Wear Properties of Wood Waste-Filled Poly(Lactic Acid) Biocomposites
Reprinted from: *Polymers* **2022**, *14*, 2230, doi:10.3390/polym14112230 137

Olga Alexeeva, Anatoliy Olkhov, Marina Konstantinova, Vyacheslav Podmasterev, Ilya Tretyakov, Tuyara Petrova, et al.
Improvement of the Structure and Physicochemical Properties of Polylactic Acid Films by Addition of Glycero-(9,10-trioxolane)-Trialeate
Reprinted from: *Polymers* **2022**, *14*, 3478, doi:10.3390/polym14173478 153

Jindi Xu, Dongying Hu, Qi Zheng, Qiulu Meng and Ning Li
The Distribution and Polymerization Mechanism of Polyfurfuryl Alcohol (PFA) with Lignin in Furfurylated Wood
Reprinted from: *Polymers* **2022**, *14*, 1071, doi:10.3390/polym14061071 173

Bruno Esteves, Pedro Aires, Umut Sen, Maria da Glória Gomes, Raquel P. F. Guiné, Idalina Domingos, et al.
Particleboard Production from *Paulownia tomentosa* (Thunb.) Steud. Grown in Portugal
Reprinted from: *Polymers* **2023**, *15*, 1158, doi:10.3390/polym15051158 187

Wilasinee Sriprom, Adilah Sirivallop, Aree Choodum, Wadcharawadee Limsakul and Worawit Wongniramaikul
Plastic/Natural Fiber Composite Based on Recycled Expanded Polystyrene Foam Waste
Reprinted from: *Polymers* **2022**, *14*, 2241, doi:10.3390/polym14112241 203

Ivan V. Nazarov, Danil P. Zarezin, Ivan A. Solomatov, Anastasya A. Danshina, Yulia V. Nelyubina, Igor R. Ilyasov and Maxim V. Bermeshev
Chiral Polymers from Norbornenes Based on Renewable Chemical Feedstocks
Reprinted from: *Polymers* **2022**, *14*, 5453, doi:10.3390/polym14245453 215

Pengcheng Yi, Jingrong Chen, Junyao Chang, Junbo Wang, Ying Lei, Ruobing Jing, et al.
Self-Healable, Strong, and Tough Polyurethane Elastomer Enabled by Carbamate-Containing Chain Extenders Derived from Ethyl Carbonate
Reprinted from: *Polymers* **2022**, *14*, 1673, doi:10.3390/polym14091673 233

Hsiao-Lin Chien, Yi-Ting Tsai, Wei-Sung Tseng, Jin-An Wu, Shin-Liang Kuo, Sheng-Lung Chang, et al.
Biodegradation of PBSA Films by Elite *Aspergillus* Isolates and Farmland Soil
Reprinted from: *Polymers* **2022**, *14*, 1320, doi:10.3390/polym14071320 247

Nussana Lehman, Akarapong Tuljittraporn, Ladawan Songtipya, Nattapon Uthaipan, Karnda Sengloyluan, Jobish Johns, et al.
Influence of Non-Rubber Components on the Properties of Unvulcanized Natural Rubber from Different Clones
Reprinted from: *Polymers* **2022**, *14*, 1759, doi:10.3390/polym14091759 269

Sok Kim, Yun Hwan Park and Yoon-E Choi
Amination of Non-Functional Polyvinyl Chloride Polymer Using Polyethyleneimine for Removal of Phosphorus from Aqueous Solution
Reprinted from: *Polymers* **2022**, *14*, 1645, doi:10.3390/polym14091645 283

Josefina Chipón, Kassandra Ramírez, José Morales and Paulo Díaz-Calderón
Rheological and Thermal Study about the Gelatinization of Different Starches (Potato, Wheat and Waxy) in Blend with Cellulose Nanocrystals
Reprinted from: *Polymers* **2022**, *14*, 1560, doi:10.3390/polym14081560 299

Mengqi Zhao, Xiaoqing Ma, Yuxi Chao, Dejun Chen and Yinnian Liao
Super-Hydrophobic Magnetic Fly Ash Coated Polydimethylsiloxane (MFA@PDMS) Sponge as an Absorbent for Rapid and Efficient Oil/Water Separation
Reprinted from: *Polymers* **2022**, *14*, 3726, doi:10.3390/polym14183726 313

About the Editors

Valentina Siracusa

Valentina Siracusa received her degree in Industrial Chemistry from the University of Catania (Italy), at 23 years old. She completed her PhD and post-PhD study working on the synthesis and characterization of innovative polyesters, used in the engineering field. After a period as a Lecturer in "Chemistry and Materials" for Engineering, since 2006, she has been an Associate Professor in Chemistry for Engineering at the University of Catania (Italy) and Invited Professor on "Life Cycle Assessment Study (LCA)" courses at the University of Bologna (Italy). She collaborates in several research projects, both for academic than industrial interest, on topics such as recycling, the environment, food packaging, graphene for packaging application, nanoparticles, and polymer drug delivery. She also collaborates with national and international research groups on biopolymers used in the field of food packaging, for the modified atmosphere packaging of fresh foods, with Life Cycle Assessment study (with SimaPro software). She is the author of more than 110 papers in high impact factor scientific journals, the author of several book chapters for Wiley, Springer, and Elsevier, and she is author of articles for a special Module of Elsevier Encyclopedia, as well as an Editorial Board Member and Lead Guest Editor of several international journals. Her research interests include the following topics: synthesis and full characterization of biodegradable and bio-based polymers; gas barrier behavior in standard as well as in moisture condition at different temperature; life cycle assessment (LCA) study of polymers; thermal and photo degradation behavior of packaging materials analyzed during food shelf-life study.

Nadia Lotti

Nadia Lotti has been a Full Professor at the University of Bologna since 2021. She conducts her research in the field of chemistry and technology of polymers. Her research interests are directed, in particular, toward the synthesis and molecular, solid and melt characterization of polycondensates and the chemical modification of commercial polymers. She currently focuses her research on synthetic plastic from renewable sources for packaging applications and on biopolymers for biomedical use. Engaged in national and international research projects, she collaborates with several European institutes. Author of numerous publications, she is a peer reviewer of several polymer science journals. She is the author/co-author of 186 articles, 4 conference papers, 4 reviews, 1 book chapter (4398 citations, h=37), and 150 communications in international and national conferences.

Michelina Soccio

Michelina Soccio has been Assistant Professor at the Civil Chemical Environmental and Materials Engineering Department of University of Bologna since 2017. Her research activity concerns the polymer science field related to the design, synthesis and characterization of new biopolymers as well as the modification of commercial polymers to address the requirements of the final applications, mainly packaging and biomedicine. She has collaborated in several national and international research projects. She is the author of 125 articles in peer-reviewed journals (h-index: 32, citations: 2768 source Scopus) and has attended over 120 national and international conferences.

Alexey L. Iordanskii

Alexey L. Iordanskii currently works at the Semenov Institute of Chemical Physics, Russian Academy of Sciences. His scientific activities include the physical concept of transport phenomena in ultrathin fibers, planar films and other micro- and nano-scaled polymer objects in biomedicine, environmental science and packaging. A.L. Iordanskii graduated from Lomonosov Moscow University and then earned a Ph.D. and DSc sequentially in the areas of polymer materials chemistry and physical chemistry of biopolymers. His current project focuses on biodegradable composites and ultrathin fibers for biomedicine and environmental applications, concerning their morphology, diffusion, and molecular dynamics. Accordingly, his research activities have been published 250+ international scientific publications.

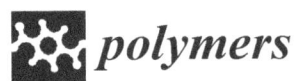

Editorial

"Polymers from Renewable Resources": Key Findings from This Topic Special Issue

Valentina Siracusa [1,*], Nadia Lotti [2], Michelina Soccio [2] and Alexey L. Iordanskii [3,*]

1. Department of Chemical Sciences, Università degli Studi di Catania, 95125 Catania, Italy
2. Department of Civil, Chemical, Environmental and Materials Engineering of the University of Bologna, 40131 Bologna, Italy; nadia.lotti@unibo.it (N.L.); m.soccio@unibo.it (M.S.)
3. N. N. Semenov Federal Research Center for Chemical Physics, Russian Academy of Sciences, 119991 Moscow, Russia
* Correspondence: vsiracus@dmfci.unict.it (V.S.); aljordan08@gmail.com (A.L.I.); Tel.: +39-3387275526 (V.S.)

Citation: Siracusa, V.; Lotti, N.; Soccio, M.; Iordanskii, A.L. "Polymers from Renewable Resources": Key Findings from This Topic Special Issue. *Polymers* 2023, 15, 3300. https://doi.org/10.3390/polym15153300

Received: 28 July 2023
Accepted: 1 August 2023
Published: 4 August 2023

Copyright: © 2023 by the authors. Licensee MDPI, Basel, Switzerland. This article is an open access article distributed under the terms and conditions of the Creative Commons Attribution (CC BY) license (https://creativecommons.org/licenses/by/4.0/).

1. Introduction

The Food and Agriculture Organization of the United Nations (FAO) has estimated that about one-third of the food produced for human consumption is currently lost or wasted, resulted in an estimated approximately USD 750 billion of direct costs for food producers every year. Of course, packaging plays a key role in reducing food waste and in food preservation and protection due to its innovative properties and eminent versatility of applications. Polymers and their composites used as packaging plastics have changed global needs and to the progress of modern technologies. However, plastic packaging is still considered in the framework of the outdated paradigm, "make, use, and dispose it", with great material and energy losses incurred after the disposal of plastics. Consequently, hundreds of millions of tons of plastics are lost due to disposal practices and are discarded throughout the environment. The most interesting challenge for the future is the use of food waste and waste in general as a renewable resource to produce new bio-based and/or biodegradable materials with well-tailored properties. The circular economy is based on a new, important concept: eliminate waste, rebuild natural capital, and create economic value by using and not consuming resources. This game-changing strategy includes the promotion of sustainable polymer technologies that are able to exclude plastics from fossil resources, use renewable resources obtained from food or natural wastes as principal feedstocks, reduce the presence of plastic wastes in the environment, and increase the quality and uptake of recycling.

2. Scope of This Special Issue

The purpose of this special collection was to cover all the topics related to biomaterials obtained from renewable resources, including innovative feedstock, polymerization processes, full characterizations, final applications and life cycle assessment analyses in order to share all the academic and industrial efforts related to new and innovative sustainable materials and technologies. Nineteen papers were published in this Topic Special Issue: one in *Sustainability*, two in *Polysaccharides*, and sixteen in *Polymers*. A brief account of the research presented in the Special Issue is provided below.

3. Overview of the Papers Included in This Topic Special Issue

Ichim and collaborators published an article in *Sustainability* [1]. They investigated the properties of composite materials that contain 50% hemp fibers as a reinforcement and 50% recycled polypropylene (rPP) as a matrix (50/50 hemp/rPP). These were produced using thermoforming in collaboration with a Romanian furniture manufacturing company. The effect of the addition of a compatibilizer, maleic anhydride, was investigated. The aim was to understand how the waste recycling could reduce the environmental impact caused by waste landfilling or incineration. Composites incorporating 50% and 100% recycled fibers were treated with 2.5% and 5% maleated polypropylene (MAPP), respectively, and compared to both the untreated composites. The addition of recycled fibers decreased the mechanical performance, while the addition of just 5% of MAPP caused an improvement in the mechanical properties of the composites containing both 50 and 100% recycled fibers. The selected materials obtained by replacing wood with composite materials allow the manufacturing of high-quality sustainable products at a low price, with significant environmental benefits for climate regulation and pollution prevention, reducing the pressure on forests, and closing the loop in the circular economy.

Two articles were published in *Polysaccharides* by authors from Brazil. The first one [2] was focused on the synthesis and characterization of antimicrobial films based on methyl-cellulose/poly (vinyl alcohol) (MC/PVOH) incorporated with cellulose nanocrystals (CNC) and natamycin (MC/PVOH/CNC/natamycin blends films) for cheese preservation. The aim was to direct the attention to the possibility of guaranteeing a final product with a high microbiological quality, improving the product's shelf life, minimizing the risks to consumers' health, as well as reducing the economic losses due to food waste. To do this, new technologies were considered, such as the production of active packaging with the incorporation of antimicrobial substances into the polymer matrix of the packaging, thereby reducing the addition of preservatives directly into the food formulation. In order to improve the use of bioplastic, CNC nanostructures obtained from a renewable source (cellulose) were used. The mechanical and optical properties and avoidance of oxygen and water vapor permeation were evaluated, while the antimicrobial activity was tested against fungi and yeasts in vitro. Despite the incorporation of CNC, the films' tensile strength was increased, and their addition did not influence the water vapor and oxygen barrier behavior. The incorporation of natamycin, an antimicrobial agent, had a negative effect on the optical and mechanical properties of the films, probably due to the lower compatibility among the antimicrobial and the polymers. The authors concluded that the films showed the potential to be applied in cheese preservation, thanks to their beneficial effect against fungi and yeasts, but more studies must be conducted in order to improve the mechanical, optical, and barrier properties of these active films.

The second paper [3] reports on the use of a totally green process based on reactive extrusion or the production of cassava starch hydrogels via a reaction with two organic crosslinking agents, citric (CA) and tartaric (TA) acids. Hydrogels are materials that can be produced using natural or synthetic polymers in different formats, including films, membranes, powders, and micro- or nanogels, with the ability of retain water or biological fluids without dissolving them. Starch is a biodegradable, non-toxic, and inexpensive biopolymer that is available worldwide. In order to obtain food-grade starch hydrogels, CA and TA non-toxic reagents were safely used in reactive extrusion, without using other reagents. As reported by the authors, reactive extrusion is a continuous process that has commercial viability; it is easy to adapt to industrial scales, offering a short reaction time (2–3 min). The physicochemical and microstructural properties of the obtained films confirmed the suitability of this green procedure in obtaining food-grade starch hydrogels with good performances, greatly reducing the processing time and effluent generation compared to those of conventional processes.

Sixteen other papers were published in *Polymers*. In particular:

- Six manuscripts described the results based on the modification of Polylactic acid (PLA), one of the most frequently used bio-based materials in the food packaging industry, which is applied for the production of disposable tableware, vegetables packaging, and fast food containers. Marano and collaborators [4] reported on the state-of-the-art barrier properties of poly(lactide) (PLA). Until now, many efforts have been made to meet precise functional requirements, such as suitable thermal, mechanical, and gas barrier properties, in order to guarantee the foods' quality and safety during the whole food shelf-life. In particular, the authors focused their attention on reviewing relevant strategies to tailor the barrier properties of PLA-based materials, with the ultimate goal of providing a general guide for the design of PLA-based packaging materials with the desired mass transfer properties for water vapor, oxygen, and/or carbon dioxide. After the revision of 295 articles, the authors concluded that several strategies could be considered in order to tailor the final PLA barrier properties, such as crystallization control and co-polymerization. Azevedo and collaborators [5] reported their results regarding the process-induced changes in the morphologies of biodegradable polybutylene adipate terephthalate (PBAT) and polylactic acid (PLA) blends modified with four multifunctional chain-extending crosslinkers (CECLs). Investigations on the PBAT/PLA blends showed that the interfacial compatibility between PLA and PBAT is poor, but it can be improved using compatibilizers in order to optimize their processability and usage performance. The CECL introduction into the blend changed the thickness of the PLA fibrils, modified the interface adhesion, and altered the deformation behavior of the PBAT matrix from brittle to ductile, proving that CECLs react selectively with PBAT, PLA, and their interface. The most synergetic effect was obtained when 1,3-phenylenebisoxazoline was used as a crosslinker. Patel and collaborators [6] described the properties of extruded Poly(lactic acid)-poly(hydroxybutyrate) (PLA-PHB)-based nanocomposite films, with bio-based additives (CNCs and ChNCs) and a oligomer lactic acid (OLA) compatibilizer included in the films at a pilot scale. The aim was to identify suitable material formulations and nanocomposite production processes for film production at a larger scale for food packaging applications. As is known, for foods sensitive to oxidation, packages with low oxygen permeability are preferred. It is well known that the crystalline phase has a significant influence on the oxygen barrier properties of a material. As a consequence, increasing the crystallinity of PLA by blending PLA with other more crystalline biopolymers, such as poly(hydroxy alkanoates) (PHAs), for example, has subsequently received a lot of attention in the food packaging industry. The most common PHA is poly(hydroxybutyrate) (PHB). In addition, the synergic effect of better-dispersed ChNCs with the assistance of OLA resulted in increased crystallinity, and, thereby, an improvement in the oxygen barrier performance modified films as compared to that of neat PLA. Rogovina and collaborators [7] analyzed the hydrolysis resistance, biodegradation-in-soil, and ion sorption behavior of film binary polyester–chitosan composites, such as polylactide (PLA)–chitosan and poly(3-hydroxybutyrate) (PHB)–chitosan. Chitosan combined with biodegradable polyesters (PLA and PHB) has been studied as a novel functional material capable of performing in aqueous environments and soil. They found that PHB-chitosan composites are more stable during acid hydrolysis, they demonstrate better degradation in soil, and have a higher capacity for iron ions sorption than PLA-chitosan composites do. Singh and collaborators [8] studied the effects of Indian rosewood waste on the thermal and dry sliding wear properties of poly(lactic acid) (PLA) biocomposites. The inclusion of some natural fibers and sustainable biocarbon was reported to enhance the wear resistance of PLA-based biocomposites. The results demonstrated that the thermal stability of the PLA biocomposites increased with an increase in the wood waste loading, the wear of the biocomposites increased with an increase in the load and sliding velocity, and a wood waste content of 46.82% was the most dominant parameter for controlling the wear of the biocomposites. Alexeeva and collaborators [9] reported a result obtained regarding films made via the introduction of Glycero-(9,10-trioxolane)-trioleate (ozonide of oleic acid triglyceride, OTOA) into polylactic acid (PLA) films. The morphological, mechanical, thermal, and water absorption properties of PLA films after OTOA addition were studied to understand their suitability for packaging as well as biomedical applications. It was found that OTOA acts as a plasticizer and leads to an increase in PLA segmental mobility, which, in turn, contributes to changes in the thermodynamic and mechanical properties of PLA films. Eventually, the obtained results evidence that the morphological, thermodynamic, and mechanical properties of PLA + OTOA films could be controlled according to the OTOA content in the films.

- Two articles described the use and properties of wood treated in order to improve its mechanical behavior. In particular, Xu and collaborators [10] described a green process used to improve fast-growing wood's durability via impregnation modification. Generally, impregnation with low-molecular-weight resins can improve the properties of wood, but these resins release free phenols and free aldehydes into the environment. In contrast, furfuryl alcohol (FA) is a green modification agent derived from pentose-rich agricultural residues and releases fewer volatile organic compounds or polycyclic aromatic hydrocarbons during the combustion and degradation of FA-modified wood. In this study, balsa wood was first immersed in an FA solution, followed by in situ polymerization to obtain furfurylated wood. The obtained samples were investigated in order to understand the mechanism of interaction (crosslinking) between FA and lignin. Esteves and collaborators [11] studied the properties of particleboards made from very young Paulownia trees from Portuguese plantations. Paulownia wood is easily air dried, without severe drying defects occurring. It is more resistant to fires than other fast-growing species are due to its high ignition temperature, high water content in the fire season, and large leaves. In this study, single-layer particleboards were made from 3-year-old Paulownia trees using different processing parameters and different board composition in order to determine the best properties, such as the mechanical properties and lower thermal conductivity, for use in dry environments. Its fast-growing rate would allow sustainable forest management to be achieved since the wood can be harvested sooner than traditional wood species can be.
- Eight articles described different materials and processes with environmentally friendly attributes. Sriprom and collaborators [12] worked on a novel, reinforced, recycled expanded polystyrene (r-EPS) foam/natural fiber composite obtained via a dissolution method. Coconut husk fiber (coir) and banana stem fiber (BSF) were used as the reinforcement materials in order to enhance the mechanical properties. Nazarov and collaborators [13] described the synthesis of different optically active polymers used as materials for dense enantioselective membranes, as well as chiral stationary phases for gas and liquid chromatography. In particular, chiral polymers were obtained from Norbornenes using renewable chemical feedstocks. As result, a series of high-molecular-weight polymers with good film-forming properties was successfully obtained. Yi and collaborators [14] described the use of two diol chain extenders, bis(2-hydroxyethyl) (1,3-pheny-lene-bis-(methylene)) dicarbamate (BDM) and bis(2-hydroxyethyl) (methylenebis(cyclohexane-4,1-diy-l)) dicarbamate (BDH), required to construct self-healing thermoplastic polyurethane elastomers (TPU). Self-healing polyurethane materials are widely applied in electronic skin, intelligent sensors, biomedical materials, and many other areas. BDM and BDH were successfully synthesized from inexpensive raw materials and incorporated into the polyurethane backbone, resulting in an innovative strategy to explore elastomers with good mechanical properties and an excellent self-healing ability. Chien and collaborators [15] studied the degradation ability of a Polybutylene succinate-co-adipate (PBSA) biodegradable polymer used for packaging and mulching. In this study, two elite fungal strains for PBSA degradation from farmlands, i.e., *Aspergillus fumigatus* L30 and *Aspergillus terreus* HC, were isolated. Additionally, the biodegradability of PBSA films buried in farmland soils was evaluated. Lehman and collaborators [16] studied the influence of non-rubber components on the properties of unvulcanized natural rubber. Natural rubber latex (NRL) is obtainable from rubber plants in the form of latex. It can be contaminated by micro-organisms because it contains various nutritious substances otherwise known as non-rubber components. For this study, the fresh natural latex from four different clones (RRIM600, RRIT251, PB235, and BPM24) was chosen. Kim and collaborators [17] described the use of a polyvinyl chloride (PVC)–polyethyleneimine (PEI) composite fiber (PEI-PVC) as a recoverable adsorbent for the removal of phosphorus from aqueous phases. The adsorptive removal of phosphorus from discharged effluents has been recognized as one of the most promising solutions

in the prevention of eutrophication. In addition, the regenerated PEI-PVC maintained a phosphorus sorption capacity almost equal to that of the first use through the desorption process, and the PEI-PVC fiber did not elute any toxic chlorines into the solution during light irradiation. Chipón and collaborators [18] analyzed the effect of cellulose nanocrystals (CNCs) on the gelatinization of different starches (potato, wheat, and waxy maize) via the characterization of the rheological and thermal properties of starch–CNC blends. Starch and cellulose are the most widely distributed polymers in nature, with starch being found in the form of granules, which are energy reservoirs for plants and cellulose and are a part of the cell walls in plant tissues. Despite the effect of CNCs on the physical properties and functionality of different starches being described in the literature, studies describing the role of CNCs during the gelatinization of starch are scarce. Therefore, this work aimed to study the gelatinization of starches from different sources in the presence of CNCs produced from cotton cellulose pulp. Zhao and collaborators [19] worked on a very interesting project. They prepared a magnetic fly ash/polydimethylsiloxane (MFA@PDMS) sponge using simple dip-coating PDMS-containing ethanol in a magnetic fly ash aqueous suspension that solidified. The presence of the PDMS matrix made the sponge super-hydrophobic, with a significant lubricating oil absorption capacity; notably, it took only 10 min for the material to adsorb six times its own weight in n-hexane (oil phase). Considering the sizable interest in the environment, the authors reported that MFA@PDMS sponge demonstrated outstanding recyclability and stability since no decline in its absorption efficiency was observed after more than eight cycles. The preparation process was simple, and the resulting magnetic sponges were super-hydrophobic and super-lipophilic. Additionally, the magnetic properties of the material make it possible to separate oil–water mixtures using an external magnetic field. The tested sponge showed good mechanical stability, oil stability, and reusability.

4. Conclusions

All the research results published in this Special Issue reported innovative procedures used to obtain new materials based on the concept of "waste to waste". Due to global industrialization, water, soil, and air pollution are among the most difficult challenges today. They are not only harmful to the natural ecosystem, but also have long-term adverse effects on human health and the economy.

Author Contributions: Writing—original draft preparation, V.S.; review and editing, V.S., N.L., M.S. and A.L.I. All authors have read and agreed to the published version of the manuscript.

Funding: This research received no external funding.

Acknowledgments: Valentina Siracusa wishes to express her most sincere gratitude to her friends and Co-Topic Editors because, without their commitment and support work, the Topic would not have been possible. Special thanks go to all the Authors for the high quality of the manuscripts they contributed. Their high-quality research work will make this Topic a valid tool for spreading knowledge on sustainable materials from renewable resources. Finally, a heartfelt thanks goes to MDPI.

Conflicts of Interest: The authors declare no conflict of interest.

References

1. Ichim, M.; Filip, I.; Stelea, L.; Lisa, G.; Muresan, E.I. Recycling of Nonwoven Waste Resulting from the Manufacturing Process of Hemp Fiber-Reinforced Recycled Polypropylene Composites for Upholstered Furniture Products. *Sustainability* **2023**, *15*, 3635. [CrossRef]
2. de Sousa, M.M.; Clemente, V.M.C.; Santos, R.M.d.S.; Oliveira, M.; Silva, J.O.R.; Batista, L.F.; Marques, C.S.; de Souza, A.L.; Medeiros, É.A.A.; Soares, N.d.F.F. Development and Characterization of Sustainable Antimicrobial Films Incorporated with Natamycin and Cellulose Nanocrystals for Cheese Preservation. *Polysaccharides* **2023**, *4*, 53–64. [CrossRef]
3. Marim, B.M.; Mantovan, J.; Gil-Giraldo, G.A.; Pereira, J.F.; Simões, B.M.; Yamashita, F.; Mali, S. Reactive Extrusion-Assisted Process to Obtain Starch Hydrogels through Reaction with Organic Acids. *Polysaccharides* **2022**, *3*, 792–803. [CrossRef]

4. Marano, S.; Laudadio, E.; Minnelli, C.; Stipa, P. Tailoring the Barrier Properties of PLA: A State-of-the-Art Review for Food Packaging Applications. *Polymers* **2022**, *14*, 1626. [CrossRef] [PubMed]
5. Azevedo, J.V.C.; Ramakers-van Dorp, E.; Grimmig, R.; Hausnerova, B.; Möginger, B. Process-Induced Morphology of Poly(Butylene Adipate Terephthalate)/Poly(Lactic Acid) Blown Extrusion Films Modified with Chain-Extending Cross-Linkers. *Polymers* **2022**, *14*, 1939. [CrossRef] [PubMed]
6. Patel, M.K.; Zaccone, M.; De Brauwer, L.; Nair, R.; Monti, M.; Martinez-Nogues, V.; Frache, A.; Oksman, K. Improvement of Poly(lactic acid)-Poly(hydroxy butyrate) Blend Properties for Use in Food Packaging: Processing, Structure Relationships. *Polymers* **2022**, *14*, 5104. [CrossRef] [PubMed]
7. Rogovina, S.; Zhorina, L.; Yakhina, A.; Shapagin, A.; Iordanskii, A.; Berlin, A. Hydrolysis, Biodegradation and Ion Sorption in Binary Biocomposites of Chitosan with Polyesters: Polylactide and Poly(3-Hydroxybutyrate). *Polymers* **2023**, *15*, 645. [CrossRef] [PubMed]
8. Singh, T.; Patnaik, A.; Ranakoti, L.; Dogossy, G.; Lendvai, L. Thermal and Sliding Wear Properties of Wood Waste-Filled Poly(Lactic Acid) Biocomposites. *Polymers* **2022**, *14*, 2230. [CrossRef] [PubMed]
9. Alexeeva, O.; Olkhov, A.; Konstantinova, M.; Podmasterev, V.; Tretyakov, I.; Petrova, T.; Koryagina, O.; Lomakin, S.; Siracusa, V.; Iordanskii, A.L. Improvement of the Structure and Physicochemical Properties of Polylactic Acid Films by Addition of Glycero-(9,10-trioxolane)-Trialeate. *Polymers* **2022**, *14*, 3478. [CrossRef] [PubMed]
10. Xu, J.; Hu, D.; Zheng, Q.; Meng, Q.; Li, N. The Distribution and Polymerization Mechanism of Polyfurfuryl Alcohol (PFA) with Lignin in Furfurylated Wood. *Polymers* **2022**, *14*, 1071. [CrossRef] [PubMed]
11. Esteves, B.; Aires, P.; Sen, U.; Gomes, M.d.G.; Guiné, R.P.F.; Domingos, I.; Ferreira, J.; Viana, H.; Cruz-Lopes, L.P. Particleboard Production from *Paulownia tomentosa* (Thunb.) Steud. Grown in Portugal. *Polymers* **2023**, *15*, 1158. [CrossRef] [PubMed]
12. Sriprom, W.; Sirivallop, A.; Choodum, A.; Limsakul, W.; Wongniramaikul, W. Plastic/Natural Fiber Composite Based on Recycled Expanded Polystyrene Foam Waste. *Polymers* **2022**, *14*, 2241. [CrossRef] [PubMed]
13. Nazarov, I.V.; Zarezin, D.P.; Solomatov, I.A.; Danshina, A.A.; Nelyubina, Y.V.; Ilyasov, I.R.; Bermeshev, M.V. Chiral Polymers from Norbornenes Based on Renewable Chemical Feedstocks. *Polymers* **2022**, *14*, 5453. [CrossRef] [PubMed]
14. Yi, P.; Chen, J.; Chang, J.; Wang, J.; Lei, Y.; Jing, R.; Liu, X.; Sun, A.; Wei, L.; Li, Y. Self-Healable, Strong, and Tough Polyurethane Elastomer Enabled by Carbamate-Containing Chain Extenders Derived from Ethyl Carbonate. *Polymers* **2022**, *14*, 1673. [CrossRef] [PubMed]
15. Chien, H.-L.; Tsai, Y.-T.; Tseng, W.-S.; Wu, J.-A.; Kuo, S.-L.; Chang, S.-L.; Huang, S.-J.; Liu, C.-T. Biodegradation of PBSA Films by Elite *Aspergillus* Isolates and Farmland Soil. *Polymers* **2022**, *14*, 1320. [CrossRef] [PubMed]
16. Lehman, N.; Tuljittraporn, A.; Songtipya, L.; Uthaipan, N.; Sengloyluan, K.; Johns, J.; Nakaramontri, Y.; Kalkornsurapranee, E. Influence of Non-Rubber Components on the Properties of Unvulcanized Natural Rubber from Different Clones. *Polymers* **2022**, *14*, 1759. [CrossRef] [PubMed]
17. Kim, S.; Park, Y.H.; Choi, Y.-E. Amination of Non-Functional Polyvinyl Chloride Polymer Using Polyethyleneimine for Removal of Phosphorus from Aqueous Solution. *Polymers* **2022**, *14*, 1645. [CrossRef] [PubMed]
18. Chipón, J.; Ramírez, K.; Morales, J.; Díaz-Calderón, P. Rheological and Thermal Study about the Gelatinization of Different Starches (Potato, Wheat and Waxy) in Blend with Cellulose Nanocrystals. *Polymers* **2022**, *14*, 1560. [CrossRef] [PubMed]
19. Zhao, M.; Ma, X.; Chao, Y.; Chen, D.; Liao, Y. Super-Hydrophobic Magnetic Fly Ash Coated Polydimethylsiloxane (MFA@PDMS) Sponge as an Absorbent for Rapid and Efficient Oil/Water Separation. *Polymers* **2022**, *14*, 3726. [CrossRef] [PubMed]

Disclaimer/Publisher's Note: The statements, opinions and data contained in all publications are solely those of the individual author(s) and contributor(s) and not of MDPI and/or the editor(s). MDPI and/or the editor(s) disclaim responsibility for any injury to people or property resulting from any ideas, methods, instructions or products referred to in the content.

Article

Recycling of Nonwoven Waste Resulting from the Manufacturing Process of Hemp Fiber-Reinforced Recycled Polypropylene Composites for Upholstered Furniture Products

Mariana Ichim [1], Ioan Filip [2], Lucia Stelea [2], Gabriela Lisa [3] and Emil Ioan Muresan [3,*]

1 Faculty of Industrial Design and Business Management, Gheorghe Asachi Technical University of Iasi, 29 Prof. Dr. Doc. D. Mangeron Blvd, 700050 Iasi, Romania
2 Taparo Company SA, 198 Borcut Street, 435600 Târgu Lăpuș, Romania
3 "Cristofor Simionescu" Faculty of Chemical Engineering and Environmental Protection, Gheorghe Asachi Technical University of Iasi, 73 Prof. Dr. Doc. D. Mangeron Blvd, 700050 Iasi, Romania
* Correspondence: eimuresan@yahoo.co.uk

Abstract: Waste recycling is a solution that reduces the environmental impact of waste landfilling or incineration. The aim of this paper is to investigate both the effect of incorporating recycled fibers obtained by defibrating 50/50 hemp/rPP nonwoven waste and the effect of the compatibilizer on the properties of composite materials. Composites incorporating 50% and 100% recycled fibers were treated with 2.5% and 5% maleated polypropylene (MAPP), respectively, and compared to both the untreated composites and the composite obtained by thermoforming from the nonwovens that generated the waste. The incorporation of 50% and 100% recycled fibers into composites decreased the tensile strength by 17.1–22.6%, the elongation at break by 12.4–20.1%, the flexural strength by 6.6–9%, and flexural modulus by 10.3–37%. The addition of 5% MAPP showed the greatest improvements in mechanical properties of composites containing 100% recycled fibers, as follows: 19.2% increase in tensile strength, 3.8% increase in flexural strength, and 14.8% increase in flexural modulus. Thermal analysis established that at temperatures ranging between 20 °C and 120 °C, the composites were thermally stable. SEM analysis revealed good coverage of the reinforcing fibers, and EDX analysis confirmed the presence of the compatibilizing agent in the structure of the composite material.

Keywords: nonwoven waste; recycling; recycled polypropylene; hemp reinforcement; composite material; compatibilization; MAPP

1. Introduction

In recent years, the usage of composite materials in the furniture industry has evolved continuously due to their benefits over traditional materials such as wood, metal, and plastics. Composite materials are very attractive for the furniture industry because they offer great flexibility in the geometrical and aesthetical design of three-dimensionally shaped products [1–4]. The combination of different possible reinforcements and matrices provides composite materials with a wide range of properties that are suitable for their end use (chairs, stools, tables, shelves, cupboards, sofa frames, racks, etc.). Composite materials for furniture applications have been obtained using different reinforcement fibers, both natural and synthetic, such as carbon [5], glass [6], hemp [7–9], flax [10], jute [11], kenaf [12], coir [13–15], oil palm [16,17], banana [18], and wood [19–21]. Nowadays, carbon and fiberglass reinforcements are increasingly being replaced by natural fibers that make composite materials environmentally friendly [22–25], sustainable [26–29], lightweight [30–32], and affordable [33–36], in addition to providing good mechanical properties [37–40].

Composite materials that contain 50% hemp fibers as reinforcement and 50% recycled polypropylene (rPP) as a matrix were produced by thermoforming by a Romanian furniture manufacturing company. The selected materials allow for the manufacturing of good-quality sustainable products for a low price. Traditionally, the structural frame of an upholstered furniture product is made of wood. Due to the environmental and economic benefits, wood is being increasingly replaced by composite material. Compared to wood, which has a long production cycle, hemp plants, which provide the reinforcing fibers for the composite material, are harvested annually. Moreover, the hemp production per hectare is higher than the average yearly wood production. Therefore, by replacing wood with composite materials in the furniture industry, the human pressure on forests can be reduced, with significant environmental benefits concerning climate regulation and pollution prevention [41].

Hemp fibers are lignocellulosic fibers extracted from the stem of the Cannabis Sativa plant through stem retting, drying, and crushing, followed by scutching and hackling of fiber bundles [42]. Hemp is an environmentally friendly renewable resource that is easily available, cheap, and biodegradable and characterized by good mechanical properties and low density (compared to synthetic fibers) [43,44]. The properties of hemp fibers confer a considerable potential for their use as a reinforcement agent in polymer matrix composite materials. The waste consisting of short hemp fibers removed in the scutching/hackling process represents a valuable source of raw material that can be used in composite material manufacturing [45,46].

Polypropylene (PP) is a recyclable thermoplastic polymer with a share of 19.4% in the demand of European plastics converters in 2019 [47]. Due to its properties, such as low density, low price, good processability, good corrosion, and impact and abrasion resistance, polypropylene is used in many applications in packaging, the textile and automotive industries, building and construction, agriculture, and the electrical and electronics industries [47]. Taking into account that polypropylene is widely used in the packaging of products with a short lifespan, a large amount of polypropylene waste is generated after the product's life cycle has ended. A part of this waste is recycled, but the major part is either discarded in landfills or incinerated for energy recovery purposes. This waste disposal method results in environmental pollution, with harmful consequences for humans and biosystems. The increase in PP waste recycling reduces the environmental impact, brings economic benefits, and ensures a circular economy. In order to obtain rPP fibers, the PP waste goes through the following steps: collection, sorting, shredding, washing, melting and extrusion into pellets/flakes/granules, remelting, and extrusion into fibers [48–50]. Numerous researchers have studied the structure and properties of rPP fibers and their use in several applications, either as matrix in composites or as reinforcement in concrete [51–61].

The problem associated with natural fiber-reinforced polypropylene composites is a low compatibility between hydrophilic fibers and the hydrophobic matrix, which leads to poor adhesion at the matrix–fiber interface and therefore to poor mechanical properties of the composite materials. In order to overcome this disadvantage, the properties of the reinforcing fibers can be modified by physical modification techniques, such as corona, plasma, and alkali treatments, or by chemical modification techniques, such as esterification-based treatments (acetylation, propionylation, or benzylation), graft polymerization, use of silane coupling agents, and treatment using isocyanates [62,63]. The polypropylene matrix can also be modified by grafting maleic anhydride (MAPP), glycidyl methacrylate, and trimethylolpropane triacrylate or by surface peroxidation with heterofunctional polyperoxides [64–70].

The manufacturing process of hemp fiber-reinforced recycled polypropylene composites generates waste in the form of nonwoven fabric scraps. This category of waste can be recycled and used within the same manufacturing process in the furniture industry or in other industries for the production of various products. The possibility of using different forms of textile waste in the production process of polymeric composites process has been investigated by several authors [71–78]. The results of these studies showed that composites that contain textile waste can be used in several applications, such as the automotive, building and construction, and furniture industries.

The aim of this research was to investigate the possibility of recycling hemp/rPP nonwoven waste into thermoformed composites for upholstered furniture products. The 50/50 hemp/rPP nonwoven waste resulted from the cutting process of nonwoven fabrics according to the pattern outline in the fabrication process of 3D composites for furniture applications. The nonwoven waste was defibrated, and the resulting recycled fibers were reintroduced in new nonwovens using reincorporation rates of 50% and 100%. The novelty of this work lies in the recycling of nonwoven waste within the same manufacturing process that generated it. Thus, the length and thickness of the reinforcing hemp fibers could be examined before and after the nonwoven recycling process. Since the length of recycled fibers is shorter than the length of original fibers, it is expected that the mechanical properties of the composite materials obtained by thermoforming of nonwovens that incorporate recycled fibers would deteriorate. To compensate for this supposed depreciation of the composite mechanical properties, an MAPP compatibilizing agent was used to improve the interaction between the recycled reinforcement fibers and the matrix. The obtained composites were evaluated in terms of mechanical and thermal properties (TG, DTG, and DSC), surface morphology, and chemical analysis (SEM and EDX).

2. Materials and Methods

2.1. Materials

In the composite materials, recycled polypropylene was used as the matrix, and hemp fibers were used as reinforcement. The rPP fibers were supplied by TAPARO Company (Târgu Lăpuș, Romania) and had the following characteristics: 9.9 dtex linear density, 25.2 cN/tex tenacity, 142.9% breaking elongation, 82.2 mm mean length, and 8.2 g/10 min melt flow index (at 230 °C/2.16 kg). Hemp fibers of 15.8 tex linear density and 38.9 cN/tex tenacity were purchased from Hempflax Europe (Pianu de Jos, Romania) and were cut to 60 mm using a cutting machine.

Recycled fibers resulting from the defibration of 50/50 hemp/rPP nonwoven scraps were reincorporated into new nonwovens from which composites were obtained. These fibers differed from the reference fibers in length, as the defibration of the nonwovens resulted in fiber shortening. Nonwoven defibration may also affect the fineness of hemp fibers because of the separation of the bundles of elementary fibers from the technical hemp fibers.

Since the characteristics of the reinforcing fibers influence the mechanical properties of the composites, the length of both raw and recycled hemp fibers was measured. Single-fiber length measurement was the technique used to determine the length distribution of hemp fibers tested before and after recycling. The fibers were stretched using tweezers along a millimeter scale. Two hundred measurements were performed for each sample. The thickness of raw and recycled hemp fibers was measured using a Mesdan Analyzer 2000 microscope (Puegnago sul Garda, Italy). Fifty fibers were analyzed. Because of variations in thickness along the fiber, multiple measurements were averaged for each fiber.

Maleic anhydride-grafted PP Polybond 3200 from Crompton Corporation (Middlebury, CT, USA) supplied by S.C Prociv SRL (Bucharest, Romania) was used to improve the compatibility between the matrix and the reinforcing fibers.

2.2. Manufacturing of the Composite Materials

The Romanian furniture manufacturing company uses thermoforming to produce 3D furniture pieces from composite materials. Thermoforming is a composite manufacturing technique that allows for a high content of reinforcing fibers and a high production rate [79]. The production process consists of the following steps: manufacturing of 50/50 hemp/rPP nonwoven fabrics, overlapping and cutting of nonwoven fabrics as per the shape of the pattern, heating of fibrous mat pieces in a hot press, transferring of the heated material to a mold, pressing, and cooling. The cutting step generates up to 20% waste in the form of nonwoven scraps that can be recycled into the same product in the same manufacturing process, making the overall process more sustainable and greener.

In the experiments, the nonwoven waste was cut on a cutting machine and defibrated using an opening machine. The recycled fibers consisting of 50% hemp and 50% rPP were reintroduced in new nonwovens proportions of 50% and 100%, respectively. The manufacturing process of nonwoven fabrics consisted of fiber opening and blending, aerodynamic web formation, and mechanical web consolidation by needle punching. Six layers of rectangular nonwovens were pressed between the heating plates of the thermoforming machine in order to obtain composite plates. When laying the nonwovens, the layers were alternated according to the direction of punching (longitudinal and transversal). The thermoforming parameters were set as follows:

- Temperature: 190 °C;
- Pressure: 0.735 MPa;
- Pressing time: 15 min;
- Cooling time: 10 min.

Since defibration of nonwoven waste affects the length of recycled fibers, a decrease in the tensile strength of the composites that contain such fibers is expected. In order to counteract the effect of recycled fiber shortening on the mechanical properties of the composites, 2.5% wt and 5% wt of MAPP compatibilizer was used. The granules were ground with a ball mill and sieved between layers 2–3 and 4–5 of the nonwoven stack (Figure 1).

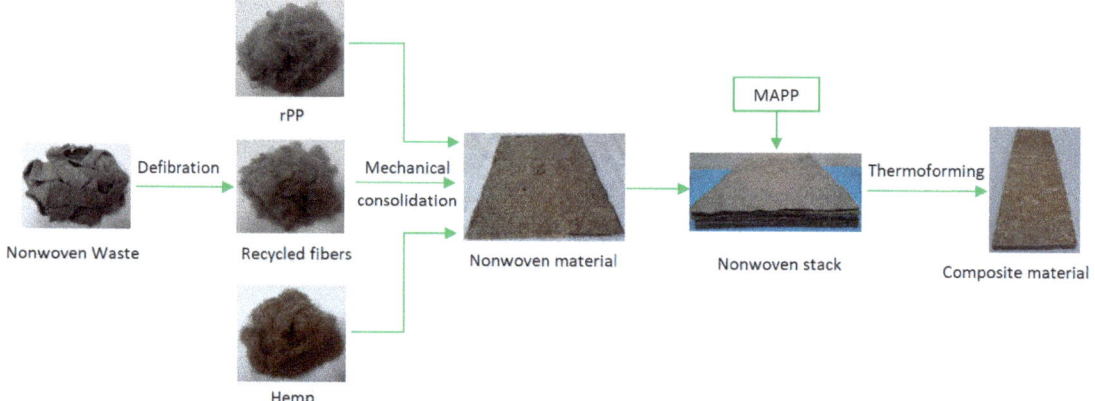

Figure 1. Flow chart of the composite manufacturing process.

Table 1 shows the composition and the coding of the manufactured composite variants.

Table 1. Experimental variants.

Variant Code	Composition
V1	50% hemp/50% rPP
V2	100% recycled fibers from V1 nonwoven waste
V2.1	97.5% V1 recycled fibers/2.5% MAPP
V2.2	95% V1 recycled fibers/5% MAPP
V3	50% V1 recycled fibers/25% hemp/25% rPP
V3.1	47.5% V1 recycled fibers/25% hemp/25% rPP/2.5% MAPP
V3.2	45% V1 recycled fibers/25% hemp/25% rPP/5% MAPP

2.3. Mechanical Properties

Both flexural and tensile tests of composite materials were carried out on an LBG testing machine (Azzano San Paolo, Italy). The length of the specimens for tensile testing was 250 mm, and the width was 25 mm, according to the EN 326-1 standard. Tensile test parameters were established according to ISO 527-4 standard as follows: 2 mm/min crosshead speed and 150 mm distance between clamps. The dimensions of the specimens for the 3-point flexural test were determined according to the specifications of the ISO 14125 standard. The specimen width was 15 mm, and the length was set depending on the specimen thickness. The cross-head speed used during the flexural tests was 2 mm/min. Five specimens were tested to determine the tensile and flexural properties of the composite materials.

2.4. Thermal Properties

The thermal degradation of rPP, hemp fibers (H), and hemp/rPP composite materials was studied by thermogravimetric analysis (TGA) using a Mettler Toledo TGA/SDTA 851 balance (Columbus, OH, USA). The weight of the samples ranged between 2 and 5 mg. The samples were subjected to heating in the temperature domain from 25 °C to 700 °C under a constant flow of nitrogen of 20 mL/min using a heating rate of 10 °C/min.

Differential scanning calorimetry (DSC) was used to determine the melting and crystallization temperatures of rPP fibers and hemp/rPP composite materials. The analysis was performed on a Mettler-Toledo DSC1 822e calorimeter (Columbus, OH, USA). The samples, weighing approximately 5 mg each, were placed in 40 µL aluminum crucibles and were scanned from −60 °C to 200 °C. The first heating stage was followed by cooling and then by a second heating stage in the same temperature range. All three thermal stages were conducted at a constant rate (10 K/min) under a constant nitrogen flow rate (150 mL/min). The heating–cooling–heating cycle of the samples allows for the determination of the crystallization and melting temperature and crystallization and melting enthalpy. The crystallinity of the composite materials (X_C(%)) was calculated using Equation (1) [80]:

$$X_c(\%) = \frac{\Delta H_f}{\Delta H_f^0} \cdot 100, \tag{1}$$

where ΔH_f is the enthalpy of fusion per unit mass of the PP calculated based on the area under the melting peak of the composite, and ΔH_f^0 is the enthalpy per unit mass of the 100% crystalline PP, with a value of 207 J/g [80].

2.5. Surface Morphology and Chemical Analysis (SEM and EDX)

The surface morphology of composite materials was analyzed through a Quanta 200 (FEI) scanning electron microscope (SEM) (Hillsboro, OR, USA) operating at an accelerating voltage of 20 kV with secondary electrons in low vacuum mode. The microscope was coupled with an energy-dispersive X-ray (EDX) system for chemical analysis and elemental analysis mapping.

3. Results and Discussions

3.1. Fiber Characteristics

Figure 2 presents the fiber length distribution of hemp fibers tested before and after recycling.

It can be seen that the length distribution of hemp fibers changed considerably after nonwoven waste recycling. The average length of hemp fibers decreased from 57.8 ± 36.7 mm (before recycling) to 24.8 ± 13.9 mm (after recycling) due to the mechanical actions to which the fibers were subjected during waste cutting and opening. This represents a significant decrease (−57.09%) in hemp fiber length after recycling. In the raw hemp fiber sample, fibers much longer than the cut length of 60 mm were found in a tangled position. After recycling, the highest frequency was recorded in the 20–25 mm length class, while before recycling the highest frequency was found in the 40–50 mm length class.

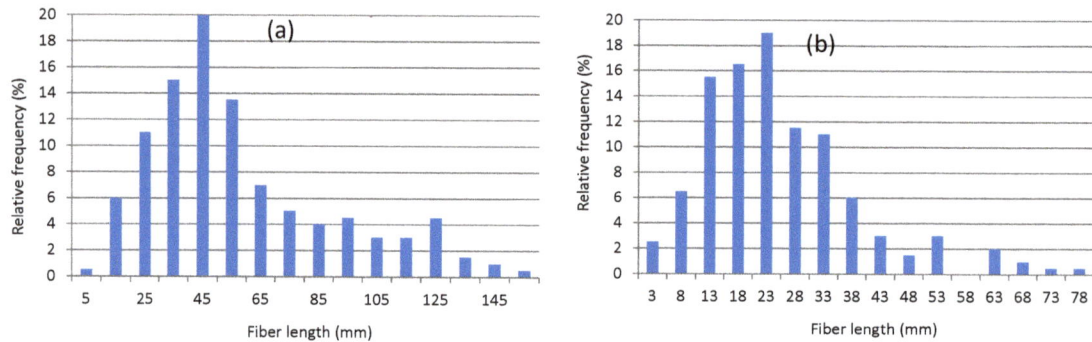

Figure 2. Frequency histograms of hemp fiber length tested (**a**) before recycling and (**b**) after recycling.

Figure 3 shows the thickness distribution of hemp fibers tested before and after recycling.

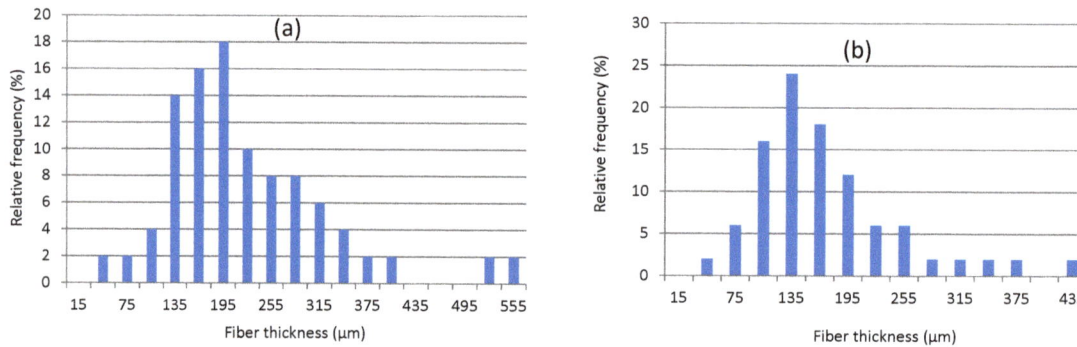

Figure 3. Frequency histograms of hemp fiber thickness tested (**a**) before recycling and (**b**) after recycling.

Technical fibers are made up of bundles of elementary fibers (cells) glued together by the middle lamella. During processing, coarse technical fibers are divided into finer fibers. This behavior explains why the average thickness of hemp fibers decreased from 223.5 ± 101.8 µm (before recycling) to 174 ± 77.5 µm (after recycling), representing a decrease of 22.15%. The recycled fibers had variable thicknesses ranging from 37 µm to 433 µm, from fine to coarse fibers. The fiber aspect ratio (length/thickness) of the recycled hemp fiber was 142.5, which is much higher than 10, which is considered the minimum value for a good transmission of stress [81].

Figure 4 shows the images of raw and recycled hemp fibers. In the raw hemp sample, compact fibers (Figure 4a), fibers with split ends (Figure 4b), and fibers with compact ends and split bundles between ends could be found (Figure 4c). The recycled hemp fibers showed the same appearance as raw hemp fibers (Figure 4d–f).

3.2. Mechanical Properties

The tensile strength and the breaking elongation of the manufactured composites are shown in Figure 5.

As shown in Figure 5a, compared to the tensile strength of composite material V1, which is the reference for comparison, the use of recycled fibers to obtain nonwoven fabrics led a reduction in the tensile strength of the composite materials. Thus, a loading of 50% recycled fibers reduced the tensile strength of the composite material by approximately 17.1% (V3), while a loading of 100% recycled fibers decreased the tensile strength of the composite by 22.6% (V2). This reduction in the tensile strength of the composites with increasing recycled fiber content can be explained by the deterioration in fiber characteristics that occurred in the opening process of the nonwoven waste. The addition of MAPP compatibilizer resulted in an increase in tensile strength of both V2 and V3 composite materials. Thus, in the case of the V2 composite material, the addition of 2.5% MAPP led to an increase in tensile strength of 6.2%, from 20.8 MPa to 22.1 MPa (V2.1), while the incorporation of 5% MAPP caused a higher increase of 19.2%, from 20.8 MPa to 24.8 MPa. This increase in tensile strength of composites treated with MAPP can be explained by the improvement in the interfacial adhesion between the reinforcing hemp fibers and the rPP matrix [82] due to the formation of covalent ester bonds between the anhydride groups of MAPP and the hydroxyl groups of the hemp surface [83].

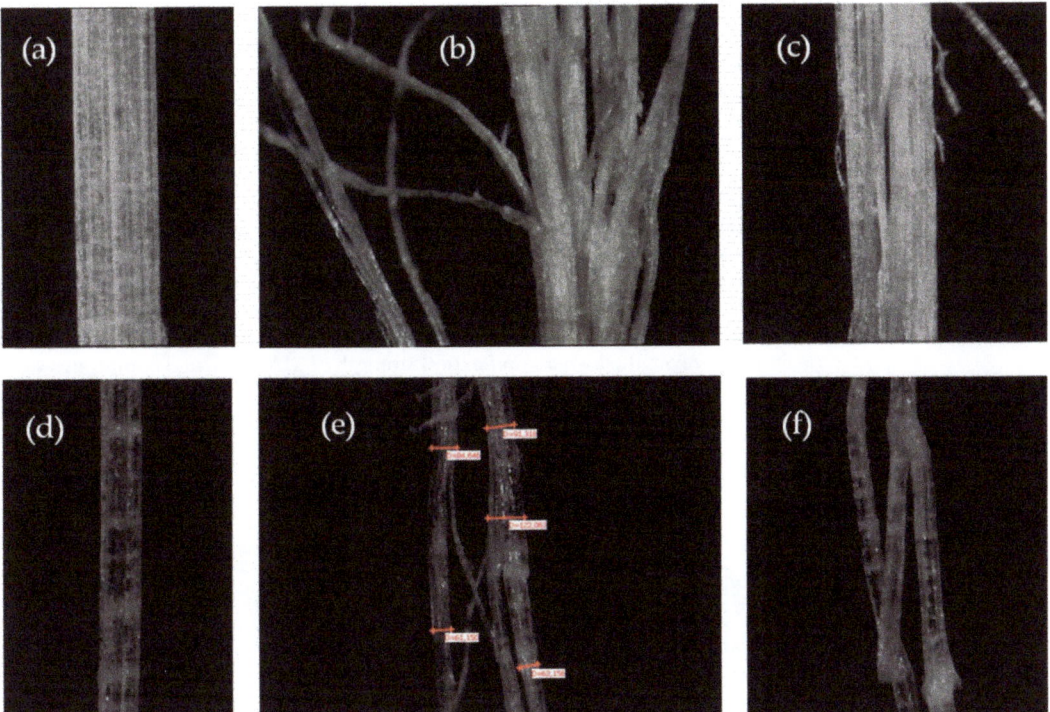

Figure 4. Images of raw hemp fibers (**a**–**c**) and recycled hemp fibers (**d**–**f**) at 128× magnification.

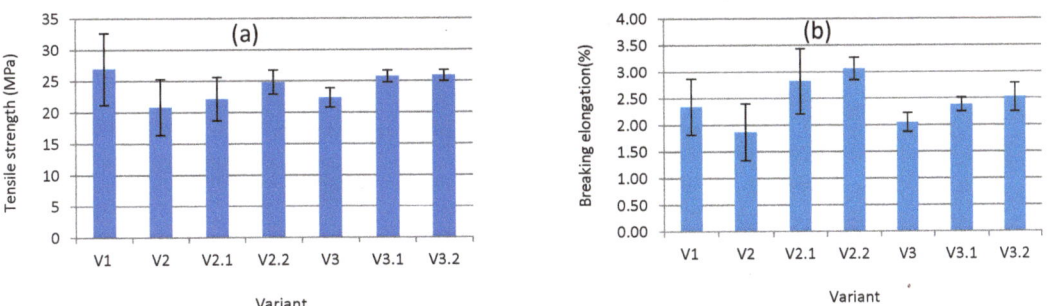

Figure 5. Tensile strength (**a**) and breaking elongation (**b**) of composite materials.

Figure 5b shows the breaking elongation of the investigated composite materials. It can be seen that the use of recycled fibers causes a decrease in the elongation at break of the composite materials. Compared to the elongation at break of the V1 composite, the elongation at break decreased by 12.4% for the V3 variant (50% recycled fibers) and by 20.1% for the V2 variant (100% recycled fibers). An explanation for this behavior may be the deterioration of the elasticity of the fibers caused by the mechanical actions to which they were subjected during defibration of the nonwoven waste [84]. The elongation at break of composite materials increased along with the increase in MAPP content. This result is consistent with the findings of Saad [83] and can be explained by the plasticizer role of MAPP [85,86]. Plasticizer separates the rPP chains and reduces the intermolecular forces between them. Thus, the rPP chains can move more easily in relation to one another [87]. Chun et al. reported a decrease in the elongation at break of composites along with an increase in the MAPP content due to the mobility restriction of the PP chains as a result of the improvement in the interfacial adhesion between the matrix and the reinforcing fibers [88]. They also noticed the plasticizing effect of MAPP at concentrations higher than 5%.

The flexural strength of the composite materials is presented in Figure 6a. The reincorporation of recycled fibers into nonwovens leads to a decrease in the flexural strength of the composites, mainly due to the reduction in the length of the hemp reinforcing fibers induced by nonwoven waste reprocessing. Thus, the incorporation of 50% recycled fibers diminished the composite flexural strength by approximately 6.6% (V3), while incorporation of 100% recycled fibers reduced the composite flexural strength by 9% (V2). Adding MAPP to the composites improved the interfacial bonding between the reinforcing hemp fibers and the polymer matrix, and this led to an increase in flexural strength of the composites. However, the increase in flexural strength of MAPP-treated composites was less than 4%.

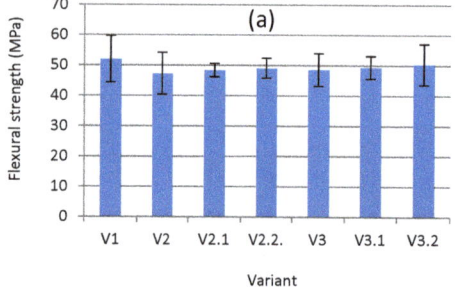 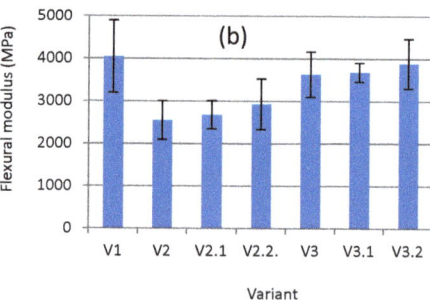

Figure 6. Flexural strength (**a**) and flexural modulus (**b**) of composite materials.

Figure 6b presents the flexural modulus of the analyzed composites. As the content of the recycled fibers into composites increases, the flexural modulus decreases due to the deterioration of fiber characteristics during reprocessing. The flexural modulus of V3 (50% recycled fibers) and V2 (100% recycled fibers) composite variants decreased by 10.3% and 37%, respectively, compared to the flexural modulus of the V1 variant. The addition of MAPP improved the flexural modulus of the composites containing recycled fibers (+14.8% maximum increase for the V2.2 variant) without reaching the value of the flexural modulus of the V1 composite material (4045 MPa).

3.3. Thermal Properties

The thermal behavior of rPP, hemp (H), and the obtained composite materials was studied by thermogravimetry. The analyzed samples were subjected to a dynamic heating cycle (according to Section 2.4). The obtained thermogravimetric curves are shown in Figures 7–9.

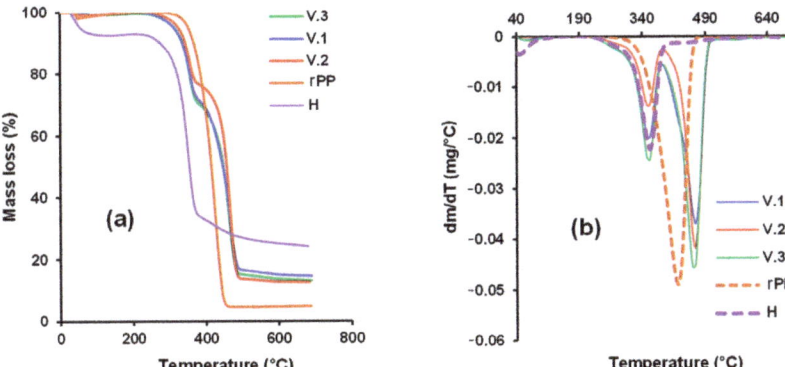

Figure 7. Graphs of thermogravimetric curves (**a**) and DTG (**b**) obtained for hemp fibers, rPP, and composite materials.

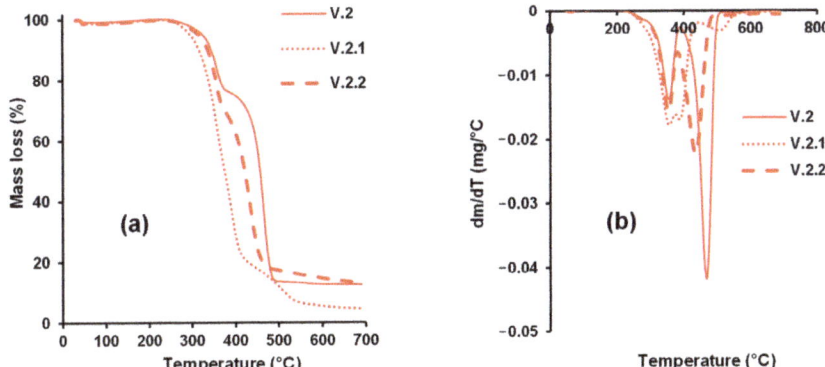

Figure 8. Graphs of thermogravimetric (**a**) and DTG (**b**) curves obtained for the composite materials from V2 variants.

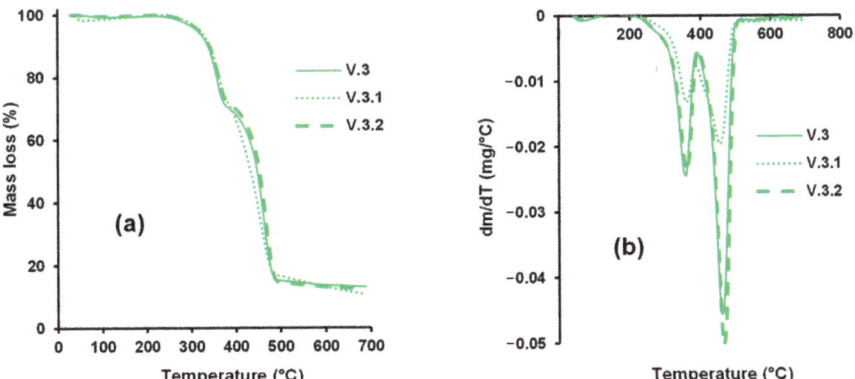

Figure 9. Graphs of thermogravimetric (**a**) and DTG (**b**) curves obtained for the composite materials from V3 variants.

T_{onset} (°C), T_{peak} (°C), and T_{endset} (°C) temperatures were automatically determined from the thermogravimetric diagrams obtained with the device using a Mettler Toledo TGA/SDTA 851.

Table 2 shows the main values extracted from these graphs, such as T_{onset} (°C), T_{peak} (°C), T_{endset} (°C), and mass loss (%).

Table 2. Thermogravimetric parameters of fibers and composite materials.

Sample	Stage	T_{onset} (°C)	T_{peak} (°C)	T_{endset} (°C)	Mass Loss (%)
Hemp, 100%	I	30	52	85.05	7.5
	II	303	359	377	54.74
	III	377	446	499	11.31
rPP, 100%	I	351.5	423.94	448.96	89.33
V1	I	31	61.79	65.81	5.8
	II	321	355	369	16.9
	III	405	465.6	483.37	65.08
V2	I	50.43	70.76	103.75	4.7
	II	325.5	355	369.4	13.72
	III	428.67	464.6	484.6	62.54
V2.1	I	70.84	75.6	111.65	4.6
	II	309.4	355.14	412.62	29.58
	III	466.97	507.33	534.31	58.22
V2.2	I	26.07	59.62	65.79	4.5
	II	330.91	353.24	365.52	21.13
	III	407.54	432.4	461.89	58.60
V3	I	27.71	69.76	113.80	4.6
	II	326.5	355.83	371.62	13.73
	III	419.4	461.7	479.7	70.68
V3.1.	I	31.81	70.10	114.40	4.42
	II	334.94	359.66	371.65	15.48
	III	406.81	455.16	478.09	65.89
V3.2.	I	33.34	61.11	103.90	4.40
	II	327.84	359.49	374.28	17.6
	III	431.29	466.72	485.66	63.29

Figures 7–9 present the TG and DTG thermograms of hemp reinforcing fibers, rPP fibers, and composite materials. The TG thermogram of hemp fiber indicates a mass loss of 7.3% for the temperature range of 41 °C to 149 °C. This mass loss can be explained by the moisture evaporation at the surface of hemp fibers. As temperature further increased between 149 °C and 303 °C, small mass losses were registered (7.06%). In the temperature domain between 303 °C and 376 °C, because of the degradation of cellulose and hemicellulose, the highest mass loss of 57.7% was recorded (maximum degradation rate at 359 °C). Above 376 °C, further mass loss was registered because of lignin degradation [89].

In the DTG curve, the peak at 55 °C was caused by hemp fiber moisture evaporation, and the steep peak at 559 °C was caused by the degradation of cellulose and lignin. The peak at 446 °C is associated with degradation of cellulose, lignin, and PP [90,91].

The TG curve of the recycled polypropylene fiber shows a maximum degradation of the fiber in the temperature range of 351 °C to 448.96 °C, with a mass loss of 98.14%. The DTG curve indicates that the maximum rate of degradation of the fiber took place at a temperature of 423.94 °C.

During the heating process, the transformations of composite materials take place in three stages. In the first stage, the moisture evaporates from the composite material. The highest mass losses were recorded for the V1 variant (5.8% in the temperature range of 31–65.8 °C, with a maximum at 61.7 °C). For the other materials, the mass losses were very close—between 4.7% (for V2 variant) and 4.4% (for V3.2 variant)—and the temperatures did not vary much, ranging between 61.79 °C (for V1) and 59.62 °C (for V2.2).

In the second stage, the temperatures at which degradation was maximum were around 355 °C for all the analyzed materials, and the mass loss was between 13.73% for the V3 variant and 29.58% for the V2.1 variant.

In the third stage, the mass loss was high, ranging between 58.22% (for the V2.1 variant, reaching a maximum at 507.33 °C) and 70.68% (for the V3 variant at the maximum temperature of 461.7 °C).

Mass losses were higher for the V3 variant (50% recycled fibers) compared to the V2 variant (100% recycled fibers) (Figures 8 and 9 and Table 2). The addition of the MAPP compatibilizer led to a slight decrease in the thermal stability of the composite material.

The analysis of DTG curves indicates that the V1, V2, and V3 composite materials showed a similar behavior. The first stage in the range of 54–120 °C corresponds to the loss of moisture, the peak at 355 °C corresponds to the degradation of hemp, the peak at 393 °C can be assigned to the degradation of lignin, and the peak at about 465 °C corresponds to rPP degradation. The temperature range indicated for the use of this composite is 20–120 °C, where mass losses are minor.

DSC analysis allowed for the determination of crystallization temperature (T_c), crystallization enthalpy (ΔH_{fc}), melting temperature (T_m), and melting enthalpy (ΔH_{fm}) of the investigated samples. The obtained data are presented graphically in Figures 10–12 and Table 3.

The total crystallinity of the composite material was calculated with Equation (1). Analysis of the data presented in Table 3 indicates that:

- The melting temperatures of the composite materials containing recycled fibers (V2 and V3) are higher than the melting temperature of the V1 composite material and lower than the melting temperature of the rPP matrix;
- With the addition of a compatibilizer, the melting temperatures decreased slightly along with the increase in the amount of MAPP;
- The crystallization temperatures of V2 and V3 composites are lower than the crystallization temperature of V1 composite material and higher than that of the 100% rPP matrix;
- The difference between the crystallization temperature of rPP and the crystallization temperatures of the obtained composite materials is approximately 4.5 °C, varying between 117.54 °C and 122.25 °C;
- With the addition of a compatibilizing agent (MAPP), the crystallization temperatures of the obtained composite materials show a slight increase compared to the untreated materials; the crystallization temperature also increases along with the increase in the concentration of the compatibilizer. This fact confirms that the hemp fibers acted as nucleating agents, and consequently, the rPP in the composite materials began to crystallize at temperatures higher than 117.54 °C, with the surface of the hemp fibers constituting crystallization centers for the polymer matrix [92].

In all analyzed cases, the crystallinity of the composite materials was lower than the crystallinity of the rPP matrix.

3.4. Surface Morphology and Chemical Analysis (SEM and EDX)

Different magnitudes were used to capture images of the composite materials. At 100× magnification, the images show a multidirectional orientation of the reinforcing hemp fibers in the composites (Figure 13a). Both the way the nonwovens were layered in the mold and the random orientation of the hemp fibers in the nonwovens give the composites fairly similar mechanical properties, regardless of the direction of the applied stress. Furthermore, as indicated by Figure 13a, the hemp fibers are uniformly distributed and well-embedded in the polymeric matrix. Some regions with poor adhesion of the matrix to the fibers could be observed, especially in the case of composites that were not treated with MAPP (Figure 13b). Such local fiber–matrix detachment may affect the structural strength of the composites.

The MAPP-treated composites show a good coverage of the reinforcing fibers and a good adhesion at the fiber–matrix interface (Figure 13c). Some images showed the presence of microvoids in the matrix structure (Figure 13d). Due to the hydrophilicity of hemp fibers, water molecules inside the microfibrils may gasify at the high temperatures of the forming process and produce voids in the composite material [93]. As shown by the TG thermogram the hemp fiber, a mass loss of 7.3% occurred due to moisture evaporation. The presence of voids can result in the degradation of the mechanical properties of the composites.

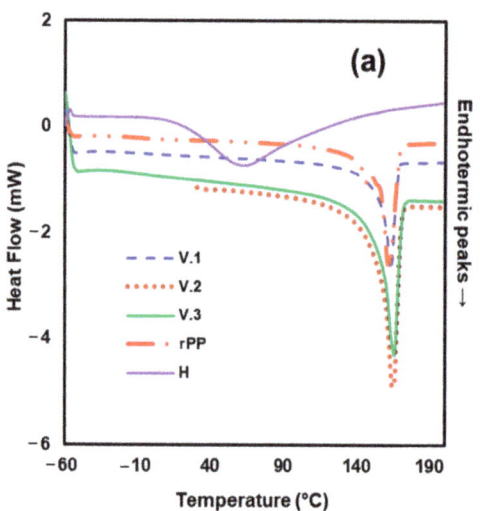
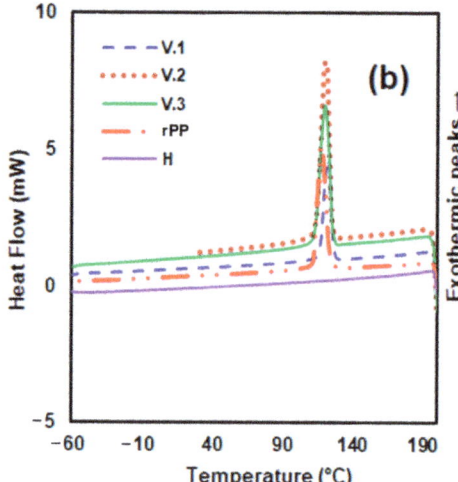

Figure 10. DSC (heating process) curves (**a**) and DSC (cooling process) curves (**b**) for hemp fibers, rPP, and composite material variants.

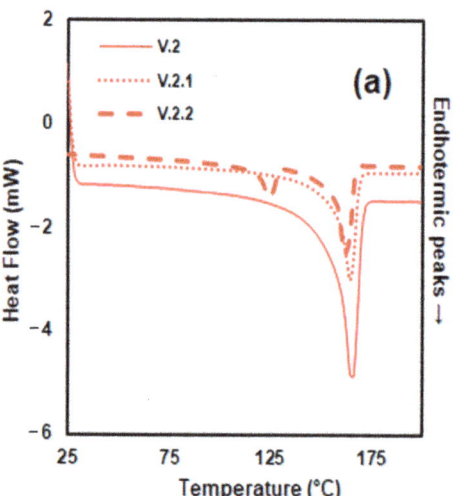
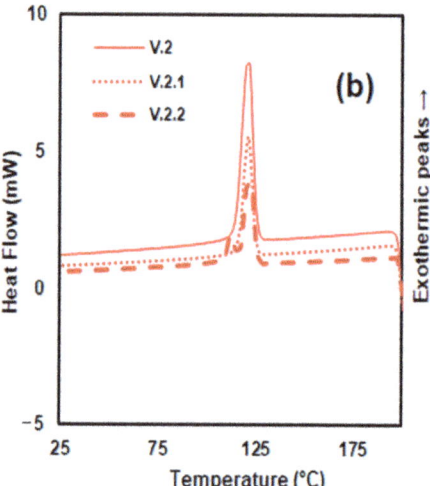

Figure 11. DSC (heating process) curves (**a**) and DSC (cooling process) curves (**b**) for V2 composite material variants.

One of the most effective ways to improve the polymer matrix–reinforcement material interactions is the use of coupling agents. Functionalization with MAPP facilitates the chemical reactions between components such as polypropylene and hemp fibers (used as reinforcements), increasing the adhesion at the interfaces and implicitly improving the mechanical and thermal properties of the composite due to the formation of new bonds between the components. [82,94]. The reactions that take place are shown schematically in Figure 14.

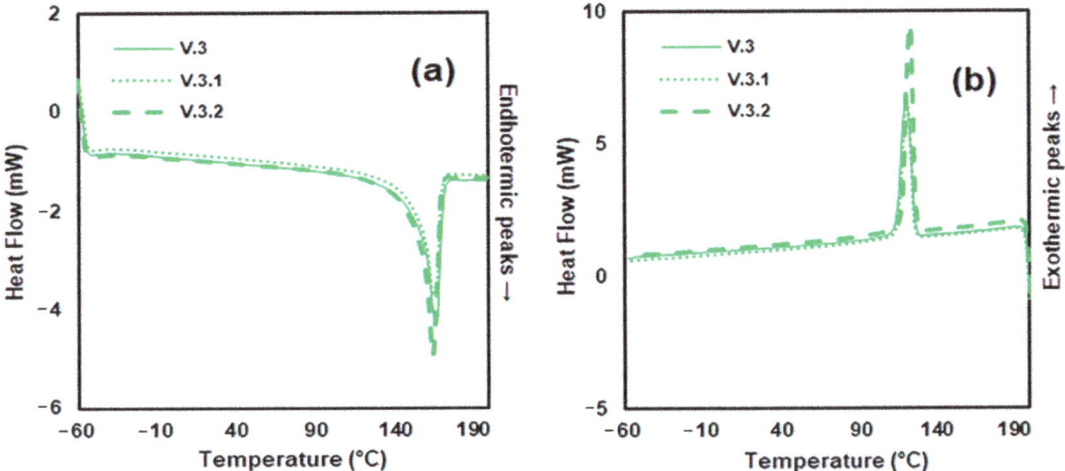

Figure 12. DSC (heating process) curves (**a**) and DSC (cooling process) curves (**b**) for V3 composite material variants.

Table 3. DSC data for rPP and hemp/rPP composite materials.

Sample	First Heating			Cooling			Second Heating		
	T_m (°C)	* ΔH_{fm} (J/g)	T_c (°C)	ΔH_{fc} (J/g)	X_C (%)	T_m (°C)	** ΔH_{fm} (J/g)	X_C (%)	
rPP 100%	156.52	49.53	117.54	93.73	45.28	168.87	81.82	39.52	
V1	164.22	39.55	122.04	40.06	19.35	163.89	42.05	20.31	
V2	164.20	47.81	120.69	57.47	27.76	165.10	51.22	24.74	
V.2.1	164.37	34.57	121.07	39.16	18.92	163.37	32.57	15.73	
V.2.2	162.89	27.87	121.69	38.04	18.76	162.22	30.07	14.53	
V3	165.47	43.05	119.94	51.96	25.10	165.81	48.97	23.65	
V.3.1	165.49	38.06	120.79	45.68	22.06	165.49	40.76	19.69	
V.3.2	165.28	48.57	122.25	57.27	27.67	164.11	51.96	25.10	

* First heating; ** second heating.

The results of the EDX analyses for the examined samples are shown in Figure 15. Five determinations were made at various points on the surface of each analyzed composite material. The average values of the weight percentages (Wt.%) and the atomic percentages (At%) of the elements from each analyzed sample were determined.

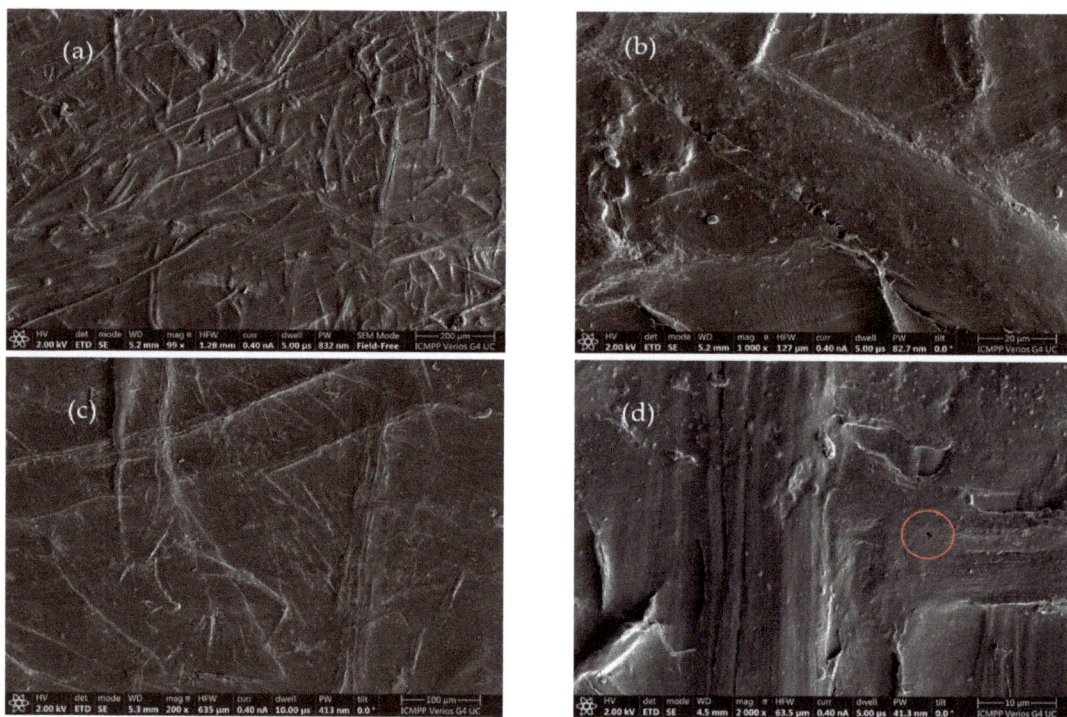

Figure 13. SEM images of (**a**) V1 composite (100×), (**b**) V2 composite (1000×), (**c**) V 3.2 composite (200×), and (**d**) V3 composite (2000×).

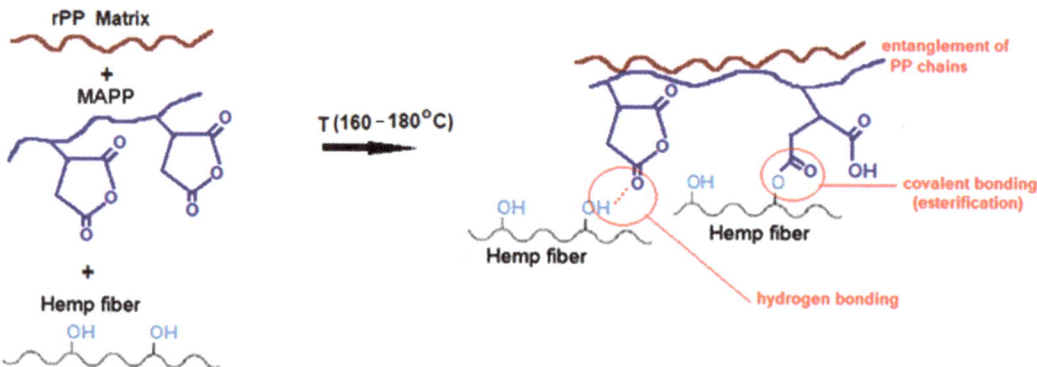

Figure 14. Scheme of the reaction between the hydroxyl groups on the surface of the hemp fiber, rPP, and MAPP.

The higher values of the O/C ratio for the samples treated with MAPP (Figure 14b,c,e,f), which increase along with the increase in MAPP concentration, confirm the presence of the compatibilizer in the analyzed composite materials.

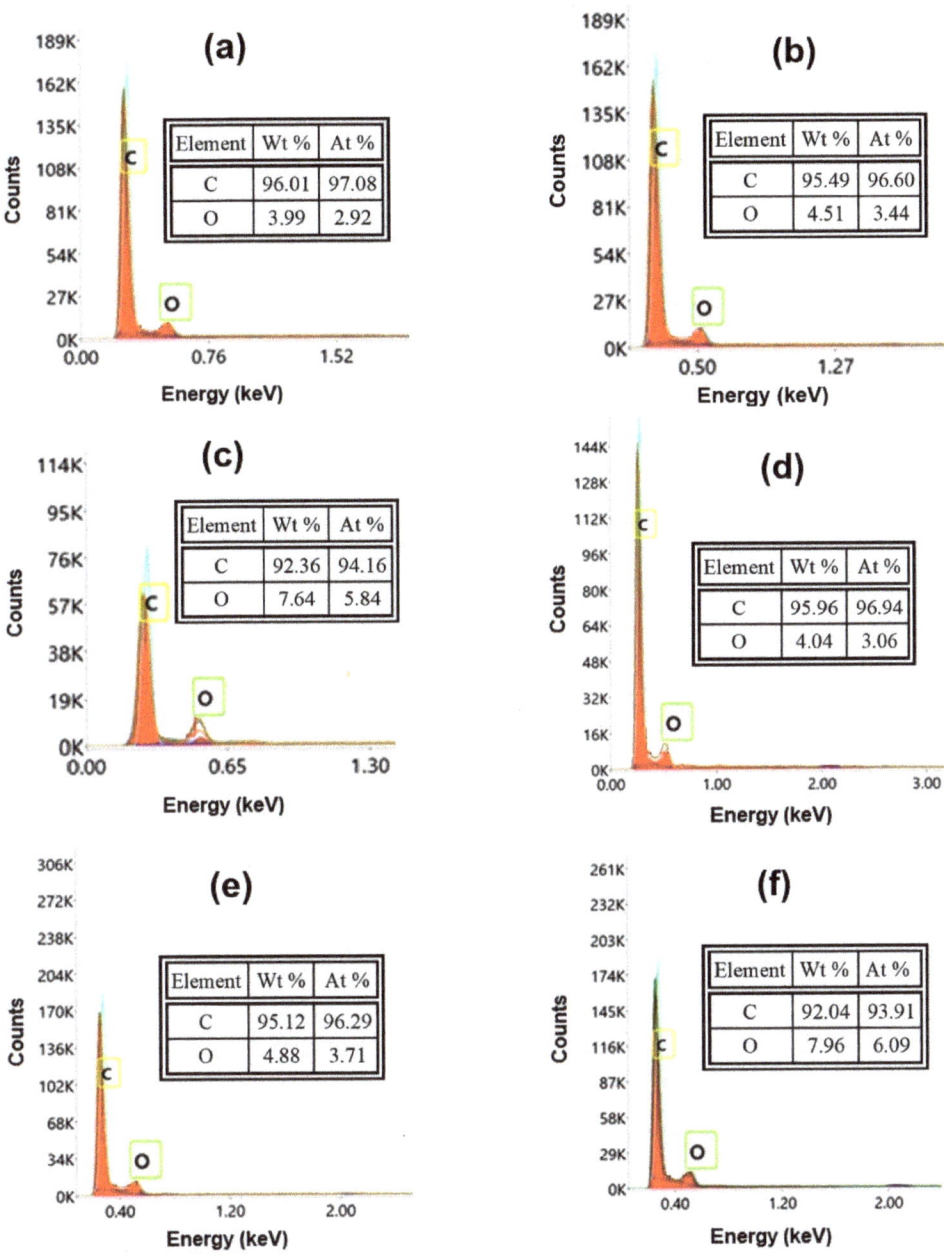

Figure 15. EDX elemental analysis of composite materials V2 (**a**), V2.1 (**b**), V2.2 (**c**), V3 (**d**), V3.1 (**e**), and V3.2 (**f**).

4. Conclusions

The increasing use of composites in the furniture industry has led to the need to consider the environmental issues generated by waste disposal in the new product development process. Waste recycling reduces the environmental impact of waste disposal, lowers the cost of products, conserves resources, and closes the loop in the circular economy.

This research investigated the possibility of recycling hemp/rPP nonwoven waste resulting from the cutting stage in the manufacturing process of 3D composites for upholstered furniture products. Nonwoven scraps were defibrated, and the resulting recycled fibers were reincorporated into new nonwovens from which composites were obtained by thermoforming. The recycling process of nonwoven waste resulted in a 57.09% reduction in hemp fiber length and a 22.15% reduction in hemp fiber thickness. The effect of recycled fiber content (50% and 100%) and MAPP content (2.5% and 5%) on the mechanical and thermal properties of composite materials was studied. This study showed that the mechanical properties of composite materials decreased with increasing recycled fiber content and increased with increasing MAPP content. Thus, the incorporation of 50% and 100% recycled fibers into composites decreased the tensile strength by 17.1–22.6%, the elongation at break by 12.4–20.1%, the flexural strength by 6.6–9%, and flexural modulus by 10.3–37%. The addition of 5% MAPP showed the greatest improvements in mechanical properties for composites containing 100% recycled fibers, as follows: 19.2% increase in tensile strength, 3.8% increase in flexural strength, and 14.8% increase in flexural modulus. Thermogravimetric analysis of the composite materials indicated that the recommended temperature range for composite applications is 20–120 °C, where mass losses are minor. The incorporation of recycled fibers from nonwoven waste into composites leads to an increase in the melting temperature and a decrease in the crystallization temperature. The crystallinity of the composite materials is lower than the crystallinity of the rPP matrix. SEM images prove a good adhesion between the rPP matrix and hemp reinforcement in the case of composite materials treated with MAPP. EDX analyses confirm the presence of the compatibilizing agent in the structure of the treated composite materials.

Composite materials containing up to 50% recycled fibers from nonwoven waste treated with 5% MAPP can be used successfully in furniture applications.

Author Contributions: All authors contributed to the research presented in this work. Their contributions are presented below. Conceptualization, methodology, and investigation: M.I., E.I.M., I.F. and L.S.; formal analysis and data curation: M.I., E.I.M., L.S. and G.L.; writing—original draft preparation: M.I. and E.I.M.; funding: I.F.; project administration: I.F.; writing—review and editing, M.I. and E.I.M. All authors have read and agreed to the published version of the manuscript.

Funding: Publication fees were financed by the Gheorghe Asachi Technical University of Iași, Romania from the Fund for University Scientific Research.

Institutional Review Board Statement: Not applicable.

Informed Consent Statement: Not applicable.

Data Availability Statement: All data are included in the published article.

Acknowledgments: This work was developed within the framework of COP Project No. 267/22.06.2020, MySMIS 121434, "Creating a center of excellence in the field of composite material at SC TAPARO SA", funded by the European Union via the European Regional Development Fund and the Romanian Government.

Conflicts of Interest: The authors declare no conflict of interest.

References

1. Shahruzzaman, M.; Biswas, S.; Islam, M.M.; Islam, M.S.; Rahman, M.S.; Haque, P.; Rahman, M.M. Furniture: Eco-Friendly Polymer Composites Applications. In *Encyclopedia of Polymer Applications*, 1st ed.; Mishra, M., Ed.; Taylor and Francis Group, CRC Press: Boca Raton, FL, USA, 2019; pp. 1517–1547. [CrossRef]
2. George, G.; Elias, L.; Luo, Z. Furniture: Polymers. In *Encyclopedia of Polymer Applications*, 1st ed.; Mishra, M., Ed.; Taylor and Francis Group, CRC Press: Boca Raton, FL, USA, 2019; pp. 1548–1560. [CrossRef]
3. Buck, L. Furniture Design with Composite Materials. Ph.D. Thesis, Brunel University, London, UK, 1997.
4. Zakriya, M.; Govindan, R. Applications of Composites in Artefact and Furniture Making. In *Natural Fibre Composites*, 1st ed.; Zakriya, M., Govindan, R., Eds.; CRC Press: Boca Raton, FL, USA, 2020; *imprint*; pp. 107–116. [CrossRef]

5. Duque Estrada, R.; Wyller, M.; Dahy, H. Aerochair Integrative design methodologies for lightweight carbon fiber furniture design. In Proceedings of the eCAADe 37/SIGraDi Conference (eCAADe: Education and Research in Computer Aided Architectural Design in Europe) and (SIGraDi: Sociedad Iberoamericana de Gráfica Digital), Porto, Portugal, 11–13 September 2019. [CrossRef]
6. Glass Fibre Reinforced Polypropylene Furniture. 2022. Available online: https://furnlink.com.au/glass-fibre-reinforced-polypropylene-furniture/ (accessed on 27 July 2022).
7. Stelea, L.; Filip, I.; Lisa, G.; Ichim, M.; Drobotă, M.; Sava, C.; Mureșan, A. Characterisation of Hemp Fibres Reinforced Composites Using Thermoplastic Polymers as Matrices. *Polymers* **2022**, *14*, 481. [CrossRef] [PubMed]
8. Ciupan, C.; Pop, E.; Filip, I.; Ciupan, E.; Câmpean, E.; Cionca, I.; Hereș, V. A new approach of the design process for replacing wooden parts of furniture. *MATEC Web. Conf.* **2017**, *137*, 06002. [CrossRef]
9. Ciupan, E.; Ciupan, C.; Câmpean, E.M.; Stelea, L.; Policsek, C.E.; Lungu, F.; Jucan, D.C. Opportunities of Sustainable Development of the Industry of Upholstered Furniture in Romania. A Case Study. *Sustainability* **2018**, *10*, 3356. [CrossRef]
10. Frulio, F. FiberFlax: The New Bio-Composite for Furniture. Available online: https://www.behance.net/gallery/16781161/FiberFlax-The-new-bio-composite-for-furniture- (accessed on 27 July 2022).
11. Zakriya, M.; Govindan, R. Jute fiber reinforced polymeric composites in moulded furniture making. *JARDCS* **2017**, *9*, 1272–1278.
12. Mahmood, A.; Sapuan, S.M.; Karmegam, K.; Abu, A.S. Design and Development of Kenaf Fiber-Reinforced Polymer Composite Polytechnic Chairs. *Asian J. Agric. Biol.* **2018**, *6*, 62–65.
13. Zaman, H. Chemically Modified Coir Fiber Reinforced Polypropylene Composites for Furniture Applications. *Int. Res. J. Mod. Eng. Technol.* **2020**, *2*, 975–982.
14. Husen, S.S.; Uday, M.L.; Chandrakant, M.B.; Madhukar, R.G.; Rupchand, D.Y. A Review on Coconut Coir Composite Board Manufacturing Machine. *Int. Res. J. Eng. Technol.* **2020**, *7*, 1317–1322.
15. Ichim, M.; Stelea, L.; Filip, I.; Lisa, G.; Muresan, E.I. Thermal and Mechanical Characterization of Coir Fibre–Reinforced Polypropylene Biocomposites. *Crystals* **2022**, *12*, 1249. [CrossRef]
16. Suhaily, S.S.; Jawaid, M.; Abdul Khalil, H.P.S.; Mohamed, A.R.; Ibrahim, F. A review of oil palm biocomposites for furniture design and applications: Potential and challenges. *BioResources* **2012**, *7*, 4400–4423.
17. Asyraf, M.R.M.; Ishak, M.R.; Syamsir, A.; Nurazzi, N.M.; Sabaruddin, F.A.; Shazleen, S.S.; Norrrahim, M.N.F.; Rafidah, M.; Ilyas, R.A.; Rashid, M.Z.A.; et al. Mechanical properties of oil palm fibre-reinforced polymer composites: A review. *J. Mater. Res. Technol.* **2022**, *17*, 33–65. [CrossRef]
18. Ikman Ishak, M.; Ismail, C.N.; Khor, C.Y.; Rosli, M.U.; RiduanJamalluddin, M.; Hazwan, M.H.M.; Nawi, M.A.M.; Mohamad Syafiq, A.K. Investigation on the Mechanical Properties of Banana Trunk Fibre–Reinforced Polymer Composites for Furniture Making Application. *IOP Conf. Ser. Mater. Sci. Eng.* **2019**, *551*, 012107. [CrossRef]
19. Mengeloglu, F.; Basboga, I.H.; Aslan, T. Selected Properties of Furniture Plant Waste Filled Thermoplastic Composites. *Pro Ligno* **2015**, *11*, 199–206.
20. Xu, K.; Du, G.; Wang, S. Wood Plastic Composites: Their Properties and Applications. In *Engineered Wood Products for Construction*; Gong, M., Ed.; IntechOpen: London, UK, 2021. [CrossRef]
21. Delviawan, A.; Suzuki, S.; Kojima, Y.; Kobori, H. The Influence of Filler Characteristics on the Physical and Mechanical Properties of Wood Plastic Composite(s). *Rev. Agric. Sci.* **2019**, *7*, 1–9. [CrossRef]
22. Wambua, P.; Ivens, J.; Verpoest, I. Natural fibres: Can they replace glass in fibre reinforced plastics? *Compos. Sci. Technol.* **2003**, *63*, 1259–1264. [CrossRef]
23. AL-Oqla, F.M.; Salit, M.S. Natural fiber composites. In *Materials Selection for Natural Fiber Composites*; AL-Oqla, F.M., Salit, M.S., Eds.; Woodhead Publishing: Cambridge, UK, 2017; pp. 23–48, ISBN 9780081009581. [CrossRef]
24. Mohammed, L.; Ansari, M.N.M.; Pua, G.; Jawaid, M.; Islam, M.S. A Review on Natural Fiber Reinforced Polymer Composite and Its Applications. *Int. J. Polym. Sci.* **2015**, *2015*, 15. [CrossRef]
25. Prakash, S.O.; Sahu, P.; Madhan, M.; Johnson Santhosh, A. A Review on Natural Fibre-Reinforced Biopolymer Composites: Properties and Applications. *Int. J. Polym. Sci.* **2022**, *2022*, 1–15. [CrossRef]
26. Jusoh, A.F.; Rejab, M.R.M.; Siregar, J.P.; Bachtiar, D. Natural Fiber Reinforced Composites: A Review on Potential for Corrugated Core of Sandwich Structures. *MATEC Web Conf.* **2016**, *74*, 33. [CrossRef]
27. Lambrache, N.; Renagi, O.; Olaru, L.; N'Drelan, B. Composite Materials with Natural Fibers. In *Fiber-Reinforced Plastics*; Masuelli, M.A., Ed.; IntechOpen: London, UK, 2022. [CrossRef]
28. Sun, Z. Progress in the research and applications of natural fiber-reinforced polymer matrix composites. *Sci. Eng. Compos. Mater.* **2018**, *25*, 835–846. [CrossRef]
29. Fuqua, M.A.; Huo, S.; Ulven, C.A. Natural Fiber Reinforced Composites. *Polym. Rev.* **2012**, *52*, 259–320. [CrossRef]
30. Azman, M.A.; Asyraf, M.R.M.; Khalina, A.; Petrů, M.; Ruzaidi, C.M.; Sapuan, S.M.; Wan Nik, W.B.; Ishak, M.R.; Ilyas, R.A.; Suriani, M.J. Natural Fiber Reinforced Composite Material for Product Design: A Short Review. *Polymers* **2021**, *13*, 1917. [CrossRef]
31. Faruk, O.; Błędzki, A.K.; Fink, H.P.; Sain, M.M. Progress Report on Natural Fiber Reinforced Composites. *Macromol. Mater Eng.* **2014**, *299*, 9–26. [CrossRef]
32. Das, P.P.; Chaudhary, V.; Motha, S.J. Fabrication and Characterization of Natural Fibre Reinforced Polymer Composites: A Review. In Proceedings of the International Conference of Advance Research & Innovation (ICARI), New Delhi, India, 19 January 2020. [CrossRef]

33. Neto, J.; Queiroz, H.; Aguiar, R.; Lima, R.; Cavalcanti, D.; Banea, M.D. A Review of Recent Advances in Hybrid Natural Fiber Reinforced Polymer Composites. *J. Renew. Mater.* **2022**, *10*, 561–589. [CrossRef]
34. Usmani, M.A.; Anas, M. Study of Natural Fibre Reinforced Composites. *IOP Conf. Ser. Mater. Sci. Eng.* **2018**, *404*, 012048. [CrossRef]
35. Seviaryna, I.; Bueno, H.G.; Maeva, E.; Tjong, J. Characterization of natural fibre-reinforced composites with advanced ultrasonic techniques. In Proceedings of the IEEE International Ultrasonics Symposium, Chicago, IL, USA, 2–6 September 2014; pp. 1428–1431. [CrossRef]
36. Mahir, F.I.; Keya, K.N.; Sarker, B.; Nahiun, K.M.; Khan, R.A. A brief review on natural fiber used as a replacement of synthetic fiber in polymer composites. *Mater. Eng. Res.* **2019**, *1*, 86–97. [CrossRef]
37. Khalid, M.Y.; Al Rashid, A.; Arif, Z.U.; Ahmed, W.; Arshad, H.; Zaidi, A.A. Natural fiber reinforced composites: Sustainable materials for emerging applications. *Results Eng.* **2021**, *11*, 100263. [CrossRef]
38. Yashas Gowda, T.G.; Sanjay, M.R.; Jyotishkumar, P.; Suchart, S. Natural Fibers as Sustainable and Renewable Resource for Development of Eco-Friendly Composites: A Comprehensive Review. *Front. Mater.* **2019**, *6*, 226. [CrossRef]
39. Di Bella, G.; Fiore, V.; Valenza, A. Natural Fibre Reinforced Composites. In *Fiber Reinforced Composites*; Cheng, Q., Ed.; Nova Science Publishers: New York, NY, USA, 2012; pp. 57–90.
40. Lotfi, A.; Li, H.; Dao, D.V.; Prusty, G. Natural fiber–reinforced composites: A review on material, manufacturing, and machinability. *J. Thermoplast. Compos. Mater.* **2021**, *34*, 238–284. [CrossRef]
41. Ciupan, M.; Ciupan, E.; Cionca, I.; Heres, V.; Muresan, C. Assessing the influence of mechanical properties of a polypropylene and hemp composite on the design of upholstered furniture. *Acta Tech. Napoc.-Ser. Appl. Math. Mech. Eng.* **2022**, *65*, 91–96.
42. Manian, A.P.; Cordin, M.; Pham, T. Extraction of cellulose fibers from flax and hemp: A review. *Cellulose* **2021**, *28*, 8275–8294. [CrossRef]
43. Ahmed, A.T.M.F.; Islam, M.Z.; Mahmud, M.S.; Sarker, M.E.; Islam, M.R. Hemp as a potential raw material toward a sustainable world: A review. *Heliyon* **2022**, *8*, e08753. [CrossRef]
44. Dhakal, H.N.; Zhang, Z. The use of hemp fibres as reinforcements in composites. In *Biofiber Reinforcements in Composite Materials*; Faruk, O., Sain, M., Eds.; Woodhead Publishing: Cambridge, UK, 2015; pp. 86–103, ISBN 9781782421221. [CrossRef]
45. Yan, Z.L.; Wang, H.; Lau, K.T.; Pather, S.; Zhang, J.C.; Lin, G.; Ding, Y. Reinforcement of polypropylene with hemp fibres. *Compos. B Eng.* **2013**, *46*, 221–226. [CrossRef]
46. Burgada, F.; Fages, E.; Quiles-Carrillo, L.; Lascano, D.; Ivorra-Martinez, J.; Arrieta, M.P.; Fenollar, O. Upgrading Recycled Polypropylene from Textile Wastes in Wood Plastic Composites with Short Hemp Fiber. *Polymers* **2021**, *13*, 1248. [CrossRef] [PubMed]
47. Plastics Europe. Plastics-the Facts. 2020. Available online: https://plasticseurope.org/knowledge-hub/plastics-the-facts-2020/ (accessed on 1 November 2022).
48. EuRIC—Plastic Recycling Factsheet. Available online: https://www.euric-aisbl.eu/facts-figures/euric-brochures (accessed on 1 November 2022).
49. Le, K. Textile Recycling Technologies, Colouring and Finishing Methods. 2018. The University of British Columbia. Available online: https://sustain.ubc.ca/about/resources/textile-recycling-technologies-colouring-and-finishing-methods (accessed on 3 September 2021).
50. Nisar, J.; Aziz, M.; Shah, A.; Shah, I.; Iqbal, M. Conversion of Polypropylene Waste into Value-Added Products: A Greener Approach. *Molecules* **2022**, *27*, 3015. [CrossRef] [PubMed]
51. Aurrekoetxea, J.; Sarrionandia, M.A.; Urrutibeascoa, I.; Maspoch, M.L. Fracture behaviour of virgin and recycled isotactic polypropylene. *J. Mater. Sci.* **2001**, *36*, 5073–5078. [CrossRef]
52. da Costa, H.M.; Ramos, V.D.; de Oliveira, M.G. Degradation of polypropylene (PP) during multiple extrusions: Thermal analysis, mechanical properties and analysis of variance. *Polym. Test.* **2007**, *26*, 676–684. [CrossRef]
53. Aurrekoetxea, J.; Sarrionandia, M.A.; Urrutibeascoa, I.; Maspoch, M.L. Efects of recycling on the microstructure and the mechanical properties of isotactic polypropylene. *J. Mater. Sci.* **2001**, *36*, 2607–2613. [CrossRef]
54. Incarnato, L.; Scarfato, P.; Acierno, D.; Milana, M.R.; Feliciani, R. Infuence of recycling and contamination on structure and transport properties of polypropylene. *J. Appl. Polym. Sci.* **2003**, *89*, 1768–1778. [CrossRef]
55. Brachet, P.; Høydal, L.T.; Hinrichsen, E.L.; Melum, F. Modifcation of mechanical properties of recycled polypropylene from post-consumer containers. *Waste Manag.* **2008**, *28*, 2456–2464. [CrossRef]
56. Khademi, F.; Ma, Y.; Ayranci, C.; Choi, K.; Duke, K. Effects of recycling on the mechanical behavior of polypropylene at room temperature through statistical analysis method. *Polym. Eng. Sci.* **2016**, *56*, 1283–1290. [CrossRef]
57. Bourmaud, A.; Le Duigou, A.; Baley, C. What is the technical and environmental interest in reusing a recycled polypropylene-hemp fibrecomposite? *Polym. Degrad. Stab.* **2011**, *96*, 1732–1739. [CrossRef]
58. Małek, M.; Jackowski, M.; Łasica, W.; Kadela, M. Characteristics of Recycled Polypropylene Fibers as an Addition to Concrete Fabrication Based on Portland Cement. *Materials* **2020**, *13*, 1827. [CrossRef] [PubMed]
59. Tuladhar, R.; Yin, S. Production of recycled polypropylene (PP) fibers from industrial plastic waste through melt spinning process. In *Use of Recycled Plastics in Eco-efficient Concrete*; Pacheco-Torgal, F., Khatib, J., Colangelo, F., Tuladhar, R., Eds.; Woodhead Publishing: Cambridge, UK, 2019; pp. 69–84, ISBN 9780081026762. [CrossRef]

60. Yin, S. *Development of Recycled Polypropylene Plastic Fibres to Reinforce Concrete*; Springer: Berlin/Heidelberg, Germany, 2015; ISBN 978-981-10-3718-4. [CrossRef]
61. Ajorloo, M.; Ghodrat, M.; Kang, W.H. Incorporation of Recycled Polypropylene and Fly Ash in Polypropylene-Based Composites for Automotive Applications. *J. Polym. Environ.* **2021**, *29*, 1298–1309. [CrossRef]
62. Luna, P.; Lizarazo Marriaga, J.; Marino, A. Compatibilization of natural fibres as reinforcement of polymeric matrices. In Proceedings of the Fifth International Conference on Sustainable Construction Materials and Technologies, Kingston University, London, UK, 14–17 July 2019. [CrossRef]
63. Karaduman, Y.; Ozdemir, H.; Karaduman, N.S.; Ozdemir, G. Interfacial Modification of Hemp Fiber–Reinforced Composites. In *Natural and Artificial Fiber-Reinforced Composites as Renewable Sources*; Günay, E., Ed.; IntechOpen: London, UK, 2018. [CrossRef]
64. Nosova, N.; Roiter, Y.; Samaryk, V.; Varvarenko, S.; Stetsyshyn, Y.; Minko, S.; Stamm, M.; Voronov, S. Polypropylene surface peroxidation with heterofunctional polyperoxides. *Macromol. Symp.* **2004**, *210*, 339–348. [CrossRef]
65. Pracella, M.; Chionna, D.; Anguillesi, I.; Kulinski, Z.; Piorkowska, E. Functionalization, compatibilization and properties of polypropylene composites with Hemp fibres. *Compos. Sci. Technol.* **2006**, *66*, 2218–2230. [CrossRef]
66. Hamour, N.; Boukerrou, A.; Djidjelli, H.; Maigret, J.-E.; Beaugrand, J. Effects of MAPP Compatibilization and Acetylation Treatment Followed by Hydrothermal Aging on Polypropylene Alfa Fiber Composites. *Int. J. Polym. Sci* **2015**, *2015*, 9. [CrossRef]
67. Mohanty, S.; Nayak, S.K.; Verma, S.K.; Tripathy, S.S. Effect of MAPP as a Coupling Agent on the Performance of Jute–PP Composites. *J. Reinf. Plast. Compos.* **2004**, *23*, 625–637. [CrossRef]
68. Khalid, M.; Salmiaton, A.; Chuah, T.G.; Ratnam, C.T.; Thomas Choong, S.Y. Effect of MAPP and TMPTA as compatibilizer on the mechanical properties of cellulose and oil palm fiber empty fruit bunch–polypropylene biocomposites. *Compos. Interfaces* **2008**, *15*, 251–262. [CrossRef]
69. Kim, H.-S.; Lee, B.-H.; Choi, S.-W.; Kim, S.; Kim, H.-J. The effect of types of maleic anhydride-grafted polypropylene (MAPP) on the interfacial adhesion properties of bio-flour-filled polypropylene composites. *Compos. Part A Appl. Sci. Manuf.* **2007**, *38*, 1473–1482. [CrossRef]
70. Sunny, T.; Pickering, K.L. Improving Polypropylene Matrix Composites Reinforced with Aligned Hemp Fibre Mats Using High Fibre Contents. *Materials* **2022**, *15*, 5587. [CrossRef]
71. Râpă, M.; Spurcaciu, B.N.; Ion, R.-M.; Grigorescu, R.M.; Darie-Niță, R.N.; Iancu, L.; Nicolae, C.-A.; Gabor, A.R.; Matei, E.; Predescu, C. Valorization of Polypropylene Waste in the Production of New Materials with Adequate Mechanical and Thermal Properties for Environmental Protection. *Materials* **2022**, *15*, 5978. [CrossRef]
72. Yalcin, I.; Sadikoglu, T.G.; Berkalp, O.B.; Bakkal, M. Utilization of various non-woven waste forms as reinforcement in polymeric composites. *Text. Res. J.* **2013**, *83*, 1551–1562. [CrossRef]
73. Sakthivel, S.; Senthil Kumar, S.; Mekala, N.; Dhanapriya, G. Development of Sound Absorbing Recycled Nonwoven Composite Materials. *IOP Conf. Ser. Mater. Sci. Eng.* **2021**, *1059*, 012023. [CrossRef]
74. Lou, C.-W.; Lin, J.-H.; Su, K.-H. Recycling Polyester and Polypropylene Nonwoven Selvages to Produce Functional Sound Absorption Composites. *Text. Res. J.* **2005**, *75*, 390–394. [CrossRef]
75. Meng, X.; Fan, W.; Ma, Y.; Wei, T.; Dou, H.; Yang, X.; Tian, H.; Yu, Y.; Zhang, T.; Gao, L. Recycling of denim fabric wastes into high-performance composites using the needle-punching nonwoven fabrication route. *Text. Res. J.* **2020**, *90*, 695–709. [CrossRef]
76. Meng, X.; Fan, W.; Wan Mahari, W.A.; Ge, S.; Xia, C.; Wu, F.; Han, L.; Wang, S.; Zhang, M.; Hu, Z.; et al. Production of three-dimensional fiber needle-punching composites from denim waste for utilization as furniture materials. *J. Clean. Prod.* **2021**, *281*, 125321. [CrossRef]
77. Renouard, N.; Merotte, J.; Kervoelen, A.; Behlouli, K.; Baley, C.; Bourmaud, A. Exploring two innovative recycling ways for poly-(propylene)-flax non wovens wastes. *Polym. Degrad. Stabil.* **2017**, *142*, 89–101. [CrossRef]
78. Ailenei, E.C.; Loghin, M.C.; Ichim, M.; Hoblea, A. New composite materials using polyester woven fabric scraps as reinforcement and thermoplastic matrix. *Ind. Text.* **2021**, *62*, 62–67. [CrossRef]
79. Ball, P. Manufacturing Processes. In *Handbook of Polymer Composites for Engineers*; Hollaway, L., Ed.; Woodhead Publishing Limited: Cambridge, UK, 1994; pp. 73–94.
80. Salazar-Cru, B.A.; Chávez-Cinco, M.I.; Morales-Cepeda, A.B.; Ramos-Galván, C.E.; Rivera-Armenta, J.L. Evaluation of Thermal Properties of Composites Prepared from Pistachio Shell Particles Treated Chemically and Polypropylene. *Molecules* **2022**, *27*, 426. [CrossRef]
81. Bourmaud, A.; Fazzini, M.; Renouard, N.; Behlouli, K.; Ouagne, P. Innovating routes for the reused of PP-flax and PP-glass non woven composites: A comparative study. *Polym. Degrad. Stab.* **2018**, *152*, 259–271. [CrossRef]
82. Liu, J.L.; Xia, R. A unified analysis of a micro-beam, droplet and CNT ring adhered on a substrate: Calculation of variation with movable boundaries. *Acta Mech. Sin.* **2013**, *29*, 62–72. [CrossRef]
83. Saad, M.J. Effect of Maleated Polypropylene (MAPP) on the Tensile, Impact and Thickness Swelling Properties of Kenaf Core—Polypropylene Composites. *J. Sci. Technol.* **2010**, *2*, 33–44.
84. Doh, S.J. A Study on the Effects of Textile Processing on Tensile Properties of Single Cotton Fibers—From Raw Cotton to Washed Garments. Ph.D. Thesis, North Carolina State University, Raleigh, NC, USA, 2004.
85. Dong, Y.; Bhattacharyya, D. Dual role of maleated polypropylene in processing and material characterisation of polypropylene/clay nanocomposites. *Mater. Sci. Eng. A* **2010**, *527*, 1617–1622. [CrossRef]

86. Dong, Y.; Bhattacharyya, D. Effect of matrix plasticisation on the characterisation of polypropylene/clay nanocomposites. *World J. Eng.* **2010**, *7*, 198–201.
87. Marcilla, A.; Beltrán, M. Mechanisms of Plasticizers Action. In *Handbook of Plasticizers*, 3rd ed.; Wypych, G., Ed.; ChemTec Publishing: Toronto, ON, Canada, 2017; pp. 119–134, ISBN 9781895198973. [CrossRef]
88. Chun, K.S.; Husseinsyah, S.; Osman, H. Tensile Properties of Polypropylene/Cocoa Pod Husk Biocomposites: Effect of Maleated Polypropylene. *Adv. Mater. Res.* **2013**, *747*, 645–648. [CrossRef]
89. Monteiro, S.N.; Calado, V.; Rodriguez, R.J.S.; Margem, F.M. Thermogravimetric behavior of natural fibres reinforced polymer compozites—An overview. *Mater. Sci. Eng. A* **2012**, *557*, 17–28. [CrossRef]
90. Joseph, P.V.; Joseph, K.; Thomas, S.; Pillai, C.K.S.; Prasad, V.S.; Groeninckx, G.; Sarkissova, M. The Thermal and Crystallization Studies of Short Sisal Fiber Reinforced Polypropylene Composites. *Mater. Sci. Eng.* **2003**, *34*, 253–266. [CrossRef]
91. Xu, H.; Liu, C.Y.; Chen, C.; Hsiao, B.S.; Zhong, G.J.; Li, Z.M. Easy alignment and effective nucleation activity of ramie fibers in injection-molded poly(lactic acid) biocomposites. *Biopolymers* **2012**, *97*, 825–839. [CrossRef]
92. Anderson, K.S.; Hillmyer, M.A. Melt preparation and nucleation efficiency of polylactide stereocomplex crystallites. *Polymer* **2006**, *47*, 2030–2035. [CrossRef]
93. Zhang, X. Manufacturing of Hemp/PP Composites and Study of its Residual Stress and Aging Behavior. Ph.D. Thesis, University of Technology of Troyes, Troyes, France, 2016.
94. Pracella, M.; Haque, M.M.-U.; Alvarez, V. Functionalization, Compatibilization and Properties of Polyolefin Composites with Natural Fibers. *Polymers* **2010**, *2*, 554–574. [CrossRef]

Disclaimer/Publisher's Note: The statements, opinions and data contained in all publications are solely those of the individual author(s) and contributor(s) and not of MDPI and/or the editor(s). MDPI and/or the editor(s) disclaim responsibility for any injury to people or property resulting from any ideas, methods, instructions or products referred to in the content.

Article

Development and Characterization of Sustainable Antimicrobial Films Incorporated with Natamycin and Cellulose Nanocrystals for Cheese Preservation

Miriane Maria de Sousa [1], Vânia Miria C. Clemente [1], Rosilene Maria de S. Santos [1], Mariane Oliveira [1], José Osvaldo Ramos Silva [1], Laís Fernanda Batista [1], Clara Suprani Marques [1,2,*], Amanda Lélis de Souza [1], Éber Antônio Alves Medeiros [1] and Nilda de Fátima Ferreira Soares [1]

1. Food Technology Department, Federal University of Viçosa, Viçosa 36570-900, Brazil
2. Department of Agrarian Sciences, Federal Institute of Minas Gerais, Campus Bambuí, Bambuí 38900-000, Brazil
* Correspondence: supraniclara@gmail.com

Abstract: Environmental pollution and food safety are both issues of global concern. In this sense, sustainable and antimicrobial nanocomposites based on cellulose/poly (vinyl alcohol) blend incorporated with natamycin and cellulose nanocrystals (CNC) were manufactured and characterized. The developed films were evaluated according to their mechanical and optical properties, and their barrier to oxygen and water vapor permeation. The antimycotic activity was evaluated in vitro against fungi and yeasts. The film's potential to act as an active packaging for Minas cheese preservation was also assessed. The incorporation of CNC increased the films' tensile strength; however, it did not influence the barrier properties to water vapor (4.12×10^{-7} g·cm·m^{-1}·h^{-1}·Pa^{-1}) and oxygen (3.64×10^{-13} g·cm·m^{-1}·h^{-1}·Pa^{-1}). The incorporation of natamycin, on the other hand, resulted in films that were more opaque (around 24%) and of a yellowish color. The active nanocomposites developed showed antimicrobial effects against all analyzed fungi and yeasts (approximately 35 mm of inhibition zone) and were able to control the growth of *S. cerevisiae* in cheese, reducing a log cycle until the 12th day of storage. Since they performed well in vitro and on food, it was concluded that the films showed potential to be applied in Minas cheese preservation.

Keywords: active packaging; cellulose nanocrystal; food packaging; food spoilage microorganisms; methyl cellulose; nanotechnology

1. Introduction

Cheese production is one of the most important activities of the dairy industry in Brazil [1]. In order to guarantee a final product with microbiological quality, improve the product shelf life, minimize risks to consumers' health, as well as to reduce economic losses due to food waste, certain precautions are required throughout the production chain, such as the implementation of good hygiene practices and food safety management systems [2,3]. Besides that, several innovative technologies have been developed and used to preserve perishable goods and to ensure food safety.

Among these new technologies, it is worth noting the active packaging with the incorporation of antimicrobial substances [4,5]. An important advantage of this kind of technology is that the addition of the preservative agents into the polymer matrix of the packaging enables reducing the addition of preservatives directly into the food formulation, which is in accordance with emerging consumer trends such as lower additive content in foods [6,7]. Moreover, since most microbial growth takes place on the food surface, the presence of antimicrobial agents in the packaging may improve their preservative action [8].

Antimicrobial packaging has been tested on different perishable foodstuffs, such as meat, fruits, vegetables, bakery goods, and dairy products, with promising results [9–13].

Regarding dairy products, preservative agents such as natamycin, lysozyme, and nisin are interesting antimicrobials to consider as active packaging additives since they are natural compounds already used in cheese production [4,13,14]. Oliveira et al. [4], for example, manufactured an active packaging incorporated with nisin, which showed relevant activity against *Staphylococcus aureus*, a pathogen occasionally linked to foodborne outbreaks [15].

Natamycin, also known as piramicin, is a natural antimycotic produced by the Gram-positive bacteria *Streptomyces natalensis*. It is considered generally recognized as safe (GRAS) by Food and Drug Administration (FDA, Silver Spring, MD, USA) and used in cheese as an antimicrobial agent, preventing product contamination by yeasts and molds [14]. Since fungal spoilage occurs mainly on the cheese surface and the current methods used to cover the food with natamycin, such as spraying and coating, have low efficiency, its incorporation into active food packaging has been considered and studied in recent years [16,17]. The active packaging developed by Fayed et al. [17] with natamycin nanoparticles was able to control the growth of *Aspergillus flavus* on Romy cheese, as well as significantly reduce toxin production. Similarly, the natamycin-based low-density polyethylene film elaborated by Anari et al. [18] exhibited activity against spoilage yeasts common in yogurt and could contribute to increasing the shelf life of this kind of product.

Another important topic that attracts the attention of researchers on a worldwide scale is the hazards regarding the accumulation of non-degradable plastics and their negative impact on the environment. In this context, there is a growing interest in the research of biodegradable polymers aiming at the production of sustainable packaging [19]. Bio-renewable sources, such as cellulose and its derivatives, have been studied as possible materials for green packaging; however, despite their desirable sustainable nature, these kinds of polymers originate films that present many limitations when compared to petroleum-based films since their barrier, mechanical and thermal properties leave something to be desired [5,20]. Low impact resistance, brittleness, and higher permeation to water vapor and other gases are characteristics usually related to bio-based packaging [5,20,21]. Furthermore, the higher hydrophilicity exhibited by several biopolymers restricts their range of applications in the food industry [21].

In order to improve the bioplastic properties, the incorporation of nanomaterials into a compatible polymeric matrix is suggested since this allows the obtainment of an improved bio-nanocomposite that might meet the mechanical and barrier requirements for food packaging [22]. Among the several nanomaterials studied, we highlight nanocrystalline cellulose (CNC): nanostructures obtained from a renewable source (cellulose) that can be added to a polymeric matrix, acting as a mechanical and barrier reinforcement without altering its biodegradable nature. In this context, this study aimed at the production of antimicrobial and sustainable packaging, which consisted of a methylcellulose (MC) and poly(vinyl alcohol) (PVOH) blend incorporated with CNC and natamycin. The mechanical, optical, and barrier properties of the films were evaluated, as well as the film's capability to preserve Minas cheese.

2. Materials and Methods

2.1. Materials

The following materials were used in the present research: natamicyn 50% (Natamax natural antimicrobial, Danisco Brazil LTDA, Brazil); CNC was extracted from softwoodpulp supplied by Klabin (São Paulo, SP, Brazil); glycerol (Sigma-Aldrich, St. Louis, MO, USA); poly(vinyl alcohol) (PVOH, PM = 85,000–124,000 g·mol^{-1}, Sigma-Aldrich, St. Louis, MO, USA); methylcellulose (MC, Sigma-Aldrich, St. Louis, MO, USA); lithium chloride; sodium chloride; potato dextrose agar (PDA, Sigma-Aldrich, St. Louis, MO, USA; Sabouraud dextrose broth (Kasvi, Padova, Italy). Minas cheese was purchased from a local market in Viçosa, Minas Gerais, transported to the laboratory, and kept under refrigeration until analyses. The yeasts *Saccharomyces cerevisiae* and *Kluyveromyces lactis*, and the food spoilage molds *Alternaria alternata*, *Rhizopus stolonifer*, *Fusarium semitectum*, and *Aspergillus niger* were obtained from the culture collection of the Food Packaging Laboratory, Federal University of Viçosa.

2.2. Methods

2.2.1. Experimental Design

The experimental design was completely randomized, and a two-factor five-level rotatable central composite design (RCCD) with five repetitions at the central point, totaling thirteen trials, was implemented to study the effect of the incorporation of CNC and natamycin in the optical, mechanical, barrier, and antimycotic properties of the MC/PVOH blend. The coded levels and their respective real values of concentrations of natamycin and CNC are displayed in Table 1.

Table 1. Composition of the two-factor five-level RCCD of the elaborated films incorporated with cellulose nanocrystals (CNC) and natamycin. Coded levels and their respective real values.

Treatments	Coded Levels		Real Values	
	X1	X2	CNC (% wt./wt.)	Natamycin (% wt./wt.)
1	−1	−1	0.70	0.70
2	+1	−1	4.30	0.70
3	−1	+1	0.70	4.30
4	+1	+1	4.30	4.30
5	−1.41	0	0	2.50
6	+1.41	0	5.00	2.50
7	0	−1.41	2.50	0
8	0	+1.41	2.50	5.00
9	0	0	2.50	2.50
10	0	0	2.50	2.50
11	0	0	2.50	2.50
12	0	0	2.50	2.50
13	0	0	2.50	2.50

The % was based on the final polymers' mass.

2.2.2. Elaboration of the Polymeric Blend Incorporated with CNC and Natamycin

Different concentrations of CNCs (Table 1, Section 2.2.1) were dispersed in 200 mL of distilled water, homogenized in ultraturrax (5 min, 20,000 rpm, model T25, IKA), and ultrasonicated (450 W, Unique Group, Rio de Janeiro, RJ, Brazil) for 5 min. After, 2.60 g of PVOH and 6.25 g of MC were added to the CNC dispersion and heated at 70 °C for 4 h under continuous stirring until solubilization, followed by gelification. Different concentrations of natamycin were added subsequently (Table 1, Section 2.2.1), followed by the incorporation of the plasticizer glycerol (30% wt./wt.). All concentrations were calculated based on the total polymer mass. The obtained dispersions were continuously stirred for 20 min and poured on glass plates (18 cm × 34 cm) until solvent evaporation.

2.2.3. Thickness and Mechanical Properties

Film thickness was measured, in µm, with a digital micrometer (model 547–401, Mitutoyo, Kawasaki, Japan). Ten specimens of each treatment were analyzed, and the measurement was realized in ten random points of each sample [9]. The films were also evaluated according to their mechanical properties of ultimate tensile strength (UTS, in MPa), elongation at break (EB, in %), and modulus of elasticity (Young's modulus, YM, in MPa) using a Universal Testing Machine (model 3367, Instron Corporation, Norwood, MA, USA) equipped with a 1 kN load cell. Five specimens (25 mm × 150 mm) of each treatment were evaluated. The initial distance of grids separation was 100 mm, and the standardized rate of separation was 50 mm·min^{-1} [23].

2.2.4. Color and Opacity Measurement

The color of the films was assessed by a Color Quest XE spectrophotometer (Hunter Lab, Reston, VA, USA), working with D65, 10° angle, and using the CIELAB scale for the L*, a* and b* coordinates. Opacity (OP), yellowness index (YI), and b* (tendency to yellow) were the parameters investigated. Two samples were analyzed for each treatment, and five measurements were performed at random points of each sample [24].

2.2.5. Water Vapor Permeability (WVP)

WVP was investigated by the gravimetric method according to ASTM E96/E96M-10 [25] with modifications [9]. The obtained films were cut into circles (Ø = 83 mm) and sealed with paraffin in poly(methyl methacrylate) cups containing a saturated solution of lithium chloride (RH = 12% ± 5% at 25 °C ± 2 °C). After, the cups were placed in desiccators containing a saturated solution of sodium chloride (RH = 75% ± 5% at 25 °C ± 2 °C). The cups were weighted periodically, enough to provide ten data points after the steady state was reached. The results were expressed as $g \cdot cm \cdot m^{-1} \cdot h^{-1} \cdot Pa^{-1}$.

2.2.6. Oxygen Permeability (O_2P)

The permeability to oxygen was verified in a VAC-V1 gas permeability tester (Labthink Instruments, Jinan, China). The experiment was carried out at 23 °C ± 2 °C and 50% RH. The exposed area of the films was 38.48 mm^2, and the analysis was conducted for 8 h. The transmission rate of oxygen through film was expressed as $g \cdot cm \cdot m^{-1} \cdot h^{-1} \cdot Pa^{-1}$.

2.2.7. In Vitro Assessment of the Antimycotic Activity of the Films

The yeasts *S. cerevisiae* and *K. lactis* were grown on Sabouraud broth, incubated for 24 h at 32 °C, plated on acidified PDA, and incubated once more for 7 days at 25 °C ± 2 °C. After, isolated colonies were selected, suspended in a 5 mL 0.85% (wt./wt.) saline solution, and adjusted to 0.5 McFarland standard (around 1.0×10^6 cells·mL^{-1}) [26]. Regarding the molds, the conidia were grown on Sabouroud broth for 7 days at 25 °C ± 2 °C, plated on acidified PDA, and incubated again at the same conditions. Subsequently, the conidia were collected and transferred to 5 mL 0.85% (wt./wt.) saline solution and adjusted to optical density (OD_{530}) of 0.10 (GBC UV/Vis 918), which represented around 5×10^4 conidia·mL^{-1} [26].

After, aliquots of 100 µL of the suspensions prepared were spread on acidified PDA, and samples of the films (Ø = 10 mm) were placed at the center of each plate. The plates inoculated with molds or yeasts were incubated for 7 days at 25 °C ± 2 °C, and the inhibition zone verified was measured in mm.

2.2.8. Assessment of the Active Films on Minas Cheese Preservation

The surface of the food sample (25 g) was decontaminated with alcohol 70% (v/v). Subsequently, aliquots of 100 µL of *S. cerevisiae* (10^4 CFU·mL^{-1}) (Section 2.2.7) were spread on cheese surface, and, right after, the samples were left to dry for 15 min [27]. The inoculated samples were packaged with the following elaborated films: film A (5% of CNC and 2% natamycin) and film B (5% of CNC and 0% of natamycin, taken as the control). The samples were stored at 4 °C for 15 days. Every three days, for a 15-day period, the samples (25 g) were homogenized with 225 mL of peptone water (0.1% v/v), diluted accordingly, and spread plated on PDA agar. The plates were incubated for 7 days at 25 °C ± 2 °C, and the results were expressed as log of colony forming unit of *S. cerevisiae* per gram of cheese (Log CFU·g^{-1}) [9].

2.2.9. Statistical Analysis

The models representing the parameter's behavior as a function of CNC and/or natamycin concentrations were obtained using analysis of variance (ANOVA) with a 5% level of significance followed by regression analysis. The software Minitab 17 was used.

3. Results and Discussion

3.1. Films' Appearance

The MC and PVOH blends incorporated with CNC and natamycin were successfully achieved, as can be seen in Figure 1. Macroscopically, they presented a smooth and homogeneous aspect. Natamycin incorporation, especially, resulted in films with a yellowish color and less transparent than films incorporated with CNC only. Complementarily, Table 2 shows the results obtained for opacity, yellow index, and tendency to yellow.

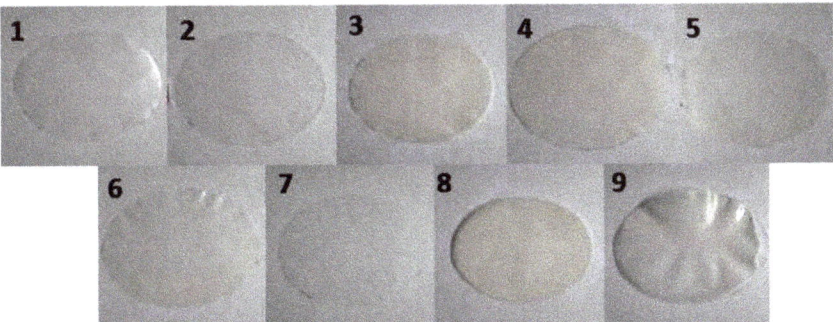

Figure 1. Photographs of the methyl cellulose and poly(vinyl alcohol) films elaborated with nanocrystalline cellulose (CNC) and natamycin: 0.7% of CNC and 0.7% of natamycin (1), 4.3% of CNC and 0.7% of natamycin (2), 0.7% of CNC and 4.3% of natamycin (3), 4.3% of CNC and 4.3% of natamycin (4), 0% of CNC and 2.5% of natamycin (5), 5% of CNC and 2.5% of natamycin (6), 2.5% of CNC and 0% of natamycin (7), 2.5% of CNC and 5% of natamycin (8), and 2.5% of CNC and 2.5% of natamycin (9).

Table 2. Results obtained for optical parameters (tendency to yellow, b*, opacity, OP, and yellow index, YI), thickness, mechanical properties (ultimate tensile strength, UTS, elongation at break, EB, Young's modulus, YM), water vapor permeability (WVP), and oxygen permeability (O_2P) of films elaborated with methyl cellulose, poly(vinyl alcohol), cellulose nanocrystals, and natamycin.

Treatment	b*	OP (%)	YI	Thickness (μm)	UTS (MPa)	EB (%)	YM (MPa)	WVP (10^{-7}) g·cm·m^{-1}·h^{-1}·Pa^{-1}	O_2P (10^{-13}) g·cm·m^{-1}·h^{-1}·Pa^{-1}
1	2.78	16.85	4.67	163	19.74	64.21	6.59	3.14	1.03
2	2.83	16.39	4.69	177	29.11	72.78	9.53	4.45	1.33
3	5.19	24.18	9.18	191	22.64	73.97	7.39	3.25	1.26
4	5.45	21.35	9.38	182	20.64	65.66	7.63	4.61	9.40
5	3.86	19.97	6.51	181	21.43	79.54	6.28	4.27	1.37
6	4.04	18.57	7.49	164	29.37	71.86	10.36	3.29	7.15
7	1.80	16.62	3.05	146	26.79	78.83	6.53	3.97	9.34
8	6.40	23.39	10.80	193	22.85	71.72	7.52	3.27	1.37
9	4.64	21.29	8.12	161	26.63	81.93	5.93	4.73	2.05
10	4.06	18.34	6.8	172	27.73	76.78	9.28	4.52	1.41
11	4.68	19.59	7.98	194	27.48	74.19	7.82	5.15	8.89
12	4.46	18.65	7.18	149	28.65	75.38	9.69	4.47	1.39
13	4.28	19.04	7.72	163	26.65	77.07	8.05	4.49	1.43

From the results for optical properties, it was possible to adjust models (non-significant lack-of-fit) that explained the behavior of the three parameters analyzed as a function of the concentration of natamycin only, as displayed in Figure 2. CNC concentration, in turn, did not impact these parameters, suggesting a good dispersion of the nanomaterial in the polymer matrix. The yellowish color of films containing natamycin is possibly due to the slightly yellow color of the powder. In addition, the increased opacity verified when higher amounts of natamycin were added into films was probably the result of the hydrophobic portion of the peptide and its poor solubility in water, which affected the antimicrobial dispersion in the hydrophilic matrix and resulted in opaquer films [14,28].

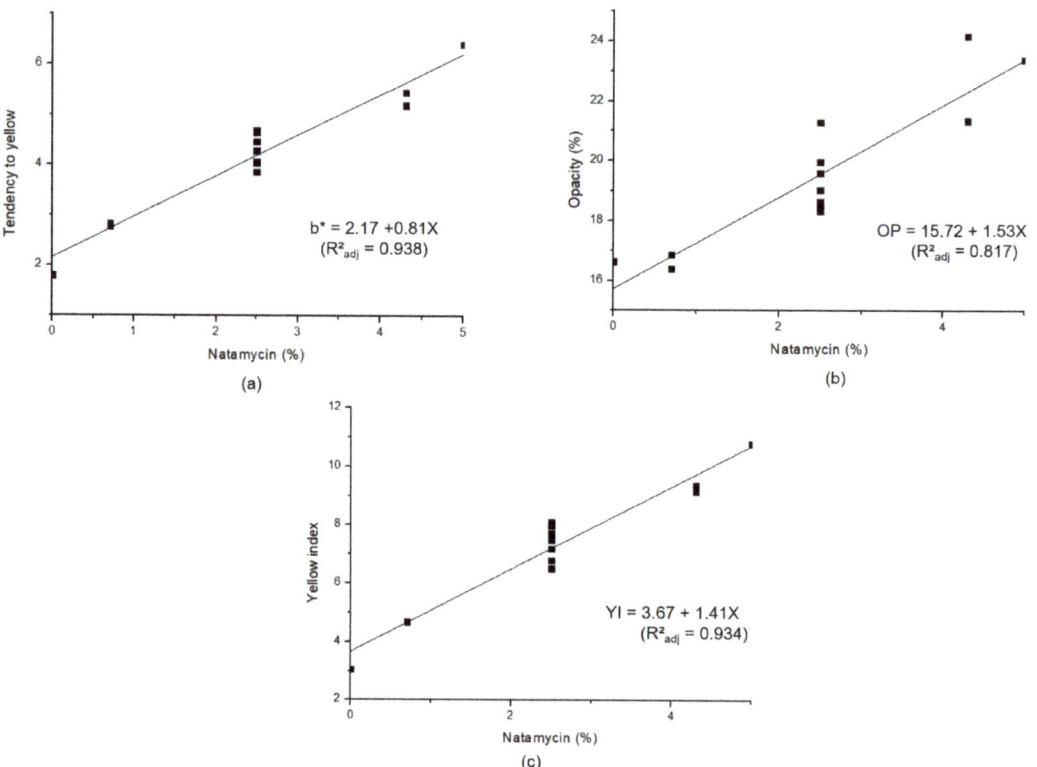

Figure 2. Plots of optical parameters as a function of natamycin concentration (% wt./wt., represented by X), their regression model, and adjusted R^2: tendency to yellow (b*, **a**), opacity (OP, **b**), and yellow index (YI, **c**).

3.2. Mechanical Properties

When developing new materials for food packaging, it is crucial to evaluate their mechanical properties since these features will allow us to infer about the packaging behavior when under certain conditions, such as mechanical stress. The results for thickness, UTS, EB, and YM are also displayed in Table 2. It was possible to adjust a regression model for only UTS, and the behavior was a function of both CNC (X_1) and natamycin (X_2) concentrations, as demonstrated in Equation (1). The other parameters (thickness, EB, and YM) were not explained by any models, therefore, we considered only the global mean value: thickness of 172 µm; 68.74% of EB; YM of 7.89 MPa.

$$UTS = 14.01 + 5.92X_1 + 4.30X_2 - 0.48X_1^2 - 0.58X_2^2 - 0.88X_1X_2 \rightarrow (R^2_{adj} = 0.910) \quad (1)$$

Complementarily to the adjusted model, Figure 3 shows the surface response obtained for UTS.

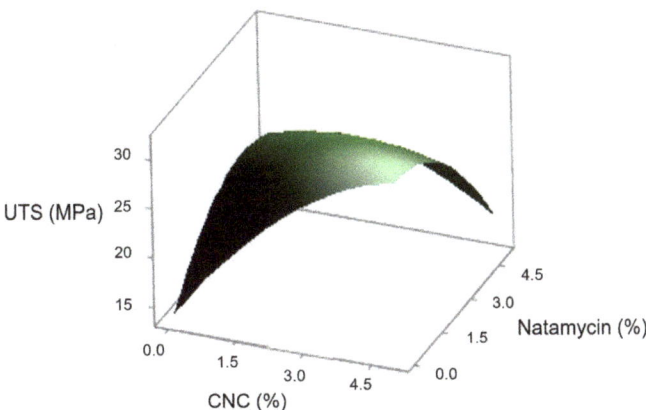

Figure 3. Three-dimensional response surface plot for ultimate tensile strength (UTS) as a function of cellulose nanocrystals (CNC, X_1) and natamycin (X_2) concentrations.

The findings indicated that both CNC and natamycin impacted the films' UTS, which assumed a parabolic shape. This suggested that, beyond certain concentrations, the additive's incorporation had a negative effect on the UTS parameter. Furthermore, the term X_1X_2 in Equation (1) suggested that, for this parameter, there was an interaction between CNC and natamycin: the UTS was likely to increase when higher concentrations of CNC and lower concentrations of natamycin were used. Other research also verified an increase in the tensile strength of sustainable films of PVOH incorporated with CNC [29,30]. This behavior may be explained by the high amount of –OH groups on the CNC structure, which are able to strongly interact with the hydrophilic polymeric matrices (MC and PVOH), increasing the compaction of the chains. This more homogenous structure may play an important role in stress redistribution all over the chains. Besides that, the high specific surface area and aspect ratio are CNCs' features that allow their application as potential reinforcing materials [31]. Natamycin, on the other hand, contributed less to the mechanical parameters, which may be due to its hydrophobic region and lower compatibility with both polymers.

3.3. Water Vapor and Oxygen Permeability

The results obtained for WVP and O_2P are presented in Table 2. In the food packaging field, it is of utmost importance to ensure that gas exchanges between the food and the external environment are minimal. While water vapor permeation can adversely affect both the product's shelf life and its sensorial quality, oxygen permeation triggers oxidation, which in turn leads to undesirable changes in flavor, texture, and appearance.

It was not possible to adjust a regression model for either WVP or O_2P as a function of CNC and natamycin concentrations (significant lack of fit, indicating not suitable models). Due to this, we considered only the global mean values: 4.12×10^{-7} g·cm·m^{-1}·h^{-1}·Pa^{-1} for WPV and 3.64×10^{-13} g·cm·m^{-1}·h^{-1}·Pa^{-1} for O_2P. Although it is said in the literature that CNC is able to improve the barrier properties of biodegradable packaging, hindering the flow of molecules through tortuous paths, contrary to our expectations, neither CNC nor natamycin influenced the gas permeability of the elaborated films [32,33].

3.4. In Vitro Antimycotic Activity

At first, the films' antimycotic activity was assessed in vitro against four molds of importance in food preservation, as well as two yeasts. Photographs of the inoculated plates after the incubation period of seven days are displayed in Figure 4.

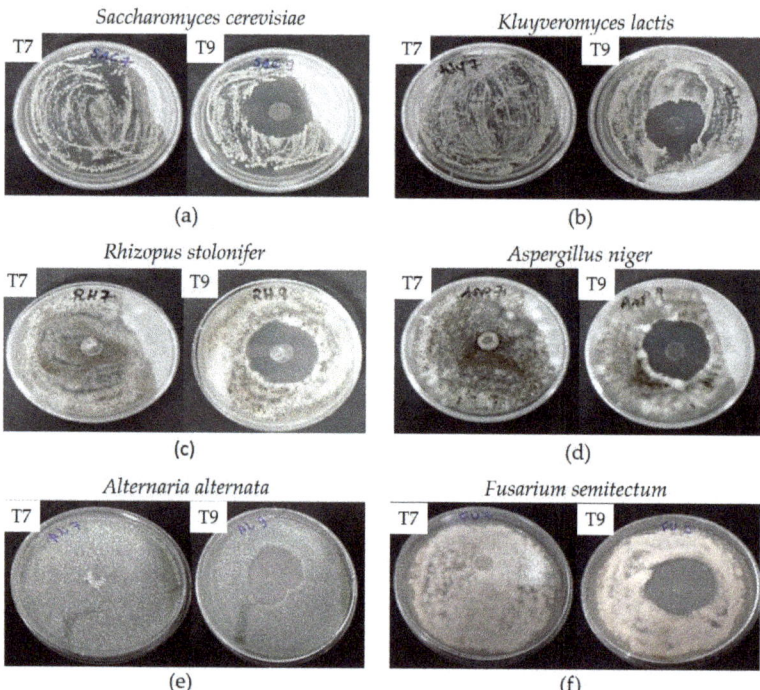

Figure 4. Photographs of plates after incubation of different yeasts (*S. cerevisiae* (**a**) and *K. lactis* (**b**)) and molds (*R. stolonifer* (**c**), *A. niger* (**d**), *A. alternate* (**e**), and *F. semitectum* (**f**)) at 25 °C for 7 days. T7: films with 2.5% of CNC and 0% of natamycin. T9: films with 2.5% of CNC and 2.5% of natamycin.

Films elaborated without natamycin (T7) did not show antimicrobial activity, as can be seen in Figure 4. When incorporated with natamycin, all the films displayed antimicrobial properties, suggesting that they could be used as active films for food preservation against these microorganisms. Regardless of the natamycin concentration, the inhibition zones verified for each mold or yeast investigated were similar, and the mean values obtained are presented in Table 3. The only exception was when the films were tested against *A. alternata*, in which it was possible to adjust a polynomial model of the inhibition zone as a function of natamycin concentration (Table 3).

Table 3. Inhibition zones obtained when the elaborated films with different concentrations of natamycin (0.7%, 2.5%, 4.3%, and 5.0%) were assessed against molds and yeasts of importance in food preservation.

Microorganism	Inhibition Zone (mm)	R^2_{adj}
Alternaria alternata	IZ = 3.12 + 17.13X − 2.37X^2	0.92
Aspergillus niger	34.7 ± 4.4	-
Rhizopus stolonifer	35.4 ± 3.0	-
Fusarium semitectum	34.2 ± 4.0	-
Saccharomyces cerevisiae	36.7 ± 3.5	-
Kluyveromyces lactis	35.0 ± 2.7	-

IZ = inhibition zone; X = natamycin concentration (wt./wt.).

Natamycin is a preservative allowed for dairy products, and it has been the object of study of several pieces of research regarding active food packaging. Similar to the present study, Chakravartula et al. (2020) [34] verified that blend films elaborated with cassava starch, chitosan, and natamycin presented potential to be used as active packaging due to the antifungal activity displayed by the films. Films incorporated with natamycin were also manufactured by Grafia et al. (2018) [16]. The authors tested the active films in vitro against *A. niger* previously isolated from onions and achieved interesting outcomes.

Literature shows that the mechanism of action of natamycin against fungi involves interference in the cells' vacuole and a preferred binding to ergosterol, a sterol molecule found in fungi cells membrane responsible for cell integrity [35–37]. Besides that, natamycin is also referred to as an efficient inhibitor of the transport of nutrients, compromising cell function [37].

3.5. Application of the Active Packaging in Minas Cheese

The central composite design allowed us to optimize the packaging system and define the film formulation that should be tested on food based on the outcomes found for the investigated properties. Minas cheese, a traditional-Brazilian product, was chosen as the food sample in this step since preliminary tests indicated that the film remained well preserved after contact with the food. The active films tested on the Minas cheese samples were elaborated with 5% of CNC and 2% of natamycin (film A), and 5% of CNC (film B, control). In this sense, cheese samples previously inoculated with *S. cerevisiae* were packaged with the films and stored for up to 15 days at low temperatures (4 °C). Regression models were adjusted for each case, and they are presented below (Equations (2) and (3)):

$$Y_A = 3.62 + 0.20T \to (R^2 adj = 0.878) \qquad (2)$$

$$Y_B = 3.79 + 0.44T - 0.014T^2 \to (R^2 adj = 0.925) \qquad (3)$$

in which Y_A is the yeast growth, in log CFU·g^{-1}, when the film A was used, Y_B is the yeast growth, also in CFU·g^{-1}, when the film B was used. For both equations, T represents the time, in days.

In Figure 5, it is possible to observe the yeast counts in Minas cheese during the storage time.

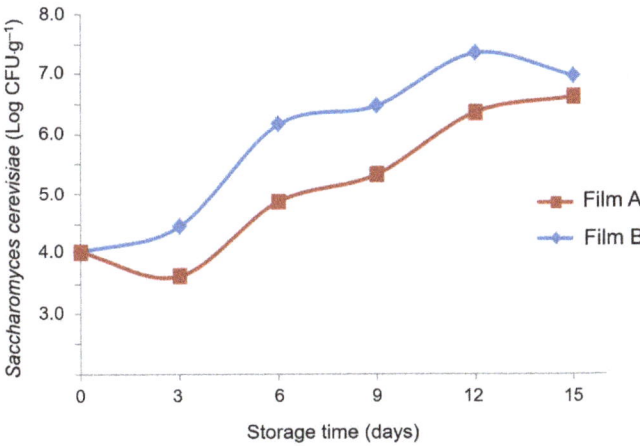

Figure 5. *Saccharomyces cerevisiae* count, in log CFU·g^{-1}, as a function of time, in Minas cheese stored at 4 °C for 15 days and packaged with methyl cellulose/poly (vinyl alcohol) blend films incorporated with 2% of natamycin and 5% of cellulose nanocrystals (Film A) and 5% of CNC (Film B).

It can be observed that the film incorporated with natamycin impacted the growth of *S. cerevisiae* in cheese in approximately one log cycle until the 12th day of refrigerated storage when compared to the film incorporated with only CNC. This means that a reduction of approximately 90% of the yeast population was achieved. Although *S. cerevisiae* is an important fermenting agent for the food industry, mostly involved in the manufacturing of bakery products and alcoholic beverages, it is considered a contaminant microorganism in cheese, usually related to product spoilage [38,39].

To date, fungal spoilage in cheese is one of the major challenges faced by the dairy industry [40]. In a study with 61 cheesemakers from the United States, 71% reported that the growth of undesirable mold on cheese surfaces was one of the most common issues in the area [41]. The occurrence of strange color and pigment development, probably caused by microbial growth, was also mentioned by around 54% of the respondents [39]. In addition, 28% of the interviewed cheesemakers mentioned that around 5 to 10% of their production was lost due to quality issues, indicating an economic loss that can be significant.

In order to minimize contaminants in dairy, it is essential to ensure the quality of the milk, as well as invest in and enforce good hygiene practices throughout the entire production chain. The adoption of novel technologies, such as active packaging, could be a complementary strategy to enhance quality and contribute to prolonging the product's shelf life. In this sense, these findings suggest that the application of a sustainable packaging MC/PVOH-based incorporated with CNC and natamycin in Minas cheese could play an important role in product preservation. Although CNC does not present antimicrobial activity, the material can be added as a nanofiller to enhance the mechanical properties of the packaging, while natamycin functions as the active component effective against molds and yeasts.

4. Conclusions

In the present study, an MC/PVOH/CNC/natamycin blend film was successfully manufactured and showed potential to be used as an antimicrobial film in Minas cheese preservation. Initially, it was hypothesized that the presence of CNC would improve the films' mechanical and barrier properties; however, only the UTS parameter was influenced by the CNC concentration. Nevertheless, this was an important finding since CNC did act as a mechanical reinforcing material. Future studies should be more focused on enhancing the barrier properties of the film since it is a significant parameter of food preservation. Furthermore, the presence of CNC did not impact the optical properties of the elaborated packaging, which was also a positive result. Natamycin, on the other hand, had a negative effect on the optical parameters and on the mechanical properties of the films, probably due to lower compatibility among the antimicrobial and the polymers. Although it enabled the film to act as an antimicrobial packaging against yeasts and molds, as well as displayed a good performance when tested on Minas cheese, its compatibility and interaction with the polymers should be studied in greater depth to allow the obtainment of active films with not only antimicrobial properties, but also better mechanical, optical, and barrier properties.

Author Contributions: Conceptualization, M.M.d.S. and N.d.F.F.S.; methodology, V.M.C.C., R.M.d.S.S., M.O. and J.O.R.S.; validation, M.M.d.S.; formal analysis, M.M.d.S.; investigation, M.M.d.S. and L.F.B.; resources, É.A.A.M. and N.d.F.F.S.; data curation, M.M.d.S.; writing—original draft preparation, M.M.d.S.; writing—review and editing, C.S.M. and A.L.d.S.; visualization, M.M.d.S. and C.S.M.; supervision, N.d.F.F.S.; project administration, N.d.F.F.S.; funding acquisition, É.A.A.M. and N.d.F.F.S. All authors have read and agreed to the published version of the manuscript.

Funding: This research was funded by CAPES (Financial code 001) and CNPq (427719/2018-6).

Informed Consent Statement: Not applicable.

Data Availability Statement: Not applicable.

Acknowledgments: The authors are grateful to Danisco Brazil LTDA and Klabin for providing Natamax and the softwood pulp, respectively.

Conflicts of Interest: The authors declare no conflict of interest.

References

1. Santos, H.C., Jr.; Maranduba, H.L.; Almeida Neto, J.A.; Rodrigues, L.B. Life cycle assessment of cheese production process in a small-sized dairy industry in Brazil. *Environ. Sci. Pollut. Res.* **2017**, *24*, 3470–3482. [CrossRef] [PubMed]
2. Carrascosa, C.; Millán, R.; Saavedra, P.; Jaber, J.R.; Raposo, A.; Sanjuán, E. Identification of the risk factors associated with cheese production to implement the hazard analysis and critical control points (HACCP) system on cheese farms. *J. Dairy Sci.* **2016**, *99*, 2606–2616. [CrossRef] [PubMed]
3. Kure, C.F.; Sakaar, I. The fungal problem in cheese industry. *Curr. Opin. Food Sci.* **2019**, *29*, 14–19. [CrossRef]
4. Oliveira, T.V.; Freitas, P.A.V.; Pola, C.C.; Terra, L.R.; Silva, J.O.R.; Badaró, A.T.; Junior, N.S.; Oliveira, M.M.; Silva, R.R.A.; Soares, N.F.F. The influence of intermolecular interactions between maleic anhydride, cellulose nanocrystal, and nisin-Z on the structural, thermal, and antimicrobial properties of starch-PVA plasticized matrix. *Polysaccharides* **2021**, *2*, 661–676. [CrossRef]
5. Marques, C.S.; Silva, R.R.A.; Arruda, T.R.; Ferreira, A.L.V.; Oliveira, T.V.; Moraes, A.R.F.; Dias, M.V.; Vanetti, M.C.D.; Soares, N.F.F. Development and investigation of zein and cellulose acetate polymer blends incorporated with garlic essential oil and β-cyclodextrin for potential food packaging application. *Polysaccharides* **2022**, *3*, 277–291. [CrossRef]
6. Motelica, L.; Ficai, D.; Ficai, A.; Oprea, O.C.; Kaya, D.A.; Andronescu, E. Biodegradable antimicrobial food packaging: Trends and Perspectives. *Foods* **2020**, *9*, 1438. [CrossRef]
7. Li, H.; Luo, J.; Li, H.; Han, S.; Fang, S.; Li, L.; Han, X.; Wu, Y. Consumer cognition analysis of food additives based on Internet public opinion in China. *Foods* **2022**, *11*, 2070. [CrossRef]
8. Yldiril, S.; Röcker, B.; Pettersen, M.K.; Nilsen-Nygaard, J.; Ayhan, Z.; Rutkaite, R.; Radusin, T.; Suminska, P.; Marcos, B.; Coma, V. Active packaging applications for food. *Compr. Rev. Food Sci. Food Saf.* **2018**, *17*, 165–199. [CrossRef]
9. Marques, C.S.; Arruda, T.R.; Silva, R.R.A.; Ferreira, A.L.V.; Oliveira, W.L.A.; Rocha, F.; Mendes, L.A.; Oliveira, T.V.; Vanetti, M.C.D.; Soares, N.F.F. Exposure to cellulose acetate films incorporated with garlic essential oil does not lead to homologous resistance in *Listeria innocua* ATCC 33090. *Food Res. Int.* **2022**, *160*, 111676. [CrossRef]
10. Takma, D.K.; Korel, F. Active packaging films as a carrier of black cumin essential oil: Development and effect on quality and shelf-life of chicken breast meat. *Food Packag. Shelf Life* **2019**, *19*, 210–217. [CrossRef]
11. Klinmalai, P.; Srisa, A.; Laorenza, Y.; Katekhong, W.; Harnkarnsujarit, N. Antifungal and plasticization effects of carvacrol in biodegradable poly(lactic acid) and poly(butylene adipate terephthalate) blend films for bakery packaging. *LWT* **2021**, *152*, 112356. [CrossRef]
12. Chiabrando, V.; Garavaglia, L.; Giacalone, G. The postharvest quality of fresh sweet cherries and strawberries with an active packaging system. *Foods* **2019**, *8*, 335. [CrossRef]
13. Irkin, R.; Esmer, O.K. Novel food packaging systems with natural antimicrobial agents. *J. Food Sci. Technol.* **2015**, *52*, 6095–6111. [CrossRef]
14. Meena, M.; Prajapati, P.; Ravichandran, C.; Sehrawat, R. Natamycin: A natural preservative for food applications—A review. *Food Sci. Technol.* **2021**, *30*, 1481–1496. [CrossRef]
15. Paramithiotis, S.; Drosinos, E.H.; Skandamis, P.N. Food recalls and warnings due to the presence of foodborne pathogens—A focus on fresh fruits, vegetables, dairy and eggs. *Curr. Opin. Food Sci.* **2017**, *18*, 71–75. [CrossRef]
16. Grafia, A.L.; Vázquez, M.B.; Bianchinotti, M.V.; Barbosa, S.E. Development of an antifungal film by polyethylene surface modification with natamycin. *Food Packag. Shelf Life* **2018**, *18*, 191–200. [CrossRef]
17. Fayed, A.; Elsayed, H.; Ali, T. Packaging fortified with natamycin nanoparticles for hindering the growth of toxigenic *Aspergillus flavus* and aflatoxin production in Romy cheese. *J. Adv. Vet. Anim. Res.* **2021**, *8*, 58–63. [CrossRef]
18. Anari, H.N.B.; Majdinasab, M.; Shaghaghian, S.; Khalesi, M. Development of a natamycin-based non-migratory antimicrobial active packaging for extending shelf-life of yogurt drink (Doogh). *Food Chem.* **2022**, *366*, 130606. [CrossRef]
19. Díaz-Montes, E. Polysaccharides: Sources, characteristics, properties, and their application in biodegradable films. *Polysaccharides* **2022**, *3*, 480–501. [CrossRef]
20. Silva, R.R.A.; Marques, C.S.; Arruda, T.R.; Teixeira, S.C.; Oliveira, T.V.; Stringheta, P.C.; Pires, A.C.S.; Soares, N.F.F. Ionic strength of methylcellulose-based films: An alternative for modulating mechanical performance and hydrophobicity for potential food packaging application. *Polysaccharides* **2022**, *3*, 426–440. [CrossRef]
21. Sid, S.; Mor, R.S.; Kishore, A.; Sharanagat, V.S. Bio-sourced polymers as alternatives to conventional food packaging materials: A review. *Trends Food Sci. Technol.* **2021**, *115*, 87–104. [CrossRef]
22. Youssef, A.M.; El-Sayed, S.M. Bionanocomposites materials for food packaging applications: Concepts and future outlook. *Carbohydr. Polym.* **2018**, *193*, 19–27. [CrossRef] [PubMed]
23. *ASTM D882-12*; Standard Test Method for Tensile Properties of Thin Plastic Sheeting. ASTM, American Society for Testing and Materials: West Conshohocken, PA, USA, 2012.
24. Ramesh, S.; Radhakrishnan, P. Cellulose nanoparticles from agro-industrial waste for the development of active packaging. *Appl. Surf. Sci.* **2019**, *484*, 1274–1281. [CrossRef]
25. *ASTM E96/E96M-10*; Standard test method for water vapor transmission of materials. ASTM, American Society for Testing and Materials: West Conshohocken, PA, USA, 2010.
26. Zomorodian, K.; Saharkhiz, J.; Pakshir, K.; Immeripour, Z.; Sadatsharifi, A. The composition, antibiofilm and antimicrobial activities of essential oil of *Ferula assa-foetida* oleo-gum-resin. *Biocatal. Agric. Biotechnol.* **2018**, *14*, 300–304. [CrossRef]

27. Marques, C.S.; Grillo, R.P.; Bravim, D.G.; Pereira, P.V.; Villanova, J.C.O.; Pinheiro, P.F.; Carneiro, J.C.S.; Bernardes, P.C. Preservation of ready-to-eat salad: A study with combination of sanitizers, ultrasound, and essential oil-containing β-cyclodextrin inclusion complex. *LWT* **2019**, *115*, 108433. [CrossRef]
28. Bierhalz, A.C.K.; Silva, M.A.; Kieckbusch, T.G. Natamycin release from alginate/pectin films for food packaging applications. *J. Food Eng.* **2012**, *110*, 18–25. [CrossRef]
29. Oyeoka, H.C.; Ewulonu, C.M.; Nwuzor, I.C.; Obele, C.M.; Nwabanne, J.T. Packaging and degradability properties of poly-vinyl alcohol/gelatin nanocomposites films filled with water hyacinth cellulose nanocrystals. *J. Bioresour. Bioprod.* **2021**, *6*, 168–185. [CrossRef]
30. Yang, W.; Qi, G.; Kenny, J.M.; Puglia, D.; Ma, P. Effect of cellulose nanocrystals and lignin nanoparticles on mechanical, antioxidant and water vapour barrier properties of glutaraldehyde crosslinked PVA films. *Polymers* **2020**, *12*, 1364. [CrossRef]
31. Azeredo, H.M.; Rosa, M.F.; Mattoso, L.H.C. Nanocellulose in biobased food packaging applications. *Ind. Crops Prod.* **2017**, *97*, 664–671. [CrossRef]
32. Yadav, M.; Liu, Y.-K.; Chiu, F.-C. Fabrication of Cellulose Nanocrystal/Silver/Alginate Bionanocomposite Films with Enhanced Mechanical and Barrier Properties for Food Packaging Application. *Nanomaterials* **2019**, *9*, 1523. [CrossRef]
33. Slavutsky, A.M.; Bertuzzi, M.A. Water barrier properties of starch films reinforced with cellulose nanocrystals obtained from sugarcane bagasse. *Carbohydr. Polym.* **2014**, *110*, 53–61. [CrossRef]
34. Chakravartula, S.S.N.; Lourenço, R.V.; Balestra, F.; Bittante, A.M.Q.B.; Sobral, P.J.A.; Rosa, M.D. Influence of pitanga (*Eugenia uniflora* L.) leaf extract and/or natamycin on properties of cassava starch/chitosan active films. *Food Packag. Shelf Life* **2020**, *24*, 100498. [CrossRef]
35. Welscher, Y.M.; Jones, L.; van Leeuwen, M.R.; Dijksterhuis, J.; Kruijff, B.; Eitzen, G.; Breukink, E. Natamycin inhibits vacuole fusion at the priming phase via a specific interaction with ergosterol. *Antimicrob. Agents Chemother.* **2010**, *54*, 2618–2625. [CrossRef]
36. Pan, H.; Zhong, C.; Xia, L.; Li, W.; Wang, Z.; Deng, L.; Li, L.; Long, C. Antifungal activity of natamycin against kiwifruit soft rot caused by *Botryosphaeria dothidea* and potential mechanisms. *Sci. Hortic.* **2022**, *305*, 111344. [CrossRef]
37. Szomek, M.; Reinholdt, P.; Walther, H.-L.; Scheidt, H.A.; Müller, P.; Obermaier, S.; Poolman, B.; Kongsted, J.; Wüstner, D. Natamycin sequesters ergosterol and interferes with substrate transport by the lysine transporter Lyp1 from yeast. *Biochim. Et Biophys. Acta (BBA) Biomembr.* **2022**, *1864*, 184012. [CrossRef]
38. Geronikou, A.; Srimahaeak, T.; Rantsiou, K.; Triantafillidis, G.; Larsen, N.; Jespersen, L. Occurrence of yeasts in white-brined cheeses: Methodologies for identification, spoilage potential and good manufacturing practices. *Front. Microbiol.* **2020**, *11*, 582778. [CrossRef]
39. Correa, F.T.; de Souza, A.C.; de Souza Júnior, E.A.; Isidoro, S.R.; Piccoli, R.H.; Dias, D.; Abreu, L.R. Effect of Brazilian green propolis on microorganism contaminants of surface of Gorgonzola-type cheese. *J. Food Sci. Technol.* **2019**, *56*, 1978–1987. [CrossRef]
40. Awasti, N.; Anand, S. The Role of Yeast and Molds in Dairy Industry: An Update. In *Dairy Processing: Advanced Research to Applications*; Minj, J., Sudhakaran, V.A., Kumari, A., Eds.; Springer: Singapore, 2020. [CrossRef]
41. Biango-Daniels, M.N.; Wolfe, B.E. American artisan cheese quality and spoilage: A survey of cheesemakers' concerns and needs. *J. Dairy Sci.* **2021**, *104*, 6283–6294. [CrossRef]

Disclaimer/Publisher's Note: The statements, opinions and data contained in all publications are solely those of the individual author(s) and contributor(s) and not of MDPI and/or the editor(s). MDPI and/or the editor(s) disclaim responsibility for any injury to people or property resulting from any ideas, methods, instructions or products referred to in the content.

Article

Reactive Extrusion-Assisted Process to Obtain Starch Hydrogels through Reaction with Organic Acids

Beatriz Marjorie Marim [1], Janaina Mantovan [1], Gina Alejandra Gil-Giraldo [1], Jéssica Fernanda Pereira [1], Bruno Matheus Simões [2], Fabio Yamashita [2] and Suzana Mali [1,*]

[1] Department of Biochemistry and Biotechnology, State University of Londrina—UEL, Londrina 86057-970, PR, Brazil
[2] Department of Food Science and Technology, State University of Londrina—UEL, Londrina 86057-970, PR, Brazil
* Correspondence: smali@uel.br; Tel.: +55-43-3371-4270; Fax: +55-43-3371-5470

Abstract: A totally green process based on reactive extrusion was used for the production of cassava starch hydrogels through reaction with two organic crosslinking agents, citric (CA) and tartaric (TA) acids. CA and TA were used at different concentrations (0, 2.5, 5.0, 10.0, 15.0, and 20.0%). Degree of substitution (DS) of hydrogels ranged from 0.023 to 0.365. Fourier transform infrared spectroscopy results showed a new band appearing at 1730 cm^{-1} associated with ester carbonyl groups. X-ray diffraction indicated that reactive extrusion resulted in the disappearance of diffraction peaks of native starch and samples with lower crystallinity indices ranging from 37% (native starch) to 8–11% in starch hydrogels. Morphology analysis showed that the original granular structure of starch was lost and replaced by a rougher and irregular structure. Water holding capacity values of starch hydrogels obtained by reactive extrusion were superior to those of native starch and the control sample (extruded without the crosslinking agents). Hydrogels obtained with the highest CA or TA concentration (20.0%) resulted in the higher DS and swelling capacities, resulting in samples with 870 and 810% of water retention, respectively. Reactive extrusion was effective in obtaining starch hydrogels by reaction with organic acids.

Keywords: citric acid; tartaric acid; crosslinking; green chemistry

Citation: Marim, B.M.; Mantovan, J.; Gil-Giraldo, G.A.; Pereira, J.F.; Simões, B.M.; Yamashita, F.; Mali, S. Reactive Extrusion-Assisted Process to Obtain Starch Hydrogels through Reaction with Organic Acids. *Polysaccharides* 2022, 3, 792–803. https://doi.org/10.3390/polysaccharides3040046

Academic Editors: Valentina Siracusa, Nadia Lotti, Michelina Soccio and Alexey Iordanskii

Received: 5 October 2022
Accepted: 2 December 2022
Published: 5 December 2022

Publisher's Note: MDPI stays neutral with regard to jurisdictional claims in published maps and institutional affiliations.

Copyright: © 2022 by the authors. Licensee MDPI, Basel, Switzerland. This article is an open access article distributed under the terms and conditions of the Creative Commons Attribution (CC BY) license (https://creativecommons.org/licenses/by/4.0/).

1. Introduction

Hydrogels are materials that can be produced from natural or synthetic polymers; these polymers are crosslinked to result in three-dimensional polymeric matrices that have the ability of retain water or biological fluids without dissolving. Hydrogels can be obtained in different formats, including films, membranes, powders, and micro or nanogels [1–3].

Starch is a biodegradable, nontoxic, and inexpensive biopolymer available worldwide with readily free hydroxyl groups with potential for functionalization [4], including crosslinking reactions [2,5]. The crosslinking for chemical modification of starch through reaction with bi- or polyfunctional agents is largely reported in the literature [6–13]. The crosslinking agents can form ether or ester intermolecular bonds between hydroxyl groups on starch molecules, resulting in a reinforced polymer matrix. Depending on the level of substitution, the resulting materials can present changes in solubility and swelling power, which are important properties for hydrogels [11,14,15].

Starch hydrogels have several potential applications; however, to obtain food-grade, biodegradable and biocompatible hydrogels, the high cost of raw materials, and the generation of toxic solvents can be considered significant disadvantages. Epichlorohydrin and glutaraldehyde have been widely used to crosslink polysaccharide materials; however, they suffer from certain toxicity [2,16,17]. In contrast, polycarboxylic acids such as CA and TA are inexpensive and nontoxic reagents that can be safely used to obtain crosslinked starches [4,18–20]. CA and other polycarboxylic acids had been described as efficient esterification and crosslinking agents of starch, and they also present a hydrolytic action depending on the processing conditions [4,15,20].

In the last few years, the use of CA and other organic acids as esterification and crosslinking agents to obtain modified starches from several sources has been reported in literature, emphasizing the efficiency, and lower environmental impact of their use [11,13,18–26].

Starch modification has been traditionally carried out in tank reactors with several reagents and long processing time, resulting in negative impacts from effluent generation; thus, the use of physical methods can be an interesting approach to minimize the excessive use of reagents [13,16,22,27]. Lipatova and Yusova [28] reported the effectiveness of mechanical activation in a rotor–stator device on starch crosslinking with CA to obtain biodegradable hydrogels.

Alternatively, the reactive extrusion process is an efficient and versatile method for modifying starch and it is considered a green technological solution since the extruder is used as a reactor, where chemical reactions such as esterification, hydroxypropylation, and crosslinking can be performed [27–30]. Reactive extrusion combines the thermomechanical energy necessary to disintegrate the native structure of the starch granule, favoring the reaction between starch and organic acids in a single process without using other reagents. Additionally, reactive extrusion is a continuous process that has commercial viability, and it is easy to adapt to industrial scales, offering short reaction time (2–3 min) [12,16,31–33].

The use of reactive extrusion to obtain esterified and crosslinked starches through reaction with organic acids is an interesting research topic that could be further investigated to contribute to knowledge in the area. Recently, Farhat et al. [22], Hong et al. [23], and Ye et al. [13] reported the use of reactive extrusion to obtain starch citrates from native corn, waxy corn, and rice starches, and they stressed that this process can be considered a promising alternative to obtain esterified and crosslinked starches.

There are few reports in the literature about the production of starch citrates by reactive extrusion [13,22,23], while starch hydrogels obtained through reaction with TA by reactive extrusion have not yet been reported in the literature. Reactive extrusion can be considered an ecofriendly process to obtain esterified and crosslinked starches, and this study intended to explore the potential of this technology for the obtainment of new biobased materials, such as starch hydrogels, contributing to the generation of knowledge in this area. Therefore, this study aimed to obtain cassava starch hydrogels by reactive extrusion using CA and TA as crosslinking and esterifying agents and to study their physicochemical and microstructural properties.

2. Materials and Methods

2.1. Materials

Cassava starch (20% amylose and 80% amylopectin) was purchased from Pinduca Co., Ltd. (Araruna, Brazil). CA and TA of analytical grade were purchased from Synthlab (Synthlab, Diadema, Brazil), similar to all other chemicals and solvents employed in this study.

2.2. Reactive Extrusion Process

CA and TA were employed at different concentrations (0, 2.5, 5.0, 10.0, 15.0, and 20.0%—g acid/100 g starch) as esterifying/crosslinking agents. Crosslinked starch hydrogels were prepared by dissolving the different concentrations of CA or TA in distilled water, which were mixed with starch to obtain samples with a moisture content of 32%. Each sample was stored in plastic sealed bags for 1 h before extrusion. A control sample (S0) was also extruded in the presence of water (without Ca or TA), resulting in a moisture content of 32%. The reactive extrusion parameters were based on previous study of Gil-Giraldo et al. [34]. The extrusion process was performed in a single screw extruder (AX Plastics, Diadema, Brazil) with a screw length/diameter ratio (L/D) of 40 and a screw diameter of 1.6 cm, and a cylindrical matrix of 0.8 cm in diameter. The temperature in all four heating zones was 100 °C, and the screw speed was 60 rpm. Each sample was collected from the extruder and air-dried (45 °C) to a constant weight and ground. Then, each sample was washed three times with absolute ethanol to remove the unreacted CA or TA, as described by Ye et al. [13] and Gil-Giraldo et al. [34]. Finally, samples were air-dried at 45 °C to a constant weight, ground, and sieved through an 80-mesh sieve to obtain powders that were employed for characterization. Starch hydrogel samples prepared by reaction with CA were labeled as SC2.5, SC5, SC10, SC15, and SC20 throughout the study. Starch hydrogel samples prepared with TA were labeled as ST2.5, ST5, ST10, ST15, and ST20 throughout the study.

2.3. Degree of Substitution (DS)

The DS of each sample was determined in triplicate employing the method described by Volkert et al. [12] and Ye et al. [13] by titration, with some modifications. Each sample (1 g) was placed in a 250 mL Erlenmeyer flask, 20 mL of deionized water and two drops of phenolphthalein were added. The solution was titrated with a sodium hydroxide solution (0.1 mol/L) until the endpoint was reached, indicating that all the free Ca or TA had been neutralized. After that, 25 mL of aqueous sodium hydroxide solution (0.5 mol/L) was added, and the sample was agitated for 60 min at room temperature. The excess alkali was back-titrated with a standard aqueous hydrochloric acid solution (0.5 mol/L) until the endpoint. A blank titration using extruded starch without CA or TA was carried out. Esterified carboxyl groups (EG, %) and DS were calculated as follows: $EG = [((v0 - v1) \times c \times M \times 100))/m]$, and $DS = (162 \times EG)/(100M - (M - 1)EG$; where EG is the content of esterified carboxyl groups (%); v0 is the volume of aqueous hydrochloric acid solution consumed by the blank (mL); v1 is the volume of aqueous hydrochloric acid solution consumed by the esterified starch sample (mL); c is the concentration of aqueous hydrochloric acid solution (mol/L); M is the molar mass of the substituent (citrate or tartarate); and m is the mass of the samples (mg).

2.4. Fourier Transform-Infrared Spectroscopy (FTIR)

Pulverized hydrogels samples were mixed with potassium bromide and compressed into pellets. The FTIR analyses were performed using a spectral resolution of 4 cm^{-1} and a spectral range of 4000–500 cm^{-1} in a Shimadzu FTIR-8300 (Kyoto, Japan) equipment.

2.5. X-ray Diffraction (XRD)

XRD analysis was carried out in a Panalytical X'Pert PRO MPD diffractometer (Almelo, The Netherlands) with copper Kα radiation (λ = 1.5418 Å) under operational conditions of 40 kV and 30 mA, with a with a ramp rate of 1°/min. The relative crystallinity index (CI) was calculated using the method described by Cheetham and Tao [35].

2.6. Scanning Electron Microscopy (SEM)

SEM analyses were performed with an FEI Quanta 200 microscope (Hillsboro, OR, USA) using an accelerating voltage of 20 kV. Each hydrogel sample was mounted for surface visualization on bronze stub, and then its surface was coated with a thin gold layer (40–50 nm).

2.7. Water Holding Capacity (WHC)

WHC was determined according to the methodology described by Butt et al. [36], with minor modifications. Approximately 1 g of each sample was weighed, and 10 mL of water was added to a pre-weighed centrifuge tube (W1). The samples were shaken for 30 min on a shaker at 25 °C and 200 rpm (Quimis Q 225M, Diadema, Brazil) and then centrifuged for 30 min at 2200 rpm (Hettich Centrifuge, Model 320R, Germany). After centrifugation, the supernatant was decanted, and tubes with residue were weighed (W2). The WHC was calculated as WHC (g/g) = ((W2 − W1) − (initial mass of sample))/(initial mass of sample).

2.8. Swelling Power at Different Times and Temperatures

The swelling power at different times was determined according to the procedure described by Yoshimura et al. [37]. Dried samples (0.5 g) were left to swell within permeable nylon tea bags (Japanese Industrial Standard, JIS K 7223). The tea bags were immersed in water at 25 °C for 1, 24, and 48 h, and then each tea bag was removed from the water, and excess water was drained for 1 min. The weight of the tea bag and samples was then measured (Wt), and the swelling was calculated according to the following equation: Swelling (g/g) = ((Wt − Wb) − Wp)/(Wp)), where Wb is the weight of a blank tea bag after water treatment, and Wp is the weight of the dry sample. Swelling power was determined in the same way at different temperatures (10, 25, and 55 °C) for 24 h.

2.9. Statistical Analysis

Statistical analysis was carried out employing the Statistica software version 7.0 (Statsoft, OK, USA) and Tukey's test was employed for mean comparison ($p \leq 0.05$).

3. Results

3.1. Degree of Substitution (DS)

Starch is composed of two polymers of D-glucopyranose, amylose, and amylopectin. Amylose is the linear fraction consisting of α-D-glucopyranose linked through α (1 → 4) linkages, and amylopectin is the branched fraction containing linear fractions linked through α (1 → 4) linkages, and ramifications at α (1 → 6); each glucopyranosyl unit in linear chains presents three reactive hydroxyl groups at position C2, C3, and C6 that have potential for functionalization [6,38]. DS is a useful parameter in the study of starch modification, and it can be defined as average number of hydroxyl groups substituted per D-glucose unit [38].

DS values of modified samples are presented in Table 1. For samples modified with CA and TA, the increase in the acid concentration resulted in higher DS values, while comparing CA- and TA-modified samples, CA-modified samples had higher DS values (Table 1). According to Seidel et al. [39], the use of different polyfunctional carboxylic acids as crosslinking and esterifying agents results in different materials, which is related to their structures. CA ($C_6H_8O_7$) is a tricarboxylic acid formed by two carboxylic groups spaced by three CH_2 groups with one OH group and one COOH group linked at the central carbon, while TA ($C_4H_6O_6$) is a dicarboxylic acid spaced by two CH_2 groups, each bearing one OH group. The difference in length of the spacer between CA and TA molecules and the kind and number of functional groups possibly resulted in a more efficient CA reaction with starch. Shen et al. [10] also stressed that carboxylic acids that have more than two carboxyl groups act more efficiently as crosslink agents than dicarboxylic acids. These authors reported that CA (tricarboxylic) was more efficient as a crosslinking agent than succinic acid (dicarboxylic).

A higher DS value was obtained for the SC20 sample (DS = 0.365), and a lower DS value was obtained for the ST2.5 sample (DS = 0.023) (Table 1). The values observed in this study were close to the values reported by Ye et al. [13] who chemically modified rice starch by reactive extrusion with CA; they reported values ranging from 0.037 to 0.138 using CA levels between 10 and 40% (g CA/100 g starch). Farhat et al. [22] reported DS values of 0.52 to 0.99 in starch citrates obtained from corn starch by reactive extrusion with CA concentrations of 40 and 100% (g CA/100 g starch). According to Ye et al. [13], reactive extrusion can be efficiently used for the obtainment of starch citrates.

The DS values of modified starch samples obtained in this study were slightly higher than those obtained for starch citrate prepared by high-temperature/long-term reactions in semidry conditions, as reported by other authors [14,25]. Mei et al. [14] reported that when CA concentration was increased from 10 to 30%, DS values increased from 0.058 to 0.178 in cassava starch citrates obtained after 8 h of reaction at room temperature.

Table 1. Degree of substitution (DS) and water holding capacity (WHC) of native starch, control sample (S0) and starches hydrogels obtained by reactive extrusion.

Samples	DS	WHC (g/g)
Native starch	0	1.84 ± 0.01 [g]
S0	0	4.99 ± 0.05 [d,e]
SC2.5	0.137 ± 0.003 [e]	3.03 ± 1.31 [f]
SC5	0.160 ± 0.011 [e]	4.62 ± 1.12 [e]
SC10	0.262 ± 0.003 [d]	5.90 ± 2.13 [c]
SC15	0.352 ± 0.002 [b]	6.22 ± 2.07 [a,b]
SC20	0.365 ± 0.001 [a]	7.02 ± 0.74 [a]
ST2.5	0.023 ± 0.001 [f]	4.99 ± 0.07 [d,e]
ST5	0.103 ± 0.004 [e]	4.92 ± 0.05 [e]
ST10	0.137 ± 0.002 [e]	6.10 ± 0.03 [b,c]
ST15	0.285 ± 0.001 [c]	6.95 ± 0.04 [a]
ST20	0.285 ± 0.004 [c]	6.66 ± 0.09 [a,b]

Different letters in the same column indicate significant difference by Tukey's test ($p \leq 0.05$).

3.2. Fourier Transform-Infrared Spectroscopy (FTIR)

The FTIR spectra of starch samples are shown in Figure 1. It was possible to identify a new important band at 1730 cm^{-1} in all hydrogel samples that indicates the stretching vibration of the carbonyl ester group, which provides clear evidence that the ester bonds were formed successfully. These results agreed with other authors' results [11,13,15,19,22,26,28] who used CA and/or TA as crosslinking and esterifying agents for starch modification. It is important to highlight that all modified samples were washed with ethanol before analysis to prevent free CA or TA from remaining in the samples; thus, the band at 1730 cm^{-1} observed in all modified samples resulted from the formation of a covalent ester bond between the CA or TA and starch.

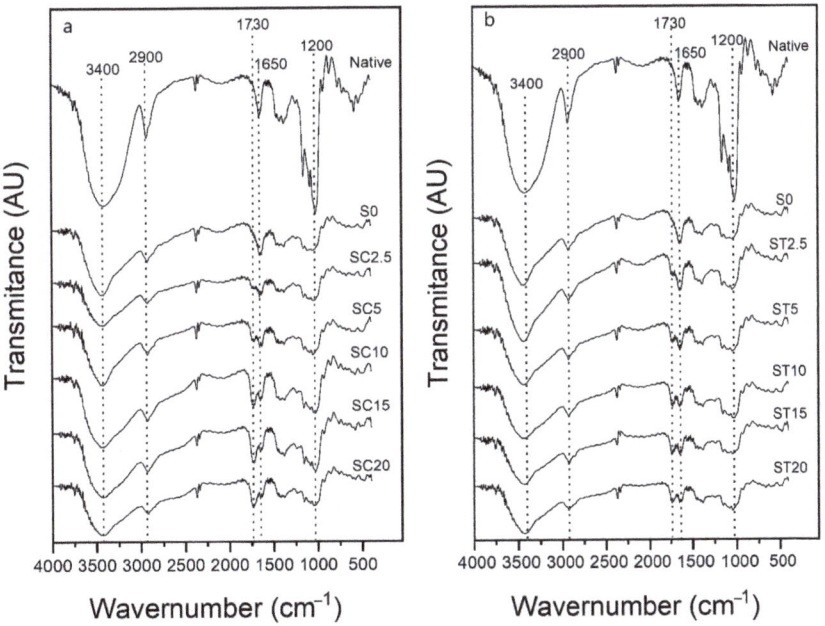

Figure 1. FTIR spectra of native starch, the control sample (extruded without reagent—S0) and starch hydrogels obtained by reactive extrusion through reaction with CA (**a**) and TA (**b**).

Other bands are also observed in the FTIR spectra of all samples: a broad band at 3400–3500 cm^{-1} that was associated with O-H elongation and hydrogen bond vibration; a band at 2900 cm^{-1} associated with C-H asymmetric elongation and vibration; a band at 1650 cm^{-1} associated with water adsorbed in amorphous regions of starch, and a band at 1200 cm^{-1} that was attributed to the stretching vibration of C-O in C-OH groups [13,15,20,22,26,28,40].

3.3. X-ray Diffraction (XRD)

The diffractograms and relative crystallinity are shown in Figure 2. Native cassava starch presented diffraction peaks at 2θ = 15.03°, 17.03°, 17.93°, and 23.02° (Figure 2), which are typical of an A-type crystallinity [41,42].

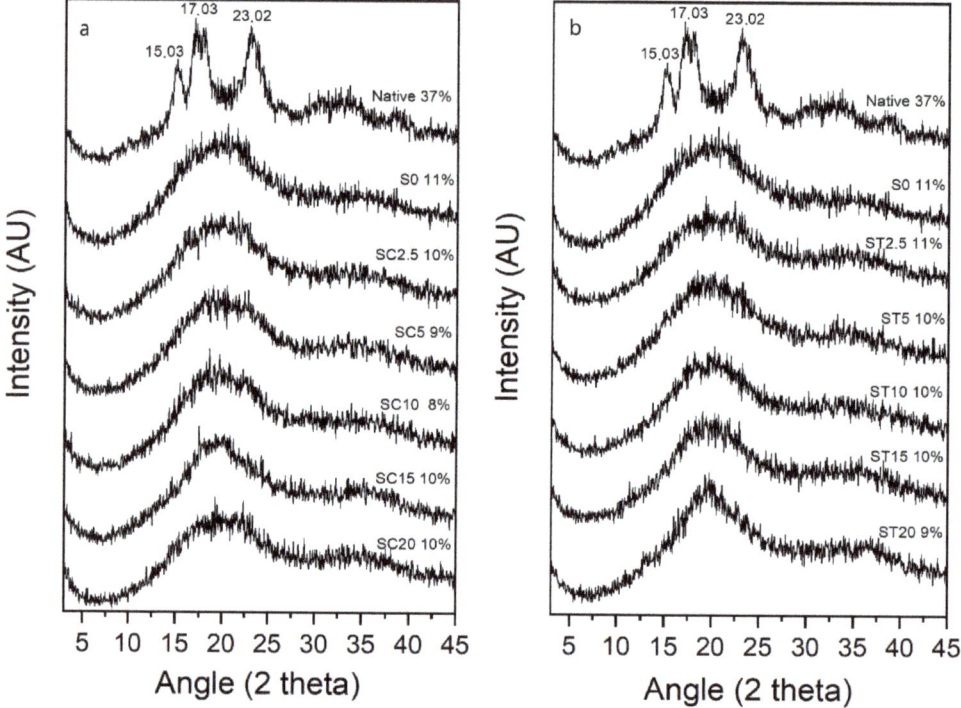

Figure 2. X-ray diffraction (XRD) patterns and relative crystallinity index (CI) of native starch, the control sample (extruded without reagent—S0) and starch hydrogels obtained by reactive extrusion through reaction with CA (**a**) and TA (**b**).

The crystallinity index (CI) of native starch was 37%, which is typical of a semicrystalline material, and a similar value (37.64%) was reported by Mei et al. [14] for native cassava starch. The CI for all extruded samples (with and without reagent) decreased, and the CI values ranged from 8 to 11% (Figure 2). During the reactive extrusion process, the starch granules were exposed to high temperatures and high shear forces, strongly affecting their crystalline structure, resulting in the disappearance of diffraction peaks and samples with low crystallinity indices (Figure 2). These results agreed with other authors that used reactive extrusion to modify starches [13,27]. Cai et al. [27] reported that rice starch crosslinked by reactive extrusion with sodium trimetaphosphate and propylene oxide had its crystalline structure almost completely disrupted, with CI values very close to those obtained in this study, ranging from 11 to 14%.

Relative crystallinity affects the physicochemical properties of starch [43]. Some authors claim that the lower crystallinity of starch hydrogels can be important for their application as carriers of various compounds, such as in a drug and chemical element delivery system and in agricultural applications [41,44,45].

3.4. Scanning Electron Microscopy (SEM)

As observed in the SEM micrograph (Figure 3), native cassava starch granules exhibited an elliptical or truncated elliptical shape, with a smooth surface without disruption. These results agreed with those reported by other authors [41,42].

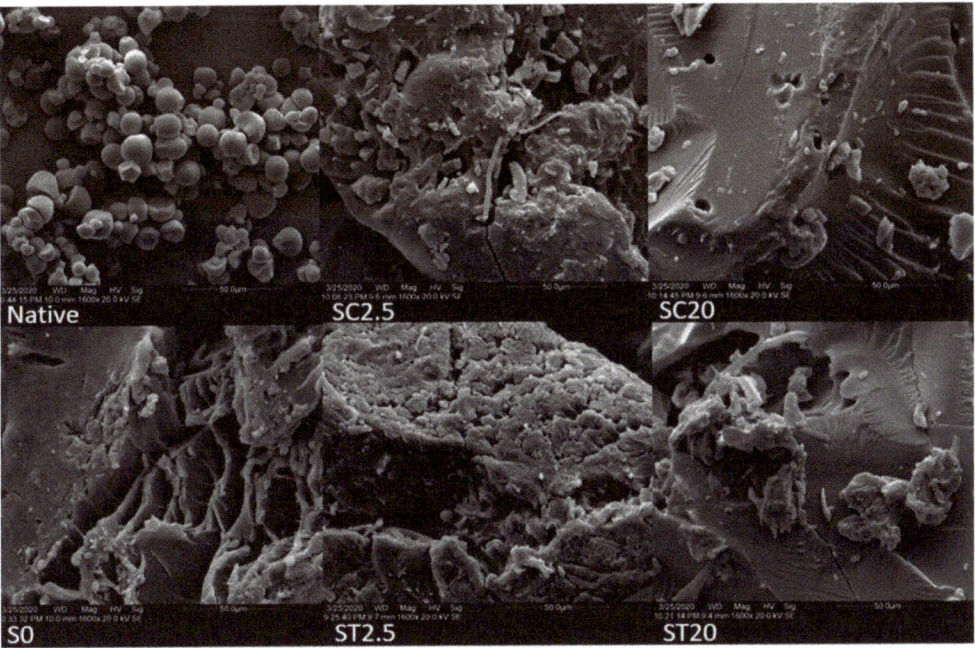

Figure 3. SEM images of native starch, the control sample (extruded without reagent—S0), and starch hydrogels obtained by reactive extrusion through reaction with CA (SC2.5 and SC20) and TA (ST2.5 and ST20).

After being subjected to extrusion, all extruded samples showed structural morphological differences in relation to native cassava starch. During reactive extrusion, the starch granules undergo changes in their structure, disintegrating totally or partially, depending on the input energy level, forming rougher structures with irregular and fragmented aggregates. SEM micrographs also demonstrated that the sample surfaces did not contain residual starch granules, which indicated that the granules were destroyed due to the thermomechanical and chemical modification effects, which agreed with the XRD results. Other authors also reported the disintegration of the granular structure of starch subjected to reactive extrusion [29,46,47].

In addition to the extrusion process, the hydrolytic action of CA and TA can also favor the breakdown of starch granules [15,20]. All samples presented several pores on their surfaces (Figure 3); however, the SC20 and ST20 samples presented more compact structures, which can occur because the higher acid concentration possibly resulted in a lower viscosity of these samples during extrusion, yielding more homogeneous structures. The presence of pores can be interesting in the case of superabsorbent hydrogels for use in agriculture because they can facilitate the entry of water into their structure. According to

Duquette et al. [20], starch hydrogels prepared by reaction with itaconic acid and CA in aqueous medium present highly irregular surfaces with the presence of pores, which can favor the diffusion of solvent and ions inside the hydrogel matrix, and additionally can favor trap particles by physical entrapment.

3.5. Water Holding Capacity (WHC)

WHC is the capacity of the sample to absorb and hold water in its polymeric network [48]. Native cassava starch presented the lowest WHC (1.84 g/g) value (Table 1). The S0 sample (extruded without reagent) and all crosslinked starch hydrogels presented significant improvements in WHC values (Table 1).

WHC depends on the presence of hydrophilic groups that can absorb and bind water molecules, leading to more swollen materials. Hydrophilic functional groups on the starch, CA and TA chains can form hydrogen bonds with water, allowing the hydrogels to hold water without dissolving. During reactive extrusion, the intramolecular bonds and the crystalline structure of starch granules were broken, resulting in the weakening of associative forces between starch chains, increasing the retention of water at low temperature by the hydrophilic groups that were exposed, which was also reported by other authors [13,33].

The sample with the higher WHC was prepared with 20.0% CA (SC20 sample, Table 1); also the sample with the higher DS, which is indicative that the starch chains were crosslinked, resulted in a more reinforced polymeric matrix capable to retain water. According to Lemos et al. [2], modified starches are reinforced by crosslinking the polymeric chains because of the covalent bonds formed, which improves their capacity to maintain water in their matrix without dissolving. Butt et al. [36] reported that the introduction of hydrophilic groups from CA resulted in increased water binding capacity of citrates due to the insertion of more hydrophilic substituent groups.

Other authors reported a contrary trend, with reductions in water holding capacity in crosslinked starch samples obtained by reactive extrusion with CA [13] and they reported that at a sufficiently high DS, it is difficult for water to penetrate into the starch citrate. Probably, in this study, the DS were not high enough to prevent the water to penetrate into the hydrogel internal structure.

The S0 sample (extruded without reagents) presented a WHC value of 4.99 g/g, a value that was higher than that presented by the crosslinked sample with CA 2.5, and similar to those presented by SC5, ST2.5 and ST5 (Table 1). During extrusion, the granular and semicrystalline structure of starch was destroyed, as observed by XRD and SEM, resulting in materials with lower crystallinity, which certainly contributed to the increase in the WHC of this sample. Heebthong and Ruttarattanamongkol [33] reported that the thermomechanical energy from the extrusion process results in disruption of the starch molecular arrangement and that water can be transferred easily into its internal structure.

3.6. Swelling Power at Different Times and Temperatures

The swelling power of samples is shown in Figure 4. It can be observed that the swelling of native starch was significantly lower than those of the S0 and samples crosslinked with CA and TA at all swelling times. The swelling of hydrogels crosslinked with CA after 1 h ranged from 3.1 (SC2.5) to 4.2 g/g (SC20), while the hydrogels crosslinked with TA ranged from 2.9 g/g (ST2.5) to 3.7 (g/g) (ST20). For all samples, the increase in time from 1 to 24 h resulted in samples with significantly higher swelling values (Figure 4), and between 24 and 48 h the swelling degree stabilized.

After 48 h of immersion in water, the higher values were 8.7 and 8.1 g/g for SC20 and ST20 samples, resulting in 870 and 810% of water retention, respectively, suggesting that crosslinked samples can swell and retain water at room temperature. Batista et al. [49] reported that superabsorbent hydrogels can absorb over 100% and up to thousands of times their dry weight in water, and they can be driven for several applications, including agriculture, drug delivery, food packaging, or as adsorbent materials to decontaminate soil and water, between others [3,18,20,49].

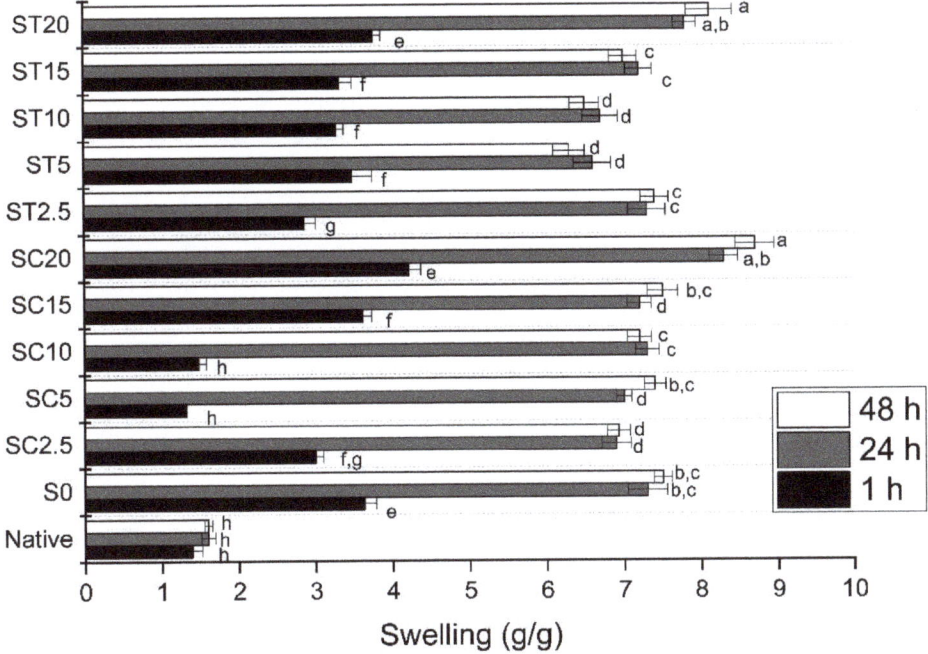

Figure 4. Effect of time on swelling of native starch, the control sample (extruded without reagent) and starch hydrogels obtained by reactive extrusion through reaction with CA and TA. Different letters indicate significant difference between means by Tukey's test ($p \leq 0.05$).

The high concentrations of CA and TA improved granule swelling due to better starch hydration, and samples SCA.20 and STA.20 also showed higher DS values (0.36 and 2.85, respectively). During reaction with 20.0% CA and TA, the starch acquires a three-dimensional matrix, which possibly results in individual chains that repel each other due to steric hindrance, as crosslinked starch derivatives try to exist in the lowest and most stable state. The repulsion between the chains can increase the capacity for expansion without dissolving, preserving the structure of the material [11,46]. Duquette et al. [20] reported that the amount of CA in starch hydrogels affects the swelling by acting as a crosslinking agent ensuring the stability of their swollen structure, and also by acting as a source of hydrophilic functional groups.

When the temperature was evaluated, all the samples presented significant increases in their swelling with increases in temperature from 10 to 55 °C (Figure 5). Hung et al. [50] reported the increase in swelling of CA-crosslinked rice starches with increasing temperature. Samples obtained through reaction with 20.0% CA and TA presented the higher swelling values at all temperatures.

After 48 h of immersion in water at 25 °C, the S0 sample presented a swelling value of 7.5 g/g, a value that was higher than those presented by the crosslinked samples with CA and TA at 2.5 and 5.0 (Figure 4), which had the lowest swelling capacity values. During extrusion, the starch granules are subjected to high temperature and pressure, and the hydrogen bonds and the crystal structure are broken, resulting in disruption of their granular structure. The free hydroxyl groups of starch chains when in contact with water can form hydrogen bonds and swell [35,51]. Similar results have also been described by Monroy et al. [42], who observed that ultrasound-modified starches lose their crystallinity and present increased swelling values.

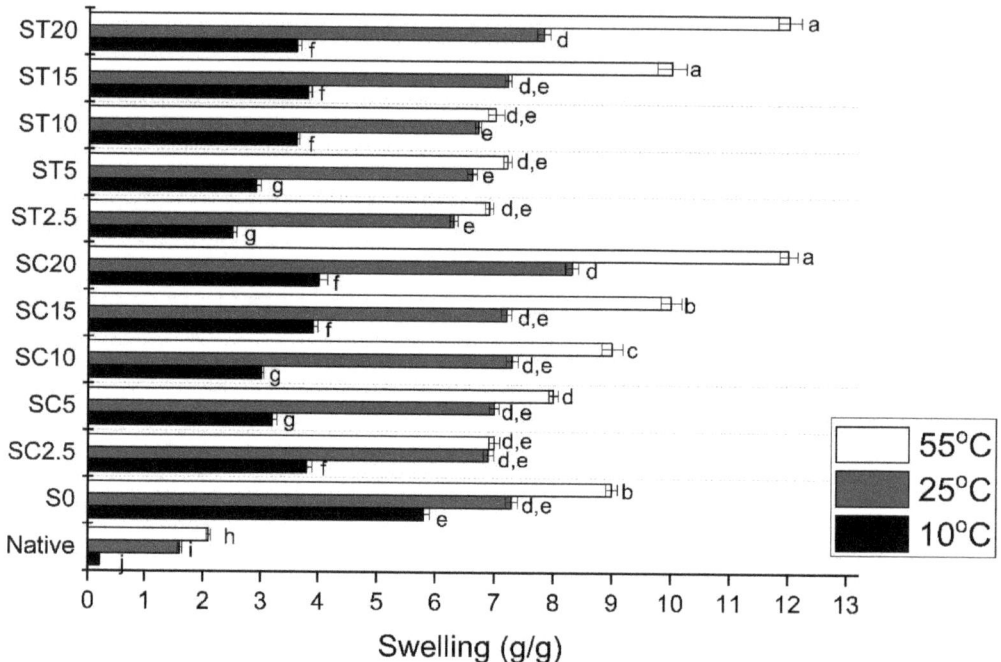

Figure 5. Effect of temperature on swelling of native starch, the control sample (extruded without reagent) and starch hydrogels obtained by reactive extrusion through reaction with CA and TA. Different letters indicate significant difference between means by Tukey's test ($p \leq 0.05$).

4. Conclusions

Reactive extrusion was effectively used for the production of cassava starch hydrogels through reaction with two crosslinking agents, CA and TA. Evidence for the occurrence of chemical modification was verified by FTIR spectroscopy with the appearance of a new band at 1730 cm^{-1}, resulting in materials with different degrees of substitution, which were higher for hydrogels obtained by crosslinking with CA. XDR showed that the starch crystalline structure was broken during reactive extrusion, and morphology analysis showed that the original granular structure of starch was lost and replaced with a rougher and more irregular structure, which resulted in hydrogels with higher water holding capacity than native starch and the control sample (extruded without the crosslinking agents). Hydrogel samples obtained with the highest CA or TA concentrations (15.0 and 20.0%) resulted in the higher DS values and water holding and swelling capacities.

Reactive extrusion was effective in obtaining starch hydrogels by reaction with organic acids, with the advantages of reduced processing time and low effluent generation when compared to conventional processes, with great possibilities to be adapted to an industrial scale, expanding the potential of starch hydrogels for several applications, including the obtainment of food-grade products.

Author Contributions: B.M.M.: experiment, writing—review and editing. J.M., G.A.G.-G., J.F.P. and B.M.S.: experiment, validation. F.Y.: resources, writing—review and editing. S.M.: conceptualization, project administration, writing—review and editing. All authors have read and agreed to the published version of the manuscript.

Funding: This work was supported by CAPES-Coordenação de Aperfeiçoamento de Pessoal de Nível Superior for the grant to the postgraduate students.

Institutional Review Board Statement: Not applicable.

Informed Consent Statement: Not applicable.

Data Availability Statement: Not applicable.

Acknowledgments: Laboratory of Spectroscopy (ESPEC), Laboratory of Electronic Microscopy and Microanalysis (LMEM) and the Laboratory of X-ray Analysis (LARX) of the State University of Londrina for the analyses.

Conflicts of Interest: The authors declare no conflict of interest.

References

1. Erdagi, S.I.; Ngwabebhoh, F.A.; Yildiz, U. Genipin crosslinked gelatin-diosgenin-nanocellulose hydrogels for potential wound dressing and healing applications. *Int. J. Biol. Macromol.* **2020**, *149*, 651–663. [CrossRef] [PubMed]
2. Lemos, P.V.F.; Marcelino, H.R.; Cardoso, L.G.; Souza, C.O.; Druzian, J.I. Starch chemical modifications applied to drug delivery systems: From fundamentals to FDA-approved raw materials. *Int. J. Biol. Macromol.* **2021**, *184*, 218–234. [CrossRef] [PubMed]
3. Liu, K.; Chen, Y.Y.; Zha, X.Q.; Li, Q.M.; Pan, L.H.; Luo, J.P. Research progress on polysaccharide/protein hydrogels: Preparation method, functional property and application as delivery systems for bioactive ingredients. *Food Res. Int.* **2021**, *147*, 110542. [CrossRef] [PubMed]
4. Golachowski, A.; Drożdż, W.; Golachowska, M.; Kapelko-Zeberska, M.; Raszewski, B. Production and properties of starch citrates—Current research. *Foods* **2020**, *9*, 1311. [CrossRef] [PubMed]
5. Elgaied-Lamouchi, D.; Descamps, N.; Lefevre, P.; Rambur, I.; Pierquin, J.-Y.; Siepmann, F.; Siepmann, J.; Muschert, S. Starch-based controlled release matrix tablets: Impact of the type of starch. *J. Drug Deliv. Sci. Technol.* **2021**, *61*, 102152. [CrossRef]
6. Dastidar, T.G.; Netravali, A.N. 'Green' crosslinking of native starches with malonic acid and their properties. *Carbohydr. Polym.* **2012**, *90*, 1620–1628. [CrossRef]
7. Liu, J.; Wang, B.; Lin, L.; Zhang, J.; Liu, W.; Xie, J.; Ding, Y. Functional, physicochemical properties and structure of cross-linked oxidized maize starch. *Food Hydrocoll.* **2014**, *36*, 45–52. [CrossRef]
8. Olsson, E.; Hedenqvist, M.S.; Johansson, C.; Järnström, L. Influence of citric acid and curing on moisture sorption, diffusion and permeability of starch films. *Carbohydr. Polym.* **2013**, *94*, 765–772. [CrossRef]
9. Otache, M.A.; Duru, R.U.; Achugasim, O.; Abayeh, O.J. Advances in the modification of starch via esterification for enhanced properties. *J. Polym. Environ.* **2020**, *29*, 1365–1379. [CrossRef]
10. Shen, L.; Xu, H.; Kong, L.; Yang, Y. Non-Toxic crosslinking of starch using polycarboxylic acids: Kinetic study and quantitative correlation of mechanical properties and crosslinking degrees. *J. Polym. Environ.* **2015**, *23*, 588–594. [CrossRef]
11. Tupa, M.V.; Altuna, L.; Herrera, M.L.; Foresti, M.L. Preparation and characterization of modified starches obtained in acetic anhydride/tartaric acid medium. *Starch-Staerke* **2020**, *72*, 1900300. [CrossRef]
12. Volkert, B.; Lehmann, A.; Greco, T.; Nejad, M.H. A comparison of different synthesis routes for starch acetates and the resulting mechanical properties. *Carbohydr. Polym.* **2010**, *79*, 571–577. [CrossRef]
13. Ye, J.; Luo, S.; Huang, A.; Chen, J.; Liu, C.; McClements, D.J. Synthesis and characterization of citric acid esterified rice starch by reactive extrusion: A new method of producing resistant starch. *Food Hydrocoll.* **2019**, *92*, 135–142. [CrossRef]
14. Mei, J.Q.; Zhou, D.N.; Jin, Z.Y.; Xu, X.M.; Chen, H.Q. Effects of citric acid esterification on digestibility, structural and physicochemical properties of cassava starch. *Food Chem.* **2015**, *187*, 378–384. [CrossRef]
15. Simões, B.M.; Cagnin, C.; Yamashita, F.; Olivato, J.B.; Garcia, P.S.; Mali, S.; Grossmann, M.V.E. Citric acid as crosslinking agent in starch/xanthan gum hydrogels produced by extrusion and thermopressing. *LWT* **2020**, *125*, 108950. [CrossRef]
16. Moad, G. Chemical modification of starch by reactive extrusion. *Prog. Polym. Sci.* **2011**, *36*, 218–237. [CrossRef]
17. Uliniuc, A.; Hamaide, T.; Popa, M.; Băcăiță, S. Modified starch-based hydrogels cross-linked with citric acid and their use as drug delivery systems for levofloxacin. *Soft Mater.* **2013**, *11*, 83–493. [CrossRef]
18. Ačkar, D.; Babić, J.; Jozinović, A.; Miličević, B.; Jokić, S.; Miličević, R.; Rajič, M.; Šubarić, D. Starch modification by organic acids and their derivatives: A review. *Molecules* **2015**, *20*, 19554–19570. [CrossRef]
19. Miskeen, S.; Hong, J.S.; Choi, H.D.; Kim, J.Y. Fabrication of citric acid-modified starch nanoparticles to improve their thermal stability and hydrophobicity. *Carbohydr. Polym.* **2021**, *253*, 117242. [CrossRef]
20. Duquette, D.; Nzediegwu, C.; Portillo-Perez, G.; Dumont, G.M.; Prasher, S. Eco-Friendly synthesis of hydrogels from starch, citric acid, and itaconic acid: Swelling capacity and metal chelation properties. *Starch-Staerke* **2020**, *72*, 1900008. [CrossRef]
21. Alimi, B.A.; Workneh, T.S. Structural and physicochemical properties of heat moisture treated and citric acid modified acha and iburu starches. *Food Hydrocoll.* **2018**, *81*, 449–455. [CrossRef]
22. Farhat, W.; Venditti, R.; Mignard, N.; Taha, M.; Becquart, F.; Ayoub, A. Polysaccharides and lignin based hydrogels with potential pharmaceutical use as a drug delivery system produced by a reactive extrusion process. *Int. J. Biol. Macromol.* **2017**, *104*, 564–575. [CrossRef] [PubMed]
23. Hong, J.S.; Chung, H.J.; Lee, B.H.; Kim, H.S. Impact of static and dynamic modes of semi-dry heat reaction on the characteristics of starch citrates. *Carbohydr. Polym.* **2020**, *233*, 115853. [CrossRef] [PubMed]
24. Kapelko-Zeberska, M.; Buksa, K.; Szumny, A.; Zieba, T.; Gryszkin, A. Analysis of molecular structure of starch citrate obtained by a well-stablished method. *LWT* **2016**, *69*, 334–341. [CrossRef]

25. Srikaeo, K.; Hao, P.T.; Lerdluksamee, C. Effects of heating temperatures and acid concentrations on physicochemical properties and starch digestibility of citric acid esterified tapioca starches. *Starch-Staerke* **2019**, *71*, 1800065. [CrossRef]
26. Zhou, J.; Tong, J.; Su, X.; Ren, L. Hydrophobic starch nanocrystals preparations through crosslinking modification using citric acid. *Int. J. Biol. Macromol.* **2016**, *91*, 1186–1193. [CrossRef]
27. Cai, C.; Wei, B.; Tian, Y.; Ma, R.; Chen, L.; Qiu, L.; Jin, Z. Structural changes of chemically modified rice starch by one-step reactive extrusion. *Food Chem.* **2019**, *288*, 354–360. [CrossRef]
28. Lipatova, I.M.; Yusova, A.A. Effect of mechanical activation on starch crosslinking with citric acid. *Int. J. Biol. Macromol.* **2021**, *185*, 688–695. [CrossRef]
29. Cai, C.; Tian, Y.; Yu, Z.; Sun, C.; Jin, Z. In vitro digestibility and predicted glycemic index of chemically modified rice starch by one-step reactive extrusion. *Starch-Staerke* **2019**, *72*, 1900012. [CrossRef]
30. Gutiérrez, T.J.; Valencia, G.A. Reactive extrusion-processed native and phosphated starch-based food packaging films governed by the hierarchical structure. *Int. J. Biol. Macromol.* **2021**, *172*, 439–451. [CrossRef]
31. González-Seligra, P.; Goyanes, S.; Famá, L. Effect of the Incorporation of Rich-Amylopectin Starch Nano/Micro Particles on the Physicochemical Properties of Starch-Based Nanocomposites Developed by Flat-Die Extrusion. *Starch-Stärke* **2022**, *74*, 2100080. [CrossRef]
32. Formela, K.; Zedler, L.; Hejna, A.; Tercjak, A. Reactive extrusion of bio-based polymer blends and composites—Current trends and future developments. *Express Polym. Lett.* **2018**, *12*, 24–57. [CrossRef]
33. Heebthong, K.; Ruttarattanamongkol, K. Physicochemical properties of cross-linked cassava starch prepared using a pilot-scale reactive twin-screw extrusion process (REX). *Starch-Staerke* **2016**, *68*, 528–540. [CrossRef]
34. Gil-Giraldo, G.A.; Mantovan, J.; Marim, B.M.; Kishima, J.O.F.; Mali, S. Surface modification of cellulose from oat hull with citric acid using ultrasonication and reactive extrusion assisted processes. *Polysaccharides* **2021**, *2*, 218–233. [CrossRef]
35. Cheetham, N.W.H.; Tao, L. Variation in crystalline type with amylose content in maize starch granules: An X-ray powder diffraction study. *Carbohydr. Polym.* **1998**, *36*, 277–284. [CrossRef]
36. Butt, N.A.; Ali, T.M.; Hasnain, A. Rice starch citrates and lactates: A comparative study on hot water and cold water swelling starches. *Int. J. Biol. Macromol.* **2019**, *127*, 107–117. [CrossRef]
37. Yoshimura, T.; Matsuo, K.; Fujioka, R. Novel biodegradable superabsorbent hydrogels derived from cotton cellulose and succinic anhydride: Synthesis and characterization. *J. Appl. Polym. Sci.* **2006**, *99*, 3251–3256. [CrossRef]
38. Namazi, H.; Dadkhah, A. Convenient method for preparation of hydrophobically modified starch nanocrystals with using fatty acids. *Carbohydr. Polym.* **2010**, *79*, 731–737. [CrossRef]
39. Seidel, C.; Kulicke, W.M.; Heß, C.; Hartmann, B.; Lechner, M.D.; Lazik, W. Influence of the cross-linking agent on the gel structure of starch derivatives. *Starch-Staerke* **2001**, *53*, 305–310. [CrossRef]
40. Hasanin, M.S. Simple, economic, ecofriendly method to extract starch nanoparticles from potato peel waste for biological applications. *Starch-Staerke* **2021**, *73*, 2100055. [CrossRef]
41. Minakawa, A.F.K.; Faria-Tischer, P.C.; Mali, S. Simple ultrasound method to obtain starch micro- and nanoparticles from cassava, corn and yam starches. *Food Chem.* **2019**, *283*, 11–18. [CrossRef] [PubMed]
42. Monroy, Y.; Rivero, S.; García, M.A. Microstructural and techno-functional properties of cassava starch modified by ultrasound. *Ultrason. Sonochem.* **2018**, *42*, 795–804. [CrossRef] [PubMed]
43. Zobel, H.F. Molecules to Granules: A Comprehensive starch review. *Starch-Stärke* **1998**, *40*, 44–50. [CrossRef]
44. Biduski, B.; Silva, W.M.F.; Colussi, R.; Halal, S.L.; El, M.; Lim, L.T.; Dias, A.R.G.; Zavareze, E. Starch hydrogels: The influence of the amylose content and gelatinization method. *Int. J. Biol. Macromol.* **2018**, *113*, 443–449. [CrossRef] [PubMed]
45. Borges, A.F.; Silva, C.; Coelho, J.F.J.; Simões, S. Oral films: Current status and future perspectives: I—Galenical development and quality attributes. *J. Control. Release* **2015**, *206*, 108–121. [CrossRef]
46. Fonseca-Florido, H.A.; Soriano-Corral, F.; Yañez-Macías, R.; González-Morones, P.; Hernández-Rodríguez, F.; Aguirre-Zurita, J.; Ávila-Orta, C.; Rodríguez-Velázquez, J. Effects of multiphase transitions and reactive extrusion on in situ thermoplasticization/succination of cassava starch. *Carbohydr. Polym.* **2019**, *225*, 115250. [CrossRef]
47. Murúa-Pagola, B.; Beristain-Guevara, C.I.; Martínez-Bustos, F. Preparation of starch derivatives using reactive extrusion and evaluation of modified starches as shell materials for encapsulation of flavoring agents by spray drying. *J. Food Eng.* **2009**, *91*, 380–386. [CrossRef]
48. Mehfooz, T.; Ali, T.M.; Ahsan, M.; Abdullah, S.; Hasnain, A. Morphological, functional and thermal characteristics of hydroxypropylated-crosslinked barley starches. *J. Food Meas. Charact.* **2021**, *15*, 237–246. [CrossRef]
49. Batista, R.A.; Espitia, P.J.P.; Quintans, J.S.S.; Freitas, M.M.; Cerqueira, M.A.; Teixeira, J.A.; Cardoso, J.C. Hydrogel as an alternative structure for food packaging systems. *Carbohydr. Polym.* **2019**, *205*, 106–116. [CrossRef]
50. Hung, P.; Van, V.; Vien, N.L.; Lan, N.T. Resistant starch improvement of rice starches under a combination of acid and heat-moisture treatments. *Food Chem.* **2016**, *191*, 67–73. [CrossRef]
51. De Graaf, R.A.; Broekroelofs, A.; Janssen, L.P.B.M. The acetylation of starch by reactive extrusion. *Starch-Staerke* **1998**, *50*, 198–205. [CrossRef]

Review

Tailoring the Barrier Properties of PLA: A State-of-the-Art Review for Food Packaging Applications

Stefania Marano [1,*], Emiliano Laudadio [1], Cristina Minnelli [2] and Pierluigi Stipa [1]

1. Department of Science and Engineering of Matter, Environment and Urban Planning, Marche Polytechnic University, 60131 Ancona, Italy; e.laudadio@staff.univpm.it (E.L.); p.stipa@staff.univpm.it (P.S.)
2. Department of Life and Environmental Sciences, Marche Polytechnic University, 60131 Ancona, Italy; c.minnelli@staff.univpm.it
* Correspondence: s.marano@staff.univpm.it

Abstract: It is now well recognized that the production of petroleum-based packaging materials has created serious ecological problems for the environment due to their resistance to biodegradation. In this context, substantial research efforts have been made to promote the use of biodegradable films as sustainable alternatives to conventionally used packaging materials. Among several biopolymers, poly(lactide) (PLA) has found early application in the food industry thanks to its promising properties and is currently one of the most industrially produced bioplastics. However, more efforts are needed to enhance its performance and expand its applicability in this field, as packaging materials need to meet precise functional requirements such as suitable thermal, mechanical, and gas barrier properties. In particular, improving the mass transfer properties of materials to water vapor, oxygen, and/or carbon dioxide plays a very important role in maintaining food quality and safety, as the rate of typical food degradation reactions (i.e., oxidation, microbial development, and physical reactions) can be greatly reduced. Since most reviews dealing with the properties of PLA have mainly focused on strategies to improve its thermal and mechanical properties, this work aims to review relevant strategies to tailor the barrier properties of PLA-based materials, with the ultimate goal of providing a general guide for the design of PLA-based packaging materials with the desired mass transfer properties.

Keywords: PLA; barrier properties; food packaging; nanoconfinement; biocomposites; clay nanoparticles; copolymers; molecular dynamics

Citation: Marano, S.; Laudadio, E.; Minnelli, C.; Stipa, P. Tailoring the Barrier Properties of PLA: A State-of-the-Art Review for Food Packaging Applications. *Polymers* **2022**, *14*, 1626. https://doi.org/10.3390/polym14081626

Academic Editors: Alexey Iordanskii, Valentina Siracusa, Michelina Soccio and Nadia Lotti

Received: 20 March 2022
Accepted: 8 April 2022
Published: 18 April 2022

Publisher's Note: MDPI stays neutral with regard to jurisdictional claims in published maps and institutional affiliations.

Copyright: © 2022 by the authors. Licensee MDPI, Basel, Switzerland. This article is an open access article distributed under the terms and conditions of the Creative Commons Attribution (CC BY) license (https://creativecommons.org/licenses/by/4.0/).

1. Introduction

The increasing use of petroleum-based plastics has raised serious environmental issues closely related to their resistance to biodegradation. In the specific case of materials used in packaging, food-contaminated plastics cannot be recycled, so significant amounts of non-degradable material are constantly accumulating in the natural environment and landfills [1]. In this context, growing research interest is directed toward a greater use of renewable resources for the production of bio-based polymers with certain desired functionalities that are, at the same time, fully biodegradable and recyclable [2]. Among a large number of biopolymer candidates, environmentally friendly thermoplastic polylactide (PLA), derived from renewable agro-resources, has certainly attracted much attention due to its promising attributes for packaging applications. These include biotic and non-biotic degradability, clarity, stiffness, low-temperature heat sealability, GRAS (Generally Recognized as Safe) status as well as suitable barrier properties to flavors and aromas, which have made PLA the most widely produced bioplastic at present for certain biological food contact applications [3]. For instance, Biota[TM] PLA-bottled water, Noble[TM] PLA-bottled juices, and Dannon[TM] yogurts are the main PLA-based packaging examples available on the market.

In this context, however, it should be noted that not all PLA grades are suitable for packaging applications, as the properties of PLA strongly depend on the physical state of the polymer, which in turn is primarily influenced by the stereochemistry, composition, and molecular weight of PLA. In fact, lactide (the cyclic dimer of the chiral lactic acid moiety) exists in three diastereoisomeric forms such as L-lactide, D-lactide, and meso-lactide, whose polymerization via a ring-opening reaction (ROP) affords several different PLA grades [4]. Optically pure PLA such as isotactic poly(L-lactide) (PLLA) and poly(D-lactide) (PDLA) are crystalline in nature with a melting point around 180 °C, while atactic poly-(meso-lactide) (PDLLA) is a fully amorphous material with a glass transition temperature of 50–57 °C. Depending on the ratio of optically active L- and D,L-monomers, it is possible to tune the amorphous and crystalline content of PLA and thus its properties. Detailed information on this subject can be found by interested readers in [5,6]. In addition, it has been widely reported that other factors such as the chain orientation and crystal packing of PLA also affect the crystallinity content, crystal thickness, spherulite size, and morphology, which in turn can strongly influence the final polymer properties [7].

PLA samples with the highest content of amorphous regions and low molecular weight (MW) are not used in packaging as they are less thermally stable and degrade much faster than their enantiomerically pure counterparts with high MW PLA (PLDA or PLLA) [8]. However, as shown in Table 1, even the highest performing PLA grades with the highest degree of crystallinity (typically PLA with 96–99% L-lactide) are not good enough to extend the applicability of PLA in packaging. They show relatively poor barrier performance as well as poor heat resistance and brittle fracture behavior compared to conventional petroleum-based plastics [9–11]. In terms of mass transfer properties, PLA exhibits moderate permeability (P) values for oxygen (O_2) and carbon dioxide (CO_2), which are higher than those of polyhydroxy butyrate-co-valerate (PHBV), polyvinylidene chloride (PVDC), ethylene vinyl alcohol (EVOH), polyethylene terephthalate (PET), and polyvinyl alcohol (PVOH), but mostly lower than those of the other polymers. However, with the exception of PVOH and EVOH, PLA has a very high P_{Water} compared to the other polymers, which also limits applications that require a high barrier to water vapor.

While most recent review articles have focused on the functional properties of PLA materials (i.e., mechanical and thermal properties) [12–16], there are no studies that have provided a comprehensive understanding of PLA barrier performance and reported the available methods to improve it. Therefore, this review provides a critical overview of recent strategies to improve the barrier properties of PLA-based plastics, focusing on the approaches that have the least environmental impact. These include crystallization and orientation, crystal modifications, blending with other impermeable biopolymers and/or nanofillers, and co-polymerization. In addition, this review presents several models for predicting the permeability of PLA-based plastic films through a molecular dynamics simulation approach (MD). To provide an overview of the strategies used by researchers, this review begins with a brief introduction to the key parameters used to characterize gas transport properties in polymer films and ends with a summary and outlook on the design of PLA-based packaging materials with the desired mass transfer properties.

Table 1. Comparison of the main barrier (permeability coefficient for O_2, water, and CO_2 in $Kg \cdot m \cdot m^{-2} \cdot s^{-1} \cdot Pa^{-1}$) and the thermal and mechanical properties of polymers used in food packaging.

Polymer *	Crystallinity Content (%)	P_{O_2} [a]	P_{Water} [b]	P_{CO_2} [c]	T_{deg} [d] Onset (°C)	Tensile Strength (MPa)	Young Modulus (GPa)	Elongation at Break (%)	Ref.
PLA	0–40	310 ± 150	161 ± 41	2811 ± 842	270	20–70	3.1–4.8	3.6–8.8	[9,12,13,17–20]
PHBV	40–60	1.1–3.2	15–24	14–40	230	35–40	3.6–5.2	4–970	[21–23]
PP	30–60	1790	312	10,500	335–450	31–48	0.2–1.4	550–1000	[20,24,25]
PET	17–40	35.9	7.8	35.9	406	45	2.7–4.1	335	[20,24–29]
PVC	10	449	16.5	247	250	4–23	2.7–3.0	200–240	[20,30–33]
PVDC	40–50	0.1–0.3	1.2–7.3	2.3–12	130	25–110	1.2–1.8	30–80	[34–36]
PVOH	15	0.7–9.5	430–840	18.2	200	31	0.08–0.7	57–122	[20,37–42]
EVOH	58–70	0.5–7.1	320–560	5.1–14.3	397	55–65	0.4–1.2	100–225	[38,43,44]
LDPE	47	3100	5.5	18,600	395	33	0.3–0.6	1075	[20,45–49]
HDPE	74	424	2.1	538	389	16–21	0.5–1.2	10.7–13.7	[20,48,50,51]

* PHBV, polyhydroxy butyrate-co-valerate; PP, polypropylene; PET, polyethylene terephthalate; PVC, polyvinyl chloride; PVDC, polyvinylidene chloride; PVOH, polyvinyl alcohol; EVOH, ethylene vinyl alcohol; LDPE, low-density polyethylene; HDPE, high-density polyethylene. [a] Oxygen permeation coefficient (P): $P \times 10^{-20} \frac{Kg\,m}{m^2 s\,Pa}$; [b] Water vapor permeation coefficient (P): $P \times 10^{-16} \frac{Kg\,m}{m^2 s\,Pa}$; [c] Carbon dioxide permeation coefficient (P): $P \times 10^{-20} \frac{Kg\,m}{m^2 s\,Pa}$; [d] Onset of degradation temperature measure by thermogravimetric analysis.

2. Concept of High Gas Barrier Material: Fundamentals of Permeation and Diffusion

The term 'barrier' refers to the inherent ability of a material to allow the exchange or permeation of low molecular weight chemical species such as gases, water vapor, and certain organic compounds (aroma molecules) [52]. This capability is extremely important in the food packaging industry, as the most important function of any packaging system is to maintain the quality and safety of the contents. In fact, foods are chemically unstable by nature and therefore need to be protected from various spoilage possibilities, lipid oxidation, and microbial contamination being the main causes of their deterioration [53]. Therefore, polymeric materials intended for use in many packaging applications must form a "high barrier" (i.e., they must prevent the penetration of substances from the packaging environment into the food and vice versa) as much as possible. The gases typically involved in food packaging are oxygen, water vapor, and carbon dioxide, and the corresponding permeability rates are known as O_2TR, WVTR, and CO_2TR, respectively.

It is well-known that permeation of low molecular weight chemicals through a non-porous polymer matrix occurs via a combination of two processes (e.g., solution and diffusion). Figure 1 shows that gas molecules are first dissolved on one side of the polymer film, followed by molecular diffusion to the other side (postulated by Thomas Graham in 1866) [54]. These processes, can therefore be described by a simple solution–diffusion mechanism using Henry and Fick's laws, which can be formally expressed in terms of permeability P, solubility S, and diffusion D, according to Equation (1):

$$P = D \cdot S = \frac{J \cdot d}{\Delta_p} = \frac{amount \cdot material\ thikness}{surface\ area \cdot time \cdot pressure\ difference} = \frac{[cm^3]\ (SATP)\ [cm]}{[cm^2]\ [s]\ [Pa]} \quad (1)$$

where J is the amount of material transported per unit time through a unit area with thickness d at standard ambient temperature and pressure (SATP; 298.15 K and 10^5 Pa) and Δ_p is the constant partial pressure difference between both sides of the polymer matrix [55]. Therefore, the magnitude of permeability is determined by the diffusion rate (D), which is a kinetic parameter, and the solubility (S), a thermodynamic parameter related to the amount of permeate sorbed by the polymer membrane. In the specific field of packaging, it is worth noting that the permeability p values at different locations in a package can vary greatly and are only an approximate estimate of the actual overall permeability. This could be attributed to different material thicknesses of the walls and seals, multilayer compositions, and/or the presence of defects (i.e., pores or leaks). Therefore, it is critical to more accurately calculate the overall permeability Q of the package using the flux J, according to Equation (2) [36]:

$$Q = \sum_i^n P_i d_i = \sum_i^n A_i J_i \quad (2)$$

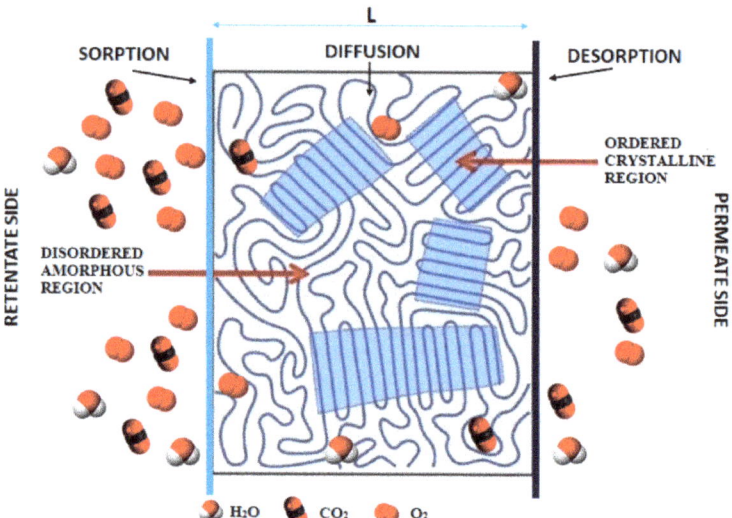

Figure 1. Schematic representation of the general mechanism of the permeation of small molecules through semicrystalline polymers.

Q is thus the sum of the permeability values P_i for each individual packaging component i relative to its wall thickness d_i. Based on Equation (1), this is equal to the sum of the corresponding flux J_i multiplied by the surface area A_i of the component. In practice, there are several methods for measuring the permeability of plastics in the form of films, sheets, laminates, co-extrusions, or plastic-coated materials, all of which are published by standards organizations such as the American Society for Testing Materials (ASTM International) and the International Organization for Standardization (ISO) [56]. These include the isostatic method [20] (also known as the continuous-flow method) and the quasi-isostatic method [57] (also known as the lag-time or constant-volume/variable-pressure method). In the latter method, a polymer film is exposed to the permeant on one side and the concentration is accumulated to values below 5 wt% on the other side. The samples are quantified at specific time intervals to produce a graph showing the amount of permeant versus time. The intercept of the x-axis is taken from the steady-state portion of the graph. This is the lag time (t_θ) used to estimate D as follows:

$$t_\theta = \frac{L^2}{6D} \qquad (3)$$

when the permeation is in a steady-state, P can be estimated from the slope of the linear part of the permeation plot.

For semicrystalline polymers (e.g., PLA), transport properties are generally evaluated using a two-phase model that identifies an impermeable crystalline phase and a permeable amorphous matrix [58]. According to this model, sorption can only occur in the amorphous regions as the denser crystalline organization makes it difficult for the permeant molecules to reach the sorption sites due to limited mobility. However, deviations from this simplified model have been reported [59–63] because, in addition to crystallinity/amorphous fraction, permeation can be influenced by other intrinsic and/or extrinsic factors such as crystal architecture, polarity, polymer microstructure, chain packing, amorphous phase morphology, presence of additives, and environmental conditions (i.e., temperature and relative humidity) [64,65].

In light of developing highly efficient and economically viable PLA-based packaging materials, the main approaches to optimize the transport properties of PLA including molecular dynamics simulations are presented below to identify the main factors that determine the permeability performance of the material.

3. PLA Morphology Modifications

3.1. Degree of Crystallinity and Crystal Polymorphism

It is well-established that the degree of crystallinity affects a variety of polymer properties including gas and/or water vapor permeability behavior. In fact, the crystalline phase is highly ordered, aligned, and denser than the amorphously oriented phase, which does not exhibit repeating patterns in the solid state. Common film-forming polymers used in packaging are semi-crystalline polymers (e.g., PET, PP, polyethylene (PE), PVDC, polyamide (PA), and EVOH), in which both the amorphous and crystalline regions coexist within the polymer matrix. Since gas diffusion through polymers is primarily controlled by the packing mode of the molecular chain segments, it is generally assumed that ordered crystalline domains should act as an effective barrier to the diffusion of gases and small molecules, making the amorphous phase the only pathway available for permeation [66]. Moreover, penetrants cannot sorb in crystalline structures because their solubility coefficients are lower compared to those of their amorphous counterparts [67,68]. For this reason, a high degree of crystallinity is particularly desirable for polymers intended for the food packaging industry [69,70]. Therefore, in a similar way to many other semicrystalline polymers, an increase in the degree of crystallinity in PLA should result in a decrease in the permeability of most low molecular weight compounds. As previously mentioned, the degree of crystallinity of PLA can be easily tuned by polymerizing a controlled mixture of the L-, D-, and meso-lactide. Depending on the isomer ratio, PLA can be fully amorphous or semicrystalline and the more optically pure polymers display higher crystallinity fractions because of higher chain symmetry. In particular, the degree in crystallinity (X_c) increases as L-lactide increases, except for PLA 50% L-lactide and 80% L-lactide, which both present 0% of X_c [9]. Alternatively, a higher degree of crystallinity can also be obtained by post-processing corona treatment and drawing (i.e., uniaxial or biaxial orientation of amorphous PLA samples) [71–73]. X_c in polymers is commonly measured using the differential scanning calorimetry (DSC) technique by dividing the enthalpy of fusion of the studied samples with the reference enthalpy value for 100% crystalline PLA (93 J/g) [12]. The gas permeability performance of several PLA grades as a function of crystallinity content has been widely investigated over the last couple of decades, and a large body of data can be found in the literature in this regard [73–79]. However, contrary to expectations, there did not appear to be a clear relationship between PLA barrier performance and its degree of crystallinity. In particular, the decrease in the gas permeability and water sorption did not occur linearly or specifically, not to the expected extent, with increasing crystallinity of PLA. For example, Tsuji and coworkers [78] conducted an in-depth investigation of the effect of the X_c of various PLA films on their water vapor permeation coefficient (P_{water}). PLA raw polymers were supplied or synthesized by ROP and the resulting samples were solution-cast to form thin films (thickness of ~50 µm). Films were subsequently made amorphous by melt quenching and recrystallized at different time and temperature to obtain samples with X_c ranging from 0 to 35%. Results showed (summarized in Table 2) that the P_{water} of PLLA films decreased monotonically from 2.18 to 1.14×10^{14} Kg/m/m^2/s/Pa as a function of increasing X_c from 0 to 20%. However, at higher X_c a plateau in the P_{water} values was reached, reporting no further reductions.

Similarly, Drieskens et al. [74] subjected compression-molded (amorphous) PLA samples to cold crystallization at different temperatures and times to obtain samples with various X_c and morphologies. They found that at low level of crystallinity (~30–35%), the oxygen permeability coefficients decreased almost linearly with increasing crystallinity, while at higher X_c (>40%), the opposite trend was observed. An analogous observation was reported in the work conducted by Guinault et al. [75], whereby the analysis of measured oxygen and helium permeability values showed that for crystallinity degrees higher than around 35%, the diffusion coefficient increased with increasing X_c, confirming the poor relationship between crystallinity and barrier properties for PLA. As a last example, Colomines et al. [79] obtained, from oxygen and helium permeation measurements, comparable permeability coefficients between amorphous PLA and semicrystalline PLA film samples prepared by compression molding, showing that the degree of crystallinity does not appear to have any effects on PLA permeability behavior.

Table 2. Water vapor permeation coefficient (P_{water}) of PLA films measured at 25 °C as a function of degree of crystallinity (X_c) and crystallization time. Data taken from [78].

X_L [a] (%)	Crystallization Time [b] (min)	X_C [c] (%)	$P \times 10^{14}$ (Kg m/m²/s/Pa)
72.2	0	0	1.90
50	0	0	1.95
99.7	0	0.6	2.08
99.4	0	0.7	2.18
99.4	0	1	1.91
99.4	5	5.7	1.90
99.4	7.5	19.1	1.14
99.4	10	32	0.99
99.4	12.5	34.9	1.04

[a] L-lactyl unit content X_L of P(LLA-DLA). [b] Crystallization time equal to 0 means that the specimens were melt-quenched. [c] Crystallinity content obtained from the DSC first run.

These studies and other similar ones suggest that the barrier properties of PLA might be less dependent on the crystalline content than expected and the effect of the resulting post-processing PLA microstructures should be further explored in relation to their different permeability behaviors. In this context, some investigations have attempted to find a correlation between the barrier properties and potential PLA crystal structure modifications. In fact, depending on processing conditions, PLA can crystallize in up to four polymorphs: α, β, γ, and the more recently reported α′ (or otherwise known as δ) form [80]. Among all, the α form, obtained by simple crystallization from the melt at high temperature, is the most stable polymorph. β and γ forms are obtained in more extreme or special conditions such as employing high energy stretching of the α form at high temperature (β form) or conducting crystallization on the hexamethylbenzene substrate (γ form) [81]. However, the α′ form reported by Zhang et al. [82] was obtained at low crystallization temperature and the resulting crystal structure was found to differ only slightly from the α structure. By simply changing the crystallization temperature from low to high values, it is therefore possible to obtain samples containing PLA in either the pure α form or α′ form, or a mixture of the two polymorphs in the same system. Differences between the α and α′ forms lie only in their chain packing mode, whereby the larger lattice dimension and the weaker interchain interactions make the α′ form somewhat more disordered compared to the α form. This peculiar difference in the packing conformation between the two crystal microstructures may affect the barrier properties of processed PLA samples. Cocca et al. [83] undertook this investigation by subjecting a series of compression-molded PLA samples to different crystallization temperatures ranging from 85 up to 165 °C. Based on the wide angle X-ray (WAXD) results, the heating treatment afforded PLA samples with different α/α′ ratios: samples in the pure α form ($T_c > 145$ °C), samples in the pure α′ ($T_c < 95$ °C) form, and samples containing a mixture of both crystal structures (105 °C < T_c < 125 °C). Figure 2A–C shows the estimated fractions of the α form in PLA compression-molded films and the related optical micrographs after cold crystallization as well as the corresponding water vapor permeability of crystallized films as a function of X_c. The water vapor permeability behavior of PLLA films showed a clear dependence on the crystal conformation. In particular, samples uniquely containing the α′ form showed the maximum permeability value, which started to progressively decrease with larger ratios of α to α′ content until the minimum value was reached for samples containing only the α form. Similar results were also observed in another following study [75]. This clearly indicates that the molecular packing mode strongly influences the permeability behavior of PLA and caution should be paid in choosing appropriate crystallization conditions to favor the formation of the more impermeable α crystal structure.

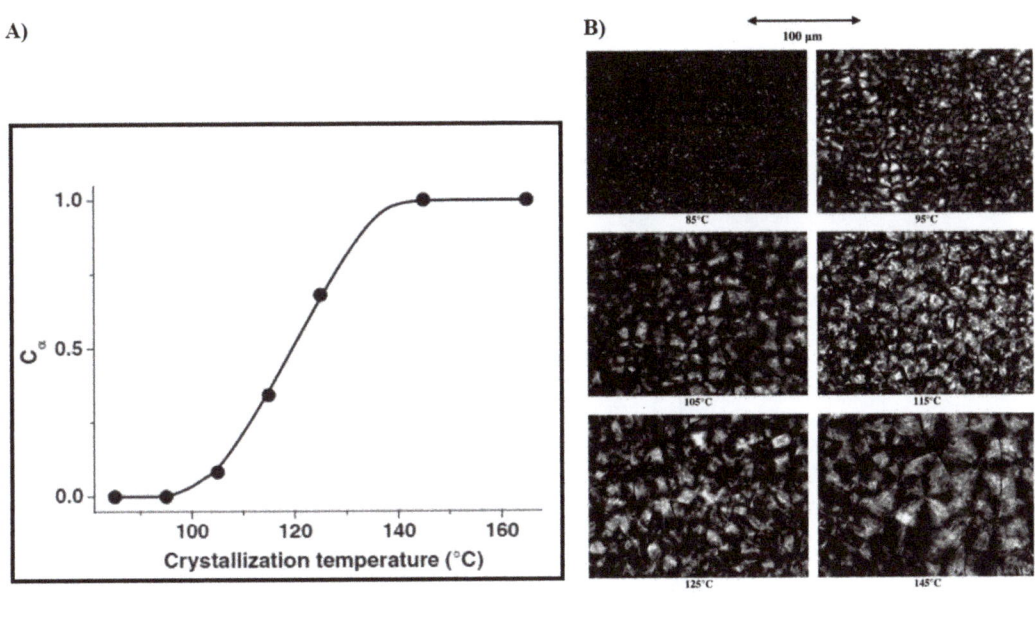

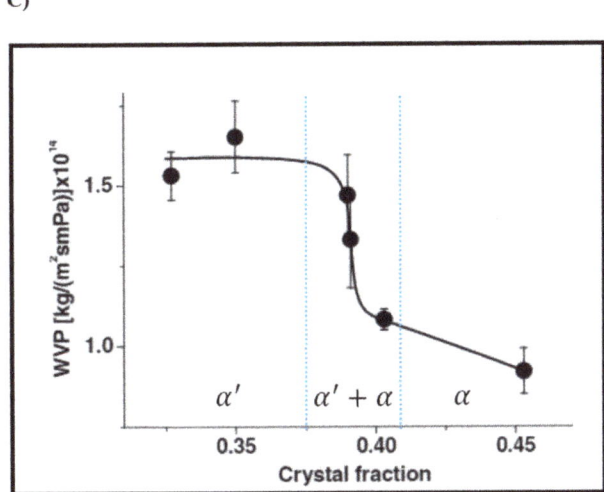

Figure 2. (**A**) Water vapor permeation estimated fraction of the α form in PLLA films; (**B**) optical micrographs of compression molded PLLA films after cold crystallization; (**C**) water vapor permeability of PLLA films crystallized as a function of degree of crystallinity. Adapted from [83] with permission from Elsevier. Copyright © 2011.

3.2. Amorphous Phase Conformation

Besides the crystallinity content and crystal conformations, the amorphous phase dynamics and particularly its degree of coupling with the crystalline phase have been reported to play an important role in the barrier properties of many polymeric materials [68,84–87]. In fact, as displayed in Figure 3, the amorphous regions in semicrystalline polymers (i.e., PLA) are typically affected by constraints imposed by the crystalline regions, which identify two different amorphous phases: the "free" mobile amorphous fraction (MAF) and the more rigid amorphous fraction (RAF) [88–90].

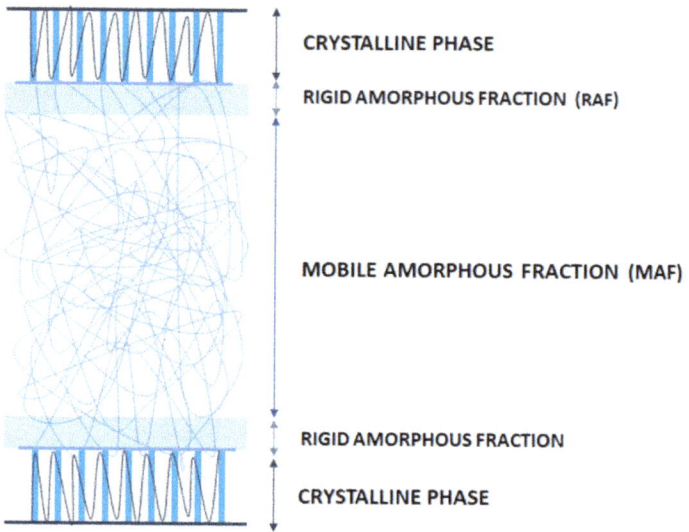

Figure 3. Schematic representation of the arrangement of crystalline, rigid amorphous and mobile amorphous fractions.

The latter does not relax as it is confined by crystalline lamellas in such a way that the amorphous phase chain mobility is greatly reduced. This creates an excess of free volume attributed to a de-densification effect [91]. For this reason, this particular constrained amorphous region appeared to have a more important role on PLA permeability behavior compared to the crystalline fraction content [61,74,77,87]. For example, Drieskens et al. [74] investigated the potential correlation between the morphology of isotropic PLA (modified by cold crystallization at different temperatures and times) and changes in PLA oxygen transport characteristics (permeability, diffusion and solubility). Amorphous PLA samples were obtained by direct quenching from the melt and subsequently annealed at the corresponding crystallization temperature (T_c) for different times (from 10 min up to 24 h). Following isothermal crystallization, different PLA crystal morphologies were reported, and the resulting microstructures characterized using DSC, electronic, and optical microscopy. In accordance with previous results, at low crystallinity increment, gas permeability was reduced due to an increase in the number and size of crystals, which in turn, increased the tortuosity of the transport path. However, at higher crystallinity contents, when the polymer matrix was completely filled with crystals, the glass transition fully shifted to higher temperatures, indicating that a constrained amorphous phase (RAF) was formed during PLA crystallization. By measuring the thermodynamic gas solubility component, it was observed that samples containing increasing amounts of the RAF fraction showed much higher gas solubility values. This was ascribed to be the major cause of plateauing in the permeability behavior at higher crystallinity levels. In line with these results, Guinault and coworkers [77] undertook an in-depth study to further elucidate the relationship between gas properties, crystallinity (ranging from ~2 to 61%), and RAF content (ranging from ~0 to 20%) in twenty PLA film samples. The permeability performance was analyzed using two model molecular probes: oxygen ('red sea mechanism' transport) and helium ('fluid like mechanism' transport). In the first mechanism, the gas transport is dependent on the volume accessible based on the microstructure of the polymer matrix, while the second one, having a much smaller size, behaves more like a "liquid" through the polymer microvoids. In both the PLLA and PDLA samples, a significant drop in helium permeability was observed upon crystallization of the fully amorphous extruded samples. However, when oxygen transport was analyzed, which is more sensitive to the polymer microstructure, there was only a small reduction in the permeability with increasing crystallinity up to approximately 35% for both sample groups. At higher crystallinity levels, the permeability to oxygen was observed to increase as a function of increasing RAF fraction. This anomalous behavior was ascribed to an increase in the overall free volume due to poor coupling between RAF and the crystalline fraction. This observation was also found in accordance with a previous study conducted by Del Rio et al. [92], whereby the evolution of the free volume was measured upon annealing of an amorphous PLLA sample using positron annihilation lifetime spectroscopy (PALS). This technique allowed for the identification of an increase in the free volume within the annealed PLLA matrix due a change in conformation of the amorphous regions occurring during crystallization: from a folded or coil conformation in the quenched samples containing exclusively MAF to a more open conformation in samples containing greater content of RAF and crystalline fraction. While this change in conformation caused a decrease in the average hole size from 95.7 to 86.5 Å^3, the number of holes significantly increased with higher RAF content, which could explain a greater overall free volume available for the diffusion of gas molecules after a certain level of crystallinity is reached in PLA samples. More recently, these results have been further investigated and clarified by other authors [61,87]. For example, Fernandes et al. [61] prepared a set of PLA samples with well-defined microstructures, spherulite size, and RAF content as a function of crystallization temperature and annealing time. It was observed that the extent of the formation of RAF in the samples was the preponderant factor governing the oxygen permeability as the solubility and diffusion coefficients were seen to increase as a function of higher RAF content. Moreover, it was interesting to note that samples that were highly nucleated prior to crystallization provided the best results in terms of oxygen barrier properties, indicating that a pre-nucleation step and the short crystallization times hinder the formation of RAF. To close the loop, Sangroniz et al. [87], via a combination of techniques (i.e., PALS, DSC, and density tester), clearly confirmed that the formation of de-densified RAF in annealed PLA samples increased the free volume as the polymer chains had a more rigid conformation than in MAF. Moreover, they found that due to the higher solubility of RAF in water, the overall free volume was further increased due to the plasticization effect of the water molecules, and therefore both factors would contribute together to the increase in the permeability values. Overall, all of these studies reported thus far indicate that an increasingly higher amount of RAF has a detrimental effect on the PLA barrier properties. To minimize the conversion of MAF into RAF with time, a pre-nucleation step and short crystallization time at high temperature may be used.

3.3. Polymer Drawing

Another common way to improve the polymeric materials' barrier properties is through molecular orientation, whereby a polymer is stretched in either one (uniaxial) or two (biaxial) predetermined directions (i.e., by extrusion or injection molding). In other words, orientation involves the use of mechanical and thermal energy to rearrange the polymer in an oriented position at the molecular level. The reduced gas and vapor permeation of oriented semicrystalline polymers is well documented [66,86,93–96], and the main mechanism involves the transformation from a spherulitic structure (the lamellar crystals propagate radially from the nucleation site) to a densely packed microfibrillar conformation (growth of lamellar crystals perpendicular to the direction of strain). This transformation induces an effective reduction in the diffusion coefficients by increasing the tortuosity of gas transport paths [66]. In the specific case of PLA, while the use of molecular orientation has been widely implemented for improving its thermal and mechanical properties, a limited number of papers could be found on how the drawing process affects its permeability behavior [62,97–99]. One relevant example is the study conducted by Delpouve and co-workers [62], whereby the effects of three different drawing modes on the water permeability properties were investigated on compression-molded PLA films (Table 3). Samples were subjected to uniaxial constant width (UCW) drawing (films are drawn only in one longitudinal direction (LD)), simultaneous biaxial (SB) drawing (films are drawn in two perpendicular directions), and sequential biaxial (SEQ) drawing (films are first drawn in LD then in the transversal direction). The permeability behavior of the drawn samples was compared to those of fully amorphous and thermally unoriented crystallized samples. As shown in Table 3, results showed that the water permeability coefficient (P) of the drawn materials were in all cases lower than the amorphous and the thermally unoriented crystallized samples, regardless of the drawing modes used. Among all of the drawn samples, a significant decrease in the permeability coefficient was obtained with the SB drawn samples, followed by the UCW and SEQ ones. In particular, the water permeability was observed to decrease from $2.15 \pm 0.07 \times 10^{-12}$ for the amorphous film (maximum p value) to $1.63 \pm 0.07 \times 10^{-12}$ mol m^{-1} s^{-1} Pa^{-1} for the SB sample (minimum p value), accounting for about 35% of reduction.

Table 3. Values related to samples' crystallinity degree (X_c), mobile amorphous phase degree (X_{am}), rigid amorphous fraction degree (X_{ar}) as well as permeability coefficient P. Data taken from [62].

PLA Sample	X_C [a] (%)	X_{am} [b] (%)	X_{ar} [c] (%)	P_{water} [d]
Amorphous film	0	100	0	2.15
Thermally crystallized film	31	41	28	2.04
UCW drawn (3 × 1)	28	66	6	2.04
SEQ drawn (3 × 3)	27	70	3	1.97
SB drawn (2 × 2)	13	86	1	1.76
SB drawn (3 × 3)	25	66	9	1.63
SB drawn (4 × 4)	31	62	7	1.63
SB drawn (3 × 3) thermos-fixed	31	59	10	1.70

[a] Crystallinity degree. [b] Mobile amorphous phase degree. [c] Rigid amorphous fraction degree. [d] Water permeability coefficient (10^{-12} mol m^{-1} s^{-1} Pa^{-1}).

As shown in Figure 4 (WAXD patterns), the best performance of SB compared to those obtained via uniaxial or sequential drawing modes was ascribed to the resulting homogeneous orthotropic structures of the SB drawn films, whereby PLA macromolecules were oriented perpendicularly to the water diffusion path. Between UCW and SEQ, the resulting worse performance of SEQ compared to UCW samples was linked to partial destruction of the crystallites because of the two sequential chain orientations. Likewise, Dong et al. [99] investigated both the oxygen and water vapor permeability of two types of uniaxial drawn PLA samples, one simply stretched with a twin-screw extruder system and the other one was first stretched and then annealed at 90 °C for two hours. All samples were stretched at different draw ratios, namely R = 1, 2, 3.5, 5, and 6.5. While the sample crystallinity content was seen to increase as a function of draw ratio, the morphological analysis of the samples' surfaces indicated that too high a stretching strength (higher than R = 3.5) promoted the formation of cracks and a porous structure with high permeability to both gas and water vapor. However, samples with draw ratios up to 3.5 showed smooth surfaces. Compared to the undrawn PLA films, both the stretched and annealed samples (at R = 3.5) showed a significant reduction (~25%) in the oxygen and water vapor permeability. A slightly better performance was obtained with the annealed samples, and this was attributed to the annealing process, which increased the density of the annealed PLLA films relative to the simply stretched one. These results confirm that the drawing process does promote higher barrier properties in PLA films compared to the corresponding undrawn samples. However, care must be taken in order not to exceed the maximum tolerated stretch, as this could have a negative impact on both film morphology and permeability performance.

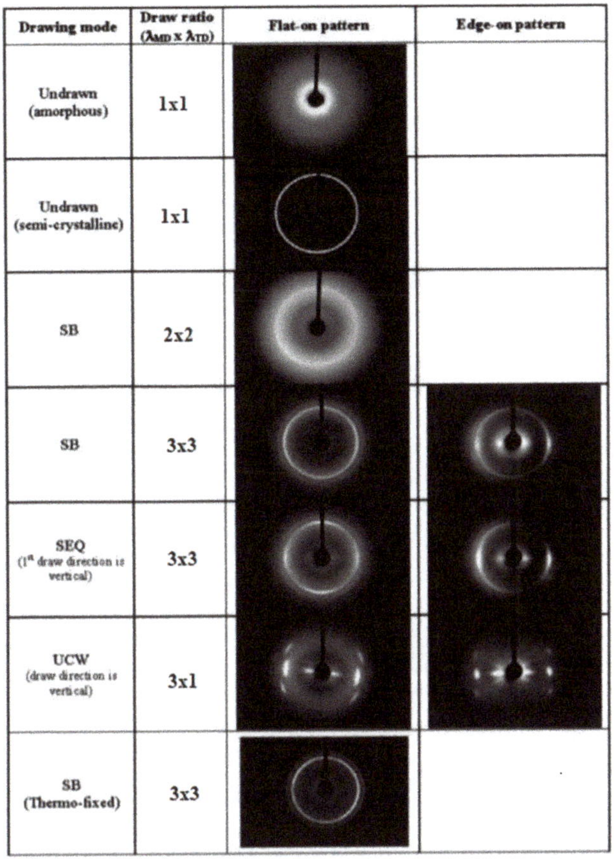

Drawing mode	Draw ratio ($\lambda_{MD} \times \lambda_{TD}$)	Flat-on pattern	Edge-on pattern
Undrawn (amorphous)	1x1		
Undrawn (semi-crystalline)	1x1		
SB	2x2		
SB	3x3		
SEQ (1st draw direction is vertical)	3x3		
UCW (draw direction is vertical)	3x1		
SB (Thermo-fixed)	3x3		

Figure 4. WAXD patterns of drawn and thermally crystallized PLA films. Adapted from [62] with permission from American Chemical Society. Copyright © 2012.

3.4. Nucleating Agents

As widely discussed in the previous paragraphs, the crystallization process and the resulting crystal size, orientation, and morphology can have a drastic effect on a wide range of polymer physical properties including the gas and water vapor permeability. PLA crystallization has been shown to be very slow [100–102], making common polymer processing operations (i.e., injection molding, extrusion, fiber spinning, melt blowing, etc.), time-consuming steps for industrial production [103]. One way to speed up these processes involves the use of effective nucleating agents, which will lower the surface free energy barrier toward nucleation and thus promote faster crystallization rates. Moreover, depending on the type, size, and aspect ratio of nucleant particles, preferential crystal orientation and/or specific crystal superstructures with tailored properties can be obtained [104–108]. Typical nucleating agents for PLA include talc, lactide, montmorillonite, boron nitride, calcium carbonate, magnesium carbonate, titanium oxide, or graphene oxide, to name a few. While these have all shown to significantly increase PLA crystallization rate to a greater or lesser extent [102,109–111], only a few of them have also been found to be particularly effective in improving its barrier properties through different mechanisms. For example, Ghassemi et al. [105] demonstrated that the addition of 3 wt% of talc resulted in a 25–30% decrease in the gas permeability and diffusion coefficients of extruded PLA films to five different gases (hydrogen, oxygen, dioxide carbon, nitrogen, and methane). This was ascribed to the plate-like structure of talc that limited gas motion within the PLA matrix (increase in the tortuosity). Likewise, Buzarovska et al. [112] demonstrated that the addition of 5 wt% of talc lowered the water vapor permeability from 6.71×10^{-12} (neat PLA) to 2.96×10^{-12} mol m/m² sPa in solution-casted PLA films, resulting in up to 55% decrease in the overall water permeability compared to neat PLA. Additionally, in this case, the resulted lower permeability was ascribed to an increase in the diffusion distance and tortuous path for the permeants and this was correlated to an even distribution of impermeable platelet talc particles within the PLA matrix. Another relevant example to tailor PLA properties is through the stereocomplex (SC) crystallite formation between enantiomeric PLLA and PDLA, which can be prepared by a simple physical blending route. As shown in Figure 5A,B, depending on the ratios of PLLA/PDLA, the resulting mix acts as a proper nucleating agent by promoting simultaneous folding of the two enantiomeric chains in triangular (for non-equimolar blends) or hexagonal (equimolar mix) shapes [113–116].

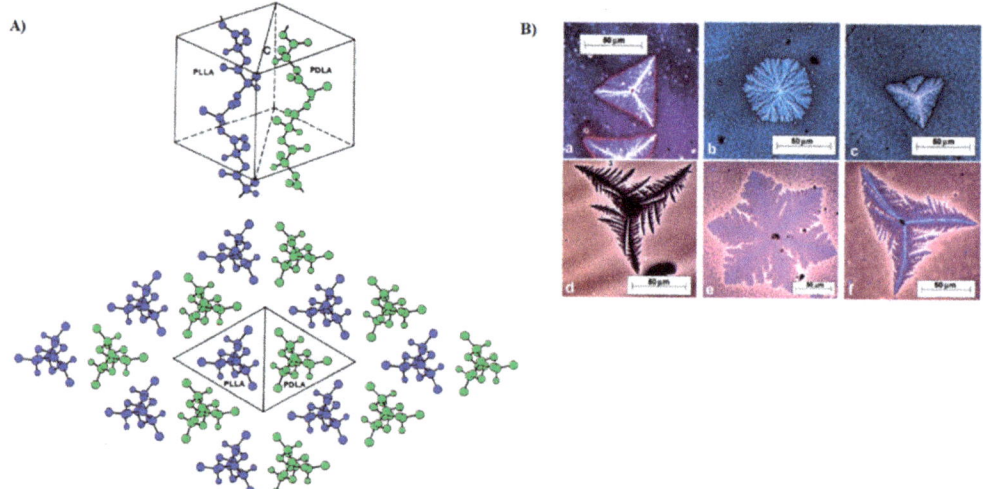

Figure 5. (**A**) SC crystalline lattice; adapted from [115] with permission from Elsevier. Copyright© 2016. (**B**) Optical microscopy micrographs of different PLLA/PDLA films with PLLA content of 75% (**a,d**), 50% (**b,e**), and 25% (**c,f**), crystallized at 200 °C; reproduced from [116] with permission from American Chemical Society. Copyright © 2010.

These peculiar conformations offer favorable positions for the polymer loops during the crystal growth, which in turn, leads to a significant reduction in the induction period for the formation of highly dense spherulites. Given the advantageous crystallization process, the potential effects of stereocomplexation on PLA barrier properties were also investigated [117–121]. For example, Tsuji and Tsuruno [117] found that the water vapor permeability values of PLLA/PDLA solution-casted stereocomplex-based films (SC-PLA) were significantly reduced compared to those of the corresponding homopolymers. In particular, the WVP of SC-PLA was reduced by 14–23% than those of pure PLLA and PDLA with X_c in the range of 0–30%, indicating that superior barrier properties were achieved, even for fully amorphous SC-PLA. In a more detailed study, Varol et al. [121] conducted an in-depth investigation into the effect of PLA stereocomplexation on the transport properties toward a wider range of permeants (water, nitrogen, oxygen, and carbon dioxide). The water permeation data revealed a drastic barrier improvement of up to 70% for SC-PLA with respect to the corresponding homopolymers with the same X_c content. Moreover, it was interesting to note that the water permeability coefficient of fully amorphous SD-PLA at 25 °C was comparable with those of pure semicrystalline PLLA and PDLA (X_c = 48%). More interestingly, the gas barrier properties of SC-PLA toward all gases were exceptionally enhanced compared to both the parent homopolymers.

For example, the permeability coefficient of pure PDLA and PLLA was seen to decrease from 339 ± 44 and 71 ± 25, respectively, up to 0.14 ± 0.02 (Barrer) in SC-PLA measured at the same conditions. Similar trends were observed toward N_2 and CO_2. The overall significant improvement in the barrier properties of SC-PLA was attributed to a change in the crystal conformation from the common α type (pseudo-orthorhombic unit cell with 10_3-helical chain conformation) of the pure homopolymers to the triangular-like crystal shape (triclinic unit cell with 3_1-helical chain conformation) for SC-PLA, which resulted in a smooth and non-porous polymeric matrix. These results clearly highlight the importance of the crystal superstructure on the barrier performance of semicrystalline PLA. In line with this research direction, many authors have reported that the addition of specific nucleating agents could manipulate the crystal superstructure of polymers [103,122–124]. In particular, it was recently shown that 1,3,5-benzenetricarboxylamide derivatives (known family of amide nucleating agents for polypropylene) can tailor the crystal superstructure of PLA, affording three distinct crystal morphologies by melt crystallization such as cone-like, shish-kebab, and needle-like structures [103]. As demonstrated by in situ polarized optical microscopy (POM) and rheological measurements, the nucleant easily dissolves in PLA melt and can self-organize into fine fibrils prior to PLA crystallization. These fibrils act as shish, from which the peculiar PLA crystal structures are formed, depending on the amount of nucleant added (0–0.5 wt%). Among all structures, the shish-kebab-like structure obtained at 0.3–0.5 wt% was further explored (by the same author) as a potential ideal conformation to enhance the barrier properties of PLA [125]. In fact, as shown in Figure 6A,B, the epitaxial growth of crystals occurred orthogonally to the long axis, which, in turn, could form a densely packed wall structure along a vertical direction of the gas diffusion path. The oxygen permeability results shown in Figure 6C display a drastic improvement in the barrier properties of PLA crystallized with increasingly higher amounts of N,N′,N″-tricyclohexyl-1,3,5-benzene-tricarboxylamide (trade name TMC-328), an active model nucleating agent belonging to the 1,3,5-benzenetricarboxylamide derivative family. The best oxygen permeability value was observed for the PLA sheet containing 0.5 wt% of nucleant ($p = 1.989 \times 10^{-20}$ $m^3 \cdot m/m^2$ sPa), exhibiting a reduction of about 300× compared to the corresponding parent PLA sheet with isotopic spherulitic crystals ($p = 5.244 \times 10^{-18}$ m^3 m/m^2 sPa). In a similar study, comparable results of enhanced barrier properties for PLA due to a shish-kebab-like structure were obtained using a different active nucleating agent belonging to the benzhydrazide family, namely octamethylene dicarboxylic dibenzoyl hydrazide (TMC-300) [126]. Like TMC-328, TMC-300 fibrils induced epitaxial growth of the PLA lamellae orthogonally to their fibrillary direction, affording "lamellae-barrier walls" stacked perpendicular to the direction of gas diffusion. By employing a series of layer-multiplying elements at the end of the extrusion setup, multilayer samples with up to 64 layers were produced. The oxygen permeability coefficient showed a gradual reduction with increasing layer number of the dense and impermeable "lamellae-barrier walls", reaching the lowest value (up to 85.4% decrease) with the 16-layer sample compared to the control one. This particularly high resistance to gas permeation was associated with the highest content of branch fibrils in the 16-layer sample that contributed to a more regular in-plane arrangement of PLA lamellae. The impact of multilayer formation on the barrier properties of PLA-based films is further discussed in the following section.

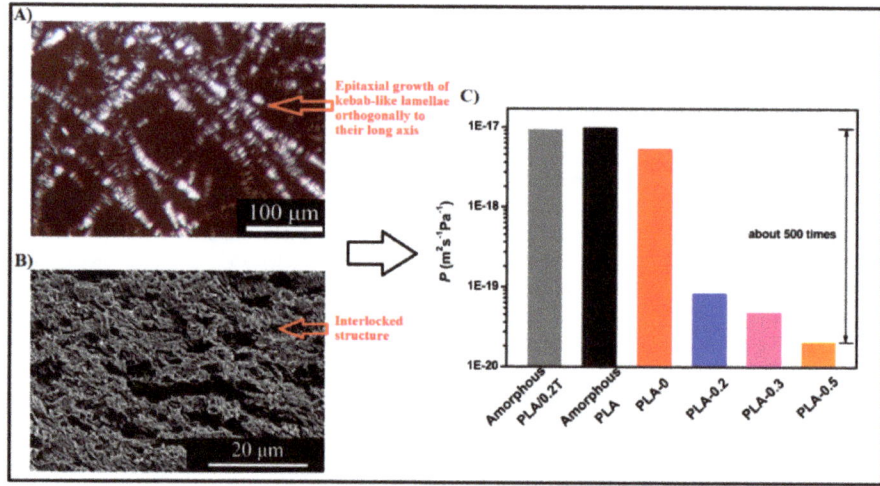

Figure 6. (**A**,**B**) POM and cross-sectional micrograph of crystal morphology for PLA with a nucleating agent showing an epitaxial growth lamellae and interlocked structure, respectively; (**C**) Oxygen permeability coefficient (*p*) values for all PLA samples as a function of nucleating agent content. Adapted from [125] with permission from American Chemical Society. Copyright © 2014.

3.5. Nanoconfinement Approach

Another way to improve the performance of packaging materials is based on the formation of coatings or multilayers [127,128]. In this context, the use of nanotechnology has recently shown that the formation of nanoscale layers (i.e., nanolayers) can result in unique crystalline morphologies that can have a profound impact on the barrier properties of the packaging films [129]. Common technologies for developing these nanostructured multilayer films include the layer-by-layer (LbL) technique [130], the electrohydrodynamic processing (EHDP) [131], and the layer-multiplying "forced assembly" coextrusion (Lm-FAC) method [132]. The latter method, introduced 40 years ago by Dow and recently updated by Baer's group, has attracted particular attention because it can control the crystallization habits of polymers and improve properties not possible with the bulk [129]. Briefly, this technique consists in combining two or three polymers into a continuous alternating layered structure with hundreds or thousands of layers of nanometric thickness. The molecular and chain organization of the polymers in a confined nanometer-scale space (ultrathin films) prevents isotropic spherulitic growth of the lamellar crystals, resulting in unique crystal orientations. As can be seen in Figure 7, these are "on-edge", where the polymer chains are oriented parallel to the substrate plane, and "in-plane", where the polymer chains are oriented perpendicular to the substrate plane [133]. Due to the high-aspect ratio and peculiar orientation of the highly ordered interlayer lamellae, the latter has led to a substantial reduction in the gas permeability by several orders of magnitude compared to their bulk film controls for a wide range of confined/confiner polymers [118,134–136] including PLA, albeit in a very limited part [137–139]. The general principles of mass transport in polymeric multilayers have been carefully summarized elsewhere, so we refer readers to the following articles: [140,141].

The orientation of crystals is undoubtedly a crucial parameter for tailoring barrier properties and is directly affected by crystallization temperature, film thickness, chain mobility, and substrate–polymer interactions. However, when PLA was chosen as a model semicrystalline polymer to study the impact of confinement on barrier properties, the results showed that the dynamics of the amorphous phase (i.e., the occurrence of RAF in the multilayer samples) also plays a crucial role in the overall barrier performance [133–137].

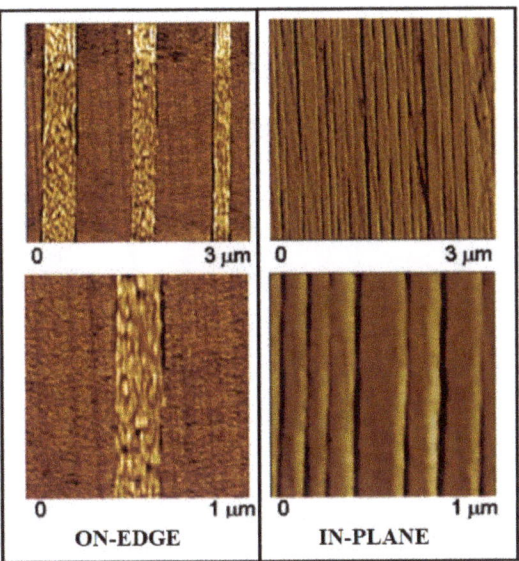

Figure 7. Example of typical AFM phase images of multilayered film cross-section with on-edge and in-plane orientations. Adapted from [133] with permission from John Wiley and Sons. Copyright © 2011.

For example, Messin et al. [137] provided new insights into the relationships between microstructure implying RAF and barrier performances of a 2049-layer film of poly(butylene succinate-co-butylene adipate) (PBSA) confined against PLA prepared using LmFAC technology. The content of the multilayer was 80 wt% PLA and 20 wt% PBSA. The continuity of the PLA/PBSA layers can be clearly seen in Figure 8A using an atomic force microscopy (AFM) image. The confinement effect caused by the PLA resulted in a slight orientation of the crystals in both the transverse and extrusion views and an increase in RAF in PBSA with densification of this fraction. As shown in Figure 8B, these structural changes significantly improved the water vapor and gas barrier properties of the PBSA layer by up to two decades in the case of CO_2 gas, mainly due to the reduction in solubility. However, it is important to note that the results of these authors contradict recent findings from other studies (see Section 3.2) that RAF is responsible for a de-densification of the amorphous phase and a decrease in the overall barrier performance. In this context, Nassar et al., in a more recent study [138], investigated the barrier properties of PLA in multinanolayer systems with two amorphous polymers (polystyrene, PS; and polycarbonate, PC), probing the effect of confinement, the compatibility between the confiner and the confined polymer, crystal orientation, and amorphous phase properties. WAXD measurements showed that the PLLA lamellae between PS layers had a mixed in-plane and on-edge orientation, while the PLLA lamellae between PC layers were clearly oriented in-plane. More importantly, the RAF content of semicrystalline PLLA was about 15% in PC/PLLA, whereas it was negligible in PS/PLLA. Oxygen permeability results showed that the occurrence of RAF in PC/PLLA samples had a detrimental impact on the barrier properties of the multilayer films, which could not be compensated by the presence of in-plane crystals. Moreover, annealing PS/PLLA films to minimize RAF content allowed for a barrier improvement of the PLLA layers by a factor of two compared with semicrystalline bulk PLLA. Overall, it can be pointed out that work conducted thus far in this regard is still limited and contradictory. Therefore, further investigation is recommended to elucidate the effects of amorphous-phase dynamics and nanoconfinement on the barrier properties of PLA-based multilayer films.

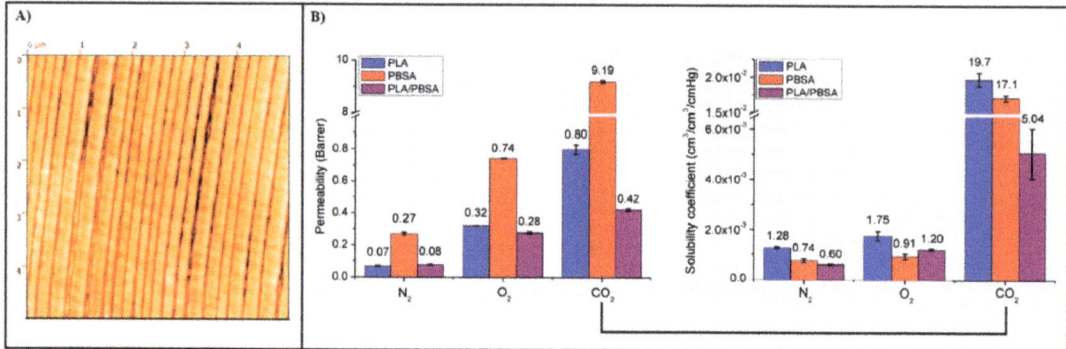

Figure 8. (**A**) AFM images of the multilayer film with PLA in light and PBSA in dark; (**B**) gas permeability and solubility coefficients for the monolayer films of PLA and PBSA and the multilayer film of PLA/PBSA. Adapted from [137] with permission from American Chemical Society. Copyright © 2017.

4. Bio-Blends and Composites

Blending is a well-established low-cost technology to develop next generation plastics with enhanced properties compared to the single components. Polymer blending is indeed a straightforward and user-friendly process, whereby materials are easily and rapidly mixed either in co-solution or in the molten state. It is therefore not surprising that this approach has been largely applied for improving PLA performance in many applications [142–144] including food packaging [12,145–147]. Among the uncountable number of co-additives employed, bio-fillers and other biodegradable polymers have recently attracted a great deal of attention, viewed in the context of minimizing environmental impacts and encouraging greater use of sustainable and renewable sources [148–153]. Those materials include natural fibers, especially as reinforcing agents (i.e., wood, cellulose, sisal, kenaf, flax, and hemp) and common biodegradable polymers (lignin, starch, polycaprolactone (PCL), polybutylene succinate (PBS), and polyhydroxyalkanoates (PHAs)). Moreover, the implementation of nano-additives (i.e., nanoclay, nanowhiskers, and 3D-isodimensional nanoparticles) for the formation of bionanocomposites in the field of food packaging have rapidly increased over the last decade due to their exceptional ability to improve PLA film properties [154–159]. However, as for all blends and composites (polymer–filler, polymer–polymer, polymer–nanomaterial), the resulting properties may vary widely as they do not depend only on the intrinsic nature of the components, but they are also highly affected by the final morphology. As shown in Figure 9, common morphologies of blends include laminar, ordered, fibrillar, droplet-type, and co-continuous. Each structure has its own advantages in terms of mechanical, thermal, and barrier properties. However, among all, laminar and co-continuous microstructures are the most desired ones for barrier improvement as the two phases are complementary reversed and the surface of each phase is an exact topological replica of the other, both contributing equally to the blend properties [160]. To obtain these structures, besides the degree of miscibility between components, the volume fraction and the choice of appropriate compatibilizers play a fundamental role in the achievement of a high degree of dispersion, which in turn, determines the overall end-product quality [161–164]. In the following paragraphs, the most relevant PLA-based bio-blends and composites are briefly reviewed in terms of the morphology–property relationship to give an update on the progress made in the improvement of PLA barrier properties.

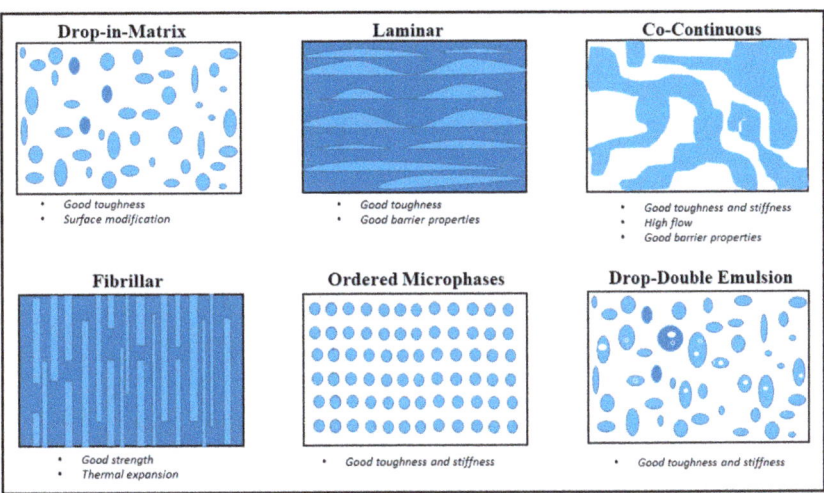

Figure 9. Types of morphology in immiscible polymer blends.

4.1. Bio-Based Reinforcing Agents

The use of bio-based reinforcing agents to produce PLA-based biocomposites with improved properties has become one of the key investment trends [165–170]. In this context, low-cost bio-renewable fibers such as cellulose, wood, kenaf, sisal, flax, and hemp have received increasing attention due to their reinforcing ability, non-toxicity, low-density, and large availability. As the name implies, these materials are commonly introduced into the polymer matrix as a reinforcement, thus enhancing its mechanical properties and stability. Concerning the barrier properties, little work has been conducted so far, as these materials are hydrophilic in nature with aa high tendency to adsorb water from the environment [171]. Moreover, the very low affinity with hydrophobic polymers (i.e., PLA) further hinders the formation of a well-dispersed system, which is one of the key factors for an efficient barrier. In light of the foregoing, Sanchez-Garcia et al. [172] investigated the morphology and transport properties of solvent-casted PLA biocomposites loaded with different ratios of purified alfa micro-cellulose (MC) fibers. Scanning electron microscope (SEM), AFM, and Raman imaging investigations showed that a good degree of dispersion was obtained only for samples with the lowest MC content (1 wt%), while at higher fiber incorporation (i.e., 10 wt%), clear presence of fiber agglomerates and phase discontinuity was reported to increase as a function of loading. As a matter of fact, water permeability data showed that the barrier properties of PLA biocomposites was only reduced by 10% in the sample containing 1 wt% of fiber content and at higher content, the permeability was seen to increase by 80%. Likewise, Luddee et al. [173] prepared a series of PLA biocomposite films containing grounded bacterial cellulose (BC) as a reinforcement and the water permeability behavior was studied as a function of filler particle size. Results showed that the incorporation of BC led to an increase in the water vapor permeability for all biocomposites compared to neat PLA. Moreover, the permeability coefficients increased linearly with the BC particle size, suggesting that BC particle sizes greatly affected the filler dispersability and their tendency to agglomerate within the PLA matrix, as also confirmed by SEM images. In order to improve compatibility between components, natural fillers (with free OH groups) can be chemically surface-modified with coupling agents. Those involve alkaline and peroxide treatment [174,175], vinyl grafting [176], acetylation [177], bleaching [170], and organosilane coating [178], to name a few. These types of treatments aim to increase the interfacial bonding strengths between the natural fibers and the polymeric matrix either through the formation of covalent bonds or mechanical interlocking. For instance, D. Kale et al. [179] carried out surface acylation of microcrystalline cellulose (MCC) to reduce the overall filler polarity. The resulted acylated MCC (AMCC) was loaded into the PLA matrix and the barrier performance of the resulting PLA biocomposites investigated. As expected, results showed that the addition of untreated MCC in the PLA matrix increased the water vapor permeability coefficients due to the low degree of dispersion and particle agglomeration compared to neat PLA. In contrast, the addition of AMCC resulted in a better filler-PLA dispersion and the overall water permeability was reduced by up to 10% compared to pure PLA. On the other hand, Tee et al. [180] investigated the barrier properties to water vapor and oxygen of the PLA biocomposite containing silane-grafted cellulose (SGC) as a filler. While water permeability coefficients increased for all biocomposites tested (with treated and untreated cellulose), oxygen permeability values decreased by up to 50% for biocomposites containing 30 wt% SGS compared to pure PLA. This was attributed to the improvement in interfacial adhesion between the filler and matrix and the higher degree of affinity after silane treatment.

4.2. Biodegradable Polymers Blends

Except for a few cases (i.e., PCL and PHAs), most of the biodegradable polymers that have been co-blended with PLA are highly polar in nature. Similar to the above case of hydrophilic reinforcing agents, blending immiscible polymers results in poor interfacial adhesion and phase-separated systems, which typically show very low overall performance; thus, appropriate compatibilization must be accomplished to obtain the desired end-properties. Among all compatibilization strategies, the use of reactive coupling agents and catalysts is considered the most viable solution for industrial application. As an example, there has been a great research interest in blending PLA with several types of starch (i.e., granular and thermoplastic) for many applications including food packaging due to its good food contact compatibility, suitable barrier properties, and low cost. However, the hydrophilic nature of starch leads to the formation of a two-phase system with very poor properties [181–183]. Many compatibilizers have been used and those include glycerol [184], polyethylene glycol [185], citric or stearic acids [186–190], maleic anhydride [187], lignocellulosic materials [188], methylenediphenyl diisocyanate [189], and adipate or citrate esters [190]. In most of the works conducted in this regard, thermoplastic starch (TPS) has been preferred over naturally granular starch as it can be deformed and dispersed to a much finer state, which in turn, greatly improves material processability. For instance, Shirai et al. [191] investigated the barrier properties of PLA/TPS biodegradable sheets with citric (CA) and adipic (AA) acids as additives (in the range of 0–1.5 wt%), prepared by a calendering–extrusion process at pilot scale. Prior to mixing, PLA and TPS were separately plasticized with 10 wt% of diisodecyl adipate (DIA) and 30 wt% of glycerol, respectively, as this pre-plasticization step has been reported to significantly improve the blend processability (extrusion). As shown in Figure 10A, all formulations could be processed continuously and the resulting sheets containing CA had a much smoother surface, homogeneous distribution, and compact structure compared to the other formulations, which, in contrast, revealed rough surfaces due to partial phase separation. The water vapor permeability studies showed that sheets containing CA exhibited the lowest WVP value, which accounted for about 70% of reduction compared to the plasticized reference sample. This was associated with a more effective interaction between starch and PLA (increases the interfacial adhesion), which in turn promoted mobility reduction over polymeric chains. The higher CA concentration (1.50 wt%) did not improve the evaluated property, suggesting that this component is efficient at lower concentrations (0.75 wt%). Comparatively, AA did not exhibit the same performance, even when mixed with CA, the WVP values were even increased in the presence of AA, probably due to partial degradation through acidolysis. In another interesting study, Muller et al. [192] studied the effect of cinnamaldehyde incorporation on the properties of the amorphous starch-PLA bilayer films intended for packaging applications. The particular compatibilizer was chosen as it is classified as GRAS (Generally Recognized As Safe) by the FDA (Food and Drug Administration) and offers antibacterial, antifungal, anti-inflammatory, and antioxidant activity to the resulting packages. The PLA/starch bilayer films were successfully obtained by compression molding followed by thermo-compression and the barrier performance studied in the presence and absence of cinnamaldehyde. Despite the lower ratio of the PLA sheet in the bilayer assembly (1/3 of the film thickness), a significant improvement in the gas barrier properties was achieved in the absence of the compatibilizer, specifically, a 96% decrease in WVP with respect to the neat starch films and a 99% decrease in OP compared to the amorphous PLA films. Surprisingly, when cinnamaldehyde was added, films exhibited lower mechanical resistance due to relevant plasticization of the amorphous regions, and WVP and OP did not notably change.

PHAs represent another relevant class of polymers that offer great potential in the food packaging industry due to good thermomechanical and barrier properties [193–202]. PHAs are derived from renewable resources and due to their bacterial origins, this class of polyesters shows good degradability features [197]. Poly(3-hydroxybutyrate) (PHB) is a homopolymer of 3-hydroxybutyrate and is the most common type of the PHA family, together with its copolymer with polyhydroxyvalerate (PHV), PHBV, which shows superior flexibility and processability. Due to the great potential of this class of polyesters, PLA-PHAs blends have attracted increasing interest in the last two decades [196,198–200]. Although both PLA and PHAs are compatible polyesters, several studies have shown that they form miscible blends only when low molecular weight (MW) fraction polymers are mixed and/or low polymer ratios (up to about 25 wt%) are incorporated into one another [21,193,201,202]. As an example, Boufarguine et al. [21] demonstrated that blending PLA with only 10 wt% PHBV using a multilayer co-extrusion process resulted in well dispersed films (up to 17 layers) and the gas barrier properties significantly improved compared with neat PLA. In particular, the helium permeability showed a reduction of about 35% with respect to pure PLA. In another study conducted by Zembouai et al. [193], it was shown that PHBV/PLA blends prepared by melt mixing in different ratios (100/0, 75/25, 50/50, 25/75, and 0/100 wt%) were not miscible, forming a two-phase system at all compositions (see Figure 10B). On the other hand, the water and oxygen barrier properties of PHBV/PLA blends were significantly improved with increasing PHBV content. At a PHBV content of 75 wt% in the blend, a reduction of about 75% and 81.5%, respectively, was achieved compared to pure PLA. However, even at a PHBV content of only 25 wt%, a reduction in the permeability coefficients for oxygen and water of about 35.3% and 22.7%, respectively, was obtained. This apparently anomalous behavior was further investigated in a comprehensive study reported by Jost and Kopitzky [202], whereby the miscibility of PLA-PHBV cast films and the resulting barrier properties were reviewed under thermodynamic aspects and correlated to their experimental results. In addition, blends were produced with different polymer molecular weight fractions and the final properties were studied in the presence and absence of selected compatibilizers. It was found that the incorporation of PHBV into the PLA matrix in the range between 20 and 35 wt% resulted in miscible blends. The reference blend (PLA:PHBV 75:25; MW: 10–40 kDa) with or without compatibilizers showed lower permeability values (to water vapor and oxygen) than the calculated values for the corresponding system. In agreement with previous findings (miscibility, crystallization, and melting of PHBV/PLA blends), this phenomenon was ascribed to the quick formation of interpenetrating PHBV spherulites, which interlock with the PLLA structures, leading to a reduction in the overall free volume. This may explain why, even for immiscible blends, the incorporation of PHBV into the PLA matrix enhances the barrier properties of PLA.

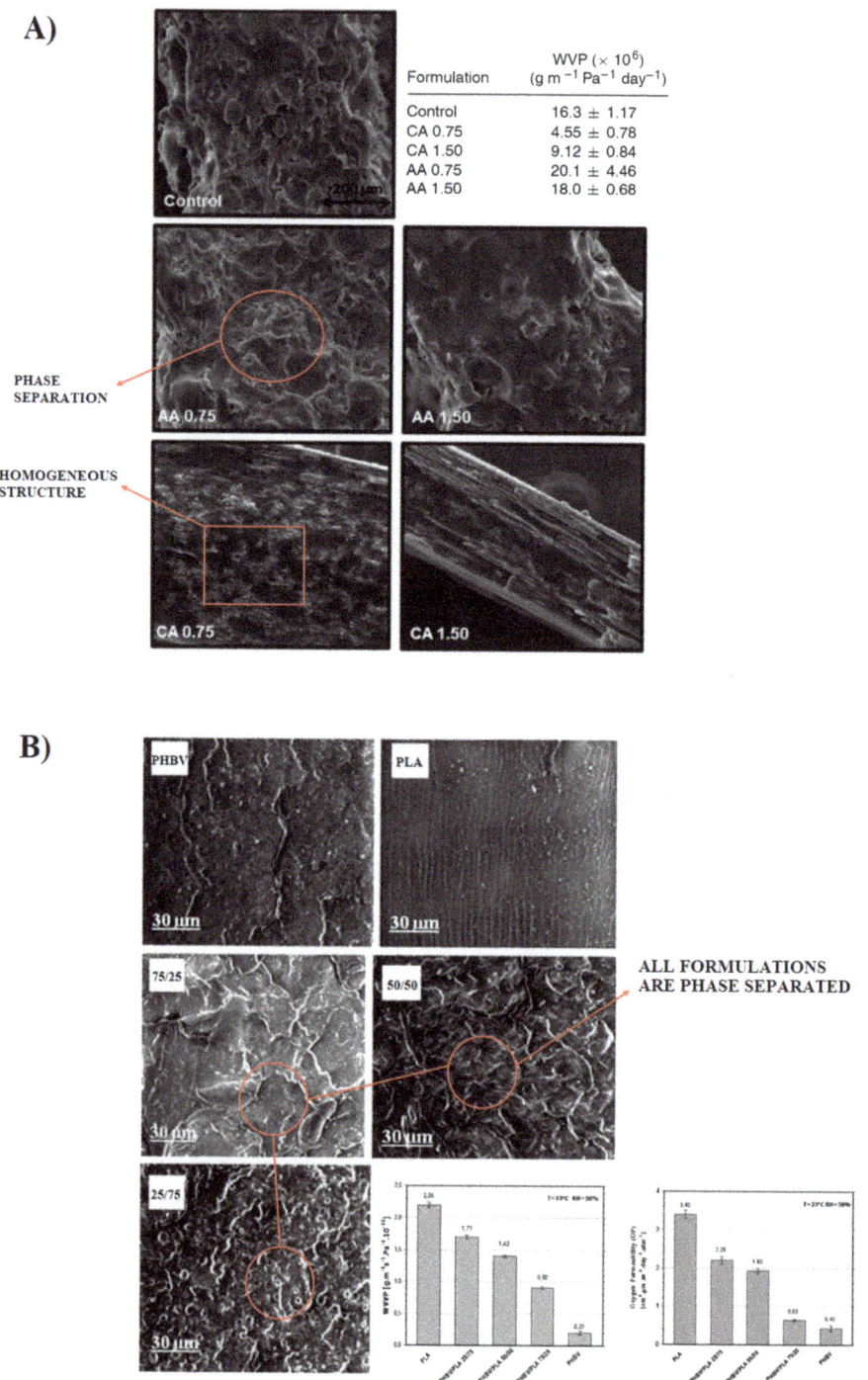

Figure 10. SEM images of (**A**) PLA/TPS sheets with CA (including WVP values), adapted from [191] and (**B**) fracture surface of neat PHBV, neat PLA, and their blends (including WVP and oxygen permeability values), adapted from [193] with permission from Elsevier. Copyright © 2013.

5. PLA-Based Nanocomposites

Nanocomposites are multicomponent systems, whereby the major constituent is typically a polymer and the minor one consists of a material with a length scale below 100 nm, referred to as nanofiller or nanoload. Nanoparticles can be classified into three major categories according to their particle geometry:

(i). layered nanoparticles that are characterized by one dimension ranging from several angstroms to several nanometers (i.e., layered silicates);
(ii). elongated particles that consist of fibrils with a diameter ranging between 1 and 100 nm and length up to several hundred nanometers (i.e., cellulose nanofibers); and
(iii). isodimensional particles that have the same size in all directions and an aspect ratio close to unity (i.e., metal oxide nanoparticles).

Due to clear evidence of their outstanding performance in many applications [203–205], it is no wonder that these systems have attracted a great deal of research interest worldwide in recent years. In the packaging field, introducing impermeable nanofillers with high aspect ratio and large surface area in the polymer matrix has appeared to be a promising approach to enhance the barrier properties of polymers [206,207]. As shown in Figure 11, this can be achieved by two specific mechanisms [208]: first, a tortuous path for gas diffusion is created though the polymer matrix as the evenly dispersed nanofillers are impermeable inorganic particles and act as an obstacle; and second, the nanoparticulate fillers may positively interact with the polymer matrix, "immobilizing" the polymer strands at the polymer–nanofiller interface and thus decreasing the overall free volume available for the gas molecules to diffuse through the polymer surface. It is therefore clear that nanocomposites offer encouraging opportunities for the food packaging industry. In the following paragraphs, the progress made in the development of PLA-based nanocomposites as high barrier materials are briefly reviewed.

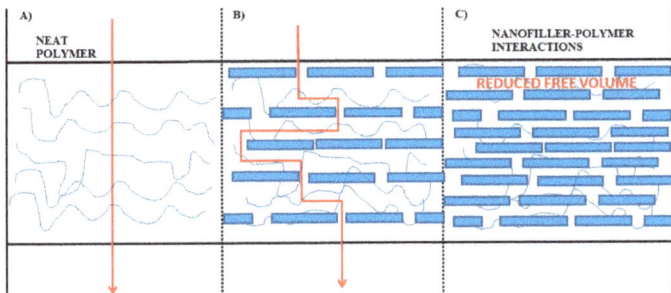

Figure 11. Simplified drawing of the "tortuous path" produced by the incorporation of exfoliated clay nanoparticles into a polymer matrix. (**A**) Neat polymer (diffusing gas molecules follow a pathway perpendicular to the film orientation); (**B**) non-interacting nanocomposite (impermeable platelets hinders direct diffusion); (**C**) interacting nanocomposite (the polymer strands are "immobilized" at the polymer–nanofiller interface and the overall free volume available is reduced.

5.1. Layered Nanofillers

Among all available nanosystems, nanoplatelets of layered silicate clay are by far the most researched nanofillers [209–212]. As shown in Figure 12, layered silicates consist of very thin films associated with counterions (exchangeable cations) [212]. Based on the types and relative content of the unit crystal lamellae, they can form three different structures:

(i). 1:1 clay types: the unit lamellar crystal consists of only one crystal sheet of silica tetrahedra in combination with an octahedral sheet, i.e., single crystal lamellae of alumina octahedra;

(ii). 2:1 clay types: consist of two crystal layers of silica tetrahedra forming the unit lamellar crystal bounded by a crystal layer of alumina octahedra located in the middle of the two layers; and

(iii). 2:2 clay types: consist of four crystal layers, alternating crystal layers of silica tetrahedra and alumina (or magnesium) octahedra.

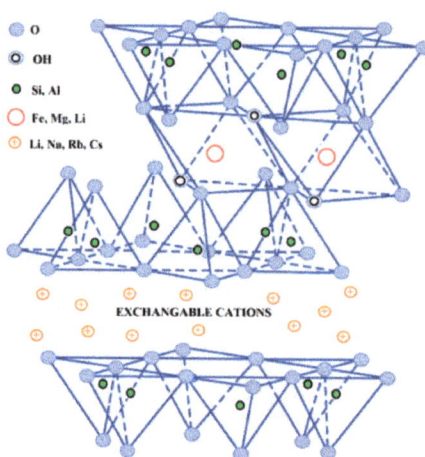

Figure 12. Example of the typical structure of the 2:1 layered silicates. Redrawn from [212] with permission of Elsevier. Copyright © 2015.

Typically, the layer thickness does not exceed 1 nm and the adjacent dimensions can vary from 300 Å up to several micrometers, depending mainly on the clay types and preparation methods [213]. For this reason, the aspect ratio (values > 1000) and the surface area (~700–900 m^2/g) are particularly high. These particular morphological features usually lead to impressive improvements in barrier properties when the nanoclays are uniformly distributed in the polymer matrices. Several studies have focused on the preparation of PLA-based layered silicate nanocomposites with improved barrier properties [148,214–217]. Among all types, the impermeable 2:1 layered phyllosilicate montmorillonite (MMT) is certainly the most commonly used prototype clay for this application [216,217]. The particular layered structure of MMT, consisting of an octahedral layer (mainly composed of Al$_4$(OH)$_{12}$) intercalated between two tetrahedral layers (composed of SiO$_4$ units), allows for the formation of multilayered polymer arrays with a high barrier. However, MMT, similar to most other clays, is inherently hydrophilic and has a high surface energy. This leads to a high segregation tendency and agglomeration of clay nanoplatelets, especially when dispersed in non-polar polymer materials (e.g., PLA) [218,219]. Agglomeration of clay platelets leads to the formation of tactoid structures with lower aspect ratios and thus lower barrier properties. To circumvent this problem, nanoclay surfaces are organically modified with cationic surfactants (usually quaternary alkylammonium compounds) by an ion exchange process with the inorganic cations naturally present in clay minerals [220–222]. Such a process reduces the surface energy of the silicates and the intercalated cationic surfactants act as compatibilizers between the hydrophilic clay and the hydrophobic polymer. The most common commercially available organically modified MMT clays include Cloisite (CH) 20A (with dimethyl dihydrogenated tallow ammonium chloride) and 30B (with methyl tallow bis-2-hydroxyethyl ammonium chloride), which have been shown to provide the greatest interlayer spacing and improved interactions with nonpolar polymers [223–225]. Moreover, to ensure homogeneous dispersion and delamination of the nanoclays in the polymer matrix, appropriate processing conditions should be applied. These may include, but are not limited to, high-pressure homogenization [226], a pre-sonication step [227] to disaggregate the clay platelets, and/or two-stage extrusion masterbatch processing to obtain a fine dispersion. The barrier performance of PLA-based nanocomposites containing CH has been studied by several authors [228–233]. Figure 13A,B shows two examples of successful CH 30B/PLA-based nanocomposites in terms of barrier performance [231,234].

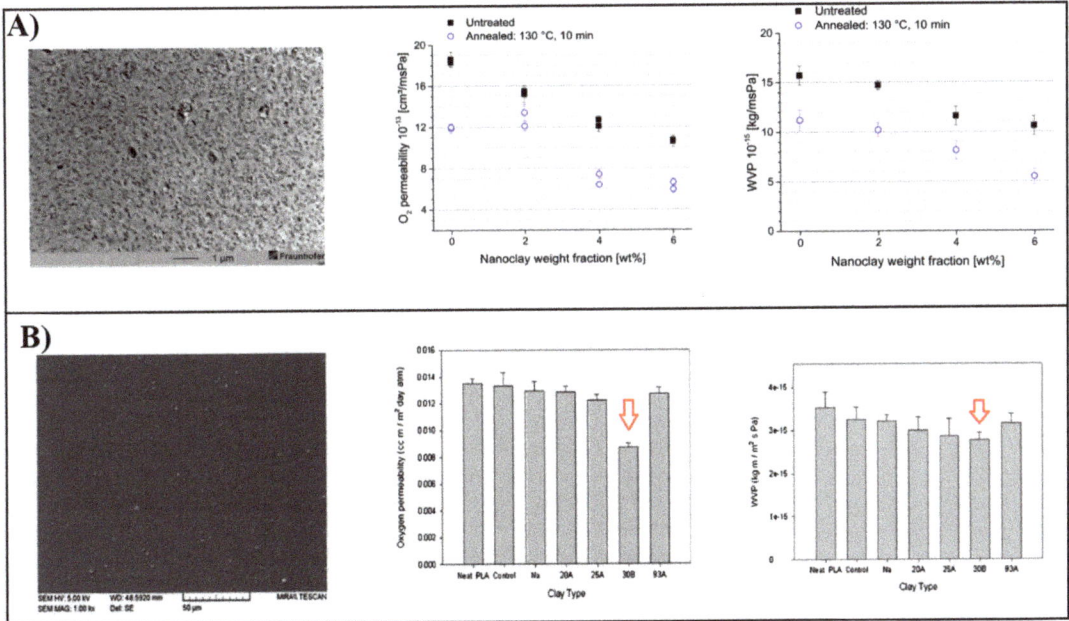

Figure 13. From left to right: (**A**) TEM image of PLA based nanocomposite films containing 6 wt% CH 30B along with oxygen and WVP values of untreated nanocomposites and annealed at 130 °C for 10 min as a function of nanoclay content. Adapted from [231] with permission of Wiley Periodicals. Copyright © 2016. (**B**) TEM and SEM images of PLA-CH 30B nanocomposites along with the relative oxygen permeability of nanocomposites with different volume fraction of clays (CNa:CH Na+, CRDP:Fyrolflex, and C30B:CH 30B). Adapted from [234] with permission from Elsevier. Copyright © 2016.

Other examples include the work of Najafi et al. [233], in which the effects of different processing conditions on the dispersion of 2 wt% CH 30B in PLA-based nanocomposites were studied and the resulting film morphologies were subjected to the oxygen permeability test. PLA–clay mixtures were prepared with and without the chain extender Joncryl® to further improve blend compatibility. Nanocomposites were prepared using a twin screw extruder with different methods. The preparation methods consisted of either simultaneous extrusion of all components together or a two-step extrusion masterbatch approach, with the chain extender added in either the first or second pass. In addition, the effect of multiple extrusion passes was also examined. According to the morphological observations conducted by SEM, TEM, and XRD, the best clay–PLA dispersions were obtained in the presence of Joncryl® when processed in the extrusion masterbatch approach, while multiple extrusion passes led to the formation of large clay aggregates due to the longer extrusion residence time. As expected, the good dispersion and distribution of clay platelets in PLA–Joncryl-based nanocomposites resulted in the lowest measured oxygen permeability, which accounted for 37% of the reduction compared to pure PLA. However, it was interesting to note that simple addition of the chain extender into pure PLA increased the oxygen permeability of the corresponding blend. This was explained by the formation of long chain branches, which reduced the crystallinity of pure PLA from ~7% to ~1%. In another study, Tenn et al. [230] investigated the incorporation of various concentrations (from 0 to 20 wt%) of CH 30B (both in the hydrated and pre-dried states) into PLA films by two-step extrusion masterbatch processing. The transport properties (water and oxygen permeability) of the resulting systems were correlated with the degree of dispersion and orientation of the nanoplatelets in the polymer matrix. To investigate the effect of the hydration state of the clays on the quality of the dispersion, CH 30B was added both in partially hydrated form (as received) and in a pre-dried state (80 °C in a vacuum oven for 12 h). For both systems of PLA–CH (hydrated and anhydrous), a decrease in water and oxygen permeability values was observed as a function of increasing nanoclay content. This is generally due to the well-known tortuosity effect, which improves the diffusion path of the permeants in the PLA matrix. However, comparing the two nanocomposites at the same nanoclay content, it was found that the system containing the hydrated form of CH had a higher barrier effect than the corresponding systems containing the dried clay form (CD). In terms of water permeability (P_w), PLA films containing 15 wt% CH showed the best performance, achieving a 95% reduction in P_w compared to the pure PLA film. The difference in water barrier performance between the two systems was ascribed to the presence of water molecules in the untreated clay component, which favored the formation of water clusters that hindered water diffusivity. Additionally, in terms of oxygen permeability, the best performance was obtained with PLA–CH systems, which achieved a reduction of up to 74% compared to the PLA-only film. The overall better performance of the PLA–CH nanocomposites was further investigated in terms of dispersion quality and it was found that in all systems, there was a coexistence of intercalated and aggregated structures depending on the clay content. However, when CH was incorporated, the TEM images showed relatively higher intercalation of clay platelets, likely due to better compatibility between CH and PLA, as observed by DSC and XRD. In addition, CH nanoplatelets were found to preferentially arrange perpendicular to the diffusion pathway, confirming the fundamental role of structural orientation in the permeation behavior of materials. These and similar studies have shown that the uniform distribution of organically modified MMT layers in the PLA matrix is the key factor for significantly improving the barrier properties of PLA.

5.2. Nanofibers or Whiskers

With the raising environmental awareness and concerns over sustainability and end-of-life disposal challenges, the interest in exploiting nanomaterials from renewable resources is rapidly increasing [235,236]. In the packaging field, nanofibers derived from natural sources have attracted a great deal of attention, not only for their environmental friendliness over traditional nanofillers, but also due to their outstanding ability to decrease the permeability of various polymeric films [237–240]. These include all the nanofibrous materials derived from cellulose, starch, chitin, and chitosan, which commonly exist in plants, animals, microorganisms, and bacteria. Among them, cellulose-based nanofillers have been more widely investigated due to advances in the production of cellulose nanofibrils (CNF) and cellulose nanowhiskers (CNW) [241–244]. CNF have elongated rod-shaped structures with large diameters and lengths variations, both ranging from a few up to 100 nm. Due to their highly crystalline nature, large surface area/aspect ratio and their ability to form a dense percolating network, CNF are known to have high barrier properties toward most gases and liquids. However, since cellulose-based materials have water-loving surfaces due to the abundance of –OH groups, blending CNF (as it is) with hydrophobic materials such as PLA is not feasible without either the addition of compatibilizing agents or appropriate chemical modifications. Several strategies have been attempted to overcome this problem such as the use of surfactants [243,245], surface acetylation [246], and reactive compatibilization [247], to name a few. For example, Espino-Pérez and co-workers [243], used an in situ surface grafting method to produce fully compatibilized PLA/CNW bionanocomposites with enhanced barrier properties. The CNW surface was grafted with a long aliphatic isocyanate chain (n-octadecyl isocyanate (ICN)) and the barrier performance of PLA/CNW and

PLA/CNW–ICN nanocomposite solution-casted films were evaluated. Visual observations of the films indicated that good filler dispersion and film transparency were obtained only with the lowest CNW and CNW–ICN content (2.5 wt%). At higher CNW concentration, aggregation of CNW in the matrix could be observed, indicating poor system compatibility. In fact, the water and oxygen permeability of both samples were not significantly reduced compared to neat PLA, likely due to the reduction in hydrogen bonds between the fibrils, which led to poor interfacial adhesion with the PLA matrix. This may suggest that while surface grafting of CNW with isocyanate is effective in enhancing the hydrophobicity of CNW, it is probably not the best compatibilizing approach to produce well-dispersed PLA–CNW systems. More recently, successful compatibilization and enhanced barrier properties of PLA–CNF nanocomposites was achieved by Nair and co-workers [248] by preparing CNF with high amounts of lignin (about 20–23 wt%) (NCFHL) from the bark of various coniferous species. Since lignin contains both polar (hydroxyl) groups and nonpolar hydrocarbon and benzene rings, its presence naturally enhances the hydrophobic nature and the barrier properties of CNF without additional modifications. As shown in Figure 14A–C, morphological examination of solution-casted PLA-NCFHL films with various NCFHL contents (5, 10, 15, and 20 wt%) revealed that with up to 10 wt% of load, fibrils were well-embedded within the PLA matrix.

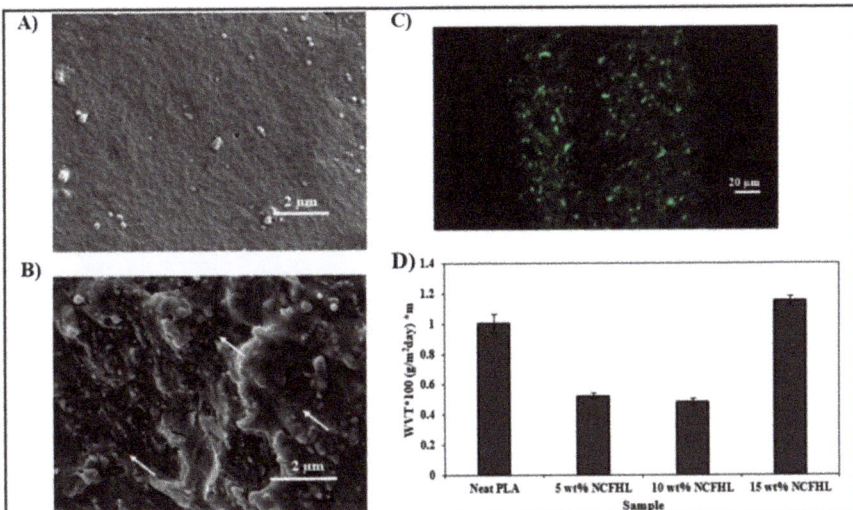

Figure 14. TEM image of (**A**) neat PLA and (**B**) the PLA–NCFHL biocomposite (with lignin) in which fibrils were well embedded within the PLA matrix (indicated by arrows); (**C**) confocal laser microscope image showing well dispersed lignin fibrils on the PLA surface; (**D**) Water vapor transmission rate of the neat PLA and PLA/lignin biocomposite. Adapted from [248] with permission from American Chemical Society. Copyright © 2018.

As expected, the good dispersion observed at 5 and 10 wt% of NCFHL loading resulted in a significant enhancement in the water permeability performance of the nanocomposites, with WVTR values reduced to about half compared to neat PLA (Figure 14D) [248]. This was ascribed to the formation of an effective interphase between the lignin and PLA, which increased the tortuous path for water vapor diffusion. In this specific context, Rigotti et al. [249] carried out an in depth-investigation on the role of microstructure of PLA layers located at the clay–matrix interface on the nanocomposites' gas transport properties. The barrier performance of solution-casted PLA films containing lauryl-functionalized cellulose nanofillers (LCNF) was examined using gas phase permeation measurements (toward CO_2, H_2, and D_2 gases) and thermal desorption spectroscopy (TDS) analysis.

Results showed that there exists a critical LCNF concentration (6.5 wt%) under which the permeation flux (P) for all gases decreases with increasing filler content, while at higher concentrations, flux increases, reaching a similar p value of neat PLA at 10 wt% of filler load. According to the SEM and TEM analyses, when filler content was lower than 6.5 wt%, LCNF appeared to be well-dispersed, while at higher concentrations, LCNF clusters and micrometric agglomerates could be observed. This is typically linked to a change in the nanostructure as a function of filler content. By analyzing experimental transport data using the permeation model of mixed matrix membranes, it was possible to conclude that at concentrations not exceeding 6.5 wt%, the interfacial regions at the filler–matrix interface were rigidified, likely due to strong lauryl–PLA chain interactions, which led to local free volume reduction. Therefore, this was suggested to be responsible to the improved barrier performance of PLA–LCNF systems, confirming that appropriate control of the nanocomposite interface properties is necessary to obtain systems with enhanced barrier capabilities.

5.3. Isodimensional Nanoparticles

Nanoparticles with three nanodimensions (less than 100 nm), also known as 3D-isodimensional nanoparticles, have gained increasing interest in the field of food packaging [250,251]. These include nanoparticles derived from most metals such as silica [252], copper [253], gold [254], silver [255], zinc [256], magnesium [257], titanium [258], and their corresponding oxides [259]. Among them, metal oxides have attracted special attention because they can be produced much more cheaply and are known for their strong antibacterial and ethylene scavenging activity [260]. These particular properties have shown that they offer an intriguing potential for the development of active nanocomposites for the packaging of fresh products, which are very sensitive to microbial spoilage. Several multifunctional PLA-based metal oxide nanocomposites have been prepared and the resulting transport properties investigated [261,262]. In a study by Lizundia et al. [261], the effect of incorporating different metal oxide nanoparticles (TiO_2, SiO_2, Fe_2O_3, and Al_2O_3) at 1 wt% on the transport properties of PLA films was investigated. TEM analysis showed that all nanoparticles had a spherical shape with similar diameters ranging from 10 to 20 nm. However, for the nanocomposites, it was interesting to note that SiO_2 and Al_2O_3 isotropically distributed in the PLA matrix, while TiO_2 and Fe_2O_3 formed large aggregates of about 100–150 nm. Unexpectedly, the most efficient metal oxide in the water permeability measurements was TiO_2, whose incorporation resulted in a reduction of about 18% compared to pure PLA, followed by Al_2O_3 and Fe_2O_3 particles, which also resulted in a relatively moderate reduction (13% and 16%, respectively). The least efficient system was PLA–SiO_2, which caused only 4% of reduction in water permeability. A different trend was observed with respect to oxygen: SiO_2 and TiO_2 improved the barrier performance of PLA to oxygen, while Al_2O_3 and Fe_2O_3 increased the overall permeability to about 5%. The author concluded that in these particular cases, the morphological characteristics (such as size, filler dispersion, and free volume) did not seem to play the most important role in the permeation process. The different barrier behavior observed in the samples was related to differences in the nature of the nanoparticles and possible filler–matrix interactions. In both cases, the best performance of TiO_2 was attributed to its large hydrophobicity, while the worst barrier performance of SiO_2 for water molecules was related to its high hydrophilicity, which favored a faster diffusion path. Moreover, PLA–ZnO antibacterial nanocomposite films were prepared in a composition range from 0.5 to 3 wt% of nanofillers by melt extrusion, and the barrier properties to water vapor were analyzed by Pantani et al. [262]. Since untreated ZnO can lead to severe degradation of the PLA matrix during the melt mixing process due to transesterification and depolymerization reactions, the surfaces of the nanoparticles were previously treated with protective agents (silane), which also acted as compatibilizers. TEM images in Figure 15A show a relatively continuous and well-dispersed rod-like ZnO distribution in the PLA matrix for all compositions [262]. However, only a slight reduction (~20%) in water vapor permeability

(6.43×10^{-7} wt%/atm × cm^2/s) was obtained with 3 wt% of filler compared to unfilled PLA (8.26×10^{-7}), which was attributed to an increased difficulty for molecules to diffuse into the polymer matrix. In a similar work, Shankar et al. [158] prepared solution-cast PLA–ZnO nanocomposite films with concentrations of less than 2 wt% (0.5, 1.0, and 1.5). Morphological analysis reported in Figure 15B shows that the ZnO nanoparticles had a cubic and rod shape with a size in the range of 50–100 nm. However, it is worth noting that the surface of the nanocomposite films was relatively rough compared to that of neat PLA and the overall roughness appeared to increase with increasing ZnO content. Nevertheless, a reduction in the water vapor permeability of up to 30% was achieved compared to unfilled PLA, even for the composite with the lowest ZnO content (0.5 wt%). No further reduction was observed at higher ZnO contents, likely due to the formation of ZnO-based microagglomerated structures. In another study, Marra et al. [263] reported that homogeneous dispersion of 1 wt% ZnO in PLA-based nanocomposites in an increase in water vapor transmission rate to about 16% compared to normal PLA. This was associated with potential changes induced by the presence of ZnO particles at PLA–filler interfaces, resulting in an increase in free volume. In the same way, Swaroop and Shukla [264] studied the barrier properties of solution-cast PLA-based films reinforced with MgO nanoparticles. At 1 wt% metal oxide content, the resulting films showed a 20% increase in water vapor permeability. Morphological studies revealed that the nanoparticles were in the form of agglomerates, resulting in a very rough film surface. This was probably the main cause of changes in free volume, absorption, and solubility at the interfaces near the "highly polar" MgO nanoparticles, which overall contributed to the high water permeability measured. These results suggest that spherical composites with metal oxide exhibit relatively poor filler–polymer compatibility because the polarity of the particles does not match that of the PLA matrix.

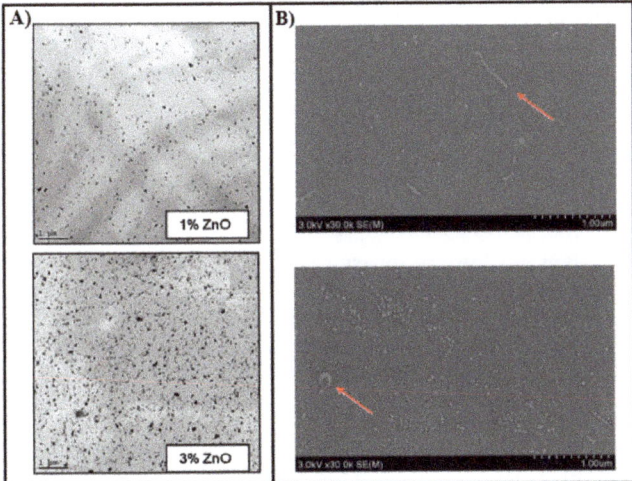

Figure 15. (**A**) TEM images of surface coated PLA nanocomposites filled with 1 and 3 wt% of ZnO nanoparticles (even dispersion), adapted from [262] with permission from Elsevier. Copyright© 2013. (**B**) SEM images of PLA nanocomposites containing 1–1.5 wt% of ZnO without any surface treatment (particle agglomeration indicated by red arrows), adapted from [158] with permission of Elsevier. Copyright © 2018.

6. Other Emerging Approaches toward High-Barrier PLA-Based Plastic

With the advent of new polymerization techniques, PLA-based block and graft copolymers with tailored properties have been synthesized based on judicious selection of comonomers and the variation of copolymer compositions [265–267]. However, due to the good biocompatibility of PLA, the great majority of these copolymers have been mostly exclusively studied for applications in the field of biomedicine [268–272] and only little attention has been paid to the design of "high barrier" PLA-based copolymers for the food-packaging sector [17,273–278]. A few random studies, mainly focusing on improving the mechanical properties of PLA by copolymerization, have also shown somewhat promising results in terms of gas permeability performance of the resulting copolymers. For example, Genovese et al. [273] synthesized PLA-based ABA triblock copolymers with a hydroxyl-terminated poly(propylene/neopentyl glycol succinate) copolymer as a midblock, through ROP followed by chain extension reaction. Depending on A/B ratio, the resulting bio-based copolymers showed enhanced mechanical and barrier properties compared to the PLA homopolymers. In particular, the oxygen permeability of the copolymer containing 33% of the midblock unit was two times lower than that of neat PLA. These results were directly related to an increase in the degree of crystallinity and the presence of the two methyl groups in the neopentyl glycol sub-unit, which was suggested to hinder the passage of small molecules. Other examples include the copolymerization of PLA with rubbery-type monomers to afford versatile thermoplastic elastomers (TPEs) with good barrier properties [277]. TPEs are commonly referred to as ABA triblock copolymers containing an incompatible A hard block (component with high Tg) and a B soft block (component with low Tg). Typical examples include petroleum-derived styrene-based TPEs (PS–TPEs), which are extensively used in the packaging field [278] due to their well-known versatile properties and low cost. In fact, depending on the ratio between the soft and hard components, these materials' properties can be easily tuned based on the required applications, ranging from slightly flexible plastics to highly elastic gums. With the increasing demand for eco-friendly alternatives, PLA-based TPEs could potentially substitute styrenic-based TPEs in this specific field. For example, Ali et al. [272] synthesized four different PLA thermoplastic polyurethane (PLAPU) copolymers with different compositions of hard PLA and soft PCL units via a chain-extension reaction. Depending on PCL content, the copolymers exhibited excellent flexibility and gas barrier properties. In particular, at the highest PCL content, the oxygen permeability was reduced by approximately 85 times compared to that of the PLA homopolymer. This behavior was ascribed to the incorporation of high molecular weight PCL segments, which led to an increase in the chain density, thus decreasing the effective path for diffusion. In a more recent study, Yuk et al. [274] prepared a series of thermoplastic superelastomers based on poly(isobutylene)-graft-poly(L-lactide) copolymers by a "grafting from" controlled polymerization in a one-pot, two-step process as reported in Figure 16A. These copolymers were based on a graft structure, which typically leads to superior physical and mechanical properties compared to linear block copolymers. As shown in Figure 16B, the oxygen barrier properties of the PLA-based graft copolymer films were evaluated and compared to those of commercial PLA and poly(styrene)−b−poly(isoprene)−b−poly(styrene) (SIS), which is one of the most widely used TPEs in food packaging applications [274]. Depending on compositions, the resulting superelastomers showed high-performance gas barrier characteristics. The oxygen permeability performance of copolymers with the highest PLA content (45 wt%) was 60-fold and 5-fold better than that of SIS and neat PLA, respectively. This was attributed to the presence of the largest fraction of homogeneously distributed semicrystalline PLA domains, which tied up the rubbery isobutylene segments, thus decreasing the channel for gas permeability through the copolymer matrix.

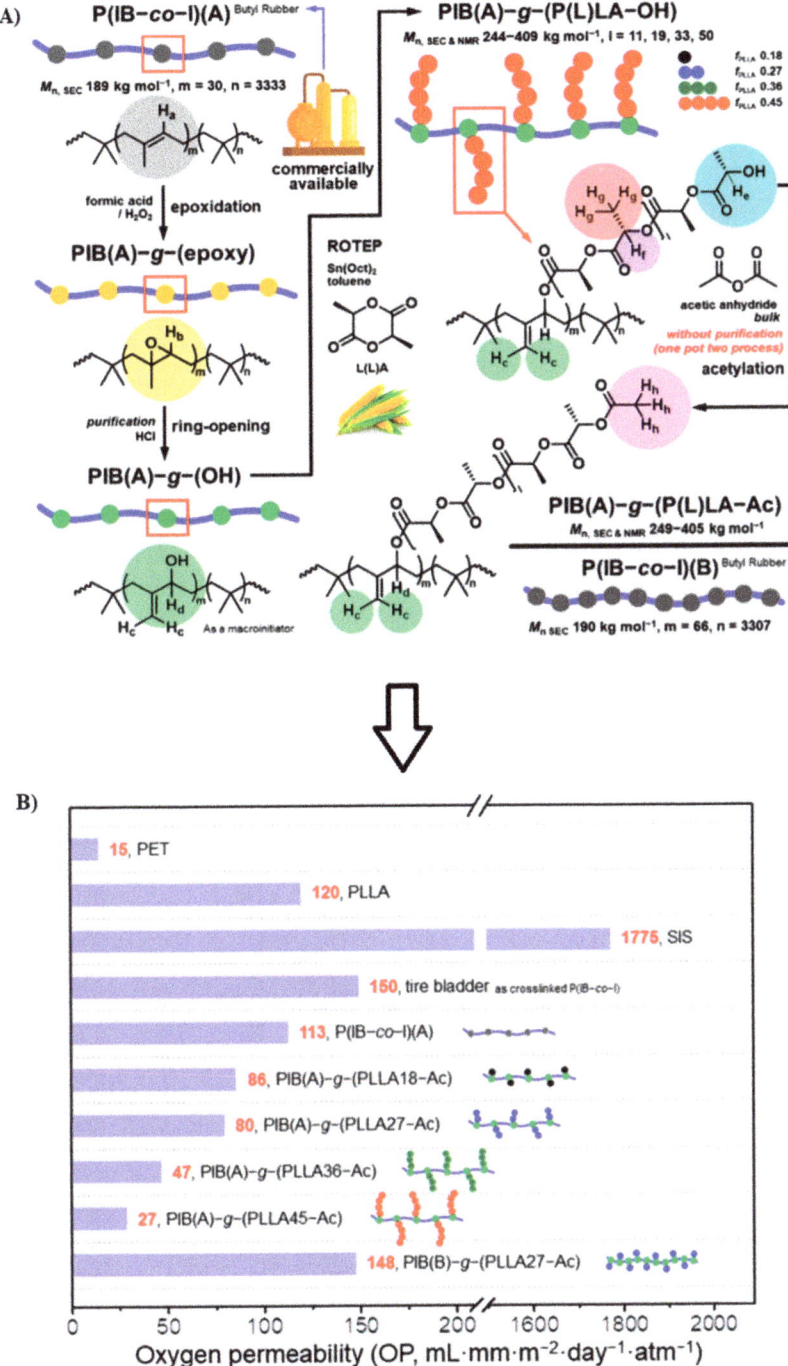

Figure 16. (**A**) Schematic synthesis of semicrystalline or amorphous poly(isobutylene)-graft-acetylated poly(lactide) (PIB-g-(P(L)LA−Ac)). (**B**) Comparison of oxygen permeability values between the newly synthesized polymers and common ones. Adapted from [274] with permission from American Chemical Society. Copyright © 2020.

7. Molecular Dynamics Simulations for Permeability Investigation of PLA-Based Materials
7.1. Most Important Tools to Study PLA-Based Plastic

In line with the growing computing power and with an ever increased number of precise and rigorous programs available to the scientific community, there have been several proposed theoretical models for the prediction of polymer permeability that are discussed in this section. Among the available approaches, molecular dynamics (MD) simulations have been successfully applied in different research areas [279] including mass transport property investigations to elucidate the sorption and diffusion mechanisms of small gas molecules in varieties of potentially useful polymers such as PLA [280]. Applying these methods, many results, difficult or impossible to detect from conventional experiments, can be obtained.

Different MD approaches can be used to investigate the permeability of PLA-based materials. Sun and Zhou [281] performed full-atomistic simulations of oxygen sorption and diffusion in amorphous PLA. The oxygen solubility coefficient (S) was calculated by fitting the dual-mode sorption model to the simulation data. The simulated S value was much higher than the experimental data obtained from the time-lag method, but slightly lower than the measurement from quartz crystal microbalance. This discrepancy was probably due to the predominant Langmuir type sorption mechanism, which holds for oxygen sorption in PLA. The time-lag method only considers oxygen molecules that are involved in the diffusion process. The allowed rotation states of successive bonds between adjacent atoms are determined from probability functions by energy consideration using the standard Monte Carlo method [282]. The solubility coefficient of gas in a glassy polymer is defined as the ratio of gas concentration to gas pressure at equilibrium, following Equation (4):

$$S = \frac{C}{p} = k_D + C_H' \frac{b}{1+bp} \qquad (4)$$

where k_D, C_H', and b can be determined by nonlinear regression fit of the oxygen sorption data that are obtained by GCMC simulations.

Information about the structural features of different PLA models at the atomic level can be provided by the radial distribution function (RDF). The RDF indicates the probability density of finding A and B atoms at a distance of r following Equation (5):

$$g_{A-B}(r) = \left(\frac{n_B}{4\pi r^2 dr}\right) / \left(\frac{N_B}{v}\right) \qquad (5)$$

where n_B is the number of B particles located at the distance r in a shell of thickness dr from particle A; N_B is the number of B particles in the system; and v is the total volume of the system. The free volume inside the polymer matrix can be obtained by rolling a spherical probe over the Connolly surface of polymer atoms. It should be noticed that the available volume for the probe to pass through is dependent on the radius of the probe [78]. Regardless of the specific computational type to use, the diffusivity of a gas in an organic solvent, polymer, or zeolite can be calculated by running a MD simulation and determining the mean square displacement (MSD) of the gas in the material. MSD is a measure of the deviation of the position of a particle with respect to a reference position over time. It is the most common measure of the spatial extent of random motion, and for this reason, it is the most useful tool to detect the small molecule diffusion from a starting state [283]. The motion pattern of penetrant gases in the host polymer can be qualitatively studied by monitoring the penetrant's displacement $|r(t) - r(0)|$ from its initial position. The diffusion coefficient D is then obtained from the slope of a plot of the MSD against time t.

H. Ebadi-Dehaghani et al. used a MD simulation to investigate and predict the gas permeability through polymer blends and nanocomposites [284]. They demonstrated that the oxygen permeation was highly dependent on blend composition, clay loading, and state of clay dispersion governed by compatibilization in the PP/PLA/clay nanocomposite film. Compared to the PLA-rich system, they found that the PP-rich films showed a greater barrier to oxygen. The lower permeability of PP-rich films was mainly due to the reduced size of dispersed domains of the oxygen barrier component (i.e., PLA in the PP matrix), leading to a more tortuous permeant path, while the higher degree of crystallinity observed in PP-rich films compared to the PLA-rich system was found to also be responsible for the higher barrier properties of the PP-rich films.

7.2. Different MD Approaches to Use for PLA Systems

The use of explicit full-atom (FA) simulation model (Figure 17A) for PLA polymers has been found to be suitable for approximately reproducing different important physical properties of amorphous PLA solids. In particular, Xiang and Anderson adopted this kind of approach to calculate material density, water sorption isotherm, and diffusion coefficient of PLA systems, thus verifying potential utilities in designing PLA based drug delivery systems, particularly for predicting drug–PLA miscibility. They combined MD simulations, the particle insertion method of Widom [285], and a theoretical sorption relation to calculate the water sorption isotherm in PLA. They found that at 0.6 (wt%) of H_2O, water molecules localized next to polar ester groups in PLA because of hydrogen bonding. Local mobility in PLA as characterized by the atomic fluctuation was sharply reduced near the Tg, decreasing further with aging at 298 K [286]. The non-Einsteinian diffusion of water was found to be correlated with the rotational β-relaxation of PLA carbonyl groups at 298 K. A relaxation–diffusion coupling model proposed by the authors provided a diffusion coefficient of 1.3×10^{-8} cm^2/s at 298 K, which is comparable to the reported experimental values [287]. In other studies, MD simulations have been performed to estimate the diffusivity coefficients of the gases CO_2, O_2, and N_2 from polypropylene (PP)/poly(lactic acid) (PLA)/clay nanocomposite films with various compositions (PP-rich and PLA-rich). Diffusion temperature dependency of the PP-rich sample for O_2 gas has been also investigated. The MSD of the gases has been calculated via MD simulation according to the Charati and Stern method [288], which consists of the generation of the initial microstructure of a polymer containing penetrant gas molecules as an amorphous cell module, and the use of a full atom approach. The diffusion coefficients of gases in PLA can also be controlled by the amount of free volume, the free-volume distribution, and the dynamics of the free volume of the polymers. Penetrant molecules reside most of the time in microcavities inside the PLA matrix, and the microcavities are elements of "free volume" (or "empty" volume) between the surrounding polymer chains. This reliable computational method showed that solubility increased with increasing temperature, which was in accordance with the experiments.

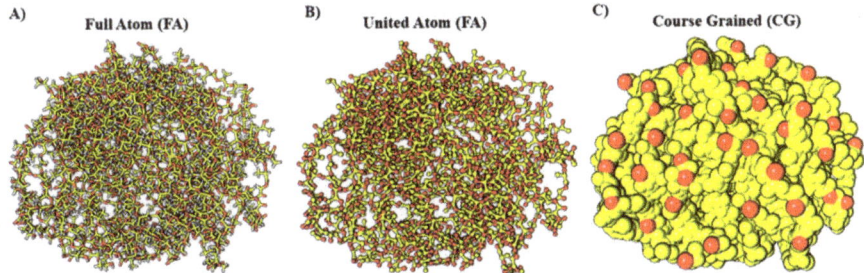

Figure 17. Representations of PLA nanoparticles using (**A**) full atoms (FA), (**B**) united atoms (UA), and (**C**) course grained (CG) methods.

Another explicit FA computational model for PLA was developed by Xiang and Anderson (2014). MD simulations of PLA glasses were carried out to explore various molecular interactions and predict certain physical properties such as material density, water sorption isotherm, and diffusion coefficient. The combined use of MD simulations, the particle insertion method of Widom, which is a statistical thermodynamic approach to the calculation of material and mixture properties [285], and a theoretical sorption relation allows us to efficiently calculate the water sorption isotherm in PLA. Weak sorption of water in amorphous PLA solids can be predicted, with results that can be similar to the experimental ones. Inspection of molecular structures of the simulated PLA glasses provided further understanding of the distribution of water in PLA polymers, which has been difficult to obtain experimentally.

The FA MD studies of fundamental properties such as water solubility, diffusivity, and distribution in PLA polymers are only a few, due to the enormous computational resources required to conduct atomistic simulations with explicit solvent models. United atom (UA) MD has also been proposed to obtain cheaper calculations than those obtained for all atom MD, retaining a high accuracy degree. This method subsumes nonpolar hydrogen atoms into their adjacent carbon atom (Figure 17B), decreasing the computational costs. While the accuracy of the united atom MD simulation has been found to be reliable for protein modeling [289], this approach was found to be unreliable in estimating the diffusion coefficients of small penetrant molecules through polymers.

Coarse-grained (CG) modeling represents a valid way to overcome the huge requirements for computing resources while maintaining high calculation efficiency. It consists of a less sensitive method than the FA and UA approaches (Figure 17C), but its accuracy degree remains high enough to describe short- and long-range phenomena at various granularity levels [290,291]. Two CG computational methods can be considered for the investigation of PLA-based material permeability: a mesoscale approach—dissipative particle dynamics (DPD) [292]—and MD simulations using the MARTINI force field in conjunction with the GROMACS package [293]. DPD simulations group atoms and molecules into fluid beads and use bead level interactions to describe the evolution of a system. For any two beads i and j, the pair wise interaction force (F^{DP}_{ij}) is the sum of the conservative force (F^C_{ij}), dissipative force (F^D_{ij}), and random force (F^R_{ij}), as shown in Equation (6):

$$F^{DP}_{ij} = F^C_{ij} + F^D_{ij} + F^R_{ij} \quad (6)$$

where F^C_{ij} is a soft repulsive force, while F^D_{ij} is a drag force or frictional force and F^R_{ij} is a random force. Of these three, the conservative force (F^C_{ij}) best describes the energy of the system. Groot and Warren (1997) established a connection between DPD beads and a real fluid by defining a relationship between the maximum repulsion between particles (a_{ij}), which is a function of the conservative force (F^C_{ij}), and the Flory–Huggins interaction parameter (x). The MARTINI force field uses, on average, a four to one mapping of non-hydrogen atoms to interaction centers (although sometimes fewer or more than four atoms are mapped on to an interaction site) and defines the interaction sites into four main types: polar (neutral water soluble atoms), non-polar (mixed groups of polar and apolar atoms), apolar (hydrophobic groups), and charged (groups bearing an ionic charge) [294].

FA and CG MD simulations represent different ways to carry out MD simulations for the permeability investigation of PLA-based materials. There is no specific method to conduct this as the choice to use one protocol with respect to the other depends on the size of the conditions of the system (PLA size, presence of solvent, number of gas molecules, and simulation time). Recent studies [295] have shown that it is possible to use both methods. This combined approach is based on starting from an all-atom MD simulation to obtain input parameters such as angles, bonds, and dihedrals of PLA chains, taking into account the crystalline and amorphous phases. After that, the MARTINI force field could be used to map PLA with various polymer segment lengths against the presence or absence of other molecules in the system.

8. Concluding Remarks and Future Outlook

With the aim of minimizing the environmental impact, several PLA-based biodegradable plastic packaging materials with improved barrier properties have been developed in the last decades. In this work, all recent strategies to improve the barrier properties of

PLA were reviewed including crystallization, orientation, stereocomplexation, blending, incorporation of nanoparticles, and copolymerization. Considering the enormous number of available approaches and the wide range of processing conditions applied, it is often difficult to establish clear relationships between the structure and barrier properties. However, the current review of the literature has identified some strategies that appear to have a greater impact on the gas permeability of PLA than others.

Starting with crystallinity content, although it has been generally demonstrated that barrier properties depend on the degree of crystallinity for many semi-crystalline polymers, increasing the crystallinity content of PLA from 0 to 40% does not always lead to a decrease in gas permeability, as expected. For PLA materials with a comparable degree of crystallinity, several studies have shown that gas permeability strongly depends on chain orientation, amorphous phase morphology, and crystalline forms (i.e., ordered α-form crystals and less ordered α'-form crystals in PLA). More importantly, further studies on the wide variety of crystalline and amorphous phase organization of PLA have revealed the importance of the crystalline superstructure and lamellar arrangement in improving gas barrier properties. In other words, the crystalline phase acts as an impermeable barrier, but its supermolecular microstructure must be properly tailored. Numerous data show that gas barrier properties can be reduced by up to two orders of magnitude if the arrangement of PLA lamellae is tuned along a perpendicular direction of the gas diffusion path. To achieve this, specific crystallization protocols must be applied to create regularly spaced lamellae in PLA packaging materials. The most successful approaches involve either molecular orientation, crystallization under nanoconfinement, addition of vertically aligned nanoplatelets, and/or use of specific fibrillar nucleating agents as templates to construct parallel-aligned shish-kebab-like crystals with well-interlocked boundaries. Consistent with these promising findings, most recent studies have focused on the development of nanocomposite structures by incorporating small amounts of nanoparticles (i.e., typically clays with at least one dimension less than 100 nm) into polymeric matrices. By acting as physical barriers to the diffusion and permeation of small molecules, such nanostructures have been shown to significantly improve the barrier properties of PLA. This phenomenon is attributed to the well-known tortuosity effect. In this context, theoretical models and empirical studies have suggested that, in addition to the aforementioned nanoparticle orientation, other key factors in achieving high-barrier materials are maximizing the aspect ratio of the nanoparticles and improving the interfacial adhesion between the nanofillers and the PLA matrix. However, despite extensive research efforts, none of these materials, with the level of gas permeability required by the end-users and at reasonable costs, have reached the market yet. As this film will be in contact with food, potential risks to human health are questioned as nanomaterials may migrate from the packages to foodstuff. In addition, the effect of nanoparticle incorporation on the biodegradability/compostability of the polymer must be thoroughly investigated. In the absence of these detailed studies, nanocomposites cannot be approved for legal utilization in industry.

Meanwhile, recent progress in polymerization methods such as controlled radical polymerization (CRP) has enabled the synthesis of interesting copolymerization products for a wide range of applications, while fitting with some of the principles of green chemistry such as the use of bio-based monomers, solvent-less protocols, and mild conditions (low temperature). Using these methods, PLA barrier properties can be enhanced by copolymerization with appropriate monomers to widen its applicability in the food packaging field. Despite the impressive developments, there are still challenges that need to be addressed. For instance, there is a huge need to discover new selective catalysts in light of expanding the range of monomers that can be used. In addition, current polymerization protocols are difficult to scale-up due to the lack of automatic systems. However, further developments are expected in the near future as growing research interest is focusing on strategies to simplify the current polymerization protocols and allow for more scalable methodologies to be implemented.

Author Contributions: Conceptualization, S.M.; Resources, P.S., E.L., and C.M.; Writing—original draft preparation, S.M.; Writing—review and editing, P.S., E.L., and C.M. All authors have read and agreed to the published version of the manuscript.

Funding: This research received no external funding.

Institutional Review Board Statement: Not applicable.

Informed Consent Statement: Not applicable.

Conflicts of Interest: The authors declare no conflict of interest.

References

1. Leal Filho, W.; Saari, U.; Fedoruk, M.; Iital, A.; Moora, H.; Klöga, M.; Voronova, V. An overview of the problems posed by plastic products and the role of extended producer responsibility in Europe. *J. Clean. Prod.* **2019**, *214*, 550–558. [CrossRef]
2. Kabir, E.; Kaur, R.; Lee, J.; Kim, K.H.; Kwon, E.E. Prospects of biopolymer technology as an alternative option for non-degradable plastics and sustainable management of plastic wastes. *J. Clean. Prod.* **2020**, *258*, 120536. [CrossRef]
3. Yadav, A.; Mangaraj, S.; Singh, R.; Das, S.K.; M, N.K.; Arora, S. Biopolymers as packaging material in food and allied industry. *Int. J. Chem. Stud.* **2018**, *6*, 2411–2418.
4. Drumright, R.E.; Gruber, P.R.; Henton, D.E. Polylactic Acid Technology. *Adv. Mater.* **2000**, *12*, 1841–1846. [CrossRef]
5. Becker, J.M.; Pounder, R.J.; Dove, A.P. Synthesis of poly(lactide)s with modified thermal and mechanical properties. *Macromol. Rapid Commun.* **2010**, *31*, 1923–1937. [CrossRef]
6. Kumar, S.; Bhatnagar, N.; Ghosh, A.K. Effect of enantiomeric monomeric unit ratio on thermal and mechanical properties of poly(lactide). *Polym. Bull.* **2016**, *73*, 2087–2104. [CrossRef]
7. Garlotta, D. A Literature Review of Poly(Lactic Acid) A Literature Review of Poly(Lactic Acid). *J. Polym. Environ.* **2019**, *9*, 63–84. [CrossRef]
8. Lunt, J. Large-scale production, properties and commercial applications of poly lactic acid polymers. *Polym. Degrad. Stab.* **1998**, *59*, 145–152. [CrossRef]
9. Auras, R.A.; Harte, B.; Selke, S.; Hernandez, R. Mechanical, physical, and barrier properties of poly(lactide) films. *J. Plast. Film Sheeting* **2003**, *19*, 123–135. [CrossRef]
10. Auras, R.A.; Singh, S.P.; Singh, J.J. Evaluation of oriented poly(lactide) polymers vs. existing PET and oriented PS for fresh food service containers. *Packag. Technol. Sci. Int. J.* **2005**, *18*, 207–216. [CrossRef]
11. Bao, L.; Dorgan, J.R.; Knauss, D.; Hait, S.; Oliveira, N.S.; Maruccho, I.M. Gas permeation properties of poly(lactic acid) revisited. *J. Memb. Sci.* **2006**, *285*, 166–172. [CrossRef]
12. Farah, S.; Anderson, D.G.; Langer, R. Physical and mechanical properties of PLA, and their functions in widespread applications—A comprehensive review. *Adv. Drug Deliv. Rev.* **2016**, *107*, 367–392. [CrossRef] [PubMed]
13. Qi, F.; Tang, M.; Chen, X.; Chen, M.; Guo, G.; Zhang, Z. Morphological structure, thermal and mechanical properties of tough poly(lactic acid) upon stereocomplexes. *Eur. Polym. J.* **2015**, *71*, 314–324. [CrossRef]
14. Nagarajan, V.; Mohanty, A.K.; Misra, M. Perspective on Polylactic Acid (PLA) based Sustainable Materials for Durable Applications: Focus on Toughness and Heat Resistance. *ACS Sustain. Chem. Eng.* **2016**, *4*, 2899–2916. [CrossRef]
15. Jin, F.L.; Hu, R.R.; Park, S.J. Improvement of thermal behaviors of biodegradable poly(lactic acid) polymer: A review. *Compos. Part B Eng.* **2019**, *164*, 287–296. [CrossRef]
16. Zhao, X.; Hu, H.; Wang, X.; Yu, X.; Zhou, W.; Peng, S. Super tough poly(lactic acid) blends: A comprehensive review. *RSC Adv.* **2020**, *10*, 13316–13368. [CrossRef]
17. Singha, S.; Hedenqvist, M.S. A Review on Barrier Properties of Poly(Lactic Acid)/Clay Nanocomposites. *Polymers* **2020**, *12*, 1095. [CrossRef]
18. Li, F.; Zhang, C.; Weng, Y. Improvement of the Gas Barrier Properties of PLA/OMMT Films by Regulating the Interlayer Spacing of OMMT and the Crystallinity of PLA. *ACS Omega* **2020**, *5*, 18675–18684. [CrossRef]
19. Najafi, N.; Heuzey, M.C.; Carreau, P.J.; Wood-Adams, P.M. Control of thermal degradation of polylactide (PLA)-clay nanocomposites using chain extenders. *Polym. Degrad. Stab.* **2012**, *97*, 554–565. [CrossRef]
20. Selke, S.E.M.; Culter, J.D.; Auras, R.A.; Rabnawaz, M. Plastics Packaging: Properties, Processing, Applications, and Regulations. In *Plastics Packaging*; Carl Hanser Verlag GmbH Co KG: Munich, Germany, 2016; pp. I–XX. [CrossRef]
21. Boufarguine, M.; Guinault, A.; Miquelard-Garnier, G.; Sollogoub, C. PLA/PHBV films with improved mechanical and gas barrier properties. *Macromol. Mater. Eng.* **2013**, *298*, 1065–1073. [CrossRef]
22. Arcos-Hernández, M.V.; Laycock, B.; Donose, B.C.; Pratt, S.; Halley, P.; Al-Luaibi, S.; Werker, A.; Lant, P.A. Physicochemical and mechanical properties of mixed culture polyhydroxyalkanoate (PHBV). *Eur. Polym. J.* **2013**, *49*, 904–913. [CrossRef]
23. Jost, V.; Langowski, H.C. Effect of different plasticisers on the mechanical and barrier properties of extruded cast PHBV films. *Eur. Polym. J.* **2015**, *68*, 302–312. [CrossRef]
24. Calafut, T. Polypropylene Films. In *Plastic Films in Food Packaging: Materials, Technology and Applications*; William Andrew: Norwich, NY, USA, 2012; pp. 17–20. [CrossRef]

25. Li, J.; Zhu, Z.; Li, T.; Peng, X.; Jiang, S.; Turng, L.S. Quantification of the Young's modulus for polypropylene: Influence of initial crystallinity and service temperature. *J. Appl. Polym. Sci.* **2020**, *137*, 48581. [CrossRef]
26. Cava, D.; Giménez, E.; Gavara, R.; Lagaron, J.M. Comparative performance and barrier properties of biodegradable thermoplastics and nanobiocomposites versus PET for food packaging applications. *J. Plast. Film Sheeting* **2006**, *22*, 265–274. [CrossRef]
27. Mohsin, M.A.; Ali, I.; Elleithy, R.H.; Al-Zahrani, S.M. Improvements in barrier properties of poly(ethylene terephthalate) films using commercially available high barrier masterbatch additives via melt blend technique. *J. Plast. Film Sheeting* **2013**, *29*, 21–38. [CrossRef]
28. Lewis, E.L.V.; Duckett, R.A.; Ward, I.M.; Fairclough, J.P.A.; Ryan, A.J. The barrier properties of polyethylene terephthalate to mixtures of oxygen, carbon dioxide and nitrogen. *Polymer* **2003**, *44*, 1631–1640. [CrossRef]
29. Dardmeh, N.; Khosrowshahi, A.; Almasi, H.; Zandi, M. Study on Effect of the Polyethylene Terephthalate/Nanoclay Nanocomposite Film on the Migration of Terephthalic Acid into the Yoghurt Drinks Simulant. *J. Food Process Eng.* **2017**, *40*, e12324. [CrossRef]
30. Awaja, F.; Pavel, D. Recycling of PET. *Eur. Polym. J.* **2005**, *41*, 1453–1477. [CrossRef]
31. Turnbull, L.; Liggat, J.J.; Macdonald, W.A. Thermal degradation chemistry of poly(ethylene naphthalate)—A study by thermal volatilisation analysis. *Polym. Degrad. Stab.* **2013**, *98*, 2244–2258. [CrossRef]
32. Pal, S.N.; Ramani, A.V.; Subramanian, N. Gas permeability studies on poly(vinyl chloride) based polymer blends intended for medical applications. *J. Appl. Polym. Sci.* **1992**, *46*, 981–990. [CrossRef]
33. Deshmukh, K.; Joshi, G.M. Thermo-mechanical properties of Poly(vinyl chloride)/graphene oxide as high performance nanocomposites. *Polym. Test.* **2014**, *34*, 211–219. [CrossRef]
34. Satapathy, S.; Palanisamy, A. Mechanical and barrier properties of polyvinyl chloride plasticized with dioctyl phthalate, epoxidized soybean oil, and epoxidized cardanol. *J. Vinyl Addit. Technol.* **2021**, *27*, 599–611. [CrossRef]
35. Yu, J.; Sun, L.; Ma, C.; Qiao, Y.; Yao, H. Thermal degradation of PVC: A review. *Waste Manag.* **2016**, *48*, 300–314. [CrossRef] [PubMed]
36. Ashley, R.J. Permeability and Plastics Packaging. In *Polymer Permeability*; Springer: Dordrecht, The Netherlands, 1985; pp. 269–308. [CrossRef]
37. Miller, K.S.; Krochta, J.M. Oxygen and aroma barrier properties of edible films: A review. *Trends Food Sci. Technol.* **1997**, *8*, 228–237. [CrossRef]
38. Emblem, A. Plastics properties for packaging materials. In *Packaging Technology*; Woodhead Publishing: Cambridge, UK, 2012; pp. 287–309. [CrossRef]
39. Klepić, M.; Setničková, K.; Lanč, M.; Žák, M.; Izák, P.; Dendisová, M.; Fuoco, A.; Jansen, J.C.; Friess, K. Permeation and sorption properties of CO_2-selective blend membranes based on polyvinyl alcohol (PVA) and 1-ethyl-3-methylimidazolium dicyanamide ([EMIM][DCA]) ionic liquid for effective CO2/H2 separation. *J. Memb. Sci.* **2020**, *597*, 117623. [CrossRef]
40. Jain, N.; Singh, V.K.; Chauhan, S. A review on mechanical and water absorption properties of polyvinyl alcohol based composites/films. *J. Mech. Behav. Mater.* **2017**, *26*, 213–222. [CrossRef]
41. Van der Schueren, B.; El Marouazi, H.; Mohanty, A.; Lévêque, P.; Sutter, C.; Romero, T.; Janowska, I. Polyvinyl alcohol-few layer graphene composite films prepared from aqueous colloids. Investigations of mechanical, conductive and gas barrier properties. *Nanomaterials* **2020**, *10*, 858. [CrossRef]
42. Holland, B.J.; Hay, J.N. The thermal degradation of poly(vinyl alcohol). *Polymer* **2001**, *42*, 6775–6783. [CrossRef]
43. Zhang, Z.; Britt, I.J.; Tung, M.A. Permeation of oxygen and water vapor through EVOH films as influenced by relative humidity. *J. Appl. Polym. Sci.* **2001**, *82*, 1866–1872. [CrossRef]
44. McKeen, L.W. Polyvinyls and Acrylics. In *Film Properties of Plastics and Elastomers*; William Andrew: Norwich, NY, USA, 2012; pp. 219–254. [CrossRef]
45. Hotta, S.; Paul, D.R. Nanocomposites formed from linear low density polyethylene and organoclays. *Polymer* **2004**, *45*, 7639–7654. [CrossRef]
46. Picard, E.; Vermogen, A.; Gérard, J.F.; Espuche, E. Influence of the compatibilizer polarity and molar mass on the morphology and the gas barrier properties of polyethylene/clay nanocomposites. *J. Polym. Sci. Part B Polym. Phys.* **2008**, *46*, 2593–2604. [CrossRef]
47. Durmuş, A.; Woo, M.; Kaşgöz, A.; Macosko, C.W.; Tsapatsis, M. Intercalated linear low density polyethylene (LLDPE)/clay nanocomposites prepared with oxidized polyethylene as a new type compatibilizer: Structural, mechanical and barrier properties. *Eur. Polym. J.* **2007**, *43*, 3737–3749. [CrossRef]
48. Ali Dadfar, S.M.; Alemzadeh, I.; Reza Dadfar, S.M.; Vosoughi, M. Studies on the oxygen barrier and mechanical properties of low density polyethylene/organoclay nanocomposite films in the presence of ethylene vinyl acetate copolymer as a new type of compatibilizer. *Mater. Des.* **2011**, *32*, 1806–1813. [CrossRef]
49. Dubdub, I.; Al-Yaari, M. Pyrolysis of low density polyethylene: Kinetic study using TGA data and ANN prediction. *Polymer* **2020**, *12*, 891. [CrossRef] [PubMed]
50. Honaker, K.; Vautard, F.; Drzal, L.T. Influence of processing methods on the mechanical and barrier properties of HDPE-GNP nanocomposites. *Adv. Compos. Hybrid Mater.* **2021**, *4*, 492–504. [CrossRef]
51. Zhao, L.; Song, P.; Cao, Z.; Fang, Z.; Guo, Z. Thermal stability and rheological behaviors of high-density polyethylene/fullerene nanocomposites. *J. Nanomater.* **2012**, *2012*, 340962. [CrossRef]

52. Lagaron, J.M.; Catalá, R.; Gavara, R. Structural characteristics defining high barrier properties in polymeric materials. *Mater. Sci. Technol.* **2004**, *20*, 1–7. [CrossRef]
53. Singh, R.K.; Singh, N. Quality of packaged foods. In *Innovations in Food Packaging*; Academic Press: Cambridge, MA, USA, 2005; pp. 24–44. [CrossRef]
54. Graham, T. The solution-diffusion model. *Lond. Edinb. Dublin Phil. Mag. Sci. Ser.* **1866**, *32*, 401–425. [CrossRef]
55. Han, J.H.; Scanlon, M.G. Mass Transfer of Gas and Solute Through Packaging Materials. In *Innovations in Food Packaging*, 2nd ed.; Academic Press: Cambridge, MA, USA, 2013; pp. 37–49. [CrossRef]
56. Fereydoon, M.; Ebnesajjad, S. Development of High-Barrier Film for Food Packaging. In *Plastic Films in Food Packaging: Materials, Technology and Applications*; William Andrew: Norwich, NY, USA, 2012; pp. 71–92. [CrossRef]
57. Ewender, J.; Welle, F. Determination and Prediction of the Lag Times of Hydrocarbons through a Polyethylene Terephthalate Film. *Packag. Technol. Sci.* **2014**, *27*, 963–974. [CrossRef]
58. Baudet, C.; Grandidier, J.C.; Cangémi, L. A two-phase model for the diffuso-mechanical behaviour of semicrystalline polymers in gaseous environment. *Int. J. Solids Struct.* **2009**, *46*, 1389–1401. [CrossRef]
59. Lin, J.; Shenogin, S.; Nazarenko, S. Oxygen solubility and specific volume of rigid amorphous fraction in semicrystalline poly(ethylene terephthalate). *Polymer* **2002**, *43*, 4733–4743. [CrossRef]
60. Delpouve, N.; Arnoult, M.; Saiter, A.; Dargent, E.; Saiter, J.M. Evidence of two mobile amorphous phases in semicrystalline polylactide observed from calorimetric investigations. *Polym. Eng. Sci.* **2014**, *54*, 1144–1150. [CrossRef]
61. Fernandes Nassar, S.; Guinault, A.; Delpouve, N.; Divry, V.; Ducruet, V.; Sollogoub, C.; Domenek, S. Multi-scale analysis of the impact of polylactide morphology on gas barrier properties. *Polymer* **2017**, *108*, 163–172. [CrossRef]
62. Delpouve, N.; Stoclet, G.; Saiter, A.; Dargent, E.; Marais, S. Water barrier properties in biaxially drawn poly(lactic acid) films. *J. Phys. Chem. B* **2012**, *116*, 4615–4625. [CrossRef]
63. Gorrasi, G.; Anastasio, R.; Bassi, L.; Pantani, R. Barrier properties of PLA to water vapour: Effect of temperature and morphology. *Macromol. Res.* **2013**, *21*, 1110–1117. [CrossRef]
64. Duan, Z.; Thomas, N.L. Water vapour permeability of poly(lactic acid): Crystallinity and the tortuous path model. *J. Appl. Phys.* **2014**, *115*, 064903. [CrossRef]
65. Sonchaeng, U.; Iñiguez-Franco, F.; Auras, R.; Selke, S.; Rubino, M.; Lim, L.T. Poly(lactic acid) mass transfer properties. *Prog. Polym. Sci.* **2018**, *86*, 85–121. [CrossRef]
66. Weinkauf, D.H.; Paul, D.R. *Effects of Structural Order on Barrier Properties*; ACS Publications: Washington, DC, USA, 1990; pp. 60–91. [CrossRef]
67. Cowling, R.; Park, G.S. Permeability, solubility and diffusion of gases in amorphous and crystalline 1,4-polybutadiene membranes. *J. Memb. Sci.* **1979**, *5*, 199–207. [CrossRef]
68. Lin, H.; Freeman, B.D. Gas solubility, diffusivity and permeability in poly(ethylene oxide). *J. Memb. Sci.* **2004**, *239*, 105–117. [CrossRef]
69. Mokwena, K.K.; Tang, J. Ethylene Vinyl Alcohol: A Review of Barrier Properties for Packaging Shelf Stable Foods. *Crit. Rev. Food Sci. Nutr.* **2012**, *52*, 640–650. [CrossRef]
70. Bekele, S. High Barrier Polyvinylidene Chloride Composition and Film. U.S. Patent 6,673,406, 6 January 2004.
71. Xie, H.; Guo, K.; Chen, J.; Wang, W.; Niu, M.; Li, X. The effect of uniaxial stretching on the crystallization behaviors of polylactic acid film. In *Proceedings of the Advanced Materials Research*; Trans Tech Publications Ltd.: Zurich, Switzerland, 2012; Volume 418, pp. 625–628. [CrossRef]
72. Stoclet, G.; Seguela, R.; Lefebvre, J.M.; Elkoun, S.; Vanmansart, C. Strain-induced molecular ordering in polylactide upon uniaxial stretching. *Macromolecules* **2010**, *43*, 1488–1498. [CrossRef]
73. Rocca-Smith, J.R.; Karbowiak, T.; Marcuzzo, E.; Sensidoni, A.; Piasente, F.; Champion, D.; Heinz, O.; Vitry, P.; Bourillot, E.; Lesniewska, E.; et al. Impact of corona treatment on PLA film properties. *Polym. Degrad. Stab.* **2016**, *132*, 109–116. [CrossRef]
74. Drieskens, M.; Peeters, R.; Mullens, J.; Franco, D.; Iemstra, P.J.; Hristova-Bogaerds, D.G. Structure versus properties relationship of poly(lactic acid). I. effect of crystallinity on barrier properties. *J. Polym. Sci. Part B Polym. Phys.* **2009**, *47*, 2247–2258. [CrossRef]
75. Guinault, A.; Sollogoub, C.; Domenek, S.; Grandmontagne, A.; Ducruet, V. Influence of crystallinity on gas barrier and mechanical properties of pla food packaging films. *Int. J. Mater. Form.* **2010**, *3*, 603–606. [CrossRef]
76. Courgneau, C.; Domenek, S.; Lebossé, R.; Guinault, A.; Avérous, L.; Ducruet, V. Effect of crystallization on barrier properties of formulated polylactide. *Polym. Int.* **2012**, *61*, 180–189. [CrossRef]
77. Guinault, A.; Sollogoub, C.; Ducruet, V.; Domenek, S. Impact of crystallinity of poly(lactide) on helium and oxygen barrier properties. *Eur. Polym. J.* **2012**, *48*, 779–788. [CrossRef]
78. Tsuji, H.; Okino, R.; Daimon, H.; Fujie, K. Water vapor permeability of poly(lactide)s: Effects of molecular characteristics and crystallinity. *J. Appl. Polym. Sci.* **2006**, *99*, 2245–2252. [CrossRef]
79. Colomines, G.; Domenek, S.; Ducruet, V.; Guinault, A. Influences of the crystallisation rate on thermal and barrier properties of polylactide acid (PLA) food packaging films. *Int. J. Mater. Form.* **2008**, *1*, 607–610. [CrossRef]
80. Di Lorenzo, M.L.; Rubino, P.; Immirzi, B.; Luijkx, R.; Hélou, M.; Androsch, R. Influence of chain structure on crystal polymorphism of poly(lactic acid). Part 2. Effect of molecular mass on the crystal growth rate and semicrystalline morphology. *Colloid Polym. Sci.* **2015**, *293*, 2459–2467. [CrossRef]

81. Pan, P.; Yang, J.; Shan, G.; Bao, Y.; Weng, Z.; Cao, A.; Yazawa, K.; Inoue, Y. Temperature-variable FTIR and solid-state 13C NMR investigations on crystalline structure and molecular dynamics of polymorphic poly(L-lactide) and poly(L-lactide)/poly(D-lactide) stereocomplex. *Macromolecules* **2012**, *45*, 189–197. [CrossRef]
82. Zhang, J.; Duan, Y.; Sato, H.; Tsuji, H.; Noda, I.; Yan, S.; Ozaki, Y. Crystal modifications and thermal behavior of poly(L-lactic acid) revealed by infrared spectroscopy. *Macromolecules* **2005**, *38*, 8012–8021. [CrossRef]
83. Cocca, M.; Di Lorenzo, M.L.; Malinconico, M.; Frezza, V. Influence of crystal polymorphism on mechanical and barrier properties of poly(L-lactic acid). *Eur. Polym. J.* **2011**, *47*, 1073–1080. [CrossRef]
84. Karayiannis, N.C.; Mavrantzas, V.G.; Theodorou, D.N. Detailed Atomistic Simulation of the Segmental Dynamics and Barrier Properties of Amorphous Poly(ethylene terephthalate) and Poly(ethylene isophthalate). *Macromolecules* **2004**, *37*, 2978–2995. [CrossRef]
85. Charlon, S.; Marais, S.; Dargent, E.; Soulestin, J.; Sclavons, M.; Follain, N. Structure-barrier property relationship of biodegradable poly(butylene succinate) and poly[(butylene succinate)-co-(butylene adipate)] nanocomposites: Influence of the rigid amorphous fraction. *Phys. Chem. Chem. Phys.* **2015**, *17*, 29918–29934. [CrossRef] [PubMed]
86. Zekriardehani, S.; Jabarin, S.A.; Gidley, D.R.; Coleman, M.R. Effect of Chain Dynamics, Crystallinity, and Free Volume on the Barrier Properties of Poly(ethylene terephthalate) Biaxially Oriented Films. *Macromolecules* **2017**, *50*, 2845–2855. [CrossRef]
87. Sangroniz, A.; Chaos, A.; Iriarte, M.; Del Río, J.; Sarasua, J.R.; Etxeberria, A. Influence of the Rigid Amorphous Fraction and Crystallinity on Polylactide Transport Properties. *Macromolecules* **2018**, *51*, 3923–3931. [CrossRef]
88. Dlubek, G.; Sen Gupta, A.; Pionteck, J.; Häßler, R.; Krause-Rehberg, R.; Kaspar, H.; Lochhaas, K.H. Glass transition and free volume in the mobile (MAF) and rigid (RAF) amorphous fractions of semicrystalline PTFE: A positron lifetime and PVT study. *Polymer* **2005**, *46*, 6075–6089. [CrossRef]
89. Ma, Q.; Georgiev, G.; Cebe, P. Constraints in semicrystalline polymers: Using quasi-isothermal analysis to investigate the mechanisms of formation and loss of the rigid amorphous fraction. *Polymer* **2011**, *52*, 4562–4570. [CrossRef]
90. Wunderlich, B. Reversible crystallization and the rigid-amorphous phase in semicrystalline macromolecules. *Prog. Polym. Sci.* **2003**, *28*, 383–450. [CrossRef]
91. Schick, C.; Dobbertin, J.; Pötter, M.; Dehne, H.; Hensel, A.; Wurm, A.; Ghoneim, A.M.; Weyer, S. Separation of components of different molecular mobility by calorimetry, dynamic mechanical and dielectric spectroscopy. *J. Therm. Anal.* **1997**, *49*, 499–511. [CrossRef]
92. Del Río, J.; Etxeberria, A.; López-Rodríguez, N.; Lizundia, E.; Sarasua, J.R. A PALS contribution to the supramolecular structure of poly(L-lactide). *Macromolecules* **2010**, *43*, 4698–4707. [CrossRef]
93. Drzal, P.L.; Halasa, A.F.; Kofinas, P. Microstructure orientation and nanoporous gas transport in semicrystalline block copolymer membranes. *Polymer* **2000**, *41*, 4671–4677. [CrossRef]
94. Hardy, L.; Espuche, E.; Seytre, G.; Stevenson, I. Gas transport properties of poly(ethylene-2,6-naphtalene dicarboxylate) films: Influence of crystallinity and orientation. *J. Appl. Polym. Sci.* **2003**, *89*, 1849–1857. [CrossRef]
95. Vermogen, A.; Picard, E.; Milan, M.L.; Masenelli-Varlot, K.; Duchet, J.; Vigier, G.; Espuche, E.; Gérard, J.F. Assessing crystalline lamellae orientation impact on the properties of semi-crystalline polymer-clay nanocomposites. *J. Polym. Sci. Part B Polym. Phys.* **2008**, *46*, 1966–1975. [CrossRef]
96. Backman, A.; Lange, J.; Hedenqvist, M.S. Transport Properties of Uniaxially Oriented Aliphatic Polyketone. *J. Polym. Sci. Part B Polym. Phys.* **2004**, *42*, 947–955. [CrossRef]
97. Lee, M.S.; Chang, K.P. Biaxially Oriented Polylactic Acid Film with High Barrier. U.S. Patent 7,951,438, 31 May 2011.
98. Guinault, A.; Menary, G.H.; Courgneau, C.; Griffith, D.; Ducruet, V.; Miri, V.; Sollogoub, C. The effect of the stretching of PLA extruded films on their crystallinity and gas barrier properties. *AIP Conf. Proc.* **2011**, *1353*, 826–831. [CrossRef]
99. Dong, T.; Yu, Z.; Wu, J.; Zhao, Z.; Yun, X.; Wang, Y.; Jin, Y.; Yang, J. Thermal and barrier properties of stretched and annealed polylactide films. *Polym. Sci. Ser. A* **2015**, *57*, 738–746. [CrossRef]
100. Saeidlou, S.; Huneault, M.A.; Li, H.; Park, C.B. Poly(lactic acid) crystallization. *Prog. Polym. Sci.* **2012**, *37*, 1657–1677. [CrossRef]
101. Nofar, M.; Zhu, W.; Park, C.B.; Randall, J. Crystallization kinetics of linear and long-chain-branched polylactide. *Ind. Eng. Chem. Res.* **2011**, *50*, 13789–13798. [CrossRef]
102. Rosely, C.V.S.; Shaiju, P.; Gowd, E.B. Poly(L-lactic acid)/Boron Nitride Nanocomposites: Influence of Boron Nitride Functionalization on the Properties of Poly(L-lactic acid). *J. Phys. Chem. B* **2019**, *123*, 8599–8609. [CrossRef]
103. Bai, H.; Zhang, W.; Deng, H.; Zhang, Q.; Fu, Q. Control of crystal morphology in poly(L-lactide) by adding nucleating agent. *Macromolecules* **2011**, *44*, 1233–1237. [CrossRef]
104. Bai, H.; Huang, C.; Xiu, H.; Zhang, Q.; Fu, Q. Enhancing mechanical performance of polylactide by tailoring crystal morphology and lamellae orientation with the aid of nucleating agent. *Polymer* **2014**, *55*, 6924–6934. [CrossRef]
105. Zhang, H.; Bai, H.; Liu, Z.; Zhang, Q.; Fu, Q. Toward high-performance poly(L-lactide) fibers via tailoring crystallization with the aid of fibrillar nucleating agent. *ACS Sustain. Chem. Eng.* **2016**, *4*, 3939–3947. [CrossRef]
106. Lan, Q.; Li, Y. Mesophase-mediated crystallization of poly(L-lactide): Deterministic pathways to nanostructured morphology and superstructure control. *Macromolecules* **2016**, *49*, 7387–7399. [CrossRef]
107. Zhang, L.; Zhao, G.; Wang, G. Investigation of the influence of pressurized CO_2 on the crystal growth of poly(L-lactic acid) by using an: In situ high-pressure optical system. *Soft Matter* **2019**, *15*, 5714–5727. [CrossRef]

108. Xiao, H.W.; Li, P.; Ren, X.; Jiang, T.; Yeh, J.T. Isothermal crystallization kinetics and crystal structure of poly(lactic acid): Effect of triphenyl phosphate and talc. *J. Appl. Polym. Sci.* **2010**, *118*, 3558–3569. [CrossRef]
109. Zaldua, N.; Mugica, A.; Zubitur, M.; Iturrospe, A.; Arbe, A.; Re, G.L.; Raquez, J.M.; Dubois, P.; Müller, A.J. The role of PLLA-g-montmorillonite nanohybrids in the acceleration of the crystallization rate of a commercial PLA. *CrystEngComm* **2016**, *18*, 9334–9344. [CrossRef]
110. Xu, P.; Luo, X.; Zhou, Y.; Yang, Y.; Ding, Y. Enhanced cold crystallization and dielectric polarization of PLA composites induced by P[MPEGMA-IL] and graphene. *Thermochim. Acta* **2017**, *657*, 156–162. [CrossRef]
111. Ghassemi, A.; Moghaddamzadeh, S.; Duchesne, C.; Rodrigue, D. Effect of annealing on gas permeability and mechanical properties of polylactic acid/talc composite films. *J. Plast. Film Sheeting* **2017**, *33*, 361–383. [CrossRef]
112. Buzarovska, A.; Bogoeva-Gaceva, G.; Fajgar, R. Effect of the talc filler on structural, water vapor barrier and mechanical properties of poly(lactic acid) composites. *J. Polym. Eng.* **2016**, *36*, 181–188. [CrossRef]
113. Brizzolara, D.; Cantow, H.J.; Diederichs, K.; Keller, E.; Domb, A.J. Mechanism of the stereocomplex formation between enantiomeric poly(lactide)s. *Macromolecules* **1996**, *29*, 191–197. [CrossRef]
114. Tsuji, H. Poly(lactide) Stereocomplexes: Formation, Structure, Properties, Degradation, and Applications. *Macromol. Biosci.* **2005**, *5*, 569–597. [CrossRef] [PubMed]
115. Tsuji, H. Poly(lactic acid) stereocomplexes: A decade of progress. *Adv. Drug Deliv. Rev.* **2016**, *107*, 97–135. [CrossRef] [PubMed]
116. Maillard, D.; Prud'homme, R.E. Differences between crystals obtained in PLLA-rich or PDLA-rich stereocomplex mixtures. *Macromolecules* **2010**, *43*, 4006–4010. [CrossRef]
117. Tsuji, H.; Tsuruno, T. Water vapor permeability of poly(L-lactide)/poly(D-lactide) stereocomplexes. *Macromol. Mater. Eng.* **2010**, *295*, 709–715. [CrossRef]
118. Xu, H.; Wu, D.; Yang, X.; Xie, L.; Hakkarainen, M. Thermostable and impermeable "nano-barrier walls" constructed by poly(lactic acid) stereocomplex crystal decorated graphene oxide nanosheets. *Macromolecules* **2015**, *48*, 2127–2137. [CrossRef]
119. Gupta, A.; Katiyar, V. Cellulose Functionalized High Molecular Weight Stereocomplex Polylactic Acid Biocomposite Films with Improved Gas Barrier, Thermomechanical Properties. *ACS Sustain. Chem. Eng.* **2017**, *5*, 6835–6844. [CrossRef]
120. Jiang, Y.; Yan, C.; Wang, K.; Shi, D.; Liu, Z.; Yang, M. Super-toughed PLA blown film with enhanced gas barrier property available for packaging and agricultural applications. *Materials* **2019**, *12*, 1663. [CrossRef]
121. Varol, N. Advanced Thermal Analysis and Transport Properties of Stereocomplex Polylactide. Ph.D. Thesis, Normandie Université, Normandie, France, 2019; p. 244.
122. Nie, M.; Han, R.; Wang, Q. Formation and alignment of hybrid shish-kebab morphology with rich beta crystals in an isotactic polypropylene pipe. *Ind. Eng. Chem. Res.* **2014**, *53*, 4142–4146. [CrossRef]
123. Xie, Q.; Han, L.; Shan, G.; Bao, Y.; Pan, P. Polymorphic Crystalline Structure and Crystal Morphology of Enantiomeric Poly(lactic acid) Blends Tailored by a Self-Assemblable Aryl Amide Nucleator. *ACS Sustain. Chem. Eng.* **2016**, *4*, 2680–2688. [CrossRef]
124. Pi, L.; Nie, M.; Wang, Q. Crystalline composition and morphology in isotactic polypropylene pipe under combining effects of rotation extrusion and fibril β-nucleating Agent. *J. Vinyl Addit. Technol.* **2019**, *25*, E195–E202. [CrossRef]
125. Bai, H.; Huang, C.; Xiu, H.; Zhang, Q.; Deng, H.; Wang, K.; Chen, F.; Fu, Q. Significantly improving oxygen barrier properties of polylactide via constructing parallel-aligned shish-kebab-like crystals with well-interlocked boundaries. *Biomacromolecules* **2014**, *15*, 1507–1514. [CrossRef]
126. Li, C.; Jiang, T.; Wang, J.; Peng, S.; Wu, H.; Shen, J.; Guo, S.; Zhang, X.; Harkin-Jones, E. Enhancing the Oxygen-Barrier Properties of Polylactide by Tailoring the Arrangement of Crystalline Lamellae. *ACS Sustain. Chem. Eng.* **2018**, *6*, 6247–6255. [CrossRef]
127. Vartiainen, J.; Vähä-Nissi, M.; Harlin, A. Biopolymer Films and Coatings in Packaging Applications—A Review of Recent Developments. *Mater. Sci. Appl.* **2014**, *5*, 708–718. [CrossRef]
128. Anukiruthika, T.; Sethupathy, P.; Wilson, A.; Kashampur, K.; Moses, J.A.; Anandharamakrishnan, C. Multilayer packaging: Advances in preparation techniques and emerging food applications. *Compr. Rev. Food Sci. Food Saf.* **2020**, *19*, 1156–1186. [CrossRef] [PubMed]
129. Carr, J.M.; Langhe, D.S.; Ponting, M.T.; Hiltner, A.; Baer, E. Confined crystallization in polymer nanolayered films: A review. *J. Mater. Res.* **2012**, *27*, 1326–1350. [CrossRef]
130. Priolo, M.A.; Holder, K.M.; Guin, T.; Grunlan, J.C. Recent advances in gas barrier thin films via layer-by-layer assembly of polymers and platelets. *Macromol. Rapid Commun.* **2015**, *36*, 866–879. [CrossRef] [PubMed]
131. Echegoyen, Y.; Fabra, M.J.; Castro-Mayorga, J.L.; Cherpinski, A.; Lagaron, J.M. High throughput electro-hydrodynamic processing in food encapsulation and food packaging applications: Viewpoint. *Trends Food Sci. Technol.* **2017**, *60*, 71–79. [CrossRef]
132. Ponting, M.; Hiltner, A.; Baer, E. Polymer nanostructures by forced assembly: Process, structure, and properties. *Macromol. Symp.* **2010**, *294*, 19–32. [CrossRef]
133. MacKey, M.; Flandin, L.; Hiltner, A.; Baer, E. Confined crystallization of PVDF and a PVDF-TFE copolymer in nanolayered films. *J. Polym. Sci. Part B Polym. Phys.* **2011**, *49*, 1750–1761. [CrossRef]
134. Wang, H.; Keum, J.K.; Hiltner, A.; Baer, E. Confined crystallization of peo in nanolayered films impacting structure and oxygen permeability. *Macromolecules* **2009**, *42*, 7055–7066. [CrossRef]
135. Ponting, M.; Lin, Y.; Keum, J.K.; Hiltner, A.; Baer, E. Effect of substrate on the isothermal crystallization kinetics of confined poly(ε-caprolactone) nanolayers. *Macromolecules* **2010**, *43*, 8619–8627. [CrossRef]

136. Zhang, G.; Baer, E.; Hiltner, A. Gas permeability of poly(4-methylpentene-1) in a confined nanolayered film system. *Polymer* **2013**, *54*, 4298–4308. [CrossRef]
137. Messin, T.; Follain, N.; Guinault, A.; Sollogoub, C.; Gaucher, V.; Delpouve, N.; Marais, S. Structure and Barrier Properties of Multinanolayered Biodegradable PLA/PBSA Films: Confinement Effect via Forced Assembly Coextrusion. *ACS Appl. Mater. Interfaces* **2017**, *9*, 29101–29112. [CrossRef]
138. Fernandes Nassar, S.; Delpouve, N.; Sollogoub, C.; Guinault, A.; Stoclet, G.; Régnier, G.; Domenek, S. Impact of Nanoconfinement on Polylactide Crystallization and Gas Barrier Properties. *ACS Appl. Mater. Interfaces* **2020**, *12*, 9953–9965. [CrossRef]
139. Messin, T.; Marais, S.; Follain, N.; Guinault, A.; Gaucher, V.; Delpouve, N.; Sollogoub, C. Biodegradable PLA/PBS multinanolayer membrane with enhanced barrier performances. *J. Memb. Sci.* **2020**, *598*, 117777. [CrossRef]
140. George, S.C.; Thomas, S. Transport phenomena through polymeric systems. *Prog. Polym. Sci.* **2001**, *26*, 985–1017. [CrossRef]
141. Yang, Y.H.; Haile, M.; Park, Y.T.; Malek, F.A.; Grunlan, J.C. Super gas barrier of all-polymer multilayer thin films. *Macromolecules* **2011**, *44*, 1450–1459. [CrossRef]
142. Hamad, K.; Kaseem, M.; Ayyoob, M.; Joo, J.; Deri, F. Polylactic acid blends: The future of green, light and tough. *Prog. Polym. Sci.* **2018**, *85*, 83–127. [CrossRef]
143. Zaaba, N.F.; Ismail, H. A review on tensile and morphological properties of Poly(lactic acid) (PLA)/thermoplastic starch (TPS) blends. *Polym. Technol. Mater.* **2019**, *58*, 1945–1964. [CrossRef]
144. Koh, J.J.; Zhang, X.; He, C. Fully biodegradable Poly(lactic acid)/Starch blends: A review of toughening strategies. *Int. J. Biol. Macromol.* **2018**, *109*, 99–113. [CrossRef] [PubMed]
145. Arrieta, M.P.; Samper, M.D.; Aldas, M.; López, J. On the use of PLA-PHB blends for sustainable food packaging applications. *Materials* **2017**, *10*, 1008. [CrossRef] [PubMed]
146. Sancheti, I.; Katkar, P.; Thorat, S.; Radhakrishnan, S.; Kulkarni, M.B. PBAT/PLA/HNT filled blends for active food packaging. *Mater. Res. Express* **2020**, *6*, 125378. [CrossRef]
147. Khosravi, A.; Fereidoon, A.; Khorasani, M.M.; Naderi, G.; Ganjali, M.R.; Zarrintaj, P.; Saeb, M.R.; Gutiérrez, T.J. Soft and hard sections from cellulose-reinforced poly(lactic acid)-based food packaging films: A critical review. *Food Packag. Shelf Life* **2020**, *23*, 100429. [CrossRef]
148. D'Amico, D.A.; Iglesias Montes, M.L.; Manfredi, L.B.; Cyras, V.P. Fully bio-based and biodegradable polylactic acid/poly(3-hydroxybutirate) blends: Use of a common plasticizer as performance improvement strategy. *Polym. Test.* **2016**, *49*, 22–28. [CrossRef]
149. Lambert, S.; Wagner, M. Environmental performance of bio-based and biodegradable plastics: The road ahead. *Chem. Soc. Rev.* **2017**, *46*, 6855–6871. [CrossRef]
150. Thakur, V.K.; Thakur, M.K.; Kessler, M.R. Handbook of composites from renewable materials. In *Handbook of Composites from Renewable Materials*; John Wiley & Sons: New York, NY, USA, 2017; pp. 1–8. [CrossRef]
151. RameshKumar, S.; Shaiju, P.; O'Connor, K.E. Bio-based and biodegradable polymers—State-of-the-art, challenges and emerging trends. *Curr. Opin. Green Sustain. Chem.* **2020**, *21*, 75–81. [CrossRef]
152. Singh, A.A.; Genovese, M.E.; Mancini, G.; Marini, L.; Athanassiou, A. Green Processing Route for Polylactic Acid-Cellulose Fiber Biocomposites. *ACS Sustain. Chem. Eng.* **2020**, *8*, 4128–4136. [CrossRef]
153. Siakeng, R.; Jawaid, M.; Asim, M.; Fouad, H.; Awad, S.; Saba, N.; Siengchin, S. Flexural and Dynamic Mechanical Properties of Alkali-Treated Coir/Pineapple Leaf Fibres Reinforced Polylactic Acid Hybrid Biocomposites. *J. Bionic Eng.* **2021**, *18*, 1430–1438. [CrossRef]
154. De Azeredo, H.M.C. Nanocomposites for food packaging applications. *Food Res. Int.* **2009**, *42*, 1240–1253. [CrossRef]
155. Yang, W.; Fortunati, E.; Dominici, F.; Giovanale, G.; Mazzaglia, A.; Balestra, G.M.; Kenny, J.M.; Puglia, D. Synergic effect of cellulose and lignin nanostructures in PLA based systems for food antibacterial packaging. *Eur. Polym. J.* **2016**, *79*, 1–12. [CrossRef]
156. Yu, H.Y.; Zhang, H.; Song, M.L.; Zhou, Y.; Yao, J.; Ni, Q.Q. From Cellulose Nanospheres, Nanorods to Nanofibers: Various Aspect Ratio Induced Nucleation/Reinforcing Effects on Polylactic Acid for Robust-Barrier Food Packaging. *ACS Appl. Mater. Interfaces* **2017**, *9*, 43920–43938. [CrossRef] [PubMed]
157. Zhang, H.; Hortal, M.; Jordá-Beneyto, M.; Rosa, E.; Lara-Lledo, M.; Lorente, I. ZnO-PLA nanocomposite coated paper for antimicrobial packaging application. *LWT—Food Sci. Technol.* **2017**, *78*, 250–257. [CrossRef]
158. Shankar, S.; Wang, L.F.; Rhim, J.W. Incorporation of zinc oxide nanoparticles improved the mechanical, water vapor barrier, UV-light barrier, and antibacterial properties of PLA-based nanocomposite films. *Mater. Sci. Eng. C* **2018**, *93*, 289–298. [CrossRef] [PubMed]
159. Mahmoodi, A.; Ghodrati, S.; Khorasani, M. High-Strength, Low-Permeable, and Light-Protective Nanocomposite Films Based on a Hybrid Nanopigment and Biodegradable PLA for Food Packaging Applications. *ACS Omega* **2019**, *4*, 14947–14954. [CrossRef]
160. Pötschke, P.; Paul, D.R. Formation of co-continuous structures in melt-mixed immiscible polymer blends. *J. Macromol. Sci. Part C Polym. Rev.* **2003**, *43*, 87–141. [CrossRef]
161. Tol, R.T.; Groeninckx, G.; Vinckier, I.; Moldenaers, P.; Mewis, J. Phase morphology and stability of co-continuous (PPE/PS)/PA6 and PS/PA6 blends: Effect of rheology and reactive compatibilization. *Polymer* **2004**, *45*, 2587–2601. [CrossRef]
162. Omonov, T.S.; Harrats, C.; Groeninckx, G.; Moldenaers, P. Anisotropy and instability of the co-continuous phase morphology in uncompatibilized and reactively compatibilized polypropylene/polystyrene blends. *Polymer* **2007**, *48*, 5289–5302. [CrossRef]

163. Zolali, A.M.; Favis, B.D. Compatibilization and toughening of co-continuous ternary blends via partially wet droplets at the interface. *Polymer* **2017**, *114*, 277–288. [CrossRef]
164. Liu, Y.; Cao, L.; Yuan, D.; Chen, Y. Design of super-tough co-continuous PLA/NR/SiO2 TPVs with balanced stiffness-toughness based on reinforced rubber and interfacial compatibilization. *Compos. Sci. Technol.* **2018**, *165*, 231–239. [CrossRef]
165. Okubo, K.; Fujii, T.; Thostenson, E.T. Multi-scale hybrid biocomposite: Processing and mechanical characterization of bamboo fiber reinforced PLA with microfibrillated cellulose. *Compos. Part A Appl. Sci. Manuf.* **2009**, *40*, 469–475. [CrossRef]
166. Mukherjee, T.; Kao, N. PLA Based Biopolymer Reinforced with Natural Fibre: A Review. *J. Polym. Environ.* **2011**, *19*, 714–725. [CrossRef]
167. Xiao, L.; Mai, Y.; He, F.; Yu, L.; Zhang, L.; Tang, H.; Yang, G. Bio-based green composites with high performance from poly(lactic acid) and surface-modified microcrystalline cellulose. *J. Mater. Chem.* **2012**, *22*, 15732–15739. [CrossRef]
168. Qiang, T.; Wang, J.; Wolcott, M.P. Facile fabrication of 100% bio-based and degradable ternary cellulose/PHBV/PLA composites. *Materials* **2018**, *11*, 330. [CrossRef] [PubMed]
169. Bajwa, D.S.; Adhikari, S.; Shojaeiarani, J.; Bajwa, S.G.; Pandey, P.; Shanmugam, S.R. Characterization of bio-carbon and lignocellulosic fiber reinforced bio-composites with compatibilizer. *Constr. Build. Mater.* **2019**, *204*, 193–202. [CrossRef]
170. Peltola, H.; Immonen, K.; Johansson, L.S.; Virkajärvi, J.; Sandquist, D. Influence of pulp bleaching and compatibilizer selection on performance of pulp fiber reinforced PLA biocomposites. *J. Appl. Polym. Sci.* **2019**, *136*, 47955. [CrossRef]
171. Mohanty, A.K.; Vivekanandhan, S.; Pin, J.M.; Misra, M. Composites from renewable and sustainable resources: Challenges and innovations. *Science* **2018**, *362*, 536–542. [CrossRef]
172. Sanchez-Garcia, M.D.; Gimenez, E.; Lagaron, J.M. Morphology and barrier properties of solvent cast composites of thermoplastic biopolymers and purified cellulose fibers. *Carbohydr. Polym.* **2008**, *71*, 235–244. [CrossRef]
173. Luddee, M.; Pivsa-Art, S.; Sirisansaneeyakul, S.; Chiravootpechyen, P. Particle size of ground bacterial cellulose affecting mechanical, thermal, and moisture barrier properties of PLA/BC biocomposites. *Energy Procedia* **2014**, *56*, 211–218. [CrossRef]
174. Li, X.; Tabil, L.G.; Panigrahi, S. Chemical treatments of natural fiber for use in natural fiber-reinforced composites: A review. *J. Polym. Environ.* **2007**, *15*, 25–33. [CrossRef]
175. Goriparthi, B.K.; Suman, K.N.S.; Mohan Rao, N. Effect of fiber surface treatments on mechanical and abrasive wear performance of polylactide/jute composites. *Compos. Part A Appl. Sci. Manuf.* **2012**, *43*, 1800–1808. [CrossRef]
176. Bakar, N.; Chee, C.Y.; Abdullah, L.C.; Ratnam, C.T.; Azowa, N. Effect of methyl methacrylate grafted kenaf on mechanical properties of polyvinyl chloride/ethylene vinyl acetate composites. *Compos. Part A Appl. Sci. Manuf.* **2014**, *63*, 45–50. [CrossRef]
177. Mukherjee, T.; Sani, M.; Kao, N.; Gupta, R.K.; Quazi, N.; Bhattacharya, S. Improved dispersion of cellulose microcrystals in polylactic acid (PLA) based composites applying surface acetylation. *Chem. Eng. Sci.* **2013**, *101*, 655–662. [CrossRef]
178. Jing, M.; Che, J.; Xu, S.; Liu, Z.; Fu, Q. The effect of surface modification of glass fiber on the performance of poly(lactic acid) composites: Graphene oxide vs. silane coupling agents. *Appl. Surf. Sci.* **2018**, *435*, 1046–1056. [CrossRef]
179. Kale, R.D.; Gorade, V.G. Preparation of acylated microcrystalline cellulose using olive oil and its reinforcing effect on poly(lactic acid) films for packaging application. *J. Polym. Res.* **2018**, *25*, 81. [CrossRef]
180. Tee, Y.B.; Talib, R.A.; Abdan, K.; Chin, N.L.; Basha, R.K.; Md Yunos, K.F. Effect of aminosilane concentrations on the properties of Poly(Lactic Acid)/Kenaf-Derived Cellulose Composites. *Polym. Polym. Compos.* **2017**, *25*, 63–76. [CrossRef]
181. Ke, T.; Sun, X. Physical properties of poly(lactic acid) and starch composites with various blending ratios. *Cereal Chem.* **2000**, *77*, 761–768. [CrossRef]
182. Ke, T.; Sun, X. Effects of Moisture Content and Heat Treatment on the Physical Properties of Starch and Poly(lactic acid) Blends. *J. Appl. Polym. Sci.* **2001**, *81*, 3069–3082. [CrossRef]
183. Li, H.; Huneault, M.A. Crystallization of PLA/Thermoplastic Starch Blends. *Int. Polym. Process.* **2008**, *23*, 412–418. [CrossRef]
184. Wu, X.S. Effect of Glycerin and Starch Crosslinking on Molecular Compatibility of Biodegradable Poly(lactic acid)-Starch Composites. *J. Polym. Environ.* **2011**, *19*, 912–917. [CrossRef]
185. Wang, J.; Zhai, W.; Zheng, W. Poly(Ethylene Glycol) Grafted Starch Introducing a Novel Interphase in Poly(Lactic Acid)/Poly(Ethylene Glycol)/Starch Ternary Composites. *J. Polym. Environ.* **2012**, *20*, 528–539. [CrossRef]
186. Wang, N.; Yu, J.; Chang, P.R.; Ma, X. Influence of citric acid on the properties of glycerol-plasticized dry starch (DTPS) and DTPS/poly(lactic acid) blends. *Starch/Staerke* **2007**, *59*, 409–417. [CrossRef]
187. Zhang, J.F.; Sun, X. Mechanical properties of poly(lactic acid)/starch composites compatibilized by maleic anhydride. *Biomacromolecules* **2004**, *5*, 1446–1451. [CrossRef]
188. Collazo-Bigliardi, S.; Ortega-Toro, R.; Chiralt, A. Using lignocellulosic fractions of coffee husk to improve properties of compatibilised starch-PLA blend films. *Food Packag. Shelf Life* **2019**, *22*, 100423. [CrossRef]
189. Wang, H.; Sun, X.; Seib, P. Mechanical properties of Poly(lactic acid) and wheat starch blends with methylenediphenyl diisocyanate. *J. Appl. Polym. Sci.* **2002**, *84*, 1257–1262. [CrossRef]
190. Shirai, M.A.; Grossmann, M.V.E.; Mali, S.; Yamashita, F.; Garcia, P.S.; Müller, C.M.O. Development of biodegradable flexible films of starch and poly(lactic acid) plasticized with adipate or citrate esters. *Carbohydr. Polym.* **2013**, *92*, 19–22. [CrossRef]
191. Shirai, M.A.; Zanela, J.; Kunita, M.H.; Pereira, G.M.; Rubira, A.F.; Müller, C.M.O.; Grossmann, M.V.E.; Yamashita, F. Influence of Carboxylic Acids on Poly(lactic acid)/Thermoplastic Starch Biodegradable Sheets Produced by Calendering–Extrusion. *Adv. Polym. Technol.* **2018**, *37*, 332–338. [CrossRef]

192. Muller, J.; González-Martínez, C.; Chiralt, A. Poly(lactic acid) (PLA) and starch bilayer films, containing cinnamaldehyde, obtained by compression moulding. *Eur. Polym. J.* **2017**, *95*, 56–70. [CrossRef]
193. Zembouai, I.; Kaci, M.; Bruzaud, S.; Benhamida, A.; Corre, Y.M.; Grohens, Y. A study of morphological, thermal, rheological and barrier properties of Poly(3-hydroxybutyrate-Co-3-Hydroxyvalerate)/polylactide blends prepared by melt mixing. *Polym. Test.* **2013**, *32*, 842–851. [CrossRef]
194. Ragaert, P.; Buntinx, M.; Maes, C.; Vanheusden, C.; Peeters, R.; Wang, S.; D'hooge, D.R.; Cardon, L. Polyhydroxyalkanoates for Food Packaging Applications. In *Reference Module in Food Science*; Elsevier: Amsterdam, The Netherlands, 2019. [CrossRef]
195. Masood, F. Polyhydroxyalkanoates in the Food Packaging Industry. In *Nanotechnology Applications in Food*; Academic Press: Cambridge, MA, USA, 2017; pp. 153–177. [CrossRef]
196. Naser, A.Z.; Deiab, I.; Darras, B.M. Poly(lactic acid) (PLA) and polyhydroxyalkanoates (PHAs), green alternatives to petroleum-based plastics: A review. *RSC Adv.* **2021**, *11*, 17151–17196. [CrossRef]
197. Urtuvia, V.; Villegas, P.; González, M.; Seeger, M. Bacterial production of the biodegradable plastics polyhydroxyalkanoates. *Int. J. Biol. Macromol.* **2014**, *70*, 208–213. [CrossRef] [PubMed]
198. Loureiro, N.C.; Esteves, J.L.; Viana, J.C.; Ghosh, S. Mechanical characterization of polyhydroxyalkanoate and poly(lactic acid) blends. *J. Thermoplast. Compos. Mater.* **2015**, *28*, 195–213. [CrossRef]
199. Ecker, J.V.; Burzic, I.; Haider, A.; Hild, S.; Rennhofer, H. Improving the impact strength of PLA and its blends with PHA in fused layer modelling. *Polym. Test.* **2019**, *78*, 105929. [CrossRef]
200. Madbouly, S.A. Bio-based polyhydroxyalkanoates blends and composites. *Phys. Sci. Rev.* **2021**. [CrossRef]
201. Modi, S.; Koelling, K.; Vodovotz, Y. Miscibility of poly(3-hydroxybutyrate-co-3-hydroxyvalerate) with high molecular weight poly(lactic acid) blends determined by thermal analysis. *J. Appl. Polym. Sci.* **2012**, *124*, 3074–3081. [CrossRef]
202. Jost, V.; Kopitzky, R. Blending of polyhydroxybutyrate-co-valerate with polylactic acid for packaging applications—Reflections on miscibility and effects on the mechanical and barrier properties. *Chem. Biochem. Eng. Q.* **2015**, *29*, 221–246. [CrossRef]
203. Rafieian, S.; Mirzadeh, H.; Mahdavi, H.; Masoumi, M.E. A review on nanocomposite hydrogels and their biomedical applications. *IEEE J. Sel. Top. Quantum Electron.* **2019**, *26*, 154–174. [CrossRef]
204. Shirdar, M.R.; Farajpour, N.; Shahbazian-Yassar, R.; Shokuhfar, T. Nanocomposite materials in orthopedic applications. *Front. Chem. Sci. Eng.* **2019**, *13*, 1–13. [CrossRef]
205. Yu, X.; Lin, K. Application of nanocomposite hydrogels in bone tissue engineering. *Chin. J. Tissue Eng. Res.* **2020**, *24*, 5441–5446. [CrossRef]
206. Kumar, N.; Kaur, P.; Bhatia, S. Advances in bio-nanocomposite materials for food packaging: A review. *Nutr. Food Sci.* **2017**, *47*, 591–606. [CrossRef]
207. Bharathi, S.K.V.; Murugesan, P.; Moses, J.A.; Anandharamakrishnan, C. Recent Trends in Nanocomposite Packaging Materials. In *Innovative Food Processing Technologies: A Comprehensive Review*; Elsevier: Amsterdam, The Netherlands, 2020; pp. 731–755. [CrossRef]
208. Tang, M.C.; Agarwal, S.; Alsewailem, F.D.; Choi, H.J.; Gupta, R.K. A model for water vapor permeability reduction in poly(lactic acid) and nanoclay nanocomposites. *J. Appl. Polym. Sci.* **2018**, *135*, 46506. [CrossRef]
209. Koh, H.C.; Park, J.S.; Jeong, M.A.; Hwang, H.Y.; Hong, Y.T.; Ha, S.Y.; Nam, S.Y. Preparation and gas permeation properties of biodegradable polymer/layered silicate nanocomposite membranes. *Desalination* **2008**, *233*, 201–209. [CrossRef]
210. Kiliaris, P.; Papaspyrides, C.D. Polymer/layered silicate (clay) nanospyrides: An overview of flame retardancy. *Prog. Polym. Sci.* **2010**, *35*, 902–958. [CrossRef]
211. Coppola, B.; Cappetti, N.; Di Maio, L.; Scarfato, P.; Incarnato, L. 3D printing of PLA/clay nanocomposites: Influence of printing temperature on printed samples properties. *Materials* **2018**, *11*, 1947. [CrossRef]
212. Gauvin, F.; Robert, M. Durability study of vinylester/silicate nanocomposites for civil engineering applications. *Polym. Degrad. Stab.* **2015**, *121*, 359–368. [CrossRef]
213. Yuan, P.; Tan, D.; Annabi-Bergaya, F. Properties and applications of halloysite nanotubes: Recent research advances and future prospects. *Appl. Clay Sci.* **2015**, *112*, 75–93. [CrossRef]
214. Bhatia, A.; Gupta, R.K.; Bhattacharya, S.N.; Choi, H.J. Effect of clay on thermal, mechanical and gas barrier properties of biodegradable poly(lactic acid)/poly(butylene succinate) (PLA/PBS) nanocomposites. *Int. Polym. Process.* **2009**, *25*, 5–14. [CrossRef]
215. Svagan, A.J.; Åkesson, A.; Cárdenas, M.; Bulut, S.; Knudsen, J.C.; Risbo, J.; Plackett, D. Transparent films based on PLA and montmorillonite with tunable oxygen barrier properties. *Biomacromolecules* **2012**, *13*, 397–405. [CrossRef]
216. Picard, E.; Espuche, E.; Fulchiron, R. Effect of an organo-modified montmorillonite on PLA crystallization and gas barrier properties. *Appl. Clay Sci.* **2011**, *53*, 58–65. [CrossRef]
217. Othman, S.H.; Ling, H.N.; Talib, R.A.; Naim, M.N.; Risyon, N.P.; Saifullah, M. PLA/MMT and PLA/halloysite bio-nanocomposite films: Mechanical, barrier, and transparency. *J. Nano Res.* **2019**, *59*, 77–93. [CrossRef]
218. Dor, M.; Levi-Kalisman, Y.; Day-Stirrat, R.J.; Mishael, Y.; Emmanuel, S. Assembly of clay mineral platelets, tactoids, and aggregates: Effect of mineral structure and solution salinity. *J. Colloid Interface Sci.* **2020**, *566*, 163–170. [CrossRef] [PubMed]
219. Jollands, M.; Gupta, R.K. Effect of mixing conditions on mechanical properties of polylactide/montmorillonite clay nanocomposites. *J. Appl. Polym. Sci.* **2010**, *118*, 1489–1493. [CrossRef]

220. Ozdemir, E.; Hacaloglu, J. Characterizations of PLA-PEG blends involving organically modified montmorillonite. *J. Anal. Appl. Pyrolysis* **2017**, *127*, 343–349. [CrossRef]
221. Ozdemir, E.; Hacaloglu, J. Polylactide/organically modified montmorillonite composite fibers. *J. Anal. Appl. Pyrolysis* **2017**, *124*, 186–194. [CrossRef]
222. Tan, L.; He, Y.; Qu, J. Structure and properties of Polylactide/Poly(butylene succinate)/Organically Modified Montmorillonite nanocomposites with high-efficiency intercalation and exfoliation effect manufactured via volume pulsating elongation flow. *Polymer* **2019**, *180*, 121656. [CrossRef]
223. Fabia, J.; Gawłowski, A.; Rom, M.; Ślusarczyk, C.; Brzozowska-Stanuch, A.; Sieradzka, M. PET fibers modified with cloisite nanoclay. *Polymer* **2020**, *12*, 774. [CrossRef]
224. Lietz, S.; Yang, J.L.; Bosch, E.; Sandler, J.K.W.; Zhang, Z.; Altstädt, V. Improvement of the mechanical properties and creep resistance of SBS block copolymers by nanoclay fillers. *Macromol. Mater. Eng.* **2007**, *292*, 23–32. [CrossRef]
225. Mallakpour, S.; Shahangi, V. Bio-modification of cloisite Na+ with chiral L-leucine and preparation of new poly(Vinyl Alcohol)/organo-nanoclay bionanocomposite films. *Synth. React. Inorg. Met. Nano-Met. Chem.* **2013**, *43*, 966–971. [CrossRef]
226. Aulin, C.; Salazar-Alvarez, G.; Lindström, T. High strength, flexible and transparent nanofibrillated cellulose-nanoclay biohybrid films with tunable oxygen and water vapor permeability. *Nanoscale* **2012**, *4*, 6622–6628. [CrossRef]
227. Planellas, M.; Sacristán, M.; Rey, L.; Olmo, C.; Aymamí, J.; Casas, M.T.; Del Valle, L.J.; Franco, L.; Puiggalí, J. Micro-molding with ultrasonic vibration energy: New method to disperse nanoclays in polymer matrices. *Ultrason. Sonochem.* **2014**, *21*, 1557–1569. [CrossRef]
228. Rhim, J.W.; Hong, S.I.; Ha, C.S. Tensile, water vapor barrier and antimicrobial properties of PLA/nanoclay composite films. *LWT-Food Sci. Technol.* **2009**, *42*, 612–617. [CrossRef]
229. Bouakaz, B.S.; Pillin, I.; Habi, A.; Grohens, Y. Synergy between fillers in organomontmorillonite/graphene-PLA nanocomposites. *Appl. Clay Sci.* **2015**, *116*, 69–77. [CrossRef]
230. Tenn, N.; Follain, N.; Soulestin, J.; Crétois, R.; Bourbigot, S.; Marais, S. Effect of nanoclay hydration on barrier properties of PLA/montmorillonite based nanocomposites. *J. Phys. Chem. C* **2013**, *117*, 12117–12135. [CrossRef]
231. Bartel, M.; Remde, H.; Bohn, A.; Ganster, J. Barrier properties of poly(lactic acid)/cloisite 30B composites and their relation between oxygen permeability and relative humidity. *J. Appl. Polym. Sci.* **2017**, *134*, 1–10. [CrossRef]
232. Kim, H.K.; Kim, S.J.; Lee, H.S.; Choi, J.H.; Jeong, C.M.; Sung, M.H.; Park, S.H.; Park, H.J. Mechanical and barrier properties of poly(lactic acid) films coated by nanoclay-ink composition. *J. Appl. Polym. Sci.* **2013**, *127*, 3823–3829. [CrossRef]
233. Najafi, N.; Heuzey, M.C.; Carreau, P.J. Polylactide (PLA)-clay nanocomposites prepared by melt compounding in the presence of a chain extender. *Compos. Sci. Technol.* **2012**, *72*, 608–615. [CrossRef]
234. Guo, Y.; Yang, K.; Zuo, X.; Xue, Y.; Marmorat, C.; Liu, Y.; Chang, C.C.; Rafailovich, M.H. Effects of clay platelets and natural nanotubes on mechanical properties and gas permeability of Poly(lactic acid) nanocomposites. *Polymer* **2016**, *83*, 246–259. [CrossRef]
235. Brayner, R.; Coradin, T.; Fiévet, F. Preface. In *Nanomaterials: A Danger or a Promise? A Chemical and Biological Perspective*; Springer: Berlin/Heidelberg, Germany, 2013; pp. v–vi. [CrossRef]
236. de Freitas, F.A.; Pessoa, W.A.G.; Lira, M.S.F.; Nobre, F.X.; Takeno, M.L. Nanomaterials obtained from renewable resources and their application as catalysts in biodiesel synthesis. *Nanomaterials* **2021**, 481–509. [CrossRef]
237. Kuswandi, B. Environmental friendly food nano-packaging. *Environ. Chem. Lett.* **2017**, *15*, 205–221. [CrossRef]
238. Othman, S.H. Bio-nanocomposite Materials for Food Packaging Applications: Types of Biopolymer and Nano-sized Filler. *Agric. Agric. Sci. Procedia* **2014**, *2*, 296–303. [CrossRef]
239. Gadhave, R.V.; Das, A.; Mahanwar, P.A.; Gadekar, P.T. Starch Based Bio-Plastics: The Future of Sustainable Packaging. *Open J. Polym. Chem.* **2018**, *8*, 21–33. [CrossRef]
240. Mohammadi, M.; Mirabzadeh, S.; Shahvalizadeh, R.; Hamishehkar, H. Development of novel active packaging films based on whey protein isolate incorporated with chitosan nanofiber and nano-formulated cinnamon oil. *Int. J. Biol. Macromol.* **2020**, *149*, 11–20. [CrossRef] [PubMed]
241. Sanchez-Garcia, M.D.; Lagaron, J.M. On the use of plant cellulose nanowhiskers to enhance the barrier properties of polylactic acid. *Cellulose* **2010**, *17*, 987–1004. [CrossRef]
242. Martínez-Sanz, M.; Lopez-Rubio, A.; Lagaron, J.M. High-barrier coated bacterial cellulose nanowhiskers films with reduced moisture sensitivity. *Carbohydr. Polym.* **2013**, *98*, 1072–1082. [CrossRef] [PubMed]
243. Espino-Pérez, E.; Bras, J.; Ducruet, V.; Guinault, A.; Dufresne, A.; Domenek, S. Influence of chemical surface modification of cellulose nanowhiskers on thermal, mechanical, and barrier properties of poly(lactide) based bionanocomposites. *Eur. Polym. J.* **2013**, *49*, 3144–3154. [CrossRef]
244. Thuy, V.T.T.; Hao, L.T.; Jeon, H.; Koo, J.M.; Park, J.; Lee, E.S.; Hwang, S.Y.; Choi, S.; Park, J.; Oh, D.X. Sustainable, self-cleaning, transparent, and moisture/oxygen-barrier coating films for food packaging. *Green Chem.* **2021**, *23*, 2658–2667. [CrossRef]
245. Bondeson, D.; Oksman, K. Dispersion and characteristics of surfactant modified cellulose whiskers nanocomposites. *Compos. Interfaces* **2007**, *14*, 617–630. [CrossRef]
246. Lee, J.H.; Park, S.H.; Kim, S.H. Surface modification of cellulose nanowhiskers and their reinforcing effect in polylactide. *Macromol. Res.* **2014**, *22*, 424–430. [CrossRef]

247. Sheng, K.; Zhang, S.; Qian, S.; Fontanillo Lopez, C.A. High-toughness PLA/Bamboo cellulose nanowhiskers bionanocomposite strengthened with silylated ultrafine bamboo-char. *Compos. Part B Eng.* **2019**, *165*, 174–182. [CrossRef]
248. Nair, S.S.; Chen, H.; Peng, Y.; Huang, Y.; Yan, N. Polylactic Acid Biocomposites Reinforced with Nanocellulose Fibrils with High Lignin Content for Improved Mechanical, Thermal, and Barrier Properties. *ACS Sustain. Chem. Eng.* **2018**, *6*, 10058–10068. [CrossRef]
249. Rigotti, D.; Checchetto, R.; Tarter, S.; Caretti, D.; Rizzuto, M.; Fambri, L.; Pegoretti, A. Polylactic acid-lauryl functionalized nanocellulose nanocomposites: Microstructural, thermo-mechanical and gas transport properties. *Express Polym. Lett.* **2019**, *13*, 858–876. [CrossRef]
250. Arrieta, M.P.; Peponi, L.; López, D.; López, J.; Kenny, J.M. An overview of nanoparticles role in the improvement of barrier properties of bioplastics for food packaging applications. *Food Packag.* **2017**, 391–424. [CrossRef]
251. Cerqueira, M.A.; Vicente, A.A.; Pastrana, L.M. Nanotechnology in Food Packaging: Opportunities and Challenges. *Nanomater. Food Packag.* **2018**, 1–11. [CrossRef]
252. Pilić, B.M.; Radusin, T.I.; Ristić, I.S.; Silvestre, C.; Lazić, V.L.; Baloš, S.S.; Duraccio, D. Uticaj dodatka hidrofobnih nanočestica silicijum(IV)-oksida na svojstva poli(mlečne kiseline). *Hem. Ind.* **2016**, *70*, 73–80. [CrossRef]
253. Bikiaris, D.N.; Triantafyllidis, K.S. HDPE/Cu-nanofiber nanocomposites with enhanced antibacterial and oxygen barrier properties appropriate for food packaging applications. *Mater. Lett.* **2013**, *93*, 1–4. [CrossRef]
254. Keith Roper, D.; Berry, K.R.; Russell, A.G.; Blake, P.A.; Keith Roper, D. Gold nanoparticles reduced in situ and dispersed in polymer thin films: Optical and thermal properties. *Nanotechnology* **2012**, *23*, 375703. [CrossRef]
255. Wu, Z.; Deng, W.; Luo, J.; Deng, D. Multifunctional nano-cellulose composite films with grape seed extracts and immobilized silver nanoparticles. *Carbohydr. Polym.* **2019**, *205*, 447–455. [CrossRef]
256. Cheng, J.; Wang, Y.; Hu, L.; Liu, N.; Xu, J.; Zhou, J. Using lantern Zn/Co-ZIF nanoparticles to provide channels for CO_2 permeation through PEO-based MMMs. *J. Memb. Sci.* **2020**, *597*, 117644. [CrossRef]
257. Bardhan, R.; Ruminski, A.M.; Brand, A.; Urban, J.J. Magnesium nanocrystal-polymer composites: A new platform for designer hydrogen storage materials. *Energy Environ. Sci.* **2011**, *4*, 4882–4895. [CrossRef]
258. Madaeni, S.S.; Badieh, M.M.S.; Vatanpour, V.; Ghaemi, N. Effect of titanium dioxide nanoparticles on polydimethylsiloxane/polyethersulfone composite membranes for gas separation. *Polym. Eng. Sci.* **2012**, *52*, 2664–2674. [CrossRef]
259. Ng, L.Y.; Mohammad, A.W.; Leo, C.P.; Hilal, N. Polymeric membranes incorporated with metal/metal oxide nanoparticles: A comprehensive review. *Desalination* **2013**, *308*, 15–33. [CrossRef]
260. Nikolic, M.V.; Vasiljevic, Z.Z.; Auger, S.; Vidic, J. Metal oxide nanoparticles for safe active and intelligent food packaging. *Trends Food Sci. Technol.* **2021**, *116*, 655–668. [CrossRef]
261. Lizundia, E.; Armentano, I.; Luzi, F.; Bertoglio, F.; Restivo, E.; Visai, L.; Torre, L.; Puglia, D. Synergic Effect of Nanolignin and Metal Oxide Nanoparticles into Poly(L-lactide) Bionanocomposites: Material Properties, Antioxidant Activity, and Antibacterial Performance. *ACS Appl. Bio Mater.* **2020**, *3*, 5263–5274. [CrossRef] [PubMed]
262. Pantani, R.; Gorrasi, G.; Vigliotta, G.; Murariu, M.; Dubois, P. PLA-ZnO nanocomposite films: Water vapor barrier properties and specific end-use characteristics. *Eur. Polym. J.* **2013**, *49*, 3471–3482. [CrossRef]
263. Marra, A.; Rollo, G.; Cimmino, S.; Silvestre, C. Assessment on the effects of ZnO and Coated ZnO particles on iPP and PLA properties for application in food packaging. *Coatings* **2017**, *7*, 29. [CrossRef]
264. Swaroop, C.; Shukla, M. Nano-magnesium oxide reinforced polylactic acid biofilms for food packaging applications. *Int. J. Biol. Macromol.* **2018**, *113*, 729–736. [CrossRef]
265. Maharana, T.; Pattanaik, S.; Routaray, A.; Nath, N.; Sutar, A.K. Synthesis and characterization of poly(lactic acid) based graft copolymers. *React. Funct. Polym.* **2015**, *93*, 47–67. [CrossRef]
266. Li, Z.; Tan, B.H.; Lin, T.; He, C. Recent advances in stereocomplexation of enantiomeric PLA-based copolymers and applications. *Prog. Polym. Sci.* **2016**, *62*, 22–72. [CrossRef]
267. Stefaniak, K.; Masek, A. Green Copolymers Based on Poly(Lactic Acid)—Short Review. *Materials* **2021**, *14*, 5254. [CrossRef]
268. Oh, J.K. Polylactide (PLA)-based amphiphilic block copolymers: Synthesis, self-assembly, and biomedical applications. *Soft Matter* **2011**, *7*, 5096–5108. [CrossRef]
269. Munim, S.A.; Raza, Z.A. Poly(lactic acid) based hydrogels: Formation, characteristics and biomedical applications. *J. Porous Mater.* **2019**, *26*, 881–901. [CrossRef]
270. Giammona, G.; Craparo, E.F. Biomedical applications of polylactide (PLA) and its copolymers. *Molecules* **2018**, *23*, 980. [CrossRef]
271. Ebrahimi, F.; Ramezani Dana, H. Poly lactic acid (PLA) polymers: From properties to biomedical applications. *Int. J. Polym. Mater. Polym. Biomater.* **2021**, 1–14. [CrossRef]
272. Ali, F.B.; Kang, D.J.; Kim, M.P.; Cho, C.H.; Kim, B.J. Synthesis of biodegradable and flexible, polylactic acid based, thermoplastic polyurethane with high gas barrier properties. *Polym. Int.* **2014**, *63*, 1620–1626. [CrossRef]
273. Genovese, L.; Soccio, M.; Lotti, N.; Gazzano, M.; Siracusa, V.; Salatelli, E.; Balestra, F.; Munari, A. Design of biobased PLLA triblock copolymers for sustainable food packaging: Thermo-mechanical properties, gas barrier ability and compostability. *Eur. Polym. J.* **2017**, *95*, 289–303. [CrossRef]
274. Yuk, J.S.; Mo, E.; Kim, S.; Jeong, H.; Gwon, H.; Kim, N.K.; Kim, Y.W.; Shin, J. Thermoplastic Superelastomers Based on Poly(isobutylene)-graft-Poly(L-lactide) Copolymers: Enhanced Thermal Stability, Tunable Tensile Strength, and Gas Barrier Property. *Macromolecules* **2020**, *53*, 2503–2515. [CrossRef]

275. Coltelli, M.B.; Mallegni, N.; Rizzo, S.; Fiori, S.; Signori, F.; Lazzeri, A. Compatibilization of poly(Lactic acid) (PLA)/plasticized cellulose acetate extruded blends through the addition of reactively extruded comb copolymers. *Molecules* **2021**, *26*, 2006. [CrossRef] [PubMed]
276. Zahir, L.; Kida, T.; Tanaka, R.; Nakayama, Y.; Shiono, T.; Kawasaki, N.; Yamano, N.; Nakayama, A. Synthesis, properties, and biodegradability of thermoplastic elastomers made from 2-methyl-1,3-propanediol, glutaric acid and lactide. *Life* **2021**, *11*, 43. [CrossRef]
277. Vinet, L.; Zhedanov, A. A 'missing' family of classical orthogonal polynomials. *J. Phys. A Math. Theor.* **2011**, *44*, 085201. [CrossRef]
278. Satterthwaite, K. Plastics Based on Styrene. In *Brydson's Plastics Materials*, 8th ed.; Butterworth-Heinemann: Oxford, UK, 2017; pp. 311–328. [CrossRef]
279. Stipa, P.; Marano, S.; Galeazzi, R.; Minnelli, C.; Mobbili, G.; Laudadio, E. Prediction of drug-carrier interactions of PLA and PLGA drug-loaded nanoparticles by molecular dynamics simulations. *Eur. Polym. J.* **2021**, *147*, 110292. [CrossRef]
280. Duez, Q.; Metwally, H.; Hoyas, S.; Lemaur, V.; Cornil, J.; De Winter, J.; Konermann, L.; Gerbaux, P. Effects of electrospray mechanisms and structural relaxation on polylactide ion conformations in the gas phase: Insights from ion mobility spectrometry and molecular dynamics simulations. *Phys. Chem. Chem. Phys.* **2020**, *22*, 4193–4204. [CrossRef] [PubMed]
281. Sun, D.; Zhou, J. Molecular simulation of oxygen sorption and diffusion in the Poly(lactic acid). *Chin. J. Chem. Eng.* **2013**, *21*, 301–309. [CrossRef]
282. Pandey, Y.N.; Doxastakis, M. Detailed atomistic Monte Carlo simulations of a polymer melt on a solid surface and around a nanoparticle. *J. Chem. Phys.* **2012**, *136*, 094901. [CrossRef] [PubMed]
283. Stipa, P.; Marano, S.; Galeazzi, R.; Minnelli, C.; Laudadio, E. Molecular dynamics simulations of quinine encapsulation into biodegradable nanoparticles: A possible new strategy against Sars-CoV-2. *Eur. Polym. J.* **2021**, *158*, 110685. [CrossRef] [PubMed]
284. Ebadi-Dehaghani, H.; Barikani, M.; Khonakdar, H.A.; Jafari, S.H.; Wagenknecht, U.; Heinrich, G. On O_2 gas permeability of PP/PLA/clay nanocomposites: A molecular dynamic simulation approach. *Polym. Test.* **2015**, *45*, 139–151. [CrossRef]
285. Wldom, B. Some topics in the theory of fluids. *J. Chem. Phys.* **1963**, *39*, 2808–2812. [CrossRef]
286. Xiang, T.X.; Anderson, B.D. Water uptake, distribution, and mobility in amorphous poly(D,L-Lactide) by molecular dynamics simulation. *J. Pharm. Sci.* **2014**, *103*, 2759–2771. [CrossRef]
287. Jarvas, G. Evaporation Models for Multicomponent Mixtures. Ph.D. Thesis, University of Pannonia, Veszprem, Hungary, 2012; p. 94.
288. Charati, S.G.; Stern, S.A. Diffusion of gases in silicone polymers: Molecular dynamics simulations. *Macromolecules* **1998**, *31*, 5529–5535. [CrossRef]
289. Guvench, O.; MacKerell, A.D. Comparison of protein force fields for molecular dynamics simulations. *Methods Mol. Biol.* **2008**, *443*, 63–88. [CrossRef]
290. Kmiecik, S.; Gront, D.; Kolinski, M.; Wieteska, L.; Dawid, A.E.; Kolinski, A. Coarse-Grained Protein Models and Their Applications. *Chem. Rev.* **2016**, *116*, 7898–7936. [CrossRef]
291. Ingólfsson, H.I.; Lopez, C.A.; Uusitalo, J.J.; de Jong, D.H.; Gopal, S.M.; Periole, X.; Marrink, S.J. The power of coarse graining in biomolecular simulations. *Wiley Interdiscip. Rev. Comput. Mol. Sci.* **2014**, *4*, 225–248. [CrossRef] [PubMed]
292. Groot, R.D.; Warren, P.B. Dissipative particle dynamics: Bridging the gap between atomistic and mesoscopic simulation. *J. Chem. Phys.* **1997**, *107*, 4423–4435. [CrossRef]
293. Marrink, S.J.; Risselada, H.J.; Yefimov, S.; Tieleman, D.P.; De Vries, A.H. The MARTINI force field: Coarse grained model for biomolecular simulations. *J. Phys. Chem. B* **2007**, *111*, 7812–7824. [CrossRef] [PubMed]
294. Maiti, A.; McGrother, S. Bead-bead interaction parameters in dissipative particle dynamics: Relation to bead-size, solubility parameter, and surface tension. *J. Chem. Phys.* **2004**, *120*, 1594–1601. [CrossRef] [PubMed]
295. Kamrani, S.M.E.; Hadizadeh, F. A coarse-grain MD (molecular dynamic) simulation of PCL–PEG and PLA–PEG aggregation as a computational model for prediction of the drug-loading efficacy of doxorubicin. *J. Biomol. Struct. Dyn.* **2019**, *37*, 4215–4221. [CrossRef]

Article

Process-Induced Morphology of Poly(Butylene Adipate Terephthalate)/Poly(Lactic Acid) Blown Extrusion Films Modified with Chain-Extending Cross-Linkers

Juliana V. C. Azevedo [1,2,3,4], Esther Ramakers-van Dorp [2], Roman Grimmig [2], Berenika Hausnerova [1,4,*] and Bernhard Möginger [2]

1. Faculty of Technology, Tomas Bata University in Zlín, Vavreckova 275, 760 01 Zlín, Czech Republic; juliana.azevedo@bio-fed.com
2. Department of Natural Sciences, University of Applied Sciences Bonn-Rhein-Sieg, von Liebig Str. 20, 53359 Rheinbach, Germany; esther.vandorp@h-brs.de (E.R.-v.D.); roman.grimmig@h-brs.de (R.G.); bernhard.moeginger@h-brs.de (B.M.)
3. BIO-FED, Branch of AKRO-PLASTIC GmbH, BioCampus Cologne, Nattermannallee 1, 50829 Köln, Germany
4. Centre of Polymer Systems, University Institute, Tomas Bata University in Zlín, Nam. T.G. Masaryka 5555, 760 01 Zlín, Czech Republic
* Correspondence: hausnerova@utb.cz

Abstract: Process-induced changes in the morphology of biodegradable polybutylene adipate terephthalate (PBAT) and polylactic acid (PLA) blends modified with various multifunctional chain-extending cross-linkers (CECLs) are presented. The morphology of unmodified and modified films produced with blown film extrusion is examined in an extrusion direction (ED) and a transverse direction (TD). While FTIR analysis showed only small peak shifts indicating that the CECLs modify the molecular weight of the PBAT/PLA blend, SEM investigations of the fracture surfaces of blown extrusion films revealed their significant effect on the morphology formed during the processing. Due to the combined shear and elongation deformation during blown film extrusion, rather spherical PLA islands were partly transformed into long fibrils, which tended to decay to chains of elliptical islands if cooled slowly. The CECL introduction into the blend changed the thickness of the PLA fibrils, modified the interface adhesion, and altered the deformation behavior of the PBAT matrix from brittle to ductile. The results proved that CECLs react selectively with PBAT, PLA, and their interface. Furthermore, the reactions of CECLs with PBAT/PLA induced by the processing depended on the deformation directions (ED and TD), thus resulting in further non-uniformities of blown extrusion films.

Keywords: poly(butylene adipate terephthalate); poly(lactic acid); chain-extending cross-linker; process-induced morphology; blown film extrusion

Citation: Azevedo, J.V.C.; Ramakers-van Dorp, E.; Grimmig, R.; Hausnerova, B.; Möginger, B. Process-Induced Morphology of Poly(Butylene Adipate Terephthalate)/Poly(Lactic Acid) Blown Extrusion Films Modified with Chain-Extending Cross-Linkers. *Polymers* 2022, 14, 1939. https://doi.org/10.3390/polym14101939

Academic Editors: Alexey Iordanskii, Valentina Siracusa, Michelina Soccio and Nadia Lotti

Received: 19 April 2022
Accepted: 3 May 2022
Published: 10 May 2022

Publisher's Note: MDPI stays neutral with regard to jurisdictional claims in published maps and institutional affiliations.

Copyright: © 2022 by the authors. Licensee MDPI, Basel, Switzerland. This article is an open access article distributed under the terms and conditions of the Creative Commons Attribution (CC BY) license (https://creativecommons.org/licenses/by/4.0/).

1. Introduction

The largest market in the plastics industry is the packaging segment, with more than 40% of plastics demand in Europe [1]. Fifty percent of all goods are packed in plastics [1]. Blown film extrusion is the most important industrial manufacturing process of polymeric films [2,3]. In this process, a molten polymer is extruded into a tube shape and subsequently drawn by nip rollers in an extrusion direction (ED). Simultaneously, the extruded melt is blown by injecting air to a substantially larger tube ratio [4–7]. As the polymer melt is drawn in both the extrusion and transverse directions (TD), the film blowing process represents a biaxial elongational flow process. The orientation of the macromolecules and the final morphology of blown films depend strongly on the chosen process parameters [2–7].

The majority of biodegradable blends are based on polybutylene adipate terephthalate (PBAT) and polylactic acid (PLA) compounds [8–11]. Their chemical reactivity is typically governed by ester, amide, and ether functional groups. PBAT is a random copolymer of butylene adipate and terephthalate, which owes its biodegradability to the butylene adipate groups and its stability and mechanical strength to the terephthalate groups [11–14]. PLA is biodegradable and entirely renewable if it originates from starch [15–18]. It often exhibits a brittle behavior, and therefore it is inappropriate for applications requiring high deformation strains [19,20]. Therefore, blending with ductile polymers such as PBAT is a reasonable approach.

Investigations on PBAT/PLA blends showed that the interfacial compatibility between PLA and PBAT is poor but can be improved by compatibilizers [21,22]. Recent developments showed that chain-extending cross-linkers (CECL) might increase melt strength, thermal stability, and phase compatibility of noncompatible polymer blends. However, only a few investigations deal with CECLs' influence on PBAT/PLA blends [16,21,23–27].

Chiu et al. [16] showed on injection molded PBAT/PLA blends that the annealed PLA has a brittle and low-deformed breaking structure, and break tracks presented a considerable acute angle. The PBAT also exhibited a brittle break, but its break angle was alleviative, and its break tracks were longer than for the PLA. The break cross-sections of PBAT/PLA (30/70) and PBAT/PLA (50/50) were similar to those of PLA. The PBAT/PLA (30/70) presented an irregular layer break cross-section with PLA spheroid-dispersed in the PBAT continuous phase, and the PBAT/PLA (50/50) showed a directive-layer break cross-section.

Wang et al. [21] found that an epoxy-terminated branched polymer (ETBP) enhances the interfacial compatibility and mechanical properties of PBAT/PLA compounds. PBAT was dispersed in the continuous PLA phase in droplet form. The phase separation structure between PLA and PBAT could be seen. The interface image between the two phases was clear and loosely bonded, showing a sea-island structure. The average size of PBAT particles in PBAT/PLA blends was 2.87 µm, indicating a typical thermodynamically immiscible system. After the addition of ETBP, the size of the dispersed PBAT particles decreased (the average size reduced by up to 0.38 µm). Moreover, the interface between PLA and PBAT became fuzzy as more PLA and PBAT were combined together. With the increase of ETBP, the tensile fracture surface became ductile, and the sea island structure nearly disappeared, indicating that the addition of more than 1.0 phr of ETBP can significantly improve the compatibility between PLA and PBAT as well as the toughness of PBAT/PLA blends [21].

Al-Itry et al. [23] modified PBAT/PLA blends with CECLs (Joncryl®, BASF, Germany) and confirmed improved thermal stability, increased molecular weight, intrinsic viscosity, and elastic modulus of the PBAT/PLA.

Dong et al. [24] investigated PBAT/PLA blends with and without two chain-extending cross-linkers. SEM analysis of the reference PBAT/PLA (20/80) blend showed that the PBAT was dispersed non-uniformly in the PLA matrix with the large domain size (1~5 µm), while adhesion between the PLA and PBAT phases was poor, as evidenced by interfacial debonding and oval cavities left by the PBAT domains after cryo-fracture. The dispersion of the PBAT domains became uniform, and the average PBAT domain size was reduced to approximately 0.5 and 1 µm after the addition of 1 wt.% of Joncryl® and 1,6-hexanediol diglycidyl ether, respectively. The interfacial adhesion between the PLA and PBAT phases improved, and results indicated that the compatibility between the PLA and PBAT was greatly enhanced by the incorporation of both CECL, which reasonably affected other properties of the blends.

To the best of our knowledge, only Arruda et al. [25] investigated PBAT/PLA blends modified with Joncryl® on the films produced via blown film extrusion. Regarding the films without chain extenders, the PLA dispersed phase presented itself as an elongated and fibrillar structure preferably arranged towards the drawn direction of the film. Arruda et al. [25] assumed that this fibrillar morphology was caused by the elongational strain derived from the film drawing process. In the films containing 0.3 and 0.6% Joncryl®, the dispersed phase appeared as ellipsoids oriented towards the film drawing. The CECLs were expected to produce the PBAT/PLA copolymer.

Pan et al. [26] studied PBAT/PLA melt compounded with methylene diphenyl diisocyanate as CECL. SEM micrographs showed that when the dispersed phase concentration reached a 1:1 ratio, complex structures, such as platelet, ribbon- or sheet-like, stratified, and co-continuous, were formed. The PBAT with a size of approximately 10 mm showed almost no wetting, with the PLA phase indicating very low compatibility of the reference blend. By the formation of a PBAT/PLA copolymer due to the addition of CECL, a decrease in the size of the PBAT phase was attained.

Phetwarotai et al. [27] investigated PLA grafted with maleic anhydride (PLA-g-MA) synthesized via reactive maleation and PBAT/PLA compounds compatibilized with toluene diisocyanate (TDI). Fracture surfaces after the tension of compression-molded films exhibited poor interfacial adhesion between PLA and PBAT phases. Upon the addition of TDI, SEM showed many elongated fibrils as the addition of the TDI enabled the strong formation of urethane and/or amide linkages between PLA and PBAT phases, which improved interfacial adhesion. Further, the SEM image of the grafted blend after tension indicated the enhanced adhesion and wettability between PLA and PBAT compared to non-grafted material. The anhydride groups of PLA-g-MA could react with the hydroxyl groups of PLA and PBAT to form the ester linkages. This strong chemical bonding was an important factor that increased the interfacial adhesion between the components.

Currently, very little is known concerning the effect of multi-functional CECLs on morphology and resulting mechanical properties of the PBAT/PLA films produced via blown film extrusion. The vast majority of the studies available were carried out on samples prepared by compression or injection molding, thus not considering the process-induced changes introduced by blown film extrusion. In our previous study [28], we showed that the chemical reactions caused by CECLs incorporation into PBAT/PLA were incomplete after compounding and that the elongation during blow film extrusion brought appropriate molecular groups into reach and thus promoted crosslinking or chain scission. Therefore, the objective of this study was to investigate in detail the effect of four CECLs on the morphology of PBAT/PLA resulting from the molding route in order to optimize their processability and usage performance.

2. Materials and Methods

Four chain-extending cross-linkers (1 wt.%) were compounded into a reference PBAT/PLA (REF) M·VERA® B5029 [29] from BIO-FED, a branch of AKRO-PLASTIC GmbH, Köln, Germany:

V1—tris(2,4-di-tert-butylphenyl)phosphite, Songnox™ 1680 (Songwon Industrial Co, Ulsan, Korea) [30];

V2—1,3-phenylenebisoxazoline, 1,3-PBO powder (Evonik, Essen, Germany) [31];

V3—aromatic polycarbodiimide, Stabaxol® P110 (Lanxess, Cologne, Germany) [32];

V4—poly(4,4-dicyclohexylmethanecarbodiimide), Carbodilite™ HMV-15CA (Nisshinbo, Tokyo, Japan) [33].

The REF compound contained 24% by weight of calcium carbonate particles (D50 1.2 µm, top cut 4 µm), PBAT represented the matrix, and PLA represented the dispersed phase. All ingredients were evenly mixed using a Mixaco CM 150-D (Mixaco Maschinenbau, Neuenrade, Germany) and compounded with a twin-screw extruder (FEL 26 MTS, Feddem GmbH, Sinzig, Germany) with a diameter of 32 mm and an L/D of 26, a screw speed of 260 rpm and an output rate of 20 kg h^{-1}.

The 25 µm thick films were produced via blown film extrusion using an LF-400 (Labtech Engineering Company, Thailand) machine with an extrusion temperature of 165 °C and a blow-up ratio (BUR) of 1:2.5. The blown film machine had a single screw with a diameter of 25 mm and an L/D of 30. From an extrusion gap of 0.8 mm, the draw ratio was estimated to be between 12 and 14. The extrusion pressures were 240 bar for the REF blend and 290 bar (V1), 159 bar (V2), 230 bar (V4), and 313 bar (V4) for the modified compounds. Storage time to testing was 24 h at 23 °C/50% r.h.

Fourier Transform Infrared Spectroscopy (FTIR) was used to identify structural changes due to chemical reactions of the CECLs with PBAT and PLA on the compounded material (granules). IR-spectra were recorded in the wavenumber range 2000 to 600 cm^{-1} using an FTIR Microscope System (Perkin Elmer Spectrum Spotlight 200, Waltham, MA, USA) with Attenuated Total Reflectance (ATR) in continuous scan mode, a spectral resolution of 16 cm^{-1} and 15 scans averages per spectrum.

Films were fractured under cryogenic conditions in extrusion (ED) and transverse (TD) directions, as seen in Figure 1, using liquid nitrogen, and sputtered with gold at 20 mA for 3 × 30 s. Afterwards, fracture surfaces of the samples were investigated using a field-emission Scanning Electron Microscope SEM (JSM-7200F, Jeol, Tokyo, Japan) at an acceleration voltage of 5.0 kV and amplifications of 3.000 and 10.000.

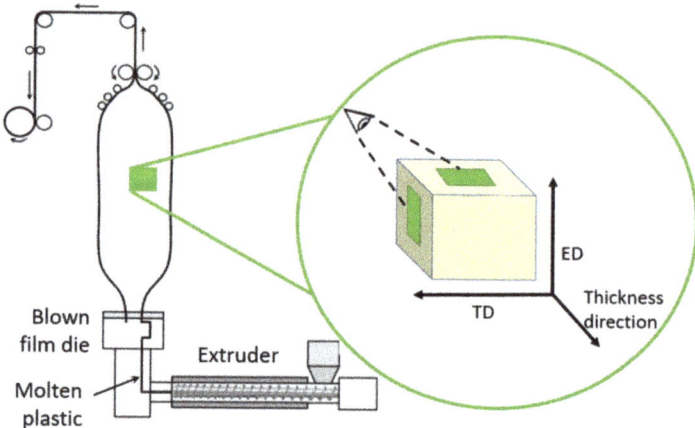

Figure 1. Sample orientation with respect to extrusion direction (ED), transverse direction (TD), and thickness directions; ED shows a TD-thickness direction-plain, and TD shows an ED-thickness direction-plain.

3. Results

PBAT and PLA are immiscible polymers [34]. Their blend morphology results from the process variables (temperature, deformation types, and rates) and the properties of the components (composition, viscosity ratio, interfacial tension, continuous phase viscosity, and elasticity of the components). Due to the blown film extrusion with different draw ratios in ED and TD, anisotropic mechanical properties and corresponding differences in a morphological structure may be expected. Recently, Azevedo et al. [28] showed (based on differential scanning calorimetry data of granules and films) that the chemical reactions were incomplete after compounding and that blown film extrusion intensified them even for rapidly cooled 25 μm films. This behavior can be explained by assuming that the CECL molecules are linked with one reactive site to polymer chains during compounding, whereas the other reactive sites remain unaffected. Depending on the kind of CECL, this may also lead to chain scission. Only the chain slip due to the elongation during blown film extrusion brings appropriate molecular groups into reach. Then, the unreacted sites can react with neighboring polymer chain segments. These reactions may lead either to further chain scission or cross-linking. The fact that the elongations at the break in extrusion direction (ED) decreased with aging and remained unaltered in the transverse direction (TD) indicates that the reactions linked to chain scission and cross-linking depended also on the introduced draw ratios during film blowing, which differed for ED and TD. Furthermore, chain scission and cross-linking altered viscosity and consequently the structure, e.g., dimensions of the dispersed phase and its geometry (spherical or fibrillar). According to dynamic mechanical analysis [28], for V1 CECLs, the T_g of PLA and PBAT phases were hardly affected, whereas for V2 to V4, an increase of T_g in the PBAT phase and a decrease in the PLA phase were observed. This means that cross-linking mainly occurred in the PBAT phase, whereas in the PLA phase, the free volume was mainly increased by partial reaction with CECL. Finalized cross-linking in the PLA phase (with corresponding T_g increases) was found for V3 and V4 compounds after the second melting.

To confirm the interactions expected from DSC and mechanical analysis, FTIR-ATR was performed on the blown films (Figure 2) as well as the granules (Supplementary Materials: Figure S1). According to Standau et al. [35] and Yuniarto et al. [36], the PLA band around 752 cm^1, together with the vibration of the α-methyl band around 864 cm^{-1}, is associated with the ester (O-CH-CH$_3$), while that many weaker peaks in the range of 1250-1050 cm^{-1} are assigned to C-O from carboxyl groups and C-O-C stretching vibrations, and the peak at 1748 cm^{-1} is associated with the carbonyl C=O stretching vibration.

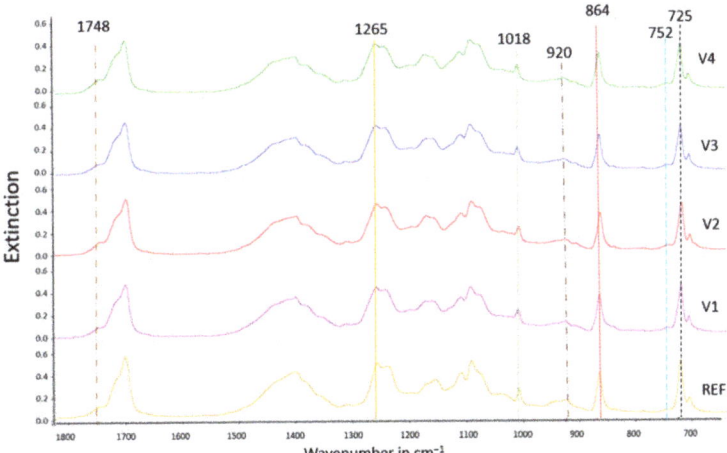

Figure 2. FTIR spectra of the blown films of unmodified PBAT/PLA (REF) and the CECL-modified samples (V1 to V4).

It has been reported [23] that there are three possible linkages in polyesters: carbon-oxygen ether linkage (β-H-C hydrogen transfer), carbonyl carbon-carbon linkage, and carbonyl carbon-oxygen, which can undergo scission. The functional groups of PBAT are described as follows (Figure 2): the peak at around 1710 cm^{-1} represents carbonyl groups (C=O) in the ester linkage, while at 1265 cm^{-1}, a peak intercepts C-O in the ester linkage, and at around 725 cm^{-1} a peak represents methylene (-CH2-) groups. Bending peaks of the benzene substitutes are located at wavenumbers between 700 and 900 cm^{-1}.

After CECL modification of the REF blend, the PLA band at 864 cm^{-1} was slightly shifted for V2, V3, and V4 to smaller wavenumbers up to 850 cm^{-1}, indicating reactions that enhance vibrations of the α-methyl group. A weak band occurring at 920 cm^{-1} is characteristic of unsaturated vinyl groups.

Overall, the spectra of V1 to V4 do not differ significantly from that obtained for REF, suggesting only small changes in the chemical structure of PBAT/PLA within the sensitivity limit of FTIR. This is in accordance with Wu et al. [37], who investigated how dicumyl peroxide (DCP) modifies the spectra of PBAT/PLA blends and found that DCP generated free radicals by thermal decomposition, initiating the formation of branching structures via hydrogen abstraction, resulting in minimal changes in the FTIR spectra.

Furthermore, FTIR performed on the granules does not show significant differences from those obtained on blown films. Therefore, regardless of its frequent usage, FTIR may not be an efficient tool to detect the CECL-attributed chemical interactions, possibly due to the strong self-interactions of PBAT and PLA [38].

SEM of fracture surfaces of blown extrusion films, as seen in Figure 3, reveals that the dimensions of the dispersed PLA phase differed among the samples and depended on the orientation of the fracture surfaces.

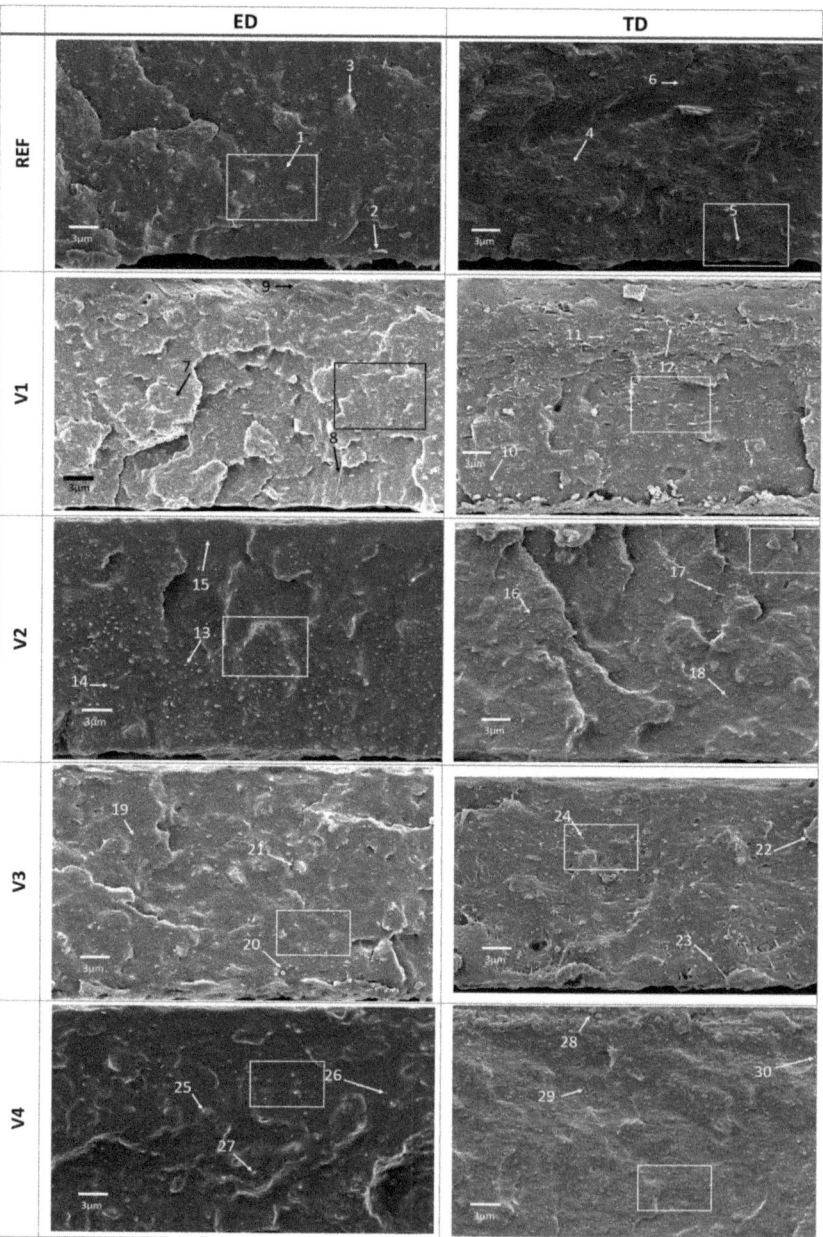

Figure 3. SEM of fracture surfaces of unmodified PBAT/PLA (REF) and CECL-modified (**V1 to V4**) films in ED and TD blown film extrusion directions. The rectangles represent the magnified areas displayed in Figure 4.

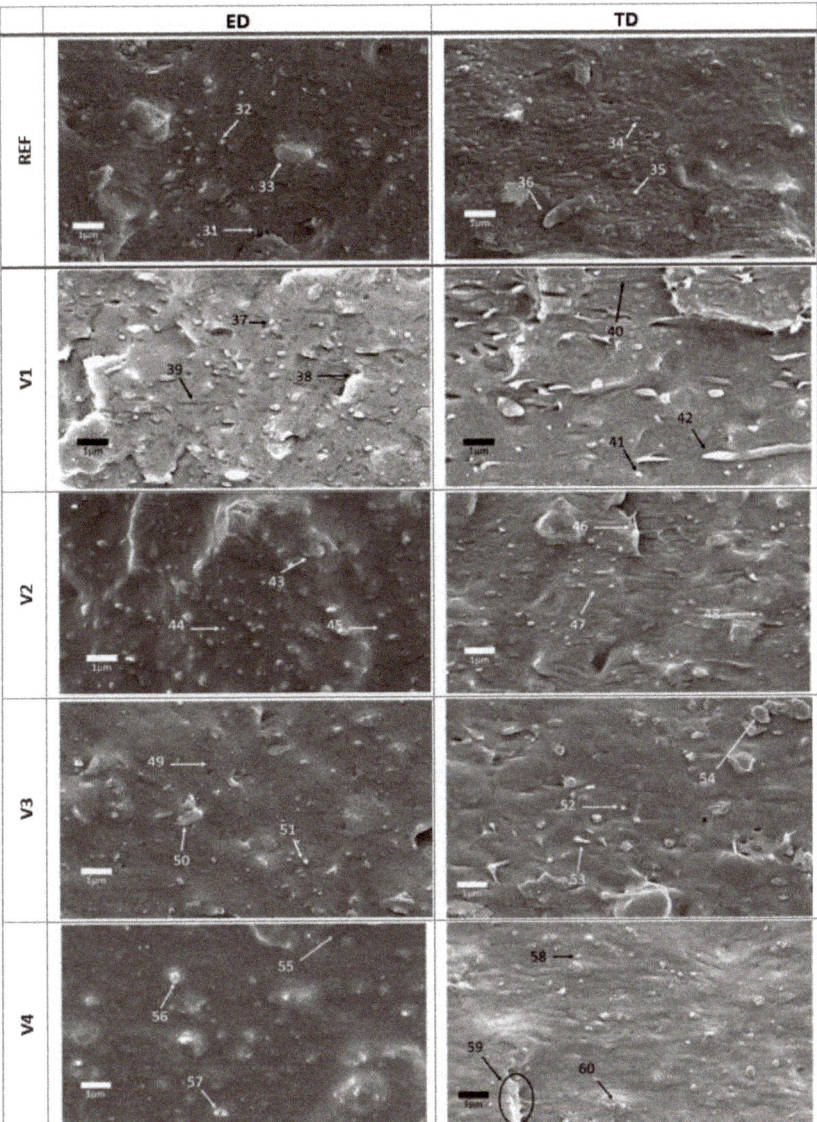

Figure 4. Detailed (magnification of 10,000) SEM of fracture surfaces of unmodified PBAT/PLA (REF) and CECL-modified (**V1 to V4**) films in ED and TD blown film extrusion directions.

REF blend exhibited a clear dispersion of PLA in the PBAT matrix with "sea-island" morphologies < 500 nm; see Figure 3—arrows 1 and 6, confirming nonhomogeneous dispersion reported in the literature [25–27]. Such morphology is associated with poor mechanical properties due to a weak interfacial adhesion between PBAT and PLA as well as internal stresses at an interface. Furthermore, the skin-core structure was found in REF consisting of coarsely dispersed PLA in the core region of the sample and fine PLA fibrils in the skin regions; see Figure 3—arrows 4 and 5 in the TD direction.

The comparison of fracture surfaces of V1 to V4 with respect to REF shows that CECLs changed the fracture appearance as follows: V1 fails brittle, V2 slightly more ductile, and

V3 to V4 significantly more ductile and tougher, as shown in Figures 3 and 4. REF and V1 to V3 exhibit fibrils, as shown in Figure 3—arrows 2, 8, 12, and 23, indicating the tough and ductile behavior of the PBAT matrix).

The calcium carbonate particles having diameters of 2 to 3 µm (D50 of 1.2 µm) were well visible on the fracture surfaces, although they were completely covered by the PBAT matrix (Figure 3—arrows 3, 5, 21, and 22), indicating a good adhesion due to a complete matrix wetting, especially for V4.

The introduction of CECLs modified the morphology of the PBAT/PLA and the structure of the fracture surfaces with respect to the dimension and geometry of the PLA islands. However, there were morphological similarities in all compounds as the structure of the dispersed PLA was circular or spherical, Figure 3—arrows 1, 6, 7, 11, 13, 18, 19, 24, 28, and 30. This and the coarser islands support the process of chain scission for V1 and V2.

The differences in ED and TD were to be expected, and PLA fibrils were found only in TD (Figure 3—arrows 5, 12, 17, 23, and 28) but not in ED. In ED, the fibrils seem to be decayed in spherical or ellipsoidal PLA islands with a maximum aspect ratio of about 3, whereas in TD, both spherical PLA islands and fibrils occur. The exception is V4 modified blend, which did not exhibit PLA fibrils in TD, suggesting a better interface adhesion, where the fibrils were completely embedded in the matrix. In addition to processing-induced PLA fibrils, fracture-induced fibrils of the PBAT matrix having 5 to 10 times smaller diameters were also found (Figure 4—arrows 31, 35, 39, 40, and 46).

The morphology of the modified blends has to be interpreted also with respect to structure formation during film blowing with a BUR of 1:2.5, a drawing ratio (DR) of 12 to 14, and temperature. During the blow phase, the elongation flow stretched and oriented the melt at the beginning in TD due to blowing and subsequently in ED due to drawing. Simultaneously, the melt cooled down the faster the film was stretched and melt viscosity increased. The occurrence of the fibrils suggests that cooling and freezing exceeded the orientation relaxation.

At temperatures below 130 °C, PLA started to crystallize, freezing in the islands in the current geometry. Due to DR >> BUR, the initially spherical PLA particles in the melt were much more stretched in ED than in TD, explaining why PLA fibrils were found only in TD. After the solidification of PLA, only the PBAT matrix could be further stretched until it had been cooled below T_g at around 60 °C. During this stretching, the calcium particles are oriented perpendicular to the thickness direction.

At larger magnifications, the fibrils due to fracturing of the PBAT matrix were well visible (Figure 4—arrows 31, 35, 39, 40, and 46). This can be explained by the lower stiffness and yield stress of PBAT ($E \approx 400$ MPa, $\sigma_y \approx 35$ MPa) [9,11] compared to PLA ($E = 3500$ MPa, $\sigma_y \approx 60$ MPa) [20], leaving the PLA in an almost nondeformed state during the fracture. Furthermore, it also enhances the fracture propagation along the PBAT/PLA interface. Rather poor adhesion is indicated with many small PLA particles ($D < 200$ nm), (Figure 4—arrows 32, 34, 37, 41, 44, 48, 51, 52, 56, and 58).

The cracks appearing on the fracture surfaces of V2 to V4 (Figure 4—arrows 45, 49, 55, and 59) were presumably not a result of the CECL modification, but of electron beam heating of the surface causing the relaxation of internal stresses and void formation.

In V1 and V3, the dispersed PLA was not covered by the PBAT matrix, indicating that the CECLs hardly affected compatibilization. The more elliptic shape of the PLA islands could have arisen due to the fact that the processing-induced liquid PLA fibrils to decay in ellipsoidal droplets, which were frozen in the current shape.

Finally, the morphological features of REF and V1 to V4, considered with respect to the degree of brittleness, structure of the PLA phase, interface adhesion of the dispersed PLA, and interface adhesion of filler particles, are summarized in Table 1.

Table 1. Summary of structure features seen on fracture surfaces of unmodified PBAT/PLA (REF) and CECL-modified (V1 to V4) films in ED and TD blown molding directions.

Cpd	Feature	ED	TD
REF	brittleness	semi-brittle	semi-tough
	dispersed PLA	circular $D \approx$ 100–400 nm no fibrillar structure	circular $D \approx$ 100–400 nm partly fibrillar in ED: $D \approx$ 100 nm, $L \approx$ 1000–2000 nm
	interface adhesion of dispersed PLA	rather poor PLA surface visible, holes of PLA phase dimensions	rather poor; PLA surface visible, holes of PLA phase dimensions; lines in fracture surface indicating poor interface adhesion of fibrils
	particle adhesion	filler particles completely covered with PBAT indicating good to excellent adhesion	
V1	brittleness	brittle	semi-brittle
	dispersed PLA	Circular/slightly elongated $D \approx$ 100–200 nm; lines in fracture surface: $D \approx$ 150 nm, $L \approx$ 1000–1500 nm	circular $D \approx$ 100–200 nm; elongated fibrils fibrils: $D \approx$ 200 nm, $L \approx$ 1000–4000 nm
	Interface adhesion of dispersed PLA	bad; PLA surface visible, holes of PLA phase dimensions	bad; well visible PLA islands and fibrils; lines in fracture surface indicating bad adhesion
	particle adhesion	filler particles completely covered with PBAT indicating good to excellent adhesion	
V2	brittleness	ductile	tough with partly fibrillated matrix
	dispersed PLA	circular $D \approx$ 100–200 nm no fibrils	circular islands $D \approx$ 100–200 nm elongated fibrils: $D \approx$ 100 nm, $L \approx$ 1000–2000 nm
	interface adhesion of dispersed PLA	poor well embedded islands with cracks in all directions, max $L \approx$ 1000 nm	bad to poor deformed islands/fibrils in fracture surface indicating some adhesion
	matrix particle adhesion	filler particles completely covered with PBAT indicating fair to excellent adhesion	
V3	brittleness	ductile to tough	ductile with fibrillated matrix
	dispersed PLA	circular/slightly elliptic $D \approx$ 200–500 nm; no fibrils	circular, partly elongated $D \approx$ 200–500 nm no fibrils
	interface adhesion of dispersed PLA	bad to mean well-embedded islands with visible surface and cracks, max $L \approx$ 500 nm	bad to mean some islands partly embedded
	particle adhesion	filler particles completely covered with PBAT, indicating fair to excellent adhesion	
V4	brittleness	ductile to tough	poor to tough with fibrillated matrix
	dispersed PLA	circular islands $D \approx$ 100–300 nm	circular islands $D \approx$ 100–300 nm no fibrils; lines with max $L \approx$ 2500 nm
	interface adhesion of dispersed PLA	poor; PLA surface partly covered with matrix	poor; no visible fibrils PLA surface partly covered with matrix
	particle adhesion	filler particles were hardly visible indicating good dispersion of the particles in the matrix	

4. Conclusions

The PBAT/PLA compound was modified with four multi-functional chain-extending cross-linkers (CECL). Previous calorimetric investigation of PBAT/PLA films revealed their selective reactions with the introduced CECL. The cross-linking effect occurred only for aromatic polycarbodiimide and poly(4,4-dicyclohexylmethanecarbodiimide, whereas chain scission was attained for modification with tris(2,4-di-tert-butylphenyl)phosphite and 1,3-phenylenebisoxazoline. FTIR did not prove to support the results of calorimetry and mechanical performance in a convincing way. Only slight shifts to lower wavenumbers were obtained for aromatic polycarbodiimide and poly(4,4-dicyclohexylmethanecarbodiimide, but also for 1,3-phenylenebisoxazoline. Morphological analysis was in accordance with the resulting mechanical properties. Tris(2,4-di-tert-butylphenyl)phosphite and aromatic polycarbodiimide showed a dispersed PLA phase that was not covered by the PBAT matrix, indicating that these two CECLs do not provide compatibilization, whereas poly(4,4-dicyclohexylmethanecarbodiimide) showed the dispersed PLA partially covered by the PBAT matrix. The most synergetic effect was obtained for 1,3-phenylenebisoxazoline, where the PLA phase was well embedded in the PBAT matrix, indicating adhesion and improved compatibilization.

Supplementary Materials: The following are available online at https://www.mdpi.com/article/10.3390/polym14101939/s1, Figure S1: FTIR spectra of the granules of unmodified PBAT/PLA (REF) and the CECL-modified samples (V1 to V4).

Author Contributions: Conceptualization, B.M.; methodology, B.M. and E.R.-v.D.; software, J.V.C.A.; validation, J.V.C.A., E.R.-v.D. and R.G.; formal analysis, J.V.C.A. and E.R.-v.D.; investigation, J.V.C.A.; resources, J.V.C.A. and B.H.; data curation, J.V.C.A.; writing—original draft preparation, J.V.C.A. and B.M.; writing—review and editing, B.H.; visualization, J.V.C.A.; supervision, B.H. and B.M.; funding acquisition, B.H. All authors have read and agreed to the published version of the manuscript.

Funding: The author B.H. acknowledges the Ministry of Education, Youth and Sports of the Czech Republic - DKRVO (RP/CPS/2022/003).

Institutional Review Board Statement: Not applicable.

Informed Consent Statement: Not applicable.

Data Availability Statement: Data are stored at the personal depository of J. Azevedo.

Acknowledgments: BIO-FED: a branch of AKRO-PLASTIC GmbH, is acknowledged for sponsoring the samples and working place of the presented work.

Conflicts of Interest: The authors declare no conflict of interest.

References

1. European Bioplastics. Bioplastics Packaging—Combining Performance with Sustainability. Available online: https://docs.european-bioplastics.org/publications/fs/EUBP_FS_Packging.pdf (accessed on 1 November 2021).
2. Suwanamornlert, P.; Kerddonfag, N.; Sane, A.; Chinsirikul, W.; Zhou, W.; Chonhencho, V. Poly(lactic acid)/poly(butylene-succinate-co-adipate) (PLA/PBSA) blend films containing thymol as alternative to synthetic preservatives for active packaging of bread. *Food Packag. Shelf Life* **2020**, *25*, 100515. [CrossRef]
3. Palai, B.; Mohanty, S.; Nayak, S.K. Synergistic effect of polylactic acid (PLA) and Poly(butylene succinate-co-adipate) (PBSA) based sustainable, reactive, super toughened eco-composite blown films for flexible packaging applications. *Polym. Test.* **2020**, *83*, 106130. [CrossRef]
4. McKeen, L.W. *Permeability Properties of Plastics and Elastomers*, 4th ed.; Plastics Design Library: Chadds Ford, PA, USA, 2016; pp. 41–60.
5. Wagner, J.R., Jr. *Multilayer Flexible Packaging*, 2nd ed.; Plastics Design Library: Chadds Ford, PA, USA, 2016; pp. 137–145.
6. Ashter, S.A. *Introduction to Bioplastics Engineering*; Plastics Design Library: Chadds Ford, PA, USA, 2016; pp. 179–209.
7. Cantor, K. *Blown Film Extrusion*, 2nd ed.; Carl Hanser Verlag GmbH & Co. KG: München, Germany, 2011.
8. Rigolin, T.R.; Costa, L.C.; Chinellato, M.A.; Muñoz, P.A.R.; Bettini, S.H.P. Chemical modification of poly(lactic acid) and its use as matrix in poly(lactic acid) poly(butylene adipate-co-terephthalate) blends. *Polym. Test.* **2017**, *63*, 542–549. [CrossRef]
9. Pietrosanto, A.; Scarfato, P.; Maio, L.D.; Incarnato, L. Development of Eco-Sustainable PBAT-Based Blown Films and Performance Analysis for Food Packaging Applications. *Materials* **2020**, *13*, 5395. [CrossRef]
10. Jiang, L.; Wolcott, M.P.; Zhang, J. Study of Biodegradable Polylactide/Poly(butylene adipate-co-terephthalate) Blends. *Biomacromolecules* **2006**, *7*, 199–207. [CrossRef]
11. Jian, J.; Xiangbin, Z.; Xianbo, H. An overview on synthesis, properties and applications of poly(butylene-adipate-co-terephthalate)-PBAT. *Adv. Ind. Eng. Polym. Res.* **2020**, *3*, 19–26. [CrossRef]
12. Lackner, M.; Ivanič, F.; Kováčová, M.; Chodák, I. Mechanical properties and structure of mixtures of poly(butylene-adipate-coterephthalate) (PBAT) with thermoplastic starch (TPS). *Int. J. Biobased Plast.* **2021**, *3*, 126–138. [CrossRef]
13. Ferreira, F.V.; Cividanes, L.S.; Gouveia, R.F.; Lona, L.M.F. An overview on properties and applications of poly(butylene adipate-co-terephthalate)–PBAT based composites. *Polym. Eng. Sci.* **2019**, *59*, 7–15. [CrossRef]
14. Ivanič, F.; Kováčová, M.; Chodák, I. The effect of plasticizer selection on properties of blends poly(butylene adipate-co-terephthalate) with thermoplastic starch. *Eur. Polym. J.* **2019**, *116*, 99–105. [CrossRef]

15. Pietrosanto, A.; Scarfato, P.; Maio, L.D.; Nobile, M.R.; Incarnato, L. Evaluation of the Suitability of Poly(Lactide)/Poly(Butylene-Adipate-co-Terephthalate) Blown Films for Chilled and Frozen Food Packaging Applications. *Polymers* **2020**, *12*, 804. [CrossRef]
16. Chiu, H.T.; Huang, S.Y.; Chen, Y.F.; Kuo, M.T.; Chiang, T.Y.; Chang, C.Y.; Wang, Y.H. Heat Treatment Effects on the Mechanical Properties and Morphologies of Poly (Lactic Acid)/Poly (Butylene Adipate-co-terephthalate) Blends. *Int. J. Polym. Sci.* **2013**, *1*, e951696. [CrossRef]
17. Hongdilokkul, P.; Keeratipinit, K.; Chawthai, S.; Hararak, B.; Seadan, M.; Suttiruengwong, S. A study on properties of PLA/PBAT from blown film process. *IOP Conf. Ser. Mater. Sci. Eng.* **2015**, *87*, e012112. [CrossRef]
18. Kijchavengkul, T.; Auras, R.; Rubino, M.; Selke, S.; Ngouajio, M.; Fernandez, R.T. Biodegradation and hydrolysis rate of aliphatic aromatic polyester. *Polym. Degrad. Stab.* **2010**, *95*, 2641–2647. [CrossRef]
19. Witt, U.; Müller, R.J.; Deckwer, R.W.-D. Biodegradation of Polyester Copolymers Containing Aromatic Compounds. *J. Macr. Sci. A* **1995**, *32*, 851–856. [CrossRef]
20. Tsuji, H. Poly(lactide) Stereocomplexes: Formation, Structure, Properties, Degradation, and Applications. *Macromol. Biosci.* **2005**, *5*, 569–597. [CrossRef] [PubMed]
21. Wang, B.; Jin, Y.; Kang, K.; Yang, N.; Weng, Y.; Huang, Z.; Men, S. Investigation on compatibility of PLA/PBAT blends modified by epoxy-terminated branched polymers through chemical micro-crosslinking. *e-Polymer* **2020**, *20*, 39–54. [CrossRef]
22. Su, S.; Duhme, M.; Kopitzky, R. Uncompatibilized PBAT/PLA Blends: Manufacturability, Miscibility and Properties. *Materials* **2020**, *13*, 4897. [CrossRef]
23. Al-Itry, R.; Amnawar, K.; Maazouz, A. Improvement of thermal stability, rheological and mechanical properties of PLA, PBAT and their blends by reactive extrusion with functionalized epoxy. *Polym. Degr. Stab.* **2012**, *97*, 1898–1914. [CrossRef]
24. Dong, W.; Zou, B.; Yan, Y.; Ma, P.; Chen, M. Effect of Chain-Extenders on the Properties and Hydrolytic Degradation Behavior of the Poly(lactide)/Poly(butylene adipate-co-terephthalate) Blends. *Int. J. Mol. Sci.* **2013**, *14*, 20189–20203. [CrossRef]
25. Arruda, L.C.; Megaton, M.; Bretas, R.E.S.; Ueki, M.N. Influence of chain extender on mechanical, thermal and morphological properties of blown films of PLA/PBAT blends. *Polym. Test.* **2015**, *43*, 27–37. [CrossRef]
26. Pan, H.; Li, Z.; Yang, J.; Li, X.; Ai, X.; Hao, Y.; Zhang, H.; Dong, L. The effect of MDI on the structure and mechanical properties of poly(lactic acid) and poly(butylene adipate-co-butylene terephthalate) blends. *RSC Adv.* **2018**, *8*, 4610–4623. [CrossRef]
27. Phetwarotai, W.; Zawong, M.; Phusunti, N.; Aht-Ong, D. Toughening and thermal characteristics of plasticized polylactide and poly(butylene adipate-co-terephthalate) blend films: Influence of compatibilization. *Int. J. Bio. Macrom.* **2021**, *183*, 346–357. [CrossRef] [PubMed]
28. Azevedo, J.V.C.; Dorp, E.R.; Hausnerová, B.; Möginger, B. The Effects of Chain-Extending Cross-Linkers on the Mechanical and Thermal Properties of Poly(butylene adipate terephthalate)/Poly(lactic acid) Blown Films. *Polymers* **2021**, *13*, 3092. [CrossRef] [PubMed]
29. BIO-FED Website. TDPG of M·VERA®B5029. Available online: https://bio-fed.com/fileadmin/bio-fed/PDFs/BIO-FED_TDPG_MVERA_B5029_B0155_2019-10-11.pdf (accessed on 27 November 2019).
30. SONGWON Website. SONGNOX™ Product Descriptions. Available online: https://www.songwon.com/products/songnox-1680 (accessed on 25 October 2019).
31. SpecialChem Webiste. Technical Datasheet of 1,3-Phenylene-bis-oxazoline. Available online: https://polymer-additives.specialchem.com/product/a-evonik-1-3-phenylene-bis-oxazoline (accessed on 25 October 2019).
32. Lanxess Website. Technical Datasheet of Stabaxol®P110. Available online: https://add.lanxess.com/fileadmin/product-import/stabaxol_p_110_en_rcr.pdf (accessed on 20 November 2020).
33. Nisshinbo ChemWebiste. Hydrolysis Stabilizer for Polyesters Including Biodegradable Resin. Available online: https://www.nisshinbo-chem.co.jp/english/products/carbodilite/poly.html (accessed on 25 October 2019).
34. Dil, E.J.; Carreau, P.J.; Favis, B.D. Morphology, Miscibility and Continuity Development in Poly(lactic acid)/Poly(butylene adipate-co-terephthalate) Blends. *Polymer* **2015**, *68*, 202–212. [CrossRef]
35. Standau, T.; Zhao, C.; Castellón, S.V.; Bonten, C.; Altstädt, V. Chemical Modification and Foam Processing of Polylactide (PLA). *Polymers* **2019**, *11*, 306. [CrossRef]
36. Yuniarto, K.; Purwanto, Y.A.; Purwanto, S.; Welt, B.A.; Purwadaria, H.K.; Sunarti, T.C. Infrared and Raman Studies on Polylactide Acid and Polyethylene Glycol-400 Blend. *AIP Conf. Proc.* **2016**, *1725*, 020101. [CrossRef]
37. Wu, A.; Huang, J.; Fan, R.; Xu, P.; Liu, G.; Li, S.Y. Effect of blending procedures and reactive compatibilizers on the properties of biodegradable poly(butylene adipate-co-terephthalate)/poly(lactic acid) blends. *J. Polym. Eng.* **2021**, *41*, 95–108. [CrossRef]
38. Bleyan, D.; Svoboda, P.; Hausnerova, B. Specific interactions of low molecular weight analogues of carnauba wax and polyethylene glycol binders of ceramic injection moulding feedstocks. *Ceram. Int.* **2015**, *41*, 3975–3982. [CrossRef]

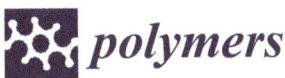

Article

Improvement of Poly(lactic acid)-Poly(hydroxy butyrate) Blend Properties for Use in Food Packaging: Processing, Structure Relationships

Mitul Kumar Patel [1], Marta Zaccone [2], Laurens De Brauwer [3], Rakesh Nair [3], Marco Monti [2], Vanesa Martinez-Nogues [4], Alberto Frache [5] and Kristiina Oksman [1,6,7,*]

1. Division of Materials Science, Department of Engineering Sciences and Mathematics, Luleå University of Technology, SE-97187 Luleå, Sweden
2. Proplast, Via Roberto di Ferro 86, 15122 Alessandria, Italy
3. Bio Base Europe Pilot Plant (BBEPP), Rodenhuizekaai 1, 9042 Gent, Belgium
4. Tecnopackaging, Polígono Industrial Empresarium, Calle Romero 12, 50720 Zaragoza, Spain
5. Department of Applied Science and Technology and Local INSTM Unit, Politecnico di Torino, Viale Teresa Michel 5, 15121 Alessandria, Italy
6. Mechanical & Industrial Engineering (MIE), University of Toronto, Toronto, ON M5S 3G8, Canada
7. Wallenberg Wood Science Center (WWSC), Luleå University of Technology, SE-97187 Luleå, Sweden
* Correspondence: kristiina.oksman@ltu.se

Citation: Patel, M.K.; Zaccone, M.; De Brauwer, L.; Nair, R.; Monti, M.; Martinez-Nogues, V.; Frache, A.; Oksman, K. Improvement of Poly(lactic acid)-Poly(hydroxy butyrate) Blend Properties for Use in Food Packaging: Processing, Structure Relationships. *Polymers* **2022**, *14*, 5104. https://doi.org/10.3390/polym14235104

Academic Editors: Alexey Iordanskii, Nadia Lotti, Valentina Siracusa and Michelina Soccio

Received: 22 October 2022
Accepted: 18 November 2022
Published: 24 November 2022

Publisher's Note: MDPI stays neutral with regard to jurisdictional claims in published maps and institutional affiliations.

Copyright: © 2022 by the authors. Licensee MDPI, Basel, Switzerland. This article is an open access article distributed under the terms and conditions of the Creative Commons Attribution (CC BY) license (https://creativecommons.org/licenses/by/4.0/).

Abstract: Poly(lactic acid)-poly(hydroxybutyrate) (PLA-PHB)-based nanocomposite films were prepared with bio-based additives (CNCs and ChNCs) and oligomer lactic acid (OLA) compatibilizer using extrusion and then blown to films at pilot scale. The aim was to identify suitable material formulations and nanocomposite production processes for film production at a larger scale targeting food packaging applications. The film-blowing process for both the PLA-PHB blend and CNC-nanocomposite was unstable and led to non-homogeneous films with wrinkles and creases, while the blowing of the ChNC-nanocomposite was stable and resulted in a smooth and homogeneous film. The optical microscopy of the blown nanocomposite films indicated well-dispersed chitin nanocrystals while the cellulose crystals were agglomerated to micrometer-size particles. The addition of the ChNCs also resulted in the improved mechanical performance of the PLA-PHB blend due to well-dispersed crystals in the nanoscale as well as the interaction between biopolymers and the chitin nanocrystals. The strength increased from 27 MPa to 37 MPa compared to the PLA-PHB blend and showed almost 36 times higher elongation at break resulting in 10 times tougher material. Finally, the nanocomposite film with ChNCs showed improved oxygen barrier performance as well as faster degradation, indicating its potential exploitation for packaging applications.

Keywords: poly(lactic acid); poly(hydroxybutyrate); film blowing; cellulose nanocrystals; chitin nanocrystals; liquid-assisted extrusion; biodegradability; oxygen barrier properties

1. Introduction

Petroleum-based plastics are currently the most popular choice for packaging applications because of their low cost, ease of processability, and tunable properties, which can be modified to meet the specific needs of the products [1]. The most popular polymers used in the packaging industry are polyethylene (PE), polypropylene (PP), polyethylene terephthalate (PET), and polystyrene (PS), which together account for more than 90% of the total volume of plastics used in food packaging [2]. Traditional thermoplastics used in packaging are often made from non-renewable fossil fuels. In addition, most fossil fuel-based polymers are non-biodegradable; as a result, poor waste management contributes to several environmental issues, including high amounts of plastic waste in the environment, the emissions of toxic gases during the incineration of polymers, and the shortage of landfill space [3]. Thus, it is essential to replace these with sustainable and biodegradable polymers.

Among biopolymers from renewable resources, polylactic acid (PLA), is a potential candidate to replace petroleum-based non-biodegradable polymers due to its good mechanical performance (stiffness and strength) [4], transparency, excellent printability [5], processability [6], and economic feasibility compared to many other biodegradable polymers. However, despite the mentioned advantages, PLA has some disadvantages, which consequently limit its use in food packaging applications. These are its inherent brittleness, which is also reflected in the difficulty of producing thin-blowing films, and its moderate gas barrier performance [7].

Gas barrier properties are critical when considering a polymer for food packaging applications because many food products are sensitive to oxidation; hence, packages with low oxygen permeability are preferred. It is well known that the crystalline phase has a significant influence on the oxygen barrier properties of a material; as a consequence, increasing the crystallinity of PLA is required for food packaging applications. To enhance the crystallinity, blending PLA with other more crystalline biopolymers, such as poly(hydroxy alkanoates) (PHAs), has subsequently received a lot of attention in the food packaging industry. The most common representative of PHAs is poly(hydroxybutyrate) (PHB) [8] which has been extensively researched for its potential applications in food packaging [9]. Our previous research demonstrated that blending PLA with 25 wt% of PHB resulted in the formation of small, finely dispersed, and highly crystalline PHB spherulites in the PLA matrix [10]. This resulted in a significant increase in the crystallinity of the material, which ultimately improved its barrier performance. Likewise, Zhang et al. [11] reported that melt blending PLA with 25 wt% of PHB showed optimal miscibility between the two polymers.

Film blowing is the most widely accepted process for manufacturing polymer-based films in the packaging industry, such as barrier films (used to protect daily meat and vegetables), frozen food packaging, and shopping bags [12,13]. However, this processing is often challenging for both PLA and PHB because of their inherent brittleness, low melt strength, and reduced elongation as compared to PP and PE [14], resulting in unstable and wrinkled bubbles that tend to collapse during the film-blowing operations [15]. In addition, most food packaging applications require ductile properties and increased flexibility, which cannot be obtained with neat PLA:PHB blends. To address these concerns, different strategies have been proposed, such as the addition of plasticizer to improve the fluidity of the polymer blend [16,17], the use of viscosity enhancers to improve the material melt strength [18], or the development of nanocomposites by the introduction of nanomaterials such as chitin nanocrystals (ChNCs) or cellulose nanocrystals (CNCs) which concurrently improve the processability and the barrier properties of the produced films [15,19]. In this context, Herrera et al. [15] demonstrated the effectiveness of ChNCs in producing PLA-based nanocomposite blown films with improved processability and mechanical properties.

Recently, cellulose and chitin nanocrystals have drawn the attention of researchers as promising reinforcing nanomaterials for producing bionanocomposites due to their large aspect ratios with good mechanical properties, as well as their biocompatible and renewable nature [20–22]. In addition to the improvement in mechanical properties, CNCs and ChNCs also act as nucleating agents for polymers, thereby enhancing the crystallization rate and increasing the crystallinity of the material. CNCs have been widely used as reinforcements to improve the thermal, mechanical, crystalline, and barrier properties of PLA [23] and PLA-PHB blends [19,24]. Singh et al. [25,26] showed that the addition of a small amount of ChNCs to PLA results in a significant improvement in PLA nucleus numbers due to faster crystallization speed and thus increased crystallinity. In addition, our previous study [10] demonstrated that the addition of 1 wt% ChNCs has a strong effect on the oxygen barrier properties of the PLA:PHB blend, which is generated simultaneously by the inherent barrier properties of ChNCs and the induction of a higher degree of crystallinity. Furthermore, ChNCs acted as an antistatic additive, resulting in less electrostatic interaction between the film surfaces, making it easier for the films to open [15]. Arrieta et al. [19] showed that the plasticized PLA-PHB-based bionanocomposite with 1 wt% functionalized CNC achieved higher elongation at break of about 150%, compared to the material without

CNCs, showing comparable values to those of commercial stretchable films used for food packaging. In addition, Jandas et al. [27] reported that reactive extrusion of a PLA-PHB blend using MA (maleic anhydride) to form an interpenetrated network structure together with modified nanoclays resulted in excellent impact strength and high elongation at break.

Although CNCs and ChNCs have great potential as mentioned above, the hydrogen groups on the surface induce high attraction between the crystals, resulting in irreversible strong hydrogen bonds between them during preparation, which makes it difficult to achieve homogeneous dispersion of nanocrystals in polymers [28]. To avoid aggregation, nanocrystals generally undergo hydrophobic surface modifications before nanocomposite preparations [24,29,30]. Unfortunately, these hydrophobic surface modifications frequently have significant economic and environmental drawbacks, which limit their industrial potential [31]. Previously, a scalable liquid-assisted extrusion approach was proposed to improve the dispersion of nanocrystals and, as a consequence, improve the final properties of the final formulations [32–34]. According to this process, nanomaterials in aqueous form along with the dispersing aid are fed into polymer melt via extrusion. OLA has been considered an effective plasticizer for PLA [35] as well as for PLA-PHB blends [36,37]; additionally, it counteracts the inherent brittleness of PLA and PHB [38] and thereby improves the processability of the biopolymer and other properties required for food packaging applications. Hence, in this study, an oligomer lactic acid (OLA) was used as a compatibilizer and dispersing aid for CNCs and ChNCs in the manufacturing of bionanocomposite films.

The present study is the final phase of a pilot-scale project for the production of PLA-PHB-based bionanocomposite films by the film-blowing process. The aim of the study was to minimize the limitations of PLA films, which include processing difficulties, stiffness, brittleness, and low barrier properties, by incorporating PHB, nanoreinforcements (cellulose and chitin nanocrystals), and OLA with a focus on developing food packaging applications. In addition, we compared two different nanomaterials in different forms to be used as reinforcement for PLA-PHB: freeze-dried CNCs, which were added via conventional melt compounding, and ChNCs in an aqueous form, which was added as liquid dispersion into the melt compounding process. The processability of the materials and the morphological, thermal, and mechanical properties were studied, and the best films were selected for evaluation of their barrier and biodegradation behavior.

2. Materials and Methods

2.1. Materials

A commercial PLA, NatureWorks Ingeo Biopolymer 2003D, for fresh food packaging and tableware was purchased from Resinex Italy srl (Mornico al Serio, BG, Italy) and was used as the main biopolymer component in the biopolymer blend. Ingeo 2003D is an extrusion grade semicrystalline polymer with a melt flow index (MFI) of 6 g/10 min (measured at 210 °C and with a load of 2.16 kg), the tensile strength of 53 MPa, 3.5 GPa tensile modulus, and 6% elongation at break, as reported by the manufacturer.

A commercial polyhydroxyalkanoate PHB, IAM NATURE B6 E15, purchased from Gruppo MAIP srl (Settimo Torinese, TO, Italy) was used as the second component in the biopolymer blend. The PHB is an extrusion-grade biopolymer with an MFI of 3 g/10 min (measured at 170 °C and with a load of 5 kg), and the tensile strength of 20 MPa was reported by the manufacturer.

A biobased and biodegradable low-molecular-weight oligomer lactic acid (OLA, (Glyplast OLA2 purchased from Condensia Quimica SA, Barcelona, Spain) was used as a compatibilizer and processing aid. The OLA is a viscous liquid having a molecular weight of 500–600 g/mol, ester content >99%, a density of 1.10 g/cm^3, and a viscosity of 90 mPa·s.

Microcrystalline cellulose (MCC) and shrimp chitin were used as starting materials for the production of the nanocrystals (CNCs and ChNCs) and were purchased from Glentham Life Sciences Ltd., Corsham, UK.

The used chemicals, sulfuric acid (H$_2$SO$_4$) and potassium phosphate (K$_3$PO$_4$), and sodium hydroxide (NaOH) used in the acid hydrolysis were purchased from Brenntag N.V. (Deerlijk, Belgium).

The CNCs were prepared at a pilot scale, the MCC was hydrolyzed with 60 wt% H$_2$SO$_4$ at 40 °C for 3 h, and the suspension was neutralized by the addition of K$_3$PO$_4$ and NaOH, followed by desalting with a ceramic ultrafiltration membrane (diameter 3 × 25 × length 1178 mm) (TAMI Industries, Nyon, France). Larger MCC particles were removed via a centrifugation step using a separation centrifuge from GEA Westfalia S1 1-01 (Oelde, Germany). The produced CNCs were concentrated at 55 °C, using a 0.5 bar steam pressure in a custom-made wiped film evaporator, a filling volume of 69 L, a maximum vacuum pressure of 1 bar, and a maximum temperature of 200 °C (GEA Canzler GmbH & Co. KG, Düren, Germany), and then lyophilized using an SP VirTis Genesis pilot freeze dryer 35 L (Barcelona, Spain) to obtain 1.4 kg of dry CNCs. The drying was performed with thermal treatment, primary drying, and secondary drying of −40 °C at 500 μbar, −40 °C to 25 °C for 44 h at 500 μbar, and 25 °C for 25 h at 80 μbar (condenser vacuum and temperature constant at 600 μbar and −45 °C), respectively. The freeze-dried CNCs are shown in Figure S1 in the Supplementary Materials.

The ChNCs were produced via sulfuric acid hydrolysis on a pilot scale by following the process reported in our earlier study [10]. Briefly, the shrimp chitin was hydrolyzed in a glass-lined reactor custom-made by Thaletec GmbH (Thale Germany) with 35 wt% H$_2$SO$_4$ at 60 °C for 2 h followed by centrifugation, dialysis, and homogenization to produce ChNCs. Isolated ChNCs had diameters and lengths ranging from 5 to 15 nm and 200 to 480 nm, respectively [10]. The supplied ChNCs (4 wt%) were then concentrated using a vacuum rotary evaporator to achieve the 18 wt% concentration.

2.2. Methods

2.2.1. Blend Processing

The compositions of the prepared formulations are shown in Table 1. A masterbatch of PLA:PHB blend was prepared by ensuring the homogeneous mixing of biopolymers in a co-rotating twin-screw extruder (Leistritz 27E, Nuremberg, Germany) with a screw diameter (D) of 27 mm and a length/diameter ratio of 40. The screw speed was kept constant at 300 rpm, and the temperature profile was set in the range of 165–175 °C. The composition of the blend was preliminarily defined, and PLA and PHB were mixed in the specific ratio of 75:25 (wt%), respectively.

Table 1. Compositions of the biopolymer blends and nanocomposites are given in weight %.

Materials	PLA-PHB (75-25)	OLA	CNC	ChNC
PLA-PHB	100	-	-	-
PLA-PHB-OLA	96	4	-	-
PLA-PHB-OLA-CNC	95	4	1	-
PLA-PHB-OLA-ChNC	95	4	-	1

2.2.2. Nanocomposite Processing

The PLA:PHB nanocomposite with OLA and CNCs was produced using conventional melt extrusion, in which the requisite amount of freeze-dried CNCs along with OLA was added to PLA-PHB blend using a twin-screw extruder (Leistritz 27E), at 300 rpm and the temperature profile in the range of 185–195 °C.

The PLA-PHB, OLA, and ChNCs were prepared via a liquid-assisted extrusion process using a co-rotating twin-screw extruder (Coperion W&P ZSK-18 MEGALab, Stuttgart, Germany). The PLA-PHB blend was fed using K-Tron gravimetric feeder (Niederlenz, Switzerland) attached to the extrusion. An aqueous suspension of ChNCs and OLA in the requisite amounts was added to produce nanocomposites. The suspension used for the liquid-assisted process was prepared by pre-dispersing a concentrated ChNC gel (18 wt%)

in ethanol at a ratio of 1:5 (water:ethanol) using magnetic stirring for 2 h, followed by the addition of OLA. The suspension was ultrasonicated prior to extrusion and preliminarily heated at 50 °C with a heating plate before dosing.

2.2.3. Film Blowing

The pellets from each formulation were dried at 80 °C for at least 4 h prior to extrusion. A MiniBlown 25 single-screw monolayer extrusion line (EUR.EX.MA Tradate, Italy) with a standard screw design (length 80 cm, diam 25 mm) attached with a circular opening die and chill rolls 380 for blow extrusion was used. For all the formulations, the velocity was set to 35 rpm and the throughput was 3.5 kg/h. The temperature profile was set at 170 °C in the barrel and 175 °C at the die. More than 2 kg of pellets was used for each formulation to ensure a steady film output. The facilities for the extrusion and film-blowing processes are shown in Figure 1.

Figure 1. Material preparation process: (**a**) melt extrusion for blending PLA-PHB, OLA, and CNCs or ChNCs; (**b**) film blowing process; (**c**) the final nanocomposite bag.

The melt flow index (MFI) of the prepared pellets was measured as an indication of the blowability of the polymers using a modular melt flow device (Instron Ceast Model MF30, Pianezza, TO, Italy). The test was performed in accordance with the ASTM D1238 standard and EN ISO 1133 methods, at 190 °C with a 2.16 kg load. The procedure was repeated at least 3 times for each sample, and the average values are reported in grams per 10 min.

The thermal stability of the films was studied using the thermogravimetric analysis (TGA) Q500 model from TA Instruments (New Castle, DE, USA). The tests were performed in a temperature range from 50 to 800 °C in N_2 atmosphere, with a heating rate of 10 °C/min (gas flow of 60 mL/min).

The thermal properties, glass transition temperature (T_g), cold crystallization temperature (T_{cc}), melting temperature (T_m), and degree of crystallinity (X_c) were studied using the differential scanning calorimetry (DSC) Q800 model from TA Instruments (New Castle, DE, USA). A single heating scan was set from −50 to 210 °C, with a heating rate of 10 °C/min, to analyze X_c, which was calculated using the following Equation (1) [39]:

$$Xc = (\Delta H_m - \Delta H_{cc})/\Delta H°_m \times (100/w) \qquad (1)$$

where ΔH_m is the melting enthalpy, ΔH_{cc} is the enthalpy of cold crystallization, $\Delta H°m = 93$ J/g for PLA, and w is the weight fraction of the measured film.

The morphology of the film fractured surface was studied by scanning electron microscopy (SEM) using a JEOL JSM-6460LV (JEOL, Tokyo, Japan) at an acceleration voltage of 5 kV. The samples were cryo-fractured in liquid nitrogen and the surface was sputter coated with platinum using an EM ACE200 Leica vacuum coater (Wetzlar, Germany) to avoid charging.

A Shimadzu AG conventional tensile tester (Kyoto, Japan) was used to test the mechanical properties of the materials. The samples were cut with a rectangular press mold with dimensions of 50 mm in length, 5.5 mm in width, and 44–56 µm in thickness. Prior to the testing, the samples were conditioned for 24 h at 25 °C in 50% RH. The test was carried out in the blowing direction (MD) using the following test parameters: a gauge length of 20 mm, strain rate of 2 mm/min, and 1 kN load cell. The tensile strength and elongation at break were directly available from the software, while the toughness and tensile modulus were calculated from the stress–strain data. The procedure was repeated for at least five samples and the average value is presented.

Oxygen permeability (OP), water vapor permeability (WVP), and biodegradability were carried out on the best material formulation based on the preliminary characterizations to determine the potential of the film used for food packaging.

OP and WVP tests were performed on the films using Multiperm 037 equipment (ExtraSolution, Fosciana, LU, Italy), according to the ASTM F2622-08 and ASTM F1249-20, respectively. The surface area of the square-formed films was 2 cm^2. The films were previously conditioned for 6 h under a continuous flux of electronically controlled anhydrous nitrogen. This preliminary step is necessary to stabilize the specimens and to remove the oxygen already present inside the sample before the beginning of the test. The duration of this phase is strongly related both to the barrier properties and to the thickness of the material under testing. The greater the thickness of the specimen, the longer the conditioning phase will be. The test was performed at 23 °C and 50% RH relative humidity. The flux of the oxygen on the surface of the film was maintained at 13.5 mL/min on average. Two specimens were tested for each formulation.

The biodegradation test under composting conditions of the selected films was performed according to the modified standard ISO 20200. It corresponds to the physical fragmentation of the organic tested material into very small parts, due to the action of temperature, humidity, and the presence of micro-organisms. A specific composting environment was established in a perforated vessel reactor (30 × 20 × 10 cm^3) by following the precise recipe specified in the standard, which includes 40 wt% sawdust, 30 wt% rabbit feed, 10 wt% ripe compost, 10 wt% corn starch, 5 wt% saccharose, 4 wt% corn seed oil, and 1 wt% urea (CAS N. 57-13-6, Merck Life Science, Rahway, NJ, USA). Around 5–6 g of each test specimen was (cut in 25 mm × 25 mm) buried at 4–6 cm depth in the vessel. The samples were incubated in a ventilated oven at 58 °C, and periodic and standard-determined monitoring of the weight of the samples was performed manually. Further, control was performed at 7, 14, 21, 28, and 45 days to control the level of disintegration. During the first week, a daily control was conducted by weighing the reactor vessel and, if necessary, adding water to restore the initial mass. After 30 days, the restoration of water was performed twice a week. After each extraction day, photographs of the films were taken for a qualitative evaluation of the physical degradation in compost over time.

3. Results

3.1. Material Processability

The MFI for PLA is 6.0 g/10 min, while the MFI of PLA-PHB blend and PLA-PHB-OLA masterbatch was 6.9 and 10.3 g/10 min, respectively. It appears that the addition of OLA affected the fluidity of the PLA-PHB blend. The higher MFI value in PLA-PHB-OLA can be attributed to the high polymer chain mobility, which decreases the melt viscosity and allows the polymer to flow easier [34]. On the other hand, the addition of CNCs and ChNCs resulted in a significant decrease in MFI to 6.4 g/10 min and 5.7 g/10 min in PLA-PHB-OLA-CNC and PLA-PHB-OLA-ChNC, respectively. This behavior is explained by the nanocrystals restricting the polymer chain mobility, resulting in an increase in polymer viscosity and melt strength [15,34].

The film blowing of the prepared material formulations was performed under stable processing conditions. The blowing process for the neat PLA-PHB blend was quite unstable due to bubble breaks and strain hardening caused by the poor melt strength and stiffness of the blend, resulting in wrinkles and creases with high thickness variation (Table 2) in the resultant film. The addition of OLA in PLA-PHB resulted in the formation of a smooth and stable process without any melt fracture during the blowing process of the PLA-PHB-OLA film because of the improved flowability. The addition of dry CNCs also caused difficulties in the blowing process and exhibited film with non-homogeneous thickness with some creases. Meanwhile, the film blowing from the nanocomposite pellet PLA-PHB-OLA-ChNC exhibited stable processing conditions resulting in a smooth and uniform film.

Table 2. The thickness of blown films produced from the compounded materials.

Materials	Film Thickness (μm)
PLA-PHB	52 ± 12
PLA-PHB-OLA	48 ± 5
PLA-PHB-OLA-CNC	48 ± 8
PLA-PHB-OLA-ChNC	47 ± 3

Figure 2 shows the visual appearance and microstructure of the prepared films. It was observed that most of the PLA-PHB blend film surface has wrinkle marks (strain hardening) in the blowing/machine directions as shown in Figure 2a1. The strain hardening occurs because of the stiffness of the polymer that is stretched in the machine direction during the process, causing fluctuation in internal bubble pressure and width of the bubble. However, this issue was resolved by the addition of OLA, resulting in an even, homogeneous, and smooth surface. On the other hand, the addition of CNCs to PLA-PHB-OLA resulted in a rough film with many white dots on the surface (Figure 2a3), which resembled large flakes when observed through a microscope (Figure 2b3). These particles are most likely CNC agglomerates that were not dispersed during nanocomposite preparation. The formation of these CNC agglomerates can be attributed to the freeze-drying step before the melt extrusion process, which resulted in an irreversible agglomeration that cannot be redispersed in the polymer matrix during the compounding process. Zaccone et al. [40] reported a similar phenomenon when dry ChNCs were fed into the PHB matrix, resulting in large ChNC agglomerates on the cast film surface. In contrast, liquid feeding of ChNCs in PLA-PHB-OLA-ChNC resulted in uniform, even, and smooth film with no evidence of nanomaterial agglomerates (Figure 2a4), indicating the positive effect of the liquid-assisted extrusion process, which preserves ChNCs in a dispersed state in the polymer matrix.

The morphology of the blown film cryo-fractured surface and the dispersion of CNCs and ChNCs in PLA-PHB-OLA-CNC and PLA-PHB-OLA-ChNC, respectively, were evaluated using SEM, with the results shown in Figure 3. It can be seen that the PLA-PHB (Figure 3a) exhibited a brittle fractured surface, while PLA-PHB-OLA (Figure 3b) showed a smoother and more homogeneous surface due to the plasticizing effect of OLA. A fractured surface of the PLA-PHB-OLA-CNC blown film exhibited a void (white arrow in Figure 3c) which was most likely caused by CNC agglomerates present during sample preparation. However, PLA-PHB-OLA-ChNC (Figure 3d) exhibited a smooth fractured surface with no evidence of ChNC agglomerates, indicating that the liquid-assisted extrusion technique combined with the OLA resulted in good ChNC dispersion.

3.2. Material Properties

The thermal properties of PLA-PHB, PLA-PHB-OLA, PLA-PHB-OLA-CNC, and PLA-PHB-OLA-ChNC films were investigated using TGA and DSC. The influence of the addition of CNC and ChNC reinforcements on the thermal properties of the PLA-PHB-OLA matrix was investigated.

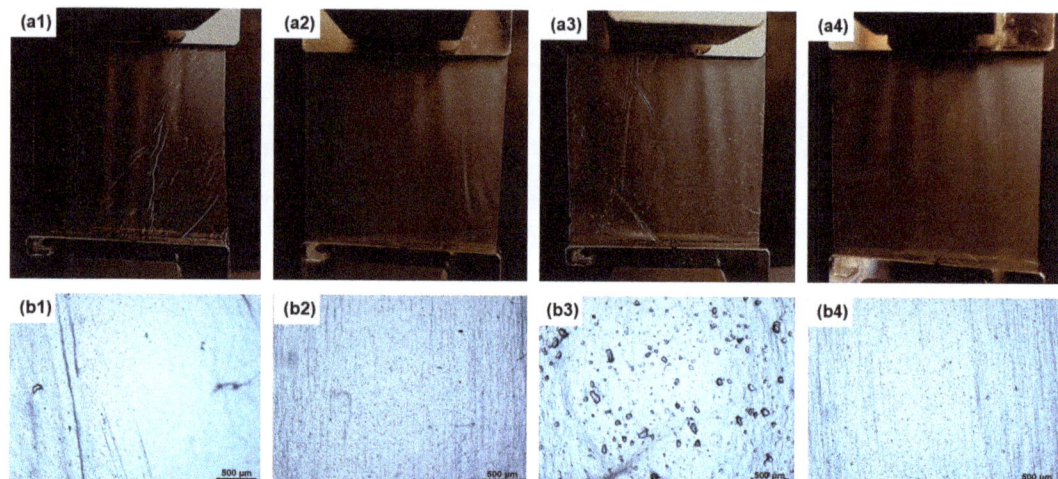

Figure 2. The visual appearance of PLA-PHB, PLA-PHB-OLA, PLA-PHB-OLA-CNC, and PLA-PHB-OLA-ChNC is shown in (**a1**–**a4**), respectively, and optical micrographs of PLA-PHB, PLA-PHB-OLA, PLA-PHB-OLA-CNC, and PLA-PHB-OLA-ChNC are shown in (**b1**–**b4**), respectively (scale bar 500 μm).

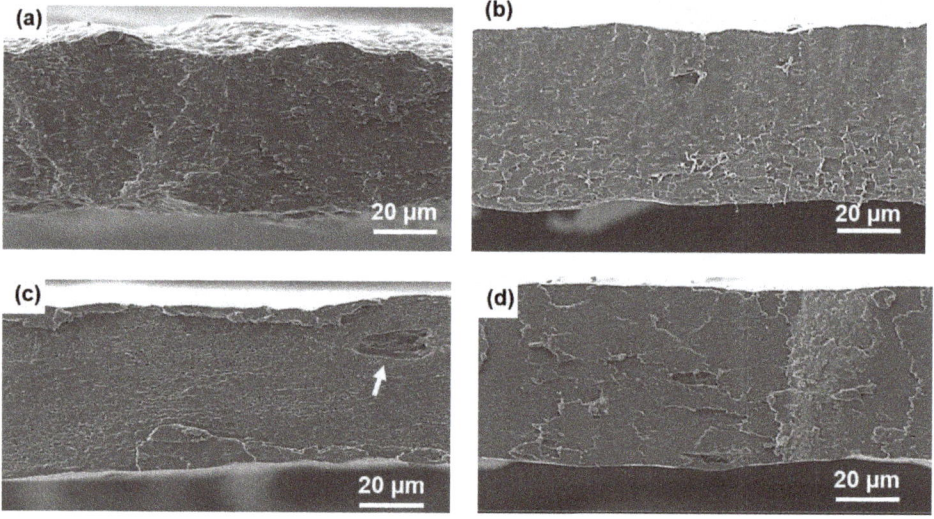

Figure 3. SEM images of the cryo-fractured surfaces of (**a**) PLA-PHB, (**b**) PLA-PHB-OLA, (**c**) PLA-PHB-OLA-CNC, and (**d**) PLA-PHB-OLA-ChNC.

TGA results of the blown film materials are summarized in Table 3, and the thermograms are shown in Figure S2 in the Supplementary Materials. The thermal degradation of the materials exhibited similar behaviors, and the decomposition process took place in a stepwise manner; the first decomposition temperature represents the PHB degradation, while the second temperature is related to PLA degradation. T_{onset}, the temperature corresponding to the initial 5 wt% weight loss, was above 200 °C for all the formulations, confirming the stability of the materials under the chosen process temperature without the risk of thermal degradation. The addition of OLA to the reference compound (PLA-PHB) did not change its thermal degradation behavior, as evidenced by the small differences in T_{onset} and T_{max} between PLA-PHB and PLA-PHB-OLA. The presence of the CNCs and ChNCs had a noticeable impact on both T_{onset} and T_{max}, which delayed the beginning and maximum thermal decomposition process in both PLA-PHB-OLA-CNC and PLA-PHB-OLA-ChNC, resulting in a slight shift toward higher temperatures compared to reference material. These phenomena were also observed in our previous studies, in which ChNCs delayed the first degradation step of TEC and GTA plasticizer in PLA-TEC-ChNC [34] and PLA-PHB-GTA-ChNC [10], respectively. The inorganic residues of all the materials, obtained at the end of the test, were somewhat similar.

Table 3. TGA of the blown films (* T_{onset} is initial 5% weight loss).

Materials	T_{onset} * (°C)	T_{max} (°C)		Residue (%)
		PHB	PLA	
PLA-PHB	282	293	364	0.43
PLA-PHB-OLA	282	295	362	0.47
PLA-PHB-OLA-CNC	288	299	366	0.53
PLA-PHB-OLA-ChNC	289	302	369	0.50

The thermal properties from the first heating DSC scan of each film are summarized in Table 4, while the thermograms are shown in Figure S3 in the Supplementary Materials. It is observed that all formulations followed a nearly identical pattern. Furthermore, all the films exhibited a single peak for glass transition temperature (T_g) and melting temperature (T_m), which corresponds to PLA. The single melting peak, according to Zhang and Thomas [8], suggests a higher degree of miscibility between PLA and PHB. In addition, it is evident that the addition of OLA had no significant effect on T_g and T_m but shifted the T_{cc} value toward lower temperatures (from 116 to 104 °C). This decrease in T_{cc} was primarily attributable to an increase in chain mobility because of OLA, which allows the material to recrystallize during the heating process, as evidenced by the increase in the Δ_{cc} value of PLA-PHB-OLA. However, the presence of CNC and ChNC restricted the chain mobility and consequently increased the T_g and T_{cc} values. The addition of OLA decreased the degree of crystallinity (X_c) in PLA-PHB-OLA, while the nanocomposite films were slightly more crystalline than the control (PLA-PHB), indicating an increase in the nucleation of spherulites due to the CNCs and ChNCs, which is consistent with our previous findings. The combination of ChNC and OLA produced a higher crystallinity degree, confirming its ability to act as a nucleating agent. The higher crystallinity results in higher melt enthalpy (ΔH_m) of the nanocomposite.

Table 4. Thermal properties of blown film materials obtained from DSC first heating scan.

Materials	T_g (°C)	T_{cc} (°C)	ΔH_{cc} (J/g)	T_m (°C)	ΔH_m (J/g)	Crystallinity (X_C%)
PLA-PHB	52	116	6.6	150	16.0	10.2
PLA-PHB-OLA	51	104	12.6	151	17.2	5.2
PLA-PHB-OLA-CNC	57	117	7.0	151	17.6	12.1
PLA-PHB-OLA-ChNC	56	110	9.0	146	21.2	13.8

In summary, the presence of CNCs and ChNCs in the formulations results in greater thermal stability, indicating that the presence of CNCs or ChNCs along with OLA has a reinforcing effect on the PLA-PHB-OLA-ChNC film.

The influence of processing and the effect of CNC and ChNC reinforcements on mechanical properties were investigated using a tensile test. The blown films were tested in the blowing direction (MD), and the mechanical properties, including Young's modulus, tensile strength, elongation at break, and work of the fracture, are summarized in Table 5, while the representative stress–strain curves for all the films are shown in Figure S4 in the Supplementary Materials.

Table 5. Mechanical properties of the blown films tested in the blowing direction compared with PLA, LDPE, and PP.

Materials	Strength (MPa)	Modulus (GPa)	Elongation at Break (%)	Toughness (MJ/m^3)
PLA-PHB	26.8 ± 3.4	2.9 ± 0.1	2 ± 1	1.8 ± 0.8
PLA-PHB-OLA	24.3 ± 2.1	2.7 ± 0.3	58 ± 6	13.6 ± 0.8
PLA-PHB-OLA-CNC	23.4 ± 0.9	2.8 ± 0.3	11 ± 0	4.9 ± 0.7
PLA-PHB-OLA-ChNC	36.8 ± 2.8	2.9 ± 0.1	71 ± 9	19.8 ± 1.3
PLA (data sheet)	53	3.5	6	-
LDPE [41]	8.3–31.4	0.17–0.28	100–650	-
PP [41]	31–41.4	1.14–1.55	100–600	-

The results indicate that blown film from the PLA-PHB blend showed reduced mechanical properties compared to the neat PLA properties reported by the manufacturer. It is possible that the wrinkled and uneven surface of the PLA-PHB films, caused by an unstable film-blowing process, hindered good stress transfer during the tensile test. This may explain the relatively lower mechanical properties. The addition of OLA to the PLA-PHB blend had minimal impact on the tensile strength (σ_{max}) and Young's modulus, whereas the flexibility of the material significantly increased to 58%, which is 29 times higher than that of the PLA-PHB blend, due to the plasticizing effect [10,34,42], resulting in a tougher material.

The addition of 1 wt% CNCs had a negative influence on the mechanical properties and resulted in a significant decrease in elongation at break (from 57.6% to 11.2%) in the PLA-PHB-OLA-CNC nanocomposite film. The decrease in flexibility of nanocomposites has been attributed to the presence of a large number of CNC agglomerates that act as stress concentrators and promote the propagation of interface-generated defects, resulting in film failure [43], which is in good agreement with optical microscopy.

However, the addition of ChNCs resulted in a substantial improvement in the mechanical performance of the PLA-PHB blend. The tensile strength increased from 24 MPa (PLA-PHB-OLA) to 37 MPa (54%), the modulus increased from 2.7 to 2.9 GPa, and the elongation at break increased from 58% to 71%, showing the best properties among all blown films. In addition, when compared to the PLA-PHB blend, the nanocomposite films were found to be more stretchable (35 times), as well as tougher (10 times); these are important properties, especially for film blowing. The reason that the addition of the ChNCs can improve the PLA-PHB blend behavior is expected to be related to the homogeneously dispersed ChNCs in PLA-PHB matrix, as well as strong interfacial adhesion and formation of a rigid ChNC percolating network within the polymer matrix which would facilitate an effective stress transfer from the polymer matrix to rigid ChNCs network, resulting in an improvement in the strength and toughness of polymer nanocomposites [24,44].

Films for food packaging are required to maintain their integrity in order to withstand the stress that occurs during shipping, handling, and storage [45]. In comparison to petroleum-based polymers such as polypropylene (PP) and low-density polyethylene (LDPE), which are frequently used in packaging applications, PLA-PHB-ChNCs nanocomposite films were shown to be stiffer, while maintaining comparable tensile strength and stretchability. Therefore, PLA-PHB-OLA-ChNC nanocomposite film, produced with liquid-assisted extrusion and the film-blowing process, could be considered the best formulation and process for food packaging applications.

The oxygen and water vapor barrier properties of PLA-PHB-OLA and PLA-PHB-OLA-ChNC films were investigated to determine the impact of the ChNCs, and the results are summarized in Table 6. Because the tests were conducted on blown films, the permeation parameter is critical and needs to be considered for both the OTR and WVTR to obtain values that are comparable between the two materials. The OTR value for PLA-PHB-OLA film was 1226 cc/m² 24 h, which is approximately 34% lower than the OTR for pure PLA. The decrease in OTR value can be attributed to the increase in crystallinity caused by blending with a high-crystallinity polymer (PHB). The addition of ChNCs has a further two-fold effect on the OTR and reduces it from 1226 to 630 cc/m² 24 h, which could be explained by a higher degree of crystallinity in the nanocomposites as well as inherent barrier properties of the ChNCs, leading to an increased tortuous path and consequently slowing down the permeation rate. It is worth noting that the OTR value of the final nanocomposite film is noticeably lower than that of commercial polymers that are commonly used for food packaging applications, such as low-density polyethylene and polypropylene. While the addition of ChNCs decreased the gas permeability, an opposite behavior is seen in water vapor permeability. Indeed, this value for PLA-PHB-OLA-ChNC is higher than that for the corresponding PLA-PHB-OLA film. The hydrophilic nature of ChNCs, which attracts more water molecules and leads to an increase in WVTR value, is one of the possible explanations for this phenomenon.

Table 6. Oxygen and water vapor barrier properties of produced materials.

Materials	Oxygen Permeability	
	OTR (cc/m² 24 h)	Permeation (cc μm /m² 24 h)
PLA-PHB-OLA	1226	67,244
PLA-PHB-OLA-ChNC	630	25,178
PLA [10]	1853	-
PE [46]	3374	-
PP [46]	1589	-

	Water vapor permeability		
Materials	WVTR (cc/m² 24 h)	Permeance (g/m² 24 h mmHg)	Permeability (g mm/m² 24 h mmHg)
PLA-PHB-OLA	33.8	2.4	0.12
PLA-PHB-OLA-ChNC	93.5	6.7	0.18

Figure 4 displays the visual appearance of the PLA-PHB-OLA and PLA-PHB-OLA-ChNC films after different time periods of disintegration in composting conditions. It is possible to observe the beginning of the disintegration of both films after 14 days as an initial cracking of the tested samples is visible. After 21 days, however, both materials exhibited strong fragmentation and a significant change in color, transitioning from a transparent appearance to yellowish color of the disintegrated particles. Nevertheless, the presence of ChNCs in PLA-PHB-OLA formulation seems to accelerate the disintegration process. These findings are in good agreement with the study reported by Arrieta et al. [47], where chitosan was found to accelerate the disintegration process of the PLA-PHB blend. Both materials are completely disintegrated after 45 days, which corresponds to the minimum amount for compliance with the adopted standard. In addition, the fragments of PLA-PHB-OLA-ChNC visible to the naked eye are smaller than the PLA-PHB-OLA ones.

The accelerating behavior of ChNCs in disintegration may be attributed to the sensitivity of this nanofiller to humidity and water vapor. These results can be linked to the poor water vapor barrier properties of the nanocomposite film. In fact, Asri et al. [48] demonstrated this kind of correlation, reporting that nanocomposites with higher water vapor permeability also exhibit better disintegration behavior. This phenomenon can be mainly explained because hydrophilic nanocomposites easily allow the microorganism to penetrate the material, accelerating the breakdown of polymeric chains and increasing the degradation rate.

Figure 4. Disintegration of produced nanocomposites under composting conditions: pictures of specimens at different test days.

4. Conclusions

A pilot-scale approach to produce PLA-PHB-based nanocomposite films reinforced with CNCs and ChNCs intended for food packaging was studied. A masterbatch of PLA-PHB blend with a 75:25 ratio was prepared using twin-screw extrusion followed by the compounding of the nanocomposites and film-blowing process. A nanocomposite with CNCs was made by feeding dry CNCs in the compounding process, while a nanocomposite with ChNCs was produced via the liquid-assisted extrusion method.

The nanocomposite produced via liquid-assisted extrusion exhibited a stable blowing process with smooth and homogeneous film, whereas the addition of dry CNCs resulted in agglomerates and non-homogeneous and rough surfaces in the produced blown film. The addition of both CNCs and ChNCs improved the thermal stability of nanocomposite films. The combination of OLA and well-dispersed ChNCs showed a reinforcing effect and simultaneously increased the tensile strength and stiffness compared to the biopolymer films and nanocomposite with CNCs.

The ChNC nanocomposite films were found to have better strength (37%), flexibility (35-fold), and toughness (10-fold) and comparable modulus when compared with the PLA-PHB blend.

The synergic effect of better-dispersed ChNCs with the assistance of OLA resulted in increased crystallinity and, thereby, an improvement in the oxygen barrier performance when compared to neat PLA.

In addition, the disintegration behavior under composting conditions of the best films was studied, and the nanocomposite films with ChNCs showed accelerated degradation.

The findings of this research indicate that the multifunctional PLA-PHB-OLA-ChNC film prepared via liquid-assisted extrusion and film blowing has great potential as a flexible film and opens a new perspective for its industrial application as short-term food packaging.

Supplementary Materials: The following supporting information can be downloaded at: https://www.mdpi.com/article/10.3390/polym14235104/s1, Figure S1: (a) SEM image of cellulose nanocrystals before freeze drying; (b) freeze-dried cellulose nanocrystals; (c) optical microscopy of dried nanocrystals showing agglomeration. Figure S2: TGA thermogram of blown films. Figure S3: DSC thermogram of the blown films. Figure S4. Tensile stress–strain graph of blown films in blowing direction.

Author Contributions: M.K.P.: conceptualization, data curation, formal analysis, investigation, methodology, validation, visualization, writing—original draft preparation, writing—review and editing; M.Z.: investigation, visualization, formal analysis, methodology, writing—review and editing; L.D.B.: investigation, visualization, formal analysis, methodology; R.N.: visualization, methodology, project administration, writing—review and editing; M.M.: methodology, investigation, writing—review and editing; V.M.-N.: methodology, investigation, visualization; A.F.: methodology, resources, supervision; K.O.: conceptualization, funding acquisition, project administration, resources, supervision, writing—review and editing. All authors have read and agreed to the published version of the manuscript.

Funding: This research was funded by Horizon 2020 BBI project NewPack (grant number: 792261) and Bio4Energy, National Strategic Research Program.

Institutional Review Board Statement: Not applicable.

Informed Consent Statement: Not applicable.

Data Availability Statement: Data are available from the corresponding author upon reasonable request.

Acknowledgments: The authors are grateful for the funding provided by the Bio4Energy National Strategic Research Program and Horizon 2020 BBI project NewPack (grant number: 792261). The Kempe Foundation is acknowledged for support of the infrastructure, Wallenberg Wood Science Center (WWSC) for resources, and LTU for open access publication.

Conflicts of Interest: The authors declare no conflict of interest.

References

1. Balakrishnan, P.; Thomas, M.S.; Pothen, L.A.; Thomas, S.; Sreekala, M.S. Polymer Films for Packaging. In *Encyclopedia of Polymeric Nanomaterials*; Kobayashi, S., Müllen, K., Eds.; Springer: Berlin/Heidelberg, Germany, 2014; pp. 1–8. [CrossRef]
2. Luzi, F.; Torre, L.; Kenny, J.M.; Puglia, D. Materials Bio-and Fossil-Based Polymeric Blends and Nanocomposites for Packaging: Structure-Property Relationship. *Materials* **2019**, *12*, 471. [CrossRef]
3. Chiara Mistretta, M.; Botta, L.; Arrigo, R.; Leto, F.; Malucelli, G.; Paolo, F.; Mantia, L. Bionanocomposite Blown Films: Insights on the Rheological and Mechanical Behavior. *Polymers* **2021**, *3*, 1167. [CrossRef] [PubMed]
4. Herrera, N.; Salaberria, A.M.; Mathew, A.P.; Oksman, K. Plasticized Polylactic Acid Nanocomposite Films with Cellulose and Chitin Nanocrystals Prepared Using Extrusion and Compression Molding with Two Cooling Rates: Effects on Mechanical, Thermal and Optical Properties. *Compos. Part A Appl. Sci. Manuf.* **2016**, *83*, 89–97. [CrossRef]
5. Fortunati, E.; Armentano, I.; Zhou, Q.; Puglia, D.; Terenzi, A.; Berglund, L.A.; Kenny, J.M. Microstructure and Nonisothermal Cold Crystallization of PLA Composites Based on Silver Nanoparticles and Nanocrystalline Cellulose. *Polym. Degrad. Stab.* **2012**, *97*, 2027–2036. [CrossRef]
6. Fortunati, E.; Armentano, I.; Iannoni, A.; Kenny, J.M. Development and Thermal Behaviour of Ternary PLA Matrix Composites. *Polym. Degrad. Stab.* **2010**, *95*, 2200–2206. [CrossRef]
7. Naser, A.Z.; Deiab, I.; Darras, B.M. Poly(Lactic Acid) (PLA) and Polyhydroxyalkanoates (PHAs), Green Alternatives to Petroleum-Based Plastics: A Review. *RSC Adv.* **2021**, *11*, 17151–17196. [CrossRef]
8. Zhang, M.; Thomas, N.L. Blending Polylactic Acid with Polyhydroxy butyrate: The Effect on Thermal, Mechanical, and Biodegradation Properties. *Adv. Polym. Technol.* **2011**, *30*, 67–79. [CrossRef]
9. Bucci, D.Z.; Tavares, L.B.B.; Sell, I. Biodegradation and Physical Evaluation of PHB Packaging. *Polym. Test.* **2007**, *26*, 908–915. [CrossRef]
10. Kumar Patel, M.; Hansson, F.; Pitkänen, O.; Geng, S.; Oksman, K. Biopolymer Blends of Poly(Lactic Acid) and Poly(Hydroxybutyrate) and Their Functionalization with Glycerol Triacetate and Chitin Nanocrystals for Food Packaging Applications. *ACS Appl. Polym. Mater.* **2022**, *4*, 6592–6601. [CrossRef] [PubMed]
11. Zhang, L.; Xiong, C.; Deng, X. Miscibility, Crystallization and Morphology of Poly(β-Hydroxybutyrate)/Poly(d,l-Lactide) Blends. *Polymer* **1996**, *37*, 235–241. [CrossRef]
12. Biron, M. Thermoplastic Processing. In *Thermoplastics and Thermoplastic Composites*, 2nd ed.; Biron, M., Ed.; Plastics Design Library; William Andrew Publishing: New York, NY, USA, 2013; pp. 715–768. [CrossRef]
13. Golebiewski, J.; Rozanski, A.; Dzwonkowski, J.; Galeski, A. Low Density Polyethylene–Montmorillonite Nanocomposites for Film Blowing. *Eur. Polym. J.* **2008**, *44*, 270–286. [CrossRef]
14. Scaffaro, R.; Sutera, F.; Botta, L. Biopolymeric Bilayer Films Produced by Co-Extrusion Film Blowing. *Polym. Test.* **2018**, *65*, 35–43. [CrossRef]
15. Herrera, N.; Roch, H.; Salaberria, A.M.; Pino-Orellana, M.A.; Labidi, J.; Fernandes, S.C.M.; Radic, D.; Leiva, A.; Oksman, K. Functionalized Blown Films of Plasticized Polylactic Acid/Chitin Nanocomposite: Preparation and Characterization. *Mater. Des.* **2016**, *92*, 846–852. [CrossRef]

16. Abdelwahab, M.A.; Flynn, A.; Chiou, B.-S.; Imam, S.; Orts, W.; Chiellini, E. Thermal, Mechanical and Morphological Characterization of Plasticized PLA-PHB Blends. *Polym. Degrad. Stab.* **2012**, *97*, 1822–1828. [CrossRef]
17. Burgos, N.; Armentano, I.; Fortunati, E.; Dominici, F.; Luzi, F.; Fiori, S.; Cristofaro, F.; Visai, L.; Jiménez, A.; Kenny, J.M. Functional Properties of Plasticized Bio-Based Poly(Lactic Acid)_Poly(Hydroxybutyrate) (PLA_PHB) Films for Active Food Packaging. *Food. Bioproc. Technol.* **2017**, *10*, 770–780. [CrossRef]
18. Lim, L.T.; Auras, R.; Rubino, M. Processing Technologies for Poly(Lactic Acid). *Prog. Polym. Sci.* **2008**, *33*, 820–852. [CrossRef]
19. Arrieta, M.P.; Fortunati, E.; Dominici, F.; López, J.; Kenny, J.M. Bionanocomposite Films Based on Plasticized PLA-PHB/Cellulose Nanocrystal Blends. *Carbohydr. Polym.* **2015**, *121*, 265–275. [CrossRef]
20. Zeng, J.B.; He, Y.S.; Li, S.L.; Wang, Y.Z. Chitin Whiskers: An Overview. *Biomacromolecules* **2012**, *13*, 1–11. [CrossRef]
21. Salaberria, A.M.; Labidi, J.; Fernandes, S.C.M. Chitin Nanocrystals and Nanofibers as Nano-Sized Fillers into Thermoplastic Starch-Based Biocomposites Processed by Melt-Mixing. *Chem. Eng. J.* **2014**, *256*, 356–364. [CrossRef]
22. Kadokawa, J. Preparation and Applications of Chitin Nanofibers/Nanowhiskers. In *Biopolymer Nanocomposites*; Dufresne, A., Thomas, S., Pothen, L.A., Eds.; John Wiley & Sons, Inc.: Hoboken, NJ, USA, 2013; pp. 131–151. [CrossRef]
23. Sanchez-Garcia, M.D.; Lagaron, J.M. On the Use of Plant Cellulose Nanowhiskers to Enhance the Barrier Properties of Polylactic Acid. *Cellulose* **2010**, *17*, 987–1004. [CrossRef]
24. Arrieta, M.P.; Fortunati, E.; Dominici, F.; Rayón, E.; López, J.; Kenny, J.M. PLA-PHB/Cellulose Based Films: Mechanical, Barrier and Disintegration Properties. *Polym. Degrad. Stab.* **2014**, *107*, 139–149. [CrossRef]
25. Singh, S.; Patel, M.; Schwendemann, D.; Zaccone, M.; Geng, S.; Maspoch, M.L.; Oksman, K. Effect of Chitin Nanocrystals on Crystallization and Properties of Poly(Lactic Acid)-Based Nanocomposites. *Polymers* **2020**, *12*, 726. [CrossRef]
26. Singh, S.; Maspoch, M.L.; Oksman, K. Crystallization of Triethyl-Citrate-Plasticized Poly(Lactic Acid) Induced by Chitin Nanocrystals. *J. Appl. Polym. Sci.* **2019**, *136*, 47936. [CrossRef]
27. Jandas, P.J.; Mohanty, S.; Nayak, S.K. Morphology and Thermal Properties of Renewable Resource-Based Polymer Blend Nanocomposites Influenced by a Reactive Compatibilizer. *ACS Sustain. Chem. Eng.* **2014**, *2*, 377–386. [CrossRef]
28. Jonoobi, M.; Harun, J.; Mathew, A.P.; Oksman, K. Mechanical Properties of Cellulose Nanofiber (CNF) Reinforced Polylactic Acid (PLA) Prepared by Twin Screw Extrusion. *Compos. Sci. Technol.* **2010**, *70*, 1742–1747. [CrossRef]
29. Wu, W.; Wang, S.; Colijn, I.; Yanat, M.; Terhaerdt, G.; Molenveld, K.; Boeriu, C.G.; Schroën, K. Chitin Nanocrystal Hydrophobicity Adjustment by Fatty Acid Esterification for Improved Polylactic Acid Nanocomposites. *Polymers* **2022**, *14*, 2619. [CrossRef]
30. Wei, L.; Luo, S.; McDonald, A.G.; Agarwal, U.P.; Hirth, K.C.; Matuana, L.M.; Sabo, R.C.; Stark, N.M. Preparation and Characterization of the Nanocomposites from Chemically Modified Nanocellulose and Poly(Lactic Acid). *J. Renew. Mater.* **2017**, *5*, 410–422. [CrossRef]
31. Dunlop, M.J.; Sabo, R.; Bissessur, R.; Acharya, B. Polylactic Acid Cellulose Nanocomposite Films Comprised of Wood and Tunicate CNCs Modified with Tannic Acid and Octadecylamine. *Polymers* **2021**, *13*, 3661. [CrossRef] [PubMed]
32. Bondeson, D.; Oksman, K. Polylactic Acid/Cellulose Whisker Nanocomposites Modified by Polyvinyl Alcohol. *Compos. Part A Appl. Sci. Manuf.* **2007**, *38*, 2486–2492. [CrossRef]
33. Oksman, K.; Mathew, A.P.; Bondeson, D.; Kvien, I. Manufacturing Process of Cellulose Whiskers/Polylactic Acid Nanocomposites. *Compos. Sci. Technol.* **2006**, *66*, 2776–2784. [CrossRef]
34. Patel, M.; Schwendemann, D.; Spigno, G.; Geng, S.; Berglund, L.; Oksman, K. Functional Nanocomposite Films of Poly(Lactic Acid) with Well-Dispersed Chitin Nanocrystals Achieved Using a Dispersing Agent and Liquid-Assisted Extrusion Process. *Molecules* **2021**, *26*, 4557. [CrossRef]
35. Burgos, N.; Tolaguera, D.; Fiori, S.; Jiménez, A. Synthesis and Characterization of Lactic Acid Oligomers: Evaluation of Performance as Poly(Lactic Acid) Plasticizers. *J. Polym. Environ.* **2014**, *22*, 227–235. [CrossRef]
36. Arrieta, M.P.; Díez García, A.; López, D.; Fiori, S.; Peponi, L. Antioxidant Bilayers Based on PHBV and Plasticized Electrospun PLA-PHB Fibers Encapsulating Catechin. *Nanomaterials* **2019**, *9*, 346. [CrossRef]
37. Luzi, F.; Dominici, F.; Armentano, I.; Fortunati, E.; Burgos, N.; Fiori, S.; Jiménez, A.; Kenny, J.M.; Torre, L. Combined Effect of Cellulose Nanocrystals, Carvacrol and Oligomeric Lactic Acid in PLA_PHB Polymeric Films. *Carbohydr. Polym.* **2019**, *223*, 115131. [CrossRef]
38. Arrieta, M.P.; Samper, M.D.; López, J.; Jiménez, A. Combined Effect of Poly(Hydroxybutyrate) and Plasticizers on Polylactic Acid Properties for Film Intended for Food Packaging. *J. Polym. Environ.* **2014**, *22*, 460–470. [CrossRef]
39. Segal, L.; Creely, J.J.; Martin, A.E.; Conrad, C.M. An Empirical Method for Estimating the Degree of Crystallinity of Native Cellulose Using the X-Ray Diffractometer. *Text. Res. J.* **1959**, *29*, 786–794. [CrossRef]
40. Zaccone, M.; Patel, M.K.; de Brauwer, L.; Nair, R.; Montalbano, M.L.; Monti, M.; Oksman, K. Influence of Chitin Nanocrystals on the Crystallinity and Mechanical Properties of Poly(Hydroxybutyrate) Biopolymer. *Polymers* **2022**, *14*, 562. [CrossRef]
41. Callister, W.D., Jr. Materials Science and Engineering—An Introduction (5th Ed.). *Anti-Corros. Methods Mater.* **2000**, *47*, 580–633. [CrossRef]
42. Herrera, N.; Mathew, A.P.; Oksman, K. Plasticized Polylactic Acid/Cellulose Nanocomposites Prepared Using Melt-Extrusion and Liquid Feeding: Mechanical, Thermal and Optical Properties. *Compos. Sci. Technol.* **2015**, *106*, 149–155. [CrossRef]
43. Orellana, J.L.; Wichhart, D.; Kitchens, C.L. Mechanical and Optical Properties of Polylactic Acid Films Containing Surfactant-Modified Cellulose Nanocrystals. *J. Nanomater.* **2018**, *2018*, 7124260. [CrossRef]

44. Chen, Y.; Gan, L.; Huang, J.; Dufresne, A. Reinforcing Mechanism of Cellulose Nanocrystals in Nanocomposites. In *Nanocellulose*; Wiley: Hoboken, NJ, USA, 2019; pp. 201–249. [CrossRef]
45. Bonilla, J.; Fortunati, E.; Atarés, L.; Chiralt, A.; Kenny, J.M. Physical, Structural and Antimicrobial Properties of Poly Vinyl Alcohol–Chitosan Biodegradable Films. *Food Hydrocoll.* **2014**, *35*, 463–470. [CrossRef]
46. Yaptenco, K.; Kim, J.; Lim, B. Gas Transmission Rates of Commercially Available Polyethylene and Polypropylene Films for Modified Atmosphere Packaging. *Philipp. Agric. Sci.* **2007**, *90*, 22–27.
47. Arrieta, M.P.; López, J.; López, D.; Kenny, J.M.; Peponi, L. Effect of Chitosan and Catechin Addition on the Structural, Thermal, Mechanical and Disintegration Properties of Plasticized Electrospun PLA-PHB Biocomposites. *Polym. Degrad. Stab.* **2016**, *132*, 145–156. [CrossRef]
48. Asri, S.E.A.M.; Zakaria, Z.; Hassan, A.; Kassim, A.H.M. Effect of Chitin Source and Content on Properties of Chitin Nanowhiskers Filled Polylactic Acid Composites. *IIUM Eng. J.* **2020**, *21*, 239–255. [CrossRef]

Article

Hydrolysis, Biodegradation and Ion Sorption in Binary Biocomposites of Chitosan with Polyesters: Polylactide and Poly(3-Hydroxybutyrate)

Svetlana Rogovina [1,*], Lubov Zhorina [1], Anastasia Yakhina [1], Alexey Shapagin [2], Alexey Iordanskii [1,*] and Alexander Berlin [1]

1. N. N. Semenov Federal Research Center for Chemical Physics Academy of Science, 119991 Moscow, Russia
2. Frumkin Institute of Physics Chemistry and Electrochemistry, Russian Academy of Science, 119071 Moscow, Russia
* Correspondence: s.rogovina@mail.ru (S.R.); aljordan08@gmail.com (A.I.)

Abstract: The film binary composites polylactide (PLA)–chitosan and poly(3-hydroxybutyrate) (PHB)–chitosan have been fabricated and their functional characteristics, such as hydrolysis resistance, biodegradation in soil, and ion sorption behavior have been explored. It was established that hydrolysis temperature and acidity of solutions are differently affected by the weight loss of these two systems. Thus, in the HCl aqueous solutions, the stability of the PHB-chitosan composites is higher than the stability of the PLA-chitosan one, while the opposite situation was observed for biodegradation in soil. The sorption capacity of both composites to Fe^{3+} ions was investigated and it was shown that, for PHB-chitosan composites, the sorption is higher than for PLA-chitosan. It was established that kinetics of sorption obeys the pseudo-first-order equation and limiting values of sorption correspond to Henry's Law formalism. By scanning electron microscopy (SEM), the comparative investigation of initial films and films containing sorbed ions was made and the change of films surface after Fe^{3+} sorption is demonstrated. The findings presented could open a new horizon in the implementation of novel functional biodegradable composites.

Keywords: polylactide; poly(3-hydroxybutyrate); chitosan; blend composite films; hydrolysis; water sorption; metal sorption

1. Introduction

Currently, the design of eco-friendly biodegradable polymeric materials, being capable of decomposing in the environment, is a key requirement for the numerous materials exploited in different fields of human activity, namely in biomedicine, packaging, for environment protection as absorbents, and others [1–3]. The use of synthetic polymers in food packaging [4] is one of the main causes of pollution in aquatic environments, including oceans, rivers, lakes, and agricultural irrigated lands [5]. At the same time, the presence of toxic ion metals harmful for human health in aqueous media requires the creation of new and efficient biodegradable sorbents based on natural polymers, which can decay after the end of their service life into the harmless substances [6].

To fulfill these requirements, the composite sorbents have been elaborated on the basis of biodegradable polyesters polylactide (PLA) [7,8] and poly(3-hydroxybutyrate) (PHB) [9] as the bio-based polymers synthesized from natural raw materials by chemical and microbiological methods, respectively, and natural polysaccharide chitosan [10]. Both polyesters have the relevant mechanical characteristics close to those of synthetic polymers with high hydrophobicity [11], while chitosan, as a highly hydrophilic biopolymer containing the amino groups, is a high effective sorbent for heavy metals [12]. However, its significant drawback is the swelling in aqueous media and has poor mechanical properties [13]. The development of the biodegradable binary composites based on these polyesters and chitosan allows one to combine mechanical characteristics of PLA and PHB with the high sorption capacity of chitosan. During exploitation, these composition materials are exposed to action of such aggressive factors as hydrolysis, oxidation, ozonolysis and also enzymatic biodegradation via numerous microorganisms [14–18].

In this connection, it is very important to separately evaluate the action of either factor on the complex of properties of investigated systems for their successful use as biodegradable sorbents of metal ions from aqueous media.

It is well known that, in the hydrophobic polyesters, the water sorption capacity is extremely poor and amounts to about a percent or even less [19]. For example, the maximal water capacity in PLA exposed in aqueous medium during a month is 1.3% [20]. By contrast, the hydrophilic chitosan is characterized by high water sorption values [21].

Evaluation of environmental disintegration of composite materials based on PLA and PHB is extremely important, since their ability to biodegrade is one of the key motivations for the use of these polyesters in the creation of green composites. It should be noted that the biodegradation rates of PLA and PHB differ essentially [22]. As shown in [23], PHB loses 14.2% of the initial mass within six months, while the mass of PLA samples remains practically unchanged over the same period of time. This fact can be explained by different mechanisms of PLA and PHB biodegradation. If the biodegradation of PHB in the soil occurs under the action of microorganisms, the biodegradation of PLA initially occurs hydrolytically, and then it proceeds enzymatically in accordance with the two-stage mechanism [24]. The non-enzymatic hydrolysis of PLA begins with the sorption of water, leading to the breaking of ester bonds and the formation of low molecular weight oligomers and the monomer of lactic acid [25]. It should be noted, however, that at elevated temperatures (more than 50 °C), the weight loss of PLA in the compost within the four weeks can reach 45 wt. % [26]. This is explained by PLA hydrolysis proceeding readily above its glass transition temperature (55–62 °C) [27]. Previously, in [28,29], we studied the features of hydrolysis and ozonolysis of PLA, PHB, and ternary compositions based on PLA, PHB, and chitosan, as well as their sorption properties with respect to metal ions [30]. The elaborated PLA–PHB binary compositions were successfully used for oil sorption from aqueous media [31]. The simultaneous use in the composition of two polyesters PLA and PHB performing a reinforcing function improved their mechanical strength, but made it difficult to estimate the effect of each of the polymers used in the process. The purpose of this work was to study the features of acid hydrolysis, soil biodegradation, and iron ion sorption of the binary polymeric composites' films, namely PLA-chitosan and PHB-chitosan as the biodegradable and eco-friendly materials potentially designated for metal sorption and packaging.

The novelty of the obtained data is connected to the establishment of different influences of PLA and PHB on the complex of properties of the investigated binary film composites polyester–chitosan.

2. Materials and Methods

2.1. Materials

PLA 4043D from Nature Works (Minnetonka, MN, USA) as pellets with diameter of 3 mm ($M_w = 2.2 \times 10^5$, $M_n = 1.65 \times 10^5$, $T_m = 155$ °C, polydespersity index D = M_w/M_n = 1.35, transparency 2.1%); PHB (Biomer, Kreilling, Germany) ($M_w = 2.05 \times 10^5$, $T_m = 175$ °C), chitosan produced by Bioprogress (Shchelkovo, Moscow region, Russia, $M_w = 4.4 \times 10^5$, degree of deacetylation 0.87), and anhydrous ferric chloride (Fluka Chemie, Buchs, Switzerland) were used.

2.2. Production of Compositions

Films were prepared by mixing PLA and PHB solutions in chloroform, in which chitosan in the form of a powder under mechanical stirring was introduced. The resulting films with a thickness of 0.2–0.3 mm were dried at room temperature.

2.3. Hydrolytic Degradation

Hydrolysis of the obtained films was proceeded in 0.005, 0.1 and 0.2 mol/L aqueous HCl solutions at temperatures 25, 40 and 70 °C. The samples were removed from HCl solution at certain intervals and placed in an oven for 2 h at a temperature of 90 °C after which they were weighed (measurement error d = 0.1 mg).

2.4. Biodegradability

The biodegradability of the obtained films was studied by modeling the processes occurring under the natural condition (ASTM D5988-12). The samples were placed in soil (pH = 7.5) for several weeks with the following measure of its weight loss at certain time intervals.

2.5. X-ray Fluorescence Analysis

For the investigation of iron ions sorption by composition films, they were placed in aqueous solutions of $FeCl_3$ of various concentrations and kept. The percentage content of sorbed iron ions in the film compositions was determined by method of X-ray fluorescence analysis on an ARL PERFORM'X X-ray Fluorescence spectrometer (Thermo Fisher Scientific, Waltham, MA, USA). Registration of spectra and all further manipulations with them were performed using the SIALMO.UQ method.

2.6. Scanning Electron Microscopy (SEM)

The morphology of the composite films PLA-chitosan and PHB-chitosan before and after the sorption of iron ions was studied by a method of scanning electron microscopy (SEM) using a Philips SEM-500 scanning electron microscope (The Netherlands) in the secondary electron mode at an accelerating voltage of 15 keV. Sample preparation consisted of thermal vacuum deposition of carbon on the film surface using a VUP-5 vacuum universal post (Russia). To obtain a continuous conductive carbon coating, the samples were rotated in the same plane during thermal spraying.

3. Results and Discussion

3.1. The Comparison of Hydrolysis Behavior for Binary Systems PLA-Chitosan and PHB-Chitosan

The durability and efficiency in using the binary composites based on the biopolyesters and chitosan as the toxic metal sorbents from aqueous media are largely determined by their ability for hydrolytic or enzymatic degradation. In this work, in the temperature range 25–70 °C, the hydrolysis of binary compositions PLA-chitosan and PHB-chitosan in aqueous hydrochloride acid solutions with various concentrations was studied. The results obtained are presented in Figures 1–3.

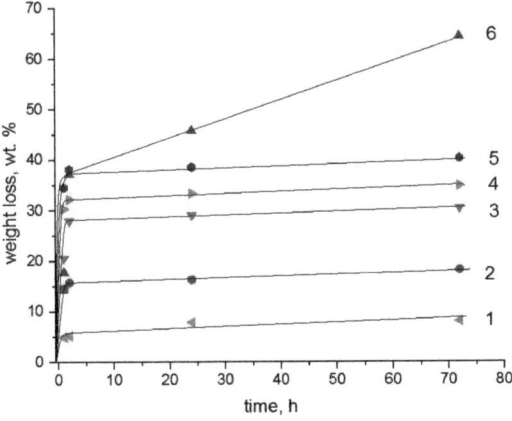

Figure 1. Weight loss kinetics for the binary composite films in the HCl solutions at 25 °C;. The numbers

show: Solution concentrations are 0.005 (1), 0.1 (3), 0.2 (5) mol/L for a PHB-chitosan system and 0.005 (2), 0.1 (4), 0.2 (6) mol/L for a PLA-chitosan system.

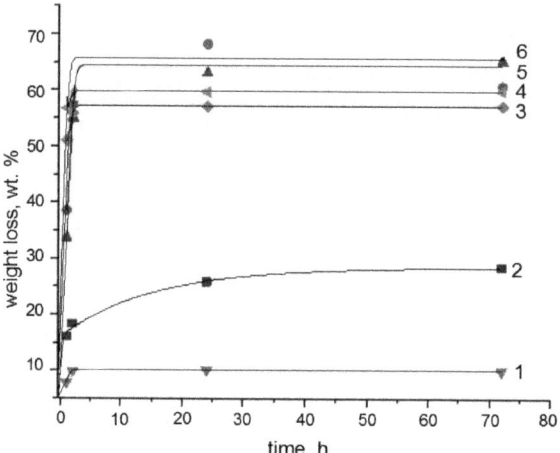

Figure 2. Weight loss kinetics for the binary composite films in the HCl solutions at 40 °C;. The numbers show: Solution concentrations are 0.005 (1), 0.1 (3), 0.2 (5) mol/L for a PHB-chitosan system and 0.005 (2), 0.1 (4), 0.2 (6) mol/L for a PLA-chitosan system.

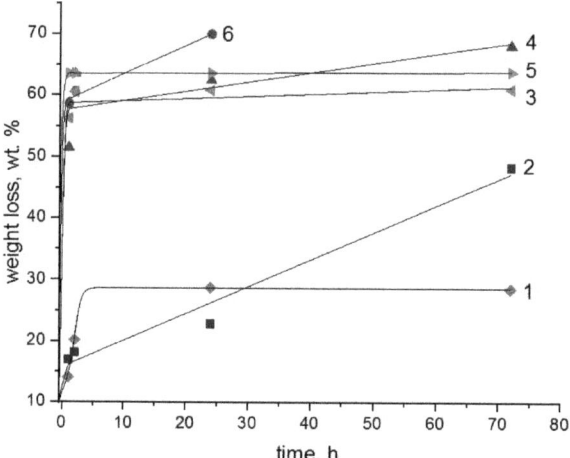

Figure 3. Weight loss kinetics for the binary composite films in the HCl solutions at 70 °C;. The numbers show: Solution concentrations are 0.005 (1), 0.1 (3), 0.2 (5) mol/L for a PHB-chitosan system and 0.005 (2), 0.1 (4), 0.2 (6) mol/L for a PLA-chitosan system.

As can be seen from the above Figures, most of the kinetic curves have two characteristic ranges. In the initial time there is a very sharp weight loss of films that takes place, after that, with the exit on the plateau or with a very slow change, weight loss is observed. The exceptions belong to the PLA-chitosan specimens exposed in extremely aggressive media, namely 70 °C; and the highest HCL concentrations. For such systems, the second kinetic stage proceeds with a remarkable rate that could be described by the linear dependence.

Based on the previous papers devoted to the study of diffusional kinetics of controlled drug release from homopolymers PHB and PLA [32–34], it is possible to propose that the initial fast stage is governed by the diffusional delivery of low molecular fraction of the polymeric components from the chitosan hydrogel matrix. The second stage is probably related to the destruction of the polyesters, PHB and PLA, that under given conditions occurs extremely slowly due to their essential crystallinity. At the given range of temperatures, all kinetic curves for PLA are located slightly higher than for PHB that again confirms somewhat higher resistance to hydrolysis of the member of the polyhydroxyalkanoates' family. It is quite clear that the hydrolysis of PLA-chitosan composition proceeds more intensively than the PHB-chitosan one, which is explained by the minor difference in the chemical structure of both polyesters, namely due to the diversity of molecular hydrophilic/hydrophobic balance in the PLA and PHB chains [35].

It should be noted that at the lowest temperature 25 °C; for the compositions PLA-chitosan and PHB-chitosan an increase in the acidity of solutions leads to the significant increases in the weight loss, but the form of the kinetic curves besides the system PLA-chitosan in 0.2 mol/L HCl solution remains the same. However, under the severest conditions at 70 °C, the weight loss kinetics curve of PLA-chitosan composition is remarkably changed in comparison with the PHB chitosan one.

3.2. Sorption of Fe^{3+} Ions by the Binary Systems PHB-Chitosan and PLA-Chitosan

The binary compositions based on biodegradable polyesters and chitosan are of great interest for their use as metal absorbents from wastewater. PLA and PHB polyesters have good mechanical properties, while chitosan, which easily swells in aqueous media, binds metal ions well. The study of electrolyte sorption capacity for the binary composite films is of great interest since such films can be easily utilized after the end of their service life without environmental violation. Recently, to estimate the feasibility of PLA–PHB–chitosan as the novel sorbent, we evaluated the sorption constants of iron and chromium ions by the above-mentioned ternary system [36]. In the present publication, to continue consideration of the ionic pollutant sorption behavior, by somewhat simplified binary systems PLA-chitosan and PHB-chitosan for which the relevant combination of good mechanical characteristics provided by the polyesters and a high ion capacity via chitosan is successfully implemented. Besides, the ion sorption comparison for these two binary composites enables the authors to reveal the role of interactions between the ester groups for the polyesters (PLA and PHB) and amino groups for chitosan. Additionally, and more important, the exploration of the binary sorbents enables the experts to estimate the difference in pollutant sorption behavior and compare the impact of both polyesters upon the chitosan capacity of Fe^{3+} ions.

The effectual recovery of ionic pollutants from the aqueous environment, primarily we say about the ions Fe^{3+}, is determined not only by the chemical structure, the morphology, and the pore pattern of biopolymer sorbent, but such key characteristics as thermodynamic affinity of the electrolyte generating the ions to functional groups of the sorbent, as well as pseudo-chemical kinetics and diffusion of the ionic pollutant [37,38]. Both kinetic processes enable the specialists to provide the complete decontamination of aqueous polluted media and to evaluate the optimal conditions for the polymer sorbent operation [39].

3.2.1. Kinetic Aspect of Fe^{3+} Sorption

The typical kinetic curves, expressing the dependences of Fe^{3+} concentration (wt. %) in the bulk compositions PLA-chitosan and PHB-chitosan (sorption, Cp) on the time of exposition (t) are presented in Figure 4a,b correspondingly. As is shown in these Figures, all the curves have well-defined limits, the values of which are determined by the concentration of the electrolyte, $FeCl_3$, in the aqueous volume (Cv). Within the interval 0.002–0.008 mol/L, with the increase in Cv, the limited values of sorption Cp_∞ are increased as well. Last finding corresponds to the positive increment of the functional dependence Cp = f(Cv) featured for most of the sorption isotherms, such as Langmuir, BET, GAB and others [40].

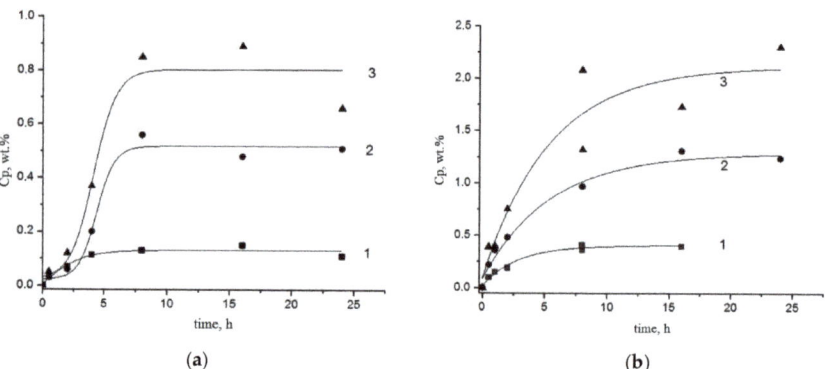

Figure 4. Kinetic curves of Fe^{3+} sorption by PLA-chitosan (**a**) and PHB-chitosan (**b**) films (50:50) wt. % from $FeCl_3$ aquatic solutions with different concentrations 0.002 (1), 0.005 (2), 0.008 (3) mol/L.

The comparison between two families of the kinetic curves for PLA-chitosan and PHB-chitosan systems showed that the limiting sorption concentration for the latter considerably exceeds the same characteristic of the former. In other words, the replacement of PLA by PHB in the chitosan-polyester composite leads to the increase in Fe^{3+} more than 100 wt. %. A reasonable cause for the observed effect may be the reduction in the ester groups interaction with amino groups, and as a consequence, the disengagement of the chitosan functional groups for the interaction with Fe^{3+} ions that essentially enhances their sorption. More specifically, the analysis of this assumption will be carried out in our next submission.

In the pseudo-first-order kinetic model of sorption, the rate of ions occupancy in the biopolymer is proportional to the amount of free active centers being available for ion immobilization [39,40]. The simplified expression of this fundamental statement in the differential form is:

$$dq/dt = k_a(q_\infty - q), \qquad (1)$$

where q is the amount of the active centers of ion sorption in the biopolymer, q_∞ is the constant characteristic corresponding to the limit value of occupied center at $t \to \infty$, t is current time of sorption and k_a is the effective constant of sorption.

The integration of Equation (1) leads readily to its own sorption kinetic equation presented in the ion concentration scale:

$$C_p = C_{p\infty}[1 - \exp(-kt)], \qquad (2)$$

where C_p and $C_{p\infty}$ are correspondingly the current and limited values of ion sorption measured at a given moment t and the limited value at $t \to \infty$ respectively; k is the effective kinetic constant of the first order differential Equation (1).

The most of ion uptake systems consisted with Equations (1) and (2) are treated in the semi-logarithmic scale to present the sorption results in the suitable form of the constants evaluation:

$$\ln[1 - (C_p/C_{p\infty})] = -kt, \qquad (3)$$

The linearity of Fe^{3+} ions sorption data in the coordinates $\ln[1-(C_p/C_{p\infty})]\sim t$ is shown in Figure 5a,b. The presented lines demonstrate the good conformity between modeling and experimental representations with higher statistical correlation coefficients. All the values of parameters for sorption Equation (3) as well as the corresponding statistical treatment of kinetic curves are presented in Table 1. The statistical coefficients (COP and Pierson coefficients calculated via "Origin2018"[C] Origin 2018 SR1 Build 9.5.1.195 software) bear evidence to the good correlation of Fe^{3+} sorption modeling with the kinetic experimental data obtained by X-ray fluorescence analysis. It is worth noting that the ion sorption kinetic constant is decreased with the volume electrolyte concentration increment.

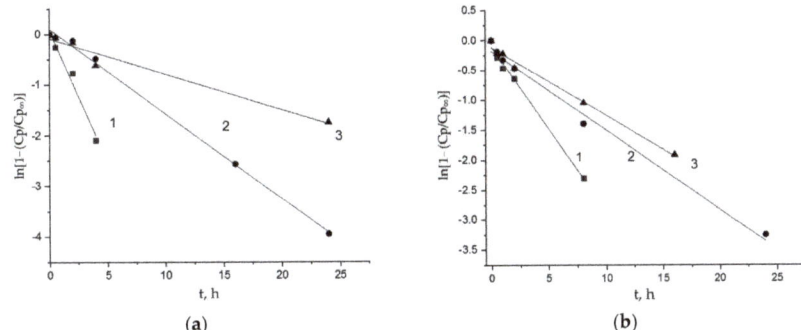

Figure 5. Kinetic curves of Fe^{3+} ions sorption by PLA-chitosan (**a**) and PHB-chitosan (**b**) in semilogarithmic coordinates. Sorption was performed from $FeCl_3$ aquatic solutions with different concentrations 0.002 (1), 0.005 (2), 0.008 (3) mol/L.

Table 1. Parameters of Fe^{3+} sorption by binary composite films PLA-chitosan and PHB-chitosan (50:50 wt. %).

$Cv(Fe^{3+})$, mol/L	k, h^{-1}	Intercept of Equation (3)	R-sq. COP	Pierson's r
		PLA-chitosan		
0.002	0.513 ± 0.053	0.051 ± 0.010	0.979	0.989
0.005	0.168 ± 0.043	0.101 ± 0.051	0.997	0.998
0.008	0.070 ± 0.081	0.089 ± 0.009	0.962	0.981
		PHB-chitosan		
0.002	0.267 ± 0.08	0.159 ± 0.032	0.998	0.993
0.005	0.129 ± 0.06	0.234 ± 0.063	0.992	0.996
0.008	0.109 ± 0.04	0.169 ± 0.036	0.996	0.998

The analogous trend was observed in our previous work [30]. It is well known that the process of sorption from aqueous medium is accompanied by transport phenomena that under determined conditions could influence the rate of contaminant recovery [41,42].

For the composites, the growth in the Fe^{3+} ion content could negatively affect the ion mobility in a given polymer matrix and hence extend the effective time of ion achievement to the active sites of sorption. For chitosan, these sites comprise predominantly –NH_2 groups and to a lesser extend –OH groups. The affinity (sorption thermodynamic impact) of ester entities belonging to PHB and PLA is essentially lower and ion sorption capacity by the polyesters can be neglected as compared to chitosan. This kinetic effect could be probably related to the concentration impact of ion diffusion in electrolyte solution (exterior diffusion) and in the biopolymer matrix (interior diffusion). We have already noted above that the process of sorption from aqueous medium is accompanied by transport phenomena that under determined conditions could influence the rate of contaminant recovery. Regularly, the increase in the concentration of electrolyte (Cv) results in the growth of its content in the biopolymer (Cp), namely in the hydrogel of chitosan as the main hydrophilic component of the composition accumulating Fe^{3+} ions. It is well known that the growth of the ion concentration leads to the enhancement of electrostatic interactions among the mobile ions and hence to the ion diffusivity decrease in a polymer matrix [43]. Consequently, owing to the diffusional hindrance, the time of iron ion access to the functional group of chitosan rises and enlarges the total time of the sorption and decreases the effective sorption constant (k).

3.2.2. Ultimate Fe^{3+} Sorption Capacity for Binary Composites: PLA-Chitosan and PHB-Chitosan and Henry's Law Model Description

The model of Fe^{3+} sorption by chitosan reinforced with PLA or PHB based on the reaction of the pseudo first order suggests the ions' immobilization on functional groups of chitosan (deacetylated-NH_2 functions). The ultimate values of Fe^{3+} sorption (Cp_∞) reflect the chitosan state when only part of its active center is occupied by the immobilized ions.

In this case, the linearity of the Cp_∞ dependence on Cv, (Figure 6) demonstrates the sorption progress in accordance with a Henry's Law formalism as the initial stage of the Langmuir isotherm [44]:

$$Cp = K_H \cdot Cv \quad (4)$$

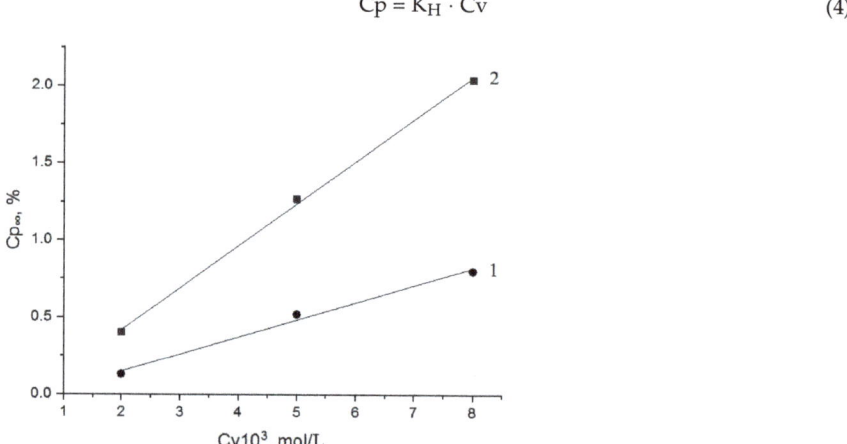

Figure 6. Ultimate Fe^{3+} sorption capacity as the function of concentration in aqueous $FeCl_3$ solutions with different molar contents (Cv) for PLA (1) and PHB (2).

The simplified character of the isotherms for both compositions testifies to the lack of interaction among ionogenic groups of the biopolymers. Regularly, such interaction should distort the form of linearity as it occurs in Langmuir or BET modeling. In the interval of $FeCl_3$ concentrations spanning the values from 2×10^{-3} to 8×10^{-3} mol/L, i.e., in the area of diluted concentrations, the ions diffuse to the sorption centers of chitosan independently without interactions with each other. In such a situation, the ion sorption should proceed independently as well, when an amino group does not cooperate with the same adjacent group in the chitosan matrix.

Besides, in the given concentration range, the number of occupied active sites for sorption under saturation (q_∞ and the corresponding values Cp_∞) are essentially far from the maximal value of the active sites in chitosan. It could also indicate that the polyesters' molecules of PHB and PLA partly screen the -NH_2 groups of chitosan and prevent intermolecular interactions among others. In given situation, we could suggest that the transport of ions is sensitive to their concentration, while the sorption process and the ion interaction with functional groups does not depend on the ion content.

3.3. Biodegradation of the Composites PLA-Chitosan and PHB-Chitosan

When studying the destruction of samples which were buried in the soil, it was found that, after exposure of the initial compositions for a month, the mass of the PHB-chitosan composition increased by 36.6 wt. % and the PLA-chitosan composition by 19.8 wt. % accordingly. The observed effect can be associated both with the ability of chitosan to absorb water from the environment and with the growth of microorganisms on the surface of the films. Since, as mentioned above, PHB is more susceptible to the action of microorganisms, it has the largest weight gain. At the same time, for both compositions containing absorbed iron ions, a decrease in the mass of samples by approximately 26 wt. % within a month takes place, that can be explained by the catalytic effect of iron ions. On Figures 7 and 8 the photographs of double films PLA-chitosan and PHB-chitosan as well as photographs of these films, containing sorbet iron ions and after exposition in soil during different times are shown.

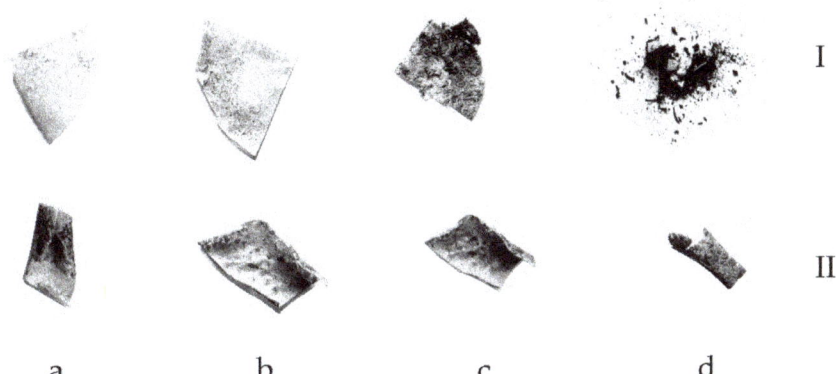

Figure 7. Photographs of PLA-chitosan films (I) and films containing sorbed iron ions (II), after exposition in soil during 2 (**a**), 4 (**b**), 7 (**c**) and 12 (**d**) weeks.

Figure 8. Photographs of PHB-chitosan films (I) and films containing sorbed iron ions (II), after exposition in soil during 2 (**a**), 4 (**b**), 7 (**c**) and 12 (**d**) weeks.

The sorption of iron ions leads to a change in the color of the samples from white to red. In addition, with increasing exposure time, their shape also changes. It should be noted that, with long-term exposure (12 weeks), in contrast to the partial destruction of the films based on PLA, the films containing PHB are completely destroyed (Figure 8), which confirms the more intensive process of biodegradation in soil for the composition containing PHB compared to films containing PLA. These data correlate with the features of degradation of compositions based on PLA and PHB described above. At the same time, it can be seen that films that do not contain iron ions are destroyed more easily than films containing iron, which is explained by hardening in the presence of metal ions.

3.4. Morphology of Binary Systems PLA-Chitosan and PHB-Chitosan

The morphology of the binary compositions was investigated by the SEM method. In Figures 9a and 10a, the photomicrographs of cross-sections of the films for PLA-chitosan and PHB-chitosan are correspondingly given. As is seen, the freeze surfaces of cross-sections of PLA and PHB are slightly different. For PLA, the surface of the fracture is more of a relief, while the surface of PHB prepared at the same conditions is smoother.

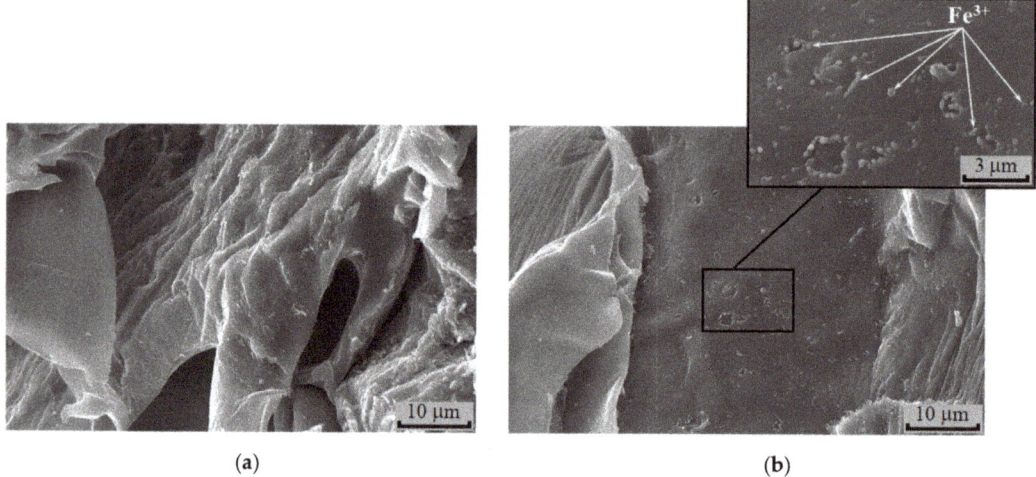

Figure 9. SEM micrographs of the initial films PLA-chitosan (50:50, wt. %) (**a**) and after sorption (**b**) from 0.3 mol/L $FeCl_3$ solution.

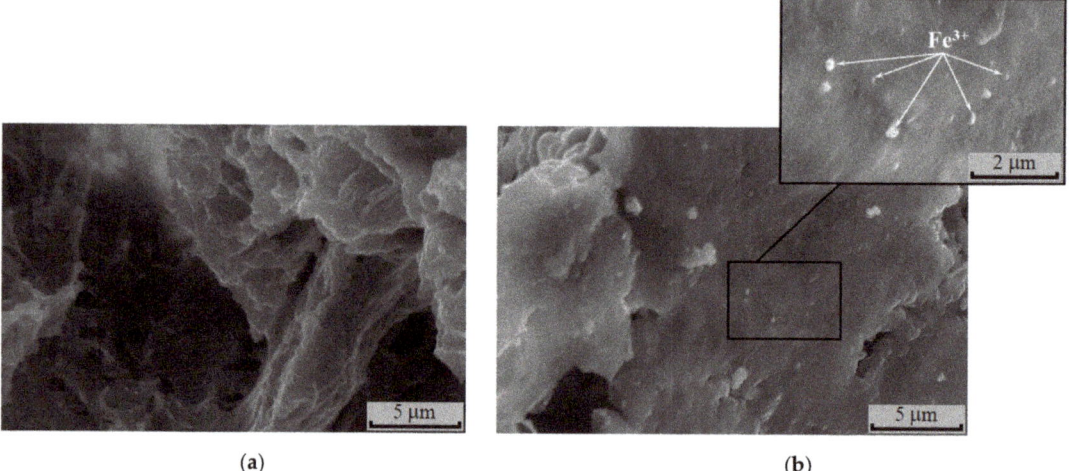

Figure 10. SEM micrographs of the initial films PHB-chitosan (50:50, wt. %) (**a**) and after sorption (**b**) from 0.3 mol/L $FeCl_3$ solution.

Due to the rather poor compatibility between polyesters and chitosan, two phases are observed. The phase separation is more reliably seen on the polymer specimens containing sorbed ions Fe^{3+} (Figures 9b and 10b). The samples exposition in the aqueous medium of $FeCl_3$ leads to chitosan structureless and the formation of areas with the fine pores. In contrast to PLA and PHB, swelling of chitosan under water action forms the smoothed fields situated between the non-swelling polyester entities. The latter should act as cross-linking agents preventing disintegration and dissolution of the chitosan phase.

4. Conclusions

The explored binary composites based on chitosan combined with the biodegradable polyesters PLA and PHB have been studied as the novel functional materials being capable to perform in aggressive media, namely aqueous environment and soil. Acid hydrolysis, biodegradation in soil and Fe^{3+} ion sorption by PLA-chitosan and PHB-chitosan composites have been studied and obtained results have been compared to understand the merits and drawbacks of the investigated systems.

It was revealed that:

1. In the temperature interval 25–70 °C in the HCl solutions (0.005–0.2) mol/L the PHB-chitosan composite stability exceeds the analogous one for PLA-chitosan composite, while the contrary situation is observed for soil biodegradation;
2. Under studying of the composites biodegradability in soil it was found that the films based on PHB degrade in a greater extent than films based on PLA, that is an additional confirmation of their different biodegradation mechanism;
3. The sorption capacity of both composites to Fe^{3+} ions was investigated and it was shown that the kinetics of sorption obeys the pseudo-first order equation and the limiting values of sorption correspond to Henry's Law formalism. The replacement of PLA by PHB in the chitosan-polyether composites leads to the increase in Fe^{3+} sorption more than on 100 wt. %.

Thus, on the basis of the obtained results, it may be concluded that the PHB-chitosan composites are more stable to the acid hydrolysis, they demonstrate better degradation in soil (may be slightly utilized after end of service life) and have a higher capacity to iron ions sorption than PLA-chitosan composites. In other words, the PHB-chitosan films are more suitable systems for Fe^3 ions sorption than PLA-chitosan ones.

Both investigated binary film composites on the base of studied biodegradable polyesters and chitosan can be considered as perspective eco-friendly biodegradable materials for use in the packing industry as well for sorbents production. Based on the obtained data, the future research should be dedicated to study thermal stability of binary composites PLA-chitosan and PHB-chitosan as well as their stability to UF action that should allow one to extend the possible areas of their application.

Author Contributions: Conceptualization, S.R.; Data curation, S.R.; Funding acquisition, A.I.; Investigation, L.Z., A.Y. and A.S.; Methodology, L.Z.; Project administration, A.B.; Supervision, A.B.; Validation, L.Z.; Visualization, L.Z.; Writing—original draft, L.Z.; Writing—review and editing, S.R. All authors have read and agreed to the published version of the manuscript.

Funding: This work was performed under financial support of the Ministry of Education and Science of the Russian Federation (contract no. FFZE-2022-0010).

Institutional Review Board Statement: Not applicable.

Data Availability Statement: The data presented in this study are available on request from the corresponding author.

Conflicts of Interest: The authors declare no conflict of interest.

References

1. Flury, M.; Narayan, R. Biodegradable plastic as an integral part of the solution to plastic waste pollution of the environment. *Curr. Opin. Green Sustain. Chem.* **2021**, *30*, 100490. [CrossRef]
2. Moshood, T.D.; Nawanir, G.; Mahmud, F.; Mohamad, F.; Ahmad, M.H.; Ghani, A.A. Biodegradable plastic applications towards sustainability: A recent innovations in the green product. *Clean. Eng. Technol.* **2022**, *6*, 100404. [CrossRef]
3. Kumar, S.R.; Shaiju, P.; O'Connor, K.E.; Babu, R. Bio-based and biodegradable polymers—State-of-the-art, challenges and emerging trends. *Curr. Opin. Green Sustain. Chem.* **2020**, *21*, 75–81. [CrossRef]
4. Khalid, M.Y.; Arif, Z.U. Novel biopolymer-based sustainable composites for food packaging applications: A narrative review. *Food Packag. Shelf Life* **2022**, *33*, 100892. [CrossRef]
5. Petersen, F.; Hubbart, J.A. The occurrence and transport of microplastics: The state of the science. *Sci. Total Environ.* **2021**, *758*, 143936. [CrossRef]

6. Arif, Z.U.; Khalid, M.Y.; Sheikh, M.F.; Zolfagharian, A.; Bodaghi, M. Biopolymers sustainable materials and their emerging applications. *J. Environ. Chem. Eng.* **2022**, *10*, 108159. [CrossRef]
7. Avérous, L. Chapter 21. Polylactic acid: Synthesis, properties and applications. In *Monomers, Polymers and Composites from Renewable Resources*; Belgacem, M.N., Gandini, A., Eds.; Elsevier: Oxford, UK; Amsterdam, The Netherlands, 2008; pp. 433–450. [CrossRef]
8. Arif, Z.U.; Khalid, M.Y.; Noroozi, R.; Sadeghianmaryan, A.; Jalalvand, M.; Hossain, M. Recent advances in 3D-printed polylactide and polycaprolactone-based biomaterials for tissue engineering applications. *Int. J. Biol. Macromol.* **2022**, *218*, 930–968. [CrossRef]
9. Palmeiro-Sanchez, T.; O'Flaherty, V.; Lens, P.N.L. Polyhydroxyalkanoate bio-production and its rise as biomaterial of the future. *J. Biotechnol.* **2022**, *348*, 10–25. [CrossRef]
10. Aranaz, I.; Alcántara, A.R.; Civera, M.C.; Arias, C.; Elorza, B.; Caballero, A.H.; Acosta, N. Chitosan: An Overview of Its Properties and Applications. *Polymers* **2021**, *13*, 3256. [CrossRef]
11. Mor, S.S.R.S.; Kishore, A.; Sharanagat, V.S. Bio-sourced polymers as alternatives to conventional food packaging materials: A review. *Trends Food Sci. Technol.* **2021**, *115*, 87–104. [CrossRef]
12. Zhang, Y.; Zhao, M.; Cheng, Q.; Wang, C.; Li, H.; Han, X.; Fan, Z.; Su, G.; Pan, D.; Li, Z. Research progress of adsorption and removal of heavy metals by chitosan and its derivatives: A review. *Chemosphere* **2021**, *279*, 130927. [CrossRef] [PubMed]
13. Srinivasa, P.C.; Ramesh, M.N.; Tharanathan, R.N. Effect of plasticizers and fatty acids on mechanical and permeability characteristics of chitosan films. *Food Hydrocoll.* **2007**, *21*, 1113–1122. [CrossRef]
14. Zhuikov, V.A.; Zhuikova, Y.V.; Makhina, T.K.; Myshkina, V.L.; Rusakov, A.; Useinov, A.; Voinova, V.V.; Bonartseva, G.A.; Berlin, A.A.; Bonartsev, A.P.; et al. Comparative Structure-Property Characterization of Poly(3-Hydroxybutyrate-Co-3-Hydroxyvalerate)s Films under Hydrolytic and Enzymatic Degradation: Finding a Transition Point in 3-Hydroxyvalerate Content. *Polymers* **2020**, *12*, 728. [CrossRef] [PubMed]
15. Olkhov, A.A.; Karpova, S.G.; Tyubaeva, P.M.; Zhulkina, A.L.; Zernova, Y.N.; Iordanskii, A.L. Effect of Ozone and Ultraviolet Radiation on Structure of Fibrous Materials Based on Poly(3-hydroxybutyrate) and Polylactide. *Inorg. Mater. Appl. Res.* **2020**, *11*, 1130–1136. [CrossRef]
16. Olkhov, A.; Karpova, S.G.; Rogovina, S.Z.; Zhorina, L.A.; Kurnosov, A.S.; Iordanskii, A.L. Effect of ozone on the structure of biodegradable matrices based on mixtures of poly(3-hydroxybutyrate), polylactide, and chitosan. *Mater. Sci.* **2022**, *4*, 35. [CrossRef]
17. Arrieta, M.P.; López, J.; Rayón, E.; Jiménez, A. Disintegrability under composting conditions of plasticized PLA–PHB blends. *Polym. Degrad. Stab.* **2014**, *108*, 307–318. [CrossRef]
18. Zhuikov, V.A.; Akoulina, E.A.; Chesnokova, D.V.; Wenhao, Y.; Makhina, T.K.; Bonartseva, G.A.; Shaitan, K.V.; Bonartsev, A.P. The Growth of 3T3 Fibroblasts on PHB, PLA and PHB/PLA Blend Films at Different Stages of Their Biodegradation In Vitro. *Polymers* **2021**, *13*, 108. [CrossRef]
19. Iordanskii, A.L.; Kamaev, P.P.; Zaikov, G.E. Water sorption and diffusion in poly(3-hydroxybutyrate) films. *Intern. J. Polym. Mater.* **1998**, *41*, 55–63. [CrossRef]
20. Valente, B.F.A.; Silvestre, A.J.D.; Neto, C.P.; Vilela, C.; Friere, C.S.R. Effect of the Micronization of Pulp Fibers on the Properties of Green Composites. *Molecules* **2021**, *26*, 5594. [CrossRef]
21. Khalid, M.N.; Agnely, F.; Yagoubi, N.; Grossiord, J.L.; Couarraze, G. Water state characterization, swelling behavior, thermal and mechanical properties of chitosan-based networks. *Eur. J. Pharm. Sci.* **2002**, *15*, 425–432. [CrossRef] [PubMed]
22. Bonartsev, A.P.; Boskhmodgiev, A.P.; Iordanskii, A.L.; Bonartseva, G.A.; Zaikov, G.E. Hydrolytic Degradation of Poly(3-hydroxybutyrate), Polylactide and their Derivatives: Kinetics, Crystallinity, and Surface Morphology. *Mol. Cryst. Liq. Cryst.* **2012**, *556*, 288–300. [CrossRef]
23. Karamanlioglu, M.; Preziosi, R.; Robson, G.D. Abiotic and biotic environment degradation of the bioplastic polymer poly(lactic acid): A review. *Polym. Degrad. Stab.* **2017**, *137*, 122–130. [CrossRef]
24. Arrieta, M.P.; Samper, M.D.; Aldas, M.; Lopez, J. On the use of PLA-PHB blends for sustainable food packing applications. *Materials* **2017**, *10*, 1008. [CrossRef]
25. Rogovina, S.; Prut, E.; Aleksanyan, K.; Krashininnikov, V.; Perepelitsyna, E.; Shaskin, D.; Ivanushkina, N.; Berlin, A. Production and investigation of structure and properties of polyethylene-polylactide composites. *J. Appl. Polym. Sci.* **2019**, *136*, 22. [CrossRef]
26. Karmanlioglu, M.; Robson, G.D. Influence of biotic and abiotic factors on the on the rate of degradation of poly(lactic) acid (PLA) coupons buried in compost and soil. *Polym. Degrad. Stab.* **2013**, *98*, 2063–2071. [CrossRef]
27. Siracusa, V.; Rocculi, P.; Romani, S.; Rosa, M.D. Biodegradable polymers for food packing: A review. *Trends Food Sci. Technol.* **2008**, *19*, 634–643. [CrossRef]
28. Rogovina, S.Z.; Zhorina, L.A.; Olkhov, A.A.; Yakhina, A.R.; Kucherenko, E.L.; Iordanskii, A.L.; Berlin, A.A. Hydrolysis of biodegradable Fibrous and Film Materials based on Polylactide and Poly(3-hydroxybutyrate) Polyesters and their Compositions with Chitosan. *Polym. Sci. Ser. D* **2022**, *15*, 447–451. [CrossRef]
29. Olkov, A.; Tyubaeva, P.; Vetcher, A.; Karpova, S.; Kurnosov, A.; Rogovina, S.; Iordanskii, A.; Berlin, A. Aggressive impacts affecting the biodegradable ultrathin fibers based on poly(3-hydroxybutyrate), polylactide and their blends: Water sorption, hydrolysis and ozonolysis. *Polymers* **2021**, *13*, 941. [CrossRef]

30. Rogovina, S.; Zhorina, L.; Iordanskii, A.; Prut, E.; Yakhina, A.; Grachev, A.; Shapagin, A.; Kuznetsova, A.; Berlin, A. New Biodegradable Absorbents Based on Polylactide, Poly(3-hydroxybutyrate) and Chitosan for Sorption of Iron and Chromium Ions. *Polym. Sci. Ser. A* **2021**, *63*, 804–814. [CrossRef]
31. Rogovina, S.; Zhorina, L.; Gatin, A.; Prut, E.; Kuznetsova, O.; Yakhina, A.; Olkhov, A.; Samoylov, N.; Grishin, M.; Iordanskii, A.; et al. Biodegradable Polylactide-Poly(3-Hydroxybutyrate) Compositions Obtained via Blending under Shear Deformations and Electrospinning: Characterization and Environmental Application. *Polymers* **2020**, *12*, 1088. [CrossRef]
32. Olkhov, A.A.; Karpova, S.G.; Bychkova, A.V.; Vetcher, A.A.; Iordanskii, A.L.C. 5 Electrospinning of fiber matrices from polyhydroxybutyrate for the controlled release drug delivery systems. In *Electrospinning*; Tański, T., Jarka, P., Eds.; IntechOpen: Rijeka, Croatia; London, UK, 2022; pp. 1–23. [CrossRef]
33. Bonartsev, A.P.; Bonartseva, G.A.; Makhina, T.K.; Iordanskii, A.L.; Myshkina, V.L.; Luchinina, E.S.; Livshits, V.A.; Boskhomdzhiev, A.P.; Markin, V.S. New poly(3-hydroxybutyrate)-based systems for controlled release of dipyridamole and indomethacin. *Appl. Biochem. Microbiol.* **2006**, *42*, 625–630. [CrossRef]
34. Wang, Y.; Qu, W.; Choi, S.H. FDA's Regulatory Science Program for Generic PLA/PLGA-Based Drug Products. American Pharmaceutical Review—The Review of American Pharmaceutical Business and Technology. *Am. Pharm. Rev.* **2016**. Available online: https://www.americanpharmaceuticalreview.com/Featured-Articles/188841-FDA-s-Regulatory-Science-Program-for-Generic-PLA-PLGA-Based-Drug-Products/ (accessed on 1 December 2022).
35. Terada, M.; Marchessault, R.H. Determination of solubility parameters for poly(3-hydroxyalkanoates). *Int. J. Biol. Macromol.* **1999**, *25*, 207–215. [CrossRef] [PubMed]
36. Zhorina, L.A.; Iordanskii, A.L.; Rogovina, S.Z.; Grachev, A.V.; Yakhina, A.V.; Prut, E.V.; Berlin, A.A. Thermal characterization and sorption of Fe III ion by ternary polylactide–poly-3-hydroxybutyrate–chitosan compositions. *Mendeleev Commun.* **2021**, *31*, 104–106. [CrossRef]
37. Brundavanam, S.; Poinern, S.G.E.J.; Fawcett, D. Kinetic and Adsorption Behaviour of Aqueous Fe^{2+}, Cu^{2+} and Zn^{2+} Using a 30 nm Hydroxyapatite Based Powder Synthesized via a Combined Ultrasound and Microwave Based Technique. *Am. J. Mater. Sci.* **2015**, *5*, 31–40.
38. Skwarczynska-Wojsa, A.L.; Chacuk, A.; Modrzejewska, Z.; Puszkarewicz, A. Sorption of calcium by chitosan hydrogel: Kinetics and equilibrium. *Desalination* **2022**, *540*, 116024. [CrossRef]
39. Beni, A.A.; Esmaeili, A. Biosorption, an efficient method for removing heavy metals from industrial effluents: A review. *Environ. Technol. Innov.* **2020**, *17*, 100503. [CrossRef]
40. Nirmala, N.; Shriniti, V.; Aasresha, K.; Arun, J.; Pugazhendhi, A. Removal of toxic metals from wastewater environment by graphene-based composites: A review on isotherm and kinetic models, recent trends, challenges and future directions. *Sci. Total Environ.* **2022**, *840*, 156564. [CrossRef]
41. Shahrin, E.W.E.S.; Narudin, N.A.H.; Shahri, N.N.M.; Nur, M.; Hobley, J.; Usman, A. A comparative study of adsorption behavior of Rifampicin, Streptomycin, and Ibuprofen contaminants from aqueous solutions onto chitosan: Dynamic interactions, kinetics, diffusions, and mechanisms. *Emerg. Contam.* **2023**, *9*, 100199. [CrossRef]
42. Asbollah, M.A.; Mahadi, A.H.; Kusrini, E.; Usman, A. Synergistic effect in concurrent removal of toxic methylene blue and acid red-1 dyes from aqueous solution by durian rind: Kinetics, isotherm, thermodynamics, and mechanism. *Int. J. Phytoremediat.* **2021**, *23*, 1432–1443. [CrossRef]
43. Iordanskii, A.L.; Shterenzon, A.L.; Moiseev, Y.V.; Zaikov, G.E. Diffusion of Electrolytes in Polymers. *Russ. Chem. Rev.* **1979**, *48*, 781. [CrossRef]
44. Czepirski, L.; Balys, M.R.; Komorowska-Czepirska, E. Some Generalization of Langmuir Adsorption Isotherm. *Int. J. Chem.* **2000**, *3*, 14.

Disclaimer/Publisher's Note: The statements, opinions and data contained in all publications are solely those of the individual author(s) and contributor(s) and not of MDPI and/or the editor(s). MDPI and/or the editor(s) disclaim responsibility for any injury to people or property resulting from any ideas, methods, instructions or products referred to in the content.

Article

Thermal and Sliding Wear Properties of Wood Waste-Filled Poly(Lactic Acid) Biocomposites

Tej Singh [1], Amar Patnaik [2], Lalit Ranakoti [3], Gábor Dogossy [4] and László Lendvai [4,*]

[1] Savaria Institute of Technology, Faculty of Informatics, ELTE Eötvös Loránd University, 9700 Szombathely, Hungary; sht@inf.elte.hu
[2] Department of Mechanical Engineering, Malaviya National Institute of Technology, Jaipur 302017, Rajasthan, India; apatnaik.mech@mnit.ac.in
[3] Mechanical Engineering Department, Graphic Era (Deemed to be University), Dehradun 248002, Uttarakhand, India; lalit_9000@yahoo.com
[4] Department of Materials Science and Engineering, Széchenyi István University, 9026 Győr, Hungary; dogossy@sze.hu
* Correspondence: lendvai.laszlo@sze.hu

Abstract: In our study, the effects of wood waste content (0, 2.5, 5, 7.5, and 10 wt.%) on thermal and dry sliding wear properties of poly(lactic acid) (PLA) biocomposites were investigated. The wear of developed composites was examined under dry contact conditions at different operating parameters, such as sliding velocity (1 m/s, 2 m/s, and 3 m/s) and normal load (10 N, 20 N, and 30 N) at a fixed sliding distance of 2000 m. Thermogravimetric analysis demonstrated that the inclusion of wood waste decreased the thermal stability of PLA biocomposites. The experimental results indicate that wear of biocomposites increased with a rise in load and sliding velocity. There was a 26–38% reduction in wear compared with pure PLA when 2.5 wt.% wood waste was added to composites. The Taguchi method with L_{25} orthogonal array was used to analyze the sliding wear behavior of the developed biocomposites. The results indicate that the wood waste content with 46.82% contribution emerged as the most crucial parameter affecting the wear of PLA biocomposites. The worn surfaces of the biocomposites were examined by scanning electron microscopy to study possible wear mechanisms and correlate them with the obtained wear results.

Keywords: poly(lactic acid); wood waste; biocomposite; sliding wear; microscopy

1. Introduction

Strict government regulations and increased environmental constraints on the burning and open-air dumping of agricultural, municipal, and industrial wastes have encouraged material scientists to develop innovative products such as biocomposites [1–3]. Several studies have highlighted that biocomposites hold the potential to be used in various applications, including automotive, infrastructure, aerospace, construction, consumer, and industrial fields [1–5]. According to reports, the forest industry was assumed to play a significant part in the economy of any country and generated a revenue of 270 billion USD worldwide in 2018 [6]. Solid wood waste is the main by-product of the forest industry, with more than 14 million tons of wood waste being generated every year, which is a significant problem when it comes to disposal [7,8]. Additionally, it was reported that more than 70 million tons of solid wood waste were generated annually during the manufacturing of wood products [9]. Furthermore, the annual production of synthetic plastics exceeds 320 million tons, which is unfortunately accumulated in productive agricultural parcels as landfills or incinerated in open-air [10]. Reducing wood waste and synthetic plastic can be achieved by replacing synthetic plastics with bioplastics to develop biocomposites. The global biocomposite market size was 16.46 billion USD in 2016, and it is estimated to reach 36.76 billion USD by 2022 with a compound annual growth rate of 14.44% [11].

Numerous scientists have examined how to improve the various properties of biocomposites, including thermal, thermomechanical, and mechanical features, by utilizing distinctive waste materials [12,13]. The potential of sugarcane bagasse and maize hull agro-wastes in thermoplastic starch-based biocomposites was studied by Dogossy and Czigany [14]. The influence of industrial wastes, namely copper slag and drill cuttings, on thermal, physical, mechanical, and antibacterial properties of poly(ε-caprolactone)-based biocomposites was investigated by Hejna et al. [15]. Kim et al. [16] studied the mechanical properties and biodegradability of rice husk and wood waste-filled poly(butylene succinate)-based biocomposites. Panaitescu et al. [17] evaluated the thermal and mechanical properties of wood waste-reinforced poly(3-hydroxybutyrate) biocomposites. Dhakal et al. [18] examined the influence of date palm fiber waste biomass on the mechanical properties of polycaprolactone-based biocomposites.

In recent years, there has been a remarkable interest in poly(lactic acid) (PLA) for many reasons, such as increasing environmental awareness, reducing product prices by capacity growth, and taking advantage of good processability [19,20]. PLA is a bio-based polymer, originating from the fermentation of corn, sugar beet, potatoes, and other agriculture-based substances. The major advantages of PLA are its biodegradability under certain temperature/pressure conditions and its non-toxic nature. It has good stiffness and strength compared with synthetic polymers, and it can be altered and adjusted for a wide range of applications, including packaging, textile, and biomedical purposes [21–26]. However, PLA has its drawbacks as well, including rapid physical aging, poor impact resistance, relatively high price, and low thermal stability. These drawbacks associated with PLA can be overcome by adopting, blending, copolymerization, or adding some materials such as filler/reinforcement [27–30].

Khan et al. [31] studied hemp hurd's impact on the mechanical properties of PLA-based biocomposites. They concluded that the evaluated tensile and flexural strength of the manufactured biocomposites decreased, whereas crystallinity and the tensile and flexural modulus improved by increasing hemp hurd content. They also pointed out that with glycidyl methacrylate grafting, the biocomposites with ≥20 wt.% hemp hurd loading demonstrated mechanical properties nearly equal to bare PLA. Orue et al. [32] explored the potential of alkali-treated walnut shells waste as filler for the PLA matrix. They incorporated alkali-treated walnut shell powder (10, 20, and 30 wt.%) into the matrix for biocomposite fabrication. A relative improvement of 50% in tensile strength was reported for treated walnut shell waste-filled biocomposites compared with the untreated counterpart. However, the tensile modulus values of treated walnut shell waste-filled biocomposites remained almost similar to unfilled PLA, and they maintained a consistent behavior with increased filler content. Boubekeur et al. [33] investigated the influence of 1–3 mm sized wood waste (a mixture of eucalyptus and Aleppo pine wood) particles on the mechanical properties of PLA-based biocomposites. The authors concluded that Young's modulus and crystallinity of the manufactured composites increased, while stress, impact strength, and elongation at break decreased with increasing wood waste percentage. Bajpai et al. [34] investigated the wear performance of natural fiber-reinforced PLA composites. Three categories of natural fibers (nettle, sisal, and *Grewia optiva*) were applied, and laminated composites were fabricated as per a hot compression process. The experimental results showed that incorporating natural fiber mats into the PLA matrix as a reinforcement remarkably enhanced the wear resistance of the neat polymer. There was a 10–44% decrease in friction coefficient and about 70% decrease in the specific wear rate of manufactured composites compared with neat PLA. Kanakannavar et al. [35] studied the effect of natural fiber 3D braided woven fabric as reinforcement in PLA composites for tribological performance. The research concluded that the fabric reinforcement decreased the specific wear rate of PLA, and about a 95% decrease was detected in the samples containing 35 wt.% reinforcement.

Although the literature is rich in research of the impacts of wood waste on the physical, mechanical, thermal, and thermo-mechanical properties of PLA-based biocomposites, the sliding wear behavior of wood waste-filled PLA-based biocomposites has not been studied so far. Moreover, the inclusion of some natural fibers and sustainable biocarbon was reported to enhance the wear resistance of PLA-based biocomposites [34–36]. Therefore, our research studied the production of PLA biocomposites using North Indian rosewood waste and investigating their thermal and dry sliding wear properties.

2. Experimental Details

2.1. Materials and Biocomposite Fabrication

PLA (Nature Works, USA, Ingeo 2003D grade) with a melt flow index of 6 g/10 min, a density of 1.24 g/cm^3, and a melting temperature of 170 °C was used in this research. North Indian rosewood waste (60 mesh) was procured from the Krishna Timber Store in Dadhol, Himachal Pradesh, India. Before use, the wood waste was treated with 2% sodium hydroxide solution for 12 h at room temperature. After that, the treated wood waste was washed with distilled water and dried in an oven for 4 h. The SEM (scanning electron microscope) micrograph of the North Indian rosewood waste particles is presented in Figure 1a. Before biocomposite manufacturing, both the wood waste and PLA were dried for 6 h in a DEGA-2500 dehumidifier at 80 °C. The melt compounding/mixing of the PLA biocomposites containing 0, 2.5, 5, 7.5, and 10 wt.% of North Indian rosewood waste was performed using an LTE 20–44 twin-screw extruder (Labtech Engineering, Samut Prakarn Thailand; L/D ratio of 44; screw diameter of 20 mm) with a screw speed of 30 rpm and a temperature profile of 155–185 °C. After melt compounding, the composites were cooled and granulated. Subsequently, the granulated samples were injection molded into dumbbell-shaped samples (Figure 1b) using an Arburg Allrounder Advance 420C (Loßburg, Germany) injection molding machine with a nozzle temperature of 195 °C. The following parameters were used for the injection molding process: injection rate of 40 cm^3/s, holding pressure at 75-65-25 MPa for 15 s, cooling time of 30 s, and mold temperature at 30 °C [37].

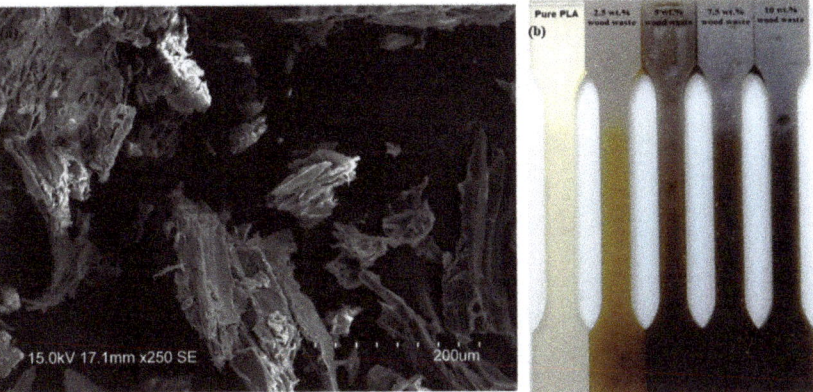

Figure 1. (a) SEM micrograph of North Indian rosewood waste, (b) fabricated biocomposites.

2.2. Thermogravimetric Analysis

The thermogravimetric tests were performed on a Shimadzu TGA-50 model scientific instrument, while the evaluation was performed using TA-60WS software. The powdered sample (~10 mg) of biocomposites was placed in a platinum pan. The thermal stability was recorded at a heating rate of 10 °C min^{-1} in a nitrogen atmosphere from 30 °C to 500 °C.

2.3. Sliding Wear Study

The sliding wear behavior of the produced biocomposites was investigated utilizing an ASTM G-99-compliant pin on disk machine (Model: TR-411, DUCOM, India). The schematic of the pin on disk machine and its detailed working principles were discussed elsewhere [38]. A 20 mm × 5 mm × 5 mm specimen was machined from the manufactured composites, and it was held stationary within the fixture, which was normal to the disk. For load-speed sensitivity, a series of trials were carried out by varying the normal load (10 N, 20 N, and 30 N) and the sliding velocity (1 m/s, 2 m/s, and 3 m/s) on the pin on disk machine for a fixed sliding distance of 2000 m. The biocomposite sample weight was measured prior to and after the wear test by utilizing an electronic weight balance (Wensar Weighing Scales Ltd., India) with an accuracy of 0.0001 g. For each sample, the wear experiment was repeated three times, and the volumetric wear in cm^3 was computed by using the following equation [37]:

$$Volumetric\ wear = \frac{\omega}{\rho} \quad (1)$$

where ω = sample weight loss (g) and ρ = sample density (g/cm^3).

The density of the manufactured biocomposites was determined by using standard water displacement method, and it was found to be 1.24 g/cm^3, 1.225 g/cm^3, 1.211 g/cm^3, 1.198 g/cm^3, and 1.183 g/cm^3 [39].

2.4. Experiment Design

In this study, the combination of control parameters for sliding wear minimization was determined using the Taguchi method. The Taguchi method is one of the most important statistical techniques used to demonstrate the influence of different control parameters with various levels. The sliding wear tests on the manufactured biocomposites were conducted under various working conditions, using four control parameters each with five levels: wood waste content (A: 0, 2.5, 5, 7.5, and 10 wt.%), normal load (B: 10, 20, 30, 40, and 50 N), sliding distance (C: 500, 100, 1500, 2000, and 2500 m) and sliding velocities (D: 0.6, 1.2, 1.8, 2.4, and 3 m/s) (as listed in Table 1).

Table 1. Levels of the control parameters used in the experiment.

Control Parameters	Levels					Units
	I	II	III	IV	V	
A: Wood waste	0	2.5	5	7.5	10	wt.%
B: Normal load	10	20	30	40	50	N
C: Sliding distance	500	1000	1500	2000	2500	m
D: Sliding velocity	0.60	1.2	1.8	2.4	3	m/s

In a full factorial design, nearly 625 (5^4) trials would be required to contemplate the impact of four control parameters, each having five levels. In contrast, the Taguchi method decreases the number of trials by utilizing orthogonal arrays, resulting in a lower number of trials with noticeable precision. Therefore, the impact of four control parameters with five levels (as presented in Table 1) was studied using L$_{25}$ orthogonal design as presented in Table 2. Further on, to assess the test results, the signal-to-noise (SN) ratio was also investigated. The Taguchi method has three categories of the SN ratio, namely lower-the-better, nominal-the-better, and higher-the-better. In this work, a 'lower-the-better' characteristic was utilized, since the intention was to minimize the wear by using the following equation [38].

$$SN\ ratio = -10\log\left[\frac{1}{n}\left(\sum_n y^2\right)\right] \quad (2)$$

where y = volumetric wear and n = number of trials.

Table 2. Experimental design.

Test Run	Control Parameters				Test Run	Control Parameters			
	A	B	C	D		A	B	C	D
1	0.0	10	500	0.6	14	5.0	40	500	1.8
2	0.0	20	1000	1.2	15	5.0	50	1000	2.4
3	0.0	30	1500	1.8	16	7.5	10	2000	1.2
4	0.0	40	2000	2.4	17	7.5	20	2500	1.8
5	0.0	50	2500	3.0	18	7.5	30	500	2.4
6	2.5	10	1000	1.8	19	7.5	40	1000	3.0
7	2.5	20	1500	2.4	20	7.5	50	1500	0.6
8	2.5	30	2000	3.0	21	10	10	2500	2.4
9	2.5	40	2500	0.6	22	10	20	500	3.0
10	2.5	50	500	1.2	23	10	30	1000	0.6
11	5.0	10	1500	3.0	24	10	40	1500	1.2
12	5.0	20	2000	0.6	25	10	50	2000	1.8
13	5.0	30	2500	1.2					

2.5. Contribution Ratio Analysis

After the SN ratio analysis, each control parameter was analyzed for their contribution ratio (ψ) towards the volumetric wear by using the following steps [40].

Step I: Calculation of the overall SN ratio mean. In this step the overall SN ratio mean ($\Re$) was computed for the 25 trials using the following equation:

$$\Re = \frac{1}{25}\sum_{n=1}^{25}(SN\ ratio) \qquad (3)$$

Step II: Level mean of the SN ratio. In this step, the level mean of the SN ratio ($\hbar_i$) was calculated for each control parameter using the following equation.

$$\hbar_i = \frac{1}{5}\sum_{j=1}^{5}(SN\ ratio)_{ij} \qquad (4)$$

where j is the level of the ith control parameter.

Step III: Sum of squares calculation. In this step, the sum of squares (λ) values were determined using the following equation:

$$\lambda = \sum_i (\hbar_i - \Re)^2 \qquad (5)$$

The ith control parameter can be determined using the following equation:

$$\lambda_i = \sum_{j=1}^{5}(\hbar_i - \Re)^2 \qquad (6)$$

where j is the level of the ith control parameter.

Step IV: Contribution ratio calculation. In the final step, the contribution ratio (ψ) of the individual control parameter was calculated by using the following equation:

$$\psi_i = \frac{\lambda_i}{\lambda} \times 100 \qquad (7)$$

2.6. Scanning Electron Microscopy

The worn surfaces of pure PLA and wood waste-filled PLA biocomposites were further examined for possible wear mechanisms using a Hitachi S-3400N scanning electron microscope (SEM; Hitachi Ltd., Tokyo, Japan). Prior to the SEM inspection, the samples were sputter-coated with a gold–palladium alloy in order to prevent charging.

3. Results and Discussion

3.1. Thermal Stability Analysis

The temperature-dependent weight loss curves and the corresponding derivatives (DTG) for pure PLA and its wood waste-filled biocomposites are illustrated in Figure 2a,b, respectively. The thermal deterioration at temperatures ranging from 30 °C to 250 °C resulted in a minor weight loss of about 2 ± 0.5%. The elimination of moisture was the primary cause of the biocomposites' weight loss at this point. A single-step decomposition process was observed both for the bare PLA and the wood waste-filled PLA biocomposites as well, in the range of (250–400 °C). The weight loss in this temperature range corresponded to the degradation of the PLA resin and the decomposition of hemicellulose, cellulose, and lignin that were present in the wood waste [41]. The temperature corresponding to the 5%, 25%, 50%, and 75% weight loss (i.e., T_5, T_{25}, T_{50}, and T_{75}) and the temperature of the maximum decomposition rate (T_{peak}) are presented in Table 3. Based on the results, the thermal degradation of wood waste-filled PLA biocomposites occurred at a lower temperature than that of pure PLA. Biopolyesters such as PLA tend to degrade at elevated temperatures as a consequence of various depolymerization processes and thermal oxidation reactions [42]. The incorporation of the less thermally stable wood waste into the polymer matrix facilitated the thermal degradation of PLA, thereby leading to an earlier decomposition of the biocomposites during the heating.

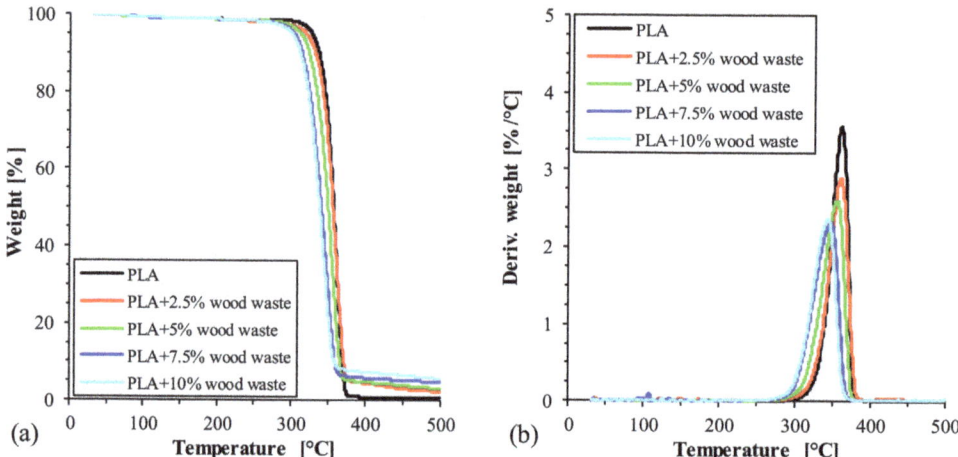

Figure 2. TGA (**a**) and DTG (**b**) curves for PLA biocomposites, respectively.

Using 5% weight loss (T_5) as the onset of the main degradation step, the temperature was 327 °C for the bare PLA but decreased to 320 °C when 2.5 wt.% wood waste was added. When the wood waste loading was increased even further (5, 7.5, and 10% wt.%), the composites' onset degradation temperature decreased to 312 °C, 304 °C, and 300 °C, respectively. Similarly, with increased wood waste content, the temperatures for 25%, 50%, and 75% weight loss and the temperatures corresponding to the highest decomposition rate also decreased considerably. Previous studies reported similar trends with various natural fiber-filled PLA composites [42–44].

Table 3. TGA and DTG results of wood waste-filled PLA biocomposites.

Biocomposite	Temperature at Different Weight Loss (°C)				T_{peak} (°C)
	T_5	T_{25}	T_{50}	T_{75}	
PLA	327	348	357	365	360
PLA+2.5 wt.% wood waste	320	346	356	365	359
PLA+5 wt.% wood waste	312	337	350	359	354
PLA+7.5 wt.% wood waste	304	327	340	351	345
PLA+10 wt.% wood waste	300	325	338	348	342

3.2. Influence of Normal Load and Sliding Velocity on Wear

Figure 3 shows the volumetric wear of composites as a function of normal load (10 N, 20 N, and 30 N) at a constant sliding velocity (1 m/s) and a 2000 m of fixed distance. Figure 3 shows that when the normal load increased, the volumetric wear of all composites increased dramatically. The volumetric wear fluctuated between 0.0342 cm^3 and 0.0660 cm^3 in pure PLA samples. Compared with pure PLA, the trend in volumetric wear for 2.5 wt.% wood waste-filled composite was modest, with an increased normal load. Adding 2.5 wt.% of wood waste reduced the volumetric wear of the PLA matrix by 26% to 34% under all loading situations. With the further addition of wood waste $\geq$ 5 wt.%, the wear of the composites increased, and it was the highest (0.0658–0.1097 cm^3) when 10 wt.% wood waste was added to the composites. The possible mechanism for the increment in volumetric wear with increased wood waste content and normal load can be explained. The lower the wood waste particle concentration, the more that the structural homogeneities remained on the higher side due to ease in the dispersion of wood waste particles within the PLA matrix. The firm embedment of the wood waste particles helped to protect the matrix in the contact zone from heat and mechanical failure, resulting in minor wear.

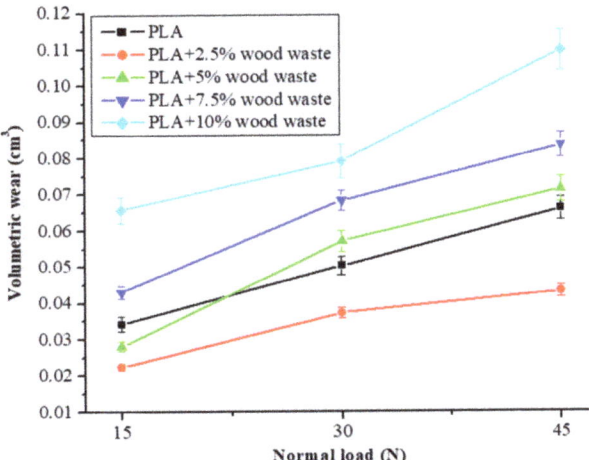

Figure 3. Volumetric wear of composite as a function of normal load.

After displaying a slight volumetric wear at 2.5 wt.% wood waste content, the volumetric wear was observed to rise when the wood waste loading was increased further. At higher concentrations, the possibilities of wood waste particles agglomerating expanded the composites and counter surface gap. As a result of the increasing distance, the adhesion between the sliding surfaces decreased, resulting in a more significant weight loss and volumetric wear. Additionally, with increased loading, the number of wood waste particles on the composite surface increased. As the normal load grew, more heat was generated during testing, and the interfacial contact temperature also increased. With this higher temperature, the bonding between the wood waste and the matrix weakened, and material removal became more accessible, increasing the wear. Similar results were reported by Bajpai et al. [34] for natural fiber-reinforced PLA composites and by Erdoğan et al. [45] for industrial waste-filled epoxy composites as well. Figure 4 shows the volumetric wear of composites as a function of sliding velocity (1 m/s, 2 m/s, and 3 m/s) at a constant normal load (30 N) and a 2000 m of fixed distance. The trend in volumetric wear for pure PLA was from 0.0502 cm^3 to 0.0977 cm^3; when wood waste was incorporated, the wear firstly decreased at 2.5 wt.% wood waste content and then increased with further wood waste loading. The volumetric wear remained at 0.0372–0.0606 cm^3 for the 2.5 wt.% wood waste-filled composite, which was 26–38% lower than that of pure PLA. In comparison, the highest volumetric wear was registered for 10 wt.% wood waste-filled composites, which fluctuated between 0.0792–0.1439 cm^3. The thermal softening of the PLA resin occurred as the sliding velocity rose due to increased heat production, resulting in increased wear with increased sliding velocity. The variation of volumetric wear with sliding velocity showed that the wear of the PLA composite increased when the sliding velocity rose higher. As the sliding velocity grew, the thermal softening of the PLA resin took place due to increased heat generation. The higher heat weakened the filler–resin bonding, and it became easier to detach the wood waste particles from the composite surface during sliding, which resulted in increased wear. Bajpai et al. [34] observed a similar mechanism for sliding wear in the case of natural fiber-reinforced PLA composites. For lower load-velocity sliding conditions, Megahed et al. [46] concluded that the generation of slight surface deformation resulted in lower wear. However, surface deformation increased at higher normal load and sliding velocity conditions, resulting in increased wear.

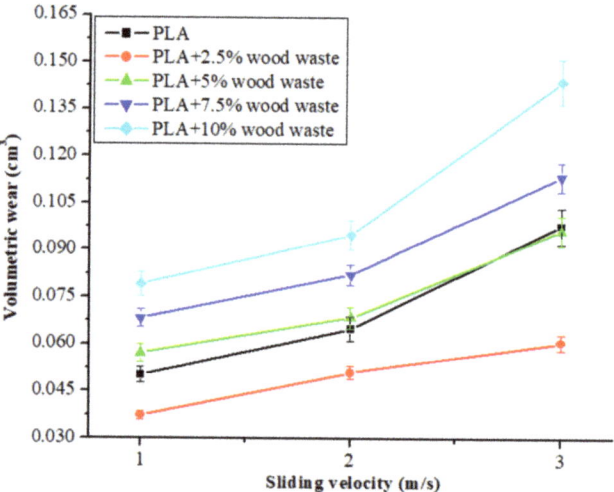

Figure 4. Volumetric wear of the composites as a function of sliding velocity.

3.3. The Taguchi Analysis for Sliding Wear Performance

According to the literature, biocomposites can be used in various applications where wear is a critical issue. The wear performance of PLA biocomposites is significantly influenced by the type and amount of reinforcement and testing parameters [28–30]. Therefore, our investigation was designed to find the most significant control parameter and combination of parameters that yielded the slightest wear during sliding. The experiments were conducted as L$_{25}$ orthogonal array design considering the impact of wood waste content, sliding distance, normal load, and sliding velocity on wear performance.

The Taguchi method suggests investigating the SN ratio by utilizing conceptual methodology that includes diagramming impacts and visually identifying the critical parameters. The results of the volumetric wear and their corresponding SN ratios are collected in Table 4. The investigation was conducted in Minitab 18. The volumetric wear obtained ranged from 0.0091 cm^3 to 0.1727 cm^3. The lowest and highest volumetric wear was obtained in test runs 6 and 25, respectively. Additionally, the influence of the selected four control parameters on the SN ratio of the volumetric wear is presented in Figure 5, while the SN ratio response is found in Table 5. As shown in Figure 5, there was a decrease in the volumetric wear of the composites upon increasing the amount of wood waste content from 0 to 2.5 wt.%; however, it started increasing above 2.5 wt.% wood waste content.

Table 4. Volumetric wear and corresponding SN ratio.

Test Run	Volumetric Wear (cm^3)	SN Ratio	Test Run	Volumetric Wear (cm^3)	SN Ratio
1	0.0094	40.5374	14	0.0663	23.5697
2	0.0365	28.7541	15	0.0960	20.3546
3	0.0508	25.8827	16	0.0592	24.5536
4	0.0686	23.2735	17	0.0780	22.1581
5	0.1332	17.5099	18	0.1169	18.6437
6	0.0091	40.8192	19	0.1189	18.4964
7	0.0218	33.2309	20	0.1033	19.7180
8	0.0606	24.3505	21	0.0830	21.6184
9	0.0368	28.6830	22	0.0785	22.1026
10	0.0353	29.0445	23	0.0916	20.7621
11	0.0515	25.7639	24	0.1471	16.6477
12	0.0258	31.7676	25	0.1727	15.2542
13	0.0600	24.4370			

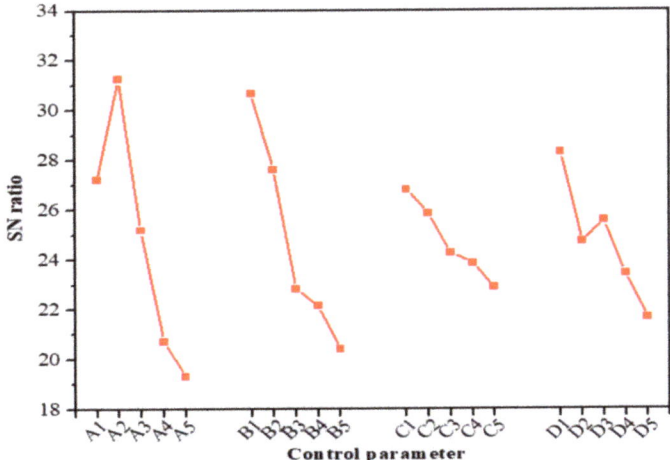

Figure 5. Main parameter effects for various SN ratio values of volumetric wear.

The minimum value of 0.0091 cm^3 for volumetric wear was obtained for the 2.5 wt.% wood waste-filled composite. The situation changed when the wood waste content started increasing. The maximum value of 0.1727 cm^3 was obtained for the biocomposite with 10 wt.% wood waste content. This significant behavior was potentially due to the agglomeration of the wood particles as a result of their poor interfacial bond with the PLA matrix. Due to the poor bonding, the wood waste particles were quickly drawn/peeled off from the PLA matrix during sliding, thus leading to the increased wear of the biocomposites. From the response displayed in Table 5, it can be assumed that among all the control parameters, wood waste content is an essential parameter, followed by normal load and sliding velocity, while sliding distance has a minimal impact on the volumetric wear of the tested biocomposite. Moreover, based on the results, it can be concluded that the combination of control parameters A_{II} (2.5 wt.% wood waste), B_I (10 N normal load), C_I (500 m sliding distance), and D_I (0.6 m/s sliding velocity) provided minimum volumetric wear. The result suggests that 2.5 wt.% wood waste-filled PLA biocomposite can be used for a low loading application of load and sliding velocity.

Table 5. SN ratio response table.

Level	A	B	C	D
I	27.19	30.66	26.78	28.29
II	31.23	27.60	25.84	24.69
III	25.18	22.82	24.25	25.54
IV	20.71	22.13	23.84	23.42
V	19.28	20.38	22.88	21.64
Delta	11.95	10.28	3.90	6.65
Rank	1	2	4	3

In addition, the influence of the most dominant control parameter (i.e., wood waste content) was analyzed on volumetric wear by drawing contour plots (Figure 6a–c) against (a) wood waste content and normal load, (b) wood waste content and sliding distance, and (c) wood waste content and sliding velocity. The contour plots demonstrate that the volumetric wear tended to increase when the wood waste content, normal load, sliding distance, and sliding velocity increased gradually. It was revealed that the lowest volumetric wear of 0.0091 cm^3 was obtained at 2.5 wt.% wood waste content and the lower value (10 N) of the normal load. In contrast, the maximum volumetric wear of 0.1727 cm^3 was obtained at 10 wt.% wood waste content and at a high level (50 N) of the normal load.

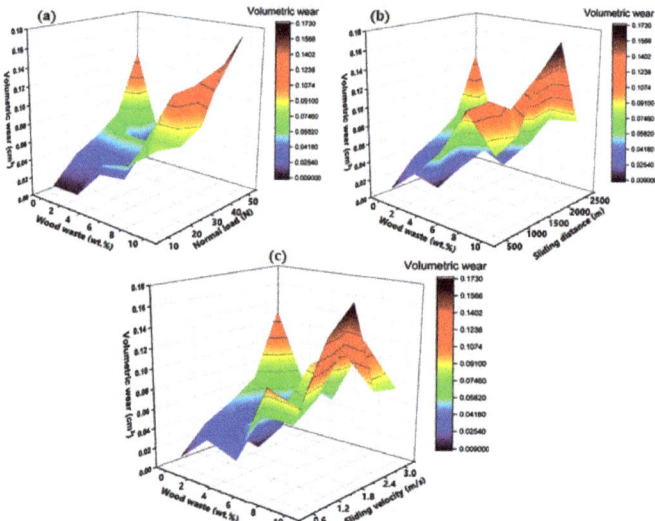

Figure 6. Contour plots of volumetric wear for wood waste content with respect to (a) normal load, (b) sliding distance, and (c) sliding velocity.

3.4. Contribution Ratio Results

The contribution ratio of each parameter for volumetric wear is listed in Table 6 and presented in Figure 7. The overall SN ratio mean value ($\Re$) for the 25 trials was determined using Equation (3) and found to be 24.72 dB. The level mean of SN ratio values for each control parameter was computed using Equation (4). The sum of squares (λ) value was determined by Equation (5), and for the individual control parameter the λ_i value was determined by Equation (6). Thereafter, the contribution ratio (ψ) for each control parameter was computed using Equation (7). The results show that wood waste, normal load, sliding distance, and sliding velocity contributed to the volumetric wear by 46.82%, 36.08%, 4.90%, and 12.20%, respectively. The contribution results indicate that the wood waste content was the most significant control parameter affecting the volumetric wear of the biocomposites, followed by the normal load.

Table 6. Contribution ratio results.

Control Parameter	$\Re$	λ_i	λ	ψ
Wood waste		94.3663		46.82
Normal load	24.72	72.7317	201.5715	36.08
Sliding distance		9.8789		4.90
Sliding velocity		24.5946		12.20

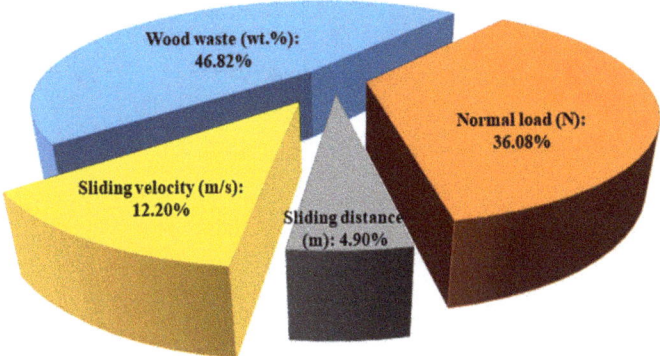

Figure 7. Contribution ratio of each control parameter to volumetric wear.

3.5. Worn Surface Morphology

The results of SEM inspections are presented in Figures 8 and 9. Figure 8a,b show the worn surfaces of bare PLA tested under 50 N load, 2500 m distance, and 3 m/s sliding velocity. In the image of lower magnification (Figure 8a), the worn surface was moderately rough, revealing possible micro-ploughing in the matrix. At a higher magnification (Figure 8b), the worn surface showed more scratches/damage to the matrix, resulting in increased material removal. As a consequence of sliding, the contact temperature was uncommonly expanded, which caused an accelerative rupture of the matrix, particularly in the interfacial zone. Accordingly, the surface damage strikingly expanded with grooves left by the matrix removal, resulting in a higher weight loss (Figure 8b). Figure 8c,d present the worn surfaces of 2.5 wt.% wood waste-filled biocomposite tested under 40 N load, 2500 m distance, and 0.6 m/s sliding velocity. In contrast with Figure 8a,b for bare PLA, the worn surfaces for 2.5 wt.% wood waste-added biocomposite was much smoother, and the matrix detachment was enormously restricted with the inclusion of wood waste particles. Even at lower magnification, the worn surface remained uniform with a lesser extent of micro-ploughing and groove formation, resulting in a slight wear of the biocomposite.

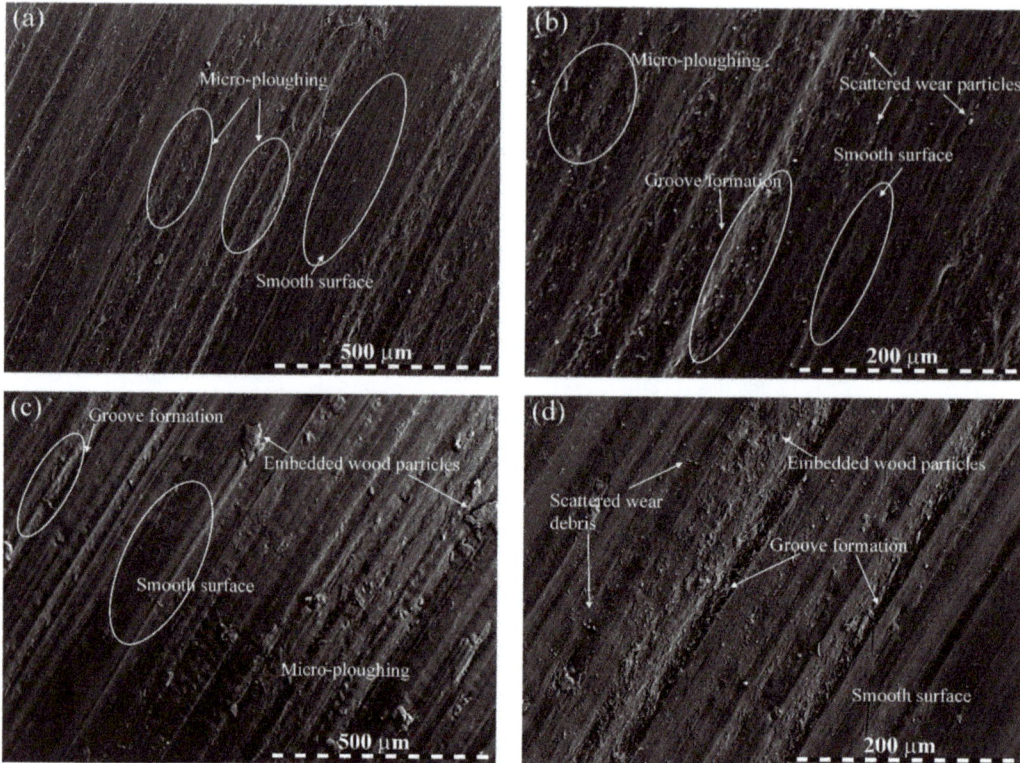

Figure 8. Worn micrographs of biocomposites: (**a**,**b**) bare PLA and (**c**,**d**) 2.5 wt.% wood waste-filled biocomposites at lower and higher magnification.

Figure 9a–f show the SEM images of the worn surfaces of 5 wt.% (under 50 N load, 1000 m distance, and at 2.4 m/s sliding velocity), 7.5 wt.% (under 40 N load, 1000 m distance, and at 3 m/s sliding velocity) and 10 wt.% (under 50 N load, 2000 m distance, and at 1.8 m/s sliding velocity) wood waste-filled PLA biocomposites. In comparison to Figure 8, the worn surfaces presented in Figure 9 were rougher with severe damage. Therefore, the worn surfaces were characterized by intense sub-surface damage due to sliding, while micro-ploughing was responsible for the heavy eradication of the surface material. The increased scattered wear particles and grooves formed by micro-ploughing contributed to the decreased wear resistance of these biocomposites. Moreover, the wood waste particles appeared to be seriously damaged, suggesting a poor filler-matrix interfacial bonding, which also resulted in elevated wear.

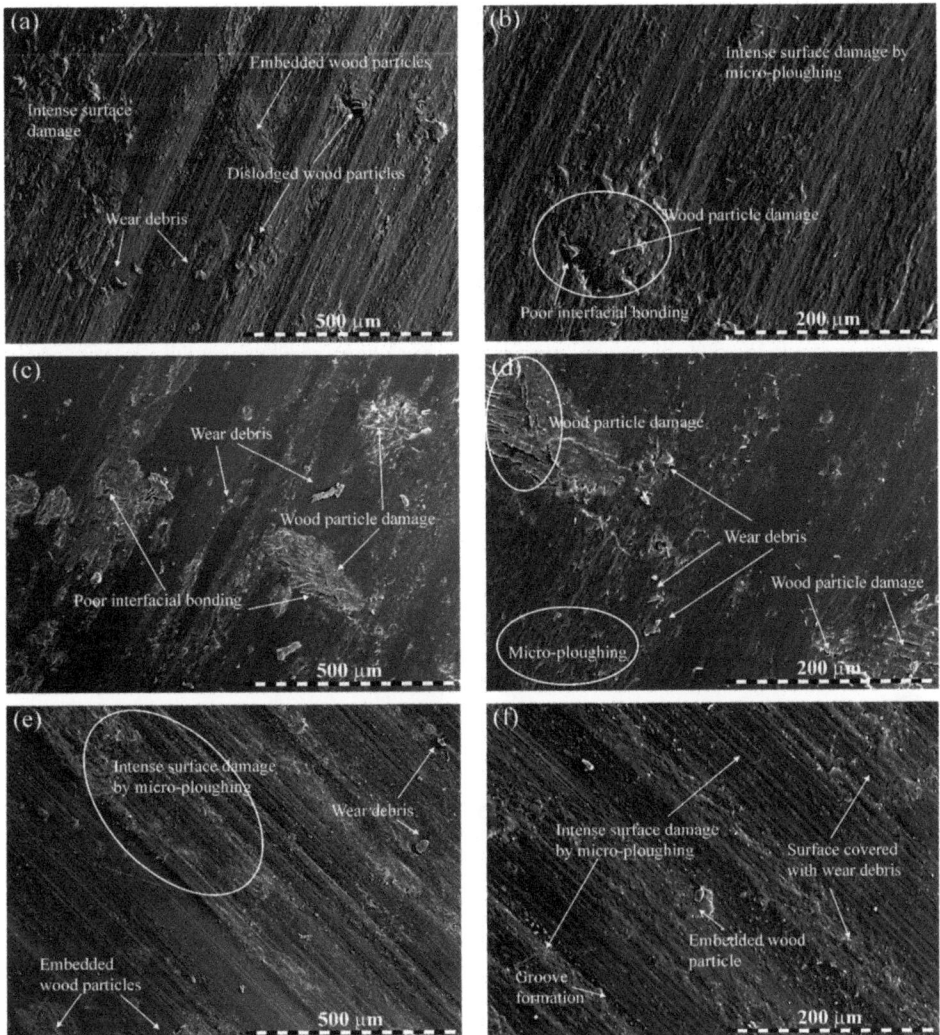

Figure 9. Worn micrographs of biocomposites: (**a**,**b**) 5 wt.%, (**c**,**d**) 7.5 wt.%, and (**e**,**f**) 10 wt.% wood waste-filled biocomposites at lower and higher magnification.

4. Conclusions

The thermal and sliding wear properties of Indian rosewood waste-filled PLA-based biocomposites were investigated. The following conclusions can be drawn:

1. The thermal stability of the PLA biocomposites increased with an increase in wood waste loading.
2. The wear of the biocomposites increased with an increase in load and sliding velocity. Compared with pure PLA, the wear in 2.5 wt.% wood waste-added biocomposites was almost 26–38% lower.
3. The Taguchi analysis demonstrated that the combination of control parameters A_{II} (wood waste of 2.5 wt.%), B_I (normal load of 10 N), C_I (sliding distance of 500 m), and D_I (sliding velocity of 0.6 m/s) offers the lowest volumetric wear for the manufactured biocomposites.

4. Wood waste content with 46.82% contribution was observed as the most dominant parameter for controlling the wear of the biocomposites, followed by the normal load, sliding velocity, and sliding distance with contributions of 36.08%, 12.20%, and 4.90%, respectively.
5. The worn surface study revealed that the micro-ploughing, grooves formation, and poor filler-matrix interfacial bonding were the possible cause of biocomposites wear.

Author Contributions: Conceptualization, T.S. and L.L.; methodology, T.S., A.P. and L.L.; software, T.S., A.P.; validation, T.S. and L.R.; formal analysis, T.S., L.L. and L.R.; investigation, T.S., L.L. and G.D.; resources, T.S., A.P. and L.L.; data curation, T.S., L.R. and L.L.; writing—original draft preparation, T.S. and L.L.; writing—review and editing, T.S., A.P. and L.L.; visualization, T.S.; supervision, A.P.; funding acquisition, L.L. All authors have read and agreed to the published version of the manuscript.

Funding: This research was funded by the Ministry of Innovation and Technology of Hungary from the National Research, Development, and Innovation Fund, financed under the TKP2021-NKTA funding scheme, grant number TKP2021-NKTA-48. L. Lendvai is grateful for the support of the János Bolyai Research Scholarship of the Hungarian Academy of Sciences.

Institutional Review Board Statement: Not applicable.

Informed Consent Statement: Not applicable.

Data Availability Statement: The data are available from the first author (T.S.) upon reasonable request.

Conflicts of Interest: The authors declare no conflict of interest.

References

1. Carvalho, J.P.; Simonassi, N.T.; Lopes, F.P.D.; Monteiro, S.N.; Vieira, C.M.F. Novel Sustainable Castor Oil-Based Polyurethane Biocomposites Reinforced with Piassava Fiber Powder Waste for High-Performance Coating Floor. *Sustainability* **2022**, *14*, 5082. [CrossRef]
2. Patti, A.; Acierno, S.; Cicala, G.; Zarrelli, M.; Acierno, D. Recovery of Waste Material from Biobags: 3D Printing Process and Thermo-Mechanical Characteristics in Comparison to Virgin and Composite Matrices. *Polymers* **2022**, *14*, 1943. [CrossRef] [PubMed]
3. Boonsiriwit, A.; Itkor, P.; Sirieawphikul, C.; Lee, Y.S. Characterization of Natural Anthocyanin Indicator Based on Cellulose Bio-Composite Film for Monitoring the Freshness of Chicken Tenderloin. *Molecules* **2022**, *27*, 2752. [CrossRef]
4. Ismail, I.; Aini, Q.; Jalil, Z.; Olaiya, N.G.; Mursal, M.; Abdullah, C.K.; Khalil Hps, A. Properties Enhancement Nano Coconut Shell Filled in Packaging Plastic Waste Bionanocomposite. *Polymers* **2022**, *14*, 772. [CrossRef] [PubMed]
5. Shaik, S.A.; Schuster, J.; Shaik, Y.P.; Kazmi, M. Manufacturing of Biocomposites for Domestic Applications Using Bio-Based Filler Materials. *J. Compos. Sci.* **2022**, *6*, 78. [CrossRef]
6. Food and Agriculture Organization of the United Nations. Forest Product Statistics. Available online: http://www.fao.org/forestry/statistics/80938/en/ (accessed on 28 June 2021).
7. Silva, C.G.; Campini, P.A.L.; Rocha, D.B.; Rosa, D.S. The influence of treated eucalyptus microfibers on the properties of PLA biocomposites. *Compos. Sci. Technol.* **2019**, *179*, 54–62. [CrossRef]
8. Etuk, S.E.; Ekpo, S.S.; Robert, U.W.; Agbasi, O.E.; Effiong, E.-A.A. Dielectric properties of Raphia Fiber from Epidermis of young Raphia Vinifera leaflet. *Acta Tech. Jaurinensis* **2022**, *15*, 91–98. [CrossRef]
9. Falk, B.; McKeever, D. Generation and recovery of solid wood waste in the U.S. *BioCycle* **2012**, *53*, 30–32.
10. Thomas, S.; Shumilova, A.; Kiselev, E.; Baranovsky, S.; Vasiliev, A.; Nemtsev, I.; Kuzmin, A.P.; Sukovatyi, A.; Avinash, R.P.; Volova, T. Thermal, mechanical and biodegradation studies of biofiller based poly-3-hydroxybutyrate biocomposites. *Int. J. Biol. Macromol.* **2019**, *155*, 1373–1384. [CrossRef]
11. Biocomposites Market by Fiber (Wood Fiber and Non-Wood Fiber), Polymer (Synthetic and Natural), Product (Hybrid and Green), End-Use Industry (Transportation, Building & Construction and Consumer Goods), and Region—Global Forecast to 2022. Available online: https://www.marketsandmarkets.com/Market-Reports/biocomposite-market258097936.html (accessed on 25 August 2021).
12. Shahavi, M.H.; Selakjani, P.P.; Abatari, M.N.; Antov, P.; Savov, V. Novel Biodegradable Poly (Lactic Acid)/Wood Leachate Composites: Investigation of Antibacterial, Mechanical, Morphological, and Thermal Properties. *Polymers* **2022**, *14*, 1227. [CrossRef]
13. Augaitis, N.; Vaitkus, S.; Członka, S.; Kairytė, A. Research of Wood Waste as a Potential Filler for Loose-Fill Building Insulation: Appropriate Selection and Incorporation into Polyurethane Biocomposite Foams. *Materials* **2020**, *13*, 5336. [CrossRef]
14. Dogossy, G.; Czigany, T. Thermoplastic starch composites reinforced by agricultural by-products: Properties, biodegradability, and application. *J. Reinf. Plast. Compos.* **2011**, *30*, 1819–1825. [CrossRef]

15. Hejna, A.; Piszcz-Karaś, K.; Filipowicz, N.; Cieśliński, H.; Namieśnik, J.; Marć, M.; Klein, M.; Formela, K. Structure and performance properties of environmentally-friendly biocomposites based on poly(ε-caprolactone) modified with copper slag and shale drill cuttings wastes. *Sci. Total Environ.* **2018**, *640–641*, 1320–1331. [CrossRef]
16. Kim, H.-S.; Yang, H.-S.; Kim, H.-J. Biodegradability and mechanical properties of agro-flour-filled polybutylene succinate biocomposites. *J. Appl. Polym. Sci.* **2005**, *97*, 1513–1521. [CrossRef]
17. Panaitescu, D.M.; Nicolae, C.A.; Gabor, A.R.; Trusca, R. Thermal and mechanical properties of poly(3-hydroxybutyrate) reinforced with cellulose fibers from wood waste. *Ind. Crop. Prod.* **2020**, *145*, 112071. [CrossRef]
18. Dhakal, H.; Bourmaud, A.; Berzin, F.; Almansour, F.; Zhang, Z.; Shah, D.; Beaugrand, J. Mechanical properties of leaf sheath date palm fibre waste biomass reinforced polycaprolactone (PCL) biocomposites. *Ind. Crop. Prod.* **2018**, *126*, 394–402. [CrossRef]
19. Freeland, B.; McCarthy, E.; Balakrishnan, R.; Fahy, S.; Boland, A.; Rochfort, K.D.; Dabros, M.; Marti, R.; Kelleher, S.M.; Gaughran, J. A Review of Polylactic Acid as a Replacement Material for Single-Use Laboratory Components. *Materials* **2022**, *15*, 2989. [CrossRef]
20. Marano, S.; Laudadio, E.; Minnelli, C.; Stipa, P. Tailoring the Barrier Properties of PLA: A State-of-the-Art Review for Food Packaging Applications. *Polymers* **2022**, *14*, 1626. [CrossRef]
21. Maleki, H.; Azimi, B.; Ismaeilimoghadam, S.; Danti, S. Poly(lactic acid)-Based Electrospun Fibrous Structures for Biomedical Applications. *Appl. Sci.* **2022**, *12*, 3192. [CrossRef]
22. Pérez-Fonseca, A.A.; Campo, A.S.M.D.; Robledo-Ortíz, J.R.; González-López, M.E. Compatibilization strategies for PLA biocomposites: A comparative study between extrusion-injection and dry blending-compression molding. *Compos. Interfaces* **2022**, *29*, 274–292. [CrossRef]
23. Siakeng, R.; Jawaid, M.; Asim, M.; Fouad, H.; Awad, S.; Saba, N.; Siengchin, S. Flexural and Dynamic Mechanical Properties of Alkali-Treated Coir/Pineapple Leaf Fibres Reinforced Polylactic Acid Hybrid Biocomposites. *J. Bionic Eng.* **2021**, *18*, 1430–1438. [CrossRef]
24. Zhao, X.; Li, K.; Wang, Y.; Tekinalp, H.; Larsen, G.; Rasmussen, D.; Ginder, R.S.; Wang, L.; Gardner, D.J.; Tajvidi, M.; et al. High-strength polylactic acid (PLA) biocomposites reinforced by epoxy-modified pine fibers. *ACS Sustain. Chem. Eng.* **2020**, *8*, 13236–13247. [CrossRef]
25. Figueroa-Velarde, V.; Diaz-Vidal, T.; Cisneros-López, E.O.; Robledo-Ortiz, J.R.; López-Naranjo, E.J.; Ortega-Gudiño, P.; Rosales-Rivera, L.C. Mechanical and Physicochemical Properties of 3D-Printed Agave Fibers/Poly(lactic) Acid Biocomposites. *Materials* **2021**, *14*, 3111. [CrossRef] [PubMed]
26. Kamarudin, S.H.; Abdullah, L.C.; Aung, M.M.; Ratnam, C.T. Thermal and Structural Analysis of Epoxidized Jatropha Oil and Alkaline Treated Kenaf Fiber Reinforced Poly(Lactic Acid) Biocomposites. *Polymers* **2020**, *12*, 2604. [CrossRef] [PubMed]
27. Yu, W.; Li, M.; Lei, W.; Pu, Y.; Sun, K.; Ma, Y. Effects of Wood Flour (WF) Pretreatment and the Addition of a Toughening Agent on the Properties of FDM 3D-Printed WF/Poly(lactic acid) Biocomposites. *Molecules* **2022**, *27*, 2985. [CrossRef]
28. Khammassi, S.; Tarfaoui, M.; Škrlová, K.; Měřínská, D.; Plachá, D.; Erchiqui, F. Poly(lactic acid) (PLA)-Based Nanocomposites: Impact of Vermiculite, Silver, and Graphene Oxide on Thermal Stability, Isothermal Crystallization, and Local Mechanical Behavior. *J. Compos. Sci.* **2022**, *6*, 112. [CrossRef]
29. Lendvai, L.; Brenn, D. Mechanical, morphological and thermal characterization of compatibilized poly(lactic acid)/thermoplastic starch blends. *Acta Technica Jaurinensis* **2020**, *13*, 1–13. [CrossRef]
30. Fekete, I.; Ronkay, F.; Lendvai, L. Highly toughened blends of poly(lactic acid) (PLA) and natural rubber (NR) for FDM-based 3D printing applications: The effect of composition and infill pattern. *Polym. Test.* **2021**, *99*, 107205. [CrossRef]
31. Khan, B.A.; Na, H.; Chevali, V.; Warner, P.; Zhu, J.; Wang, H. Glycidyl methacrylate-compatibilized poly(lactic acid)/hemp hurd biocomposites: Processing, crystallization, and thermo-mechanical response. *J. Mater. Sci. Technol.* **2018**, *34*, 387–397. [CrossRef]
32. Orue, A.; Eceiza, A.; Arbelaiz, A. The use of alkali treated walnut shells as filler in plasticized poly(lactic acid) matrix composites. *Ind. Crop. Prod.* **2019**, *145*, 111993. [CrossRef]
33. Boubekeur, B.; Belhaneche-Bensemra, N.; Massardier, V. Low-density polyethylene/poly(lactic acid) blends reinforced by waste wood flour. *J. Vinyl Addit. Technol.* **2020**, *26*, 443–451. [CrossRef]
34. Bajpai, P.K.; Singh, I.; Madaan, J. Tribological behavior of natural fiber reinforced PLA composites. *Wear* **2013**, *297*, 829–840. [CrossRef]
35. Kanakannavar, S.; Pitchaimani, J.; Ramesh, M.R. Tribological behaviour of natural fibre 3D braided woven fabric reinforced PLA composites. *Proc. Inst. Mech. Eng. Part J J. Eng. Tribol.* **2020**, *235*, 1353–1364. [CrossRef]
36. Snowdon, M.R.; Wu, F.; Mohanty, A.K.; Misra, M. Comparative study of the extrinsic properties of poly(lactic acid)-based biocomposites filled with talc versus sustainable biocarbon. *RSC Adv.* **2019**, *9*, 6752–6761. [CrossRef]
37. Lendvai, L.; Singh, T.; Fekete, G.; Patnaik, A.; Dogossy, G. Utilization of Waste Marble Dust in Poly(Lactic Acid)-Based Biocomposites: Mechanical, Thermal and Wear Properties. *J. Polym. Environ.* **2021**, *29*, 2952–2963. [CrossRef]
38. Singh, T.; Gangil, B.; Singh, B.; Verma, S.K.; Biswas, D.; Fekete, G. Natural-synthetic fiber reinforced homogeneous and functionally graded vinylester composites: Effect of bagasse-Kevlar hybridization on wear behavior. *J. Mater. Res. Technol.* **2019**, *8*, 5961–5971. [CrossRef]
39. Singh, T.; Lendvai, L.; Dogossy, G.; Fekete, G. Physical, mechanical and thermal properties of *Dalbergia sissoo* wood waste filled poly(lactic acid) composites. *Polym. Compos.* **2021**, *42*, 4380–4389. [CrossRef]

40. Chauhan, R.; Singh, T.; Kumar, N.; Patnaik, A.; Thakur, N. Experimental investigation and optimization of impinging jet solar thermal collector by Taguchi method. *Appl. Therm. Eng.* **2017**, *116*, 100–109. [CrossRef]
41. Rajeshkumar, G.; Seshadri, S.A.; Devnani, G.; Sanjay, M.; Siengchin, S.; Maran, J.P.; Al-Dhabi, N.A.; Karuppiah, P.; Mariadhas, V.A.; Sivarajasekar, N.; et al. Environment friendly, renewable and sustainable poly lactic acid (PLA) based natural fiber reinforced composites—A comprehensive review. *J. Clean. Prod.* **2021**, *310*, 127483. [CrossRef]
42. Nyambo, C.; Mohanty, A.K.; Misra, M. Effect of Maleated Compatibilizer on Performance of PLA/Wheat Straw-Based Green Composites. *Macromol. Mater. Eng.* **2011**, *296*, 710–718. [CrossRef]
43. Huda, M.S.; Drzal, A.L.T.; Misra, M.; Mohanty, A.K.; Williams, K.; Mielewski, D.F. A Study on Biocomposites from Recycled Newspaper Fiber and Poly(lactic acid). *Ind. Eng. Chem. Res.* **2005**, *44*, 5593–5601. [CrossRef]
44. Tao, Y.; Wang, H.; Li, Z.; Li, P.; Shi, S.Q. Development and Application of Wood Flour-Filled Polylactic Acid Composite Filament for 3D Printing. *Materials* **2017**, *10*, 339. [CrossRef] [PubMed]
45. Erdoğan, A.; Gök, M.S.; Koç, V.; Günen, A. Friction and wear behavior of epoxy composite filled with industrial wastes. *J. Clean. Prod.* **2019**, *237*, 117588. [CrossRef]
46. Megahed, A.A.; Agwa, M.; Megahed, M. Improvement of Hardness and Wear Resistance of Glass Fiber-Reinforced Epoxy Composites by the Incorporation of Silica/Carbon Hybrid Nanofillers. *Polym. Technol. Eng.* **2017**, *57*, 251–259. [CrossRef]

Article

Improvement of the Structure and Physicochemical Properties of Polylactic Acid Films by Addition of Glycero-(9,10-trioxolane)-Trialeate

Olga Alexeeva [1,*], Anatoliy Olkhov [1,2,3], Marina Konstantinova [1], Vyacheslav Podmasterev [1], Ilya Tretyakov [2], Tuyara Petrova [2], Olga Koryagina [1], Sergey Lomakin [1,2], Valentina Siracusa [4,*] and Alexey L. Iordanskii [2,*]

[1] N.M. Emanuel Institute of Biochemical Physics, Russian Academy of Sciences, 119334 Moscow, Russia
[2] N.N. Semenov Federal Research Center for Chemical Physics, Russian Academy of Sciences, 119991 Moscow, Russia
[3] Academic Department of Innovational Materials and Technologies Chemistry, Plekhanov Russian University of Economics, 117997 Moscow, Russia
[4] Department of Chemical Science (DSC), University of Catania, Viale A. Doria 6, 95125 Catania, Italy
* Correspondence: alexol@yandex.ru (O.A.); vsiracus@dmfci.unict.it (V.S.); aljordan08@gmail.com (A.L.I.)

Abstract: Glycero-(9,10-trioxolane)-trioleate (ozonide of oleic acid triglyceride, OTOA) was introduced into polylactic acid (PLA) films in amounts of 5, 10, 30, 50, and 70% w/w. The morphological, mechanical, thermal, and water absorption properties of PLA films after the OTOA addition were studied. The morphological analysis of the films showed that the addition of OTOA increased the diameter of PLA spherulites and, as a consequence, increased the proportion of amorphous regions in PLA films. A study of the thermodynamic properties of PLA films by differential scanning calorimetry (DSC) demonstrated a decrease in the glass transition temperature of the films with an increase in the OTOA content. According to DSC and XRD data, the degree of crystallinity of the PLA films showed a tendency to decrease with an increase in the OTOA content in the films, which could be accounted for the plasticizing effect of OTOA. The PLA film with 10% OTOA content was characterized by good smoothness, hydrophobicity, and optimal mechanical properties. Thus, while maintaining high tensile strength of 21 MPa, PLA film with 10% OTOA showed increased elasticity with 26% relative elongation at break, as compared to the 2.7% relative elongation for pristine PLA material. In addition, DMA method showed that PLA film with 10% OTOA exhibits increased strength characteristics in the dynamic load mode. The resulting film materials based on optimized PLA/OTOA compositions could be used in various packaging and biomedical applications.

Keywords: polylactic acid; polymer films; ozonide; trioleate; biocompatibility

Citation: Alexeeva, O.; Olkhov, A.; Konstantinova, M.; Podmasterev, V.; Tretyakov, I.; Petrova, T.; Koryagina, O.; Lomakin, S.; Siracusa, V.; Iordanskii, A.L. Improvement of the Structure and Physicochemical Properties of Polylactic Acid Films by Addition of Glycero-(9,10-trioxolane)-Trialeate. *Polymers* 2022, 14, 3478. https://doi.org/10.3390/polym14173478

Academic Editor: Luminita Marin

Received: 5 August 2022
Accepted: 22 August 2022
Published: 25 August 2022

Publisher's Note: MDPI stays neutral with regard to jurisdictional claims in published maps and institutional affiliations.

Copyright: © 2022 by the authors. Licensee MDPI, Basel, Switzerland. This article is an open access article distributed under the terms and conditions of the Creative Commons Attribution (CC BY) license (https://creativecommons.org/licenses/by/4.0/).

1. Introduction

In recent decades, large volumes of non-degradable polymer packaging materials have been produced by petrochemical industry, making significant contribution to the global problem of environmental pollution [1]. One of the promising ways to address this problem is the production of packaging materials using biodegradable polymers [2]. Today, several biodegradable polymers, such as polylactic acid (PLA), polyhydroxybutyrate, and poly(ε-caprolactone) are well known [3] and are widely used in medicine [4], tissue engineering [5], production of disposable tableware and packaging materials [6].

Polylactic acid (PLA) is a compostable polymer derived from natural raw materials, which is synthesized from lactic acid monomers by catalytic ring-opening polymerization [7]. PLA could be formed via electrospinning into ultrafine fiber materials [8], extruded into films and molded into diverse shapes [9]. PLA is one of the most frequently used biobased materials in the food packaging industry, which is applied for production of disposable tableware, vegetables packaging, and fast-food containers [10]. Though PLA is a biodegradable polymer, its degradation can occur only in hydrolytic and enzymatic media [11]. PLA degradation takes place in several stages: diffusion of water into the matrix, hydrolysis of ester bonds, reduction in molecular weight, and absorption of residues by microorganisms. The rate of hydrolysis depends on the temperature and water content and is catalyzed by the free carboxyl groups of the hydrolyzed ends of the PLA [12].

The successful application of biodegradable polymer materials in packaging should be based on good mechanical and barrier properties, improved elasticity, biodegradability, etc. [13]. Widespread use of PLA films in packaging applications is restricted by their poor ductility and barrier properties, relatively low thermostability [14]. Considerable efforts have been made to improve the physicochemical properties of PLA films in order to accelerate their employment in packaging industry, such as the addition of modifiers, copolymerization or blending [15,16]. PLA blending with other polymers, such as poly(hydroxybutyrate) (PHB), could significantly improve the final properties of polymer films [17–19]. It should be noted that blending of PLA with other polymers requires the use of plasticizers such as limonene [20] and polyethylene glycol [21], in order to increase the miscibility of the polymers in the blend and provide the necessary ductility. Due to strict requirements in food packaging applications, biocompatible and non-toxic plasticizers should be provided in order to improve the properties of the PLA-based films.

It should be noted that modification of PLA materials with functional additives could not only improve the physicochemical properties of PLA films and fiber materials but could also provide additional functionality for various applications [22]. Inclusion of various antimicrobial substances in the polymer matrix could impart antibacterial properties to the material, prevent the growth of microbes inside the package, and also improve the mechanical properties [23]. The use of packaging materials based on biodegradable PLA with antimicrobial additives can reduce environmental impact and provide product protection from physical, chemical, and microbiological factors, which is of great interest in terms of sustainability, adding functional properties to the packaging materials and reducing environmental risks [24,25].

According to the literature, ozonides of vegetable oils (olive, sunflower, etc.), which are the products of ozone reaction with the C=C double bonds in the molecules of unsaturated fatty acids [26], possess good antibacterial properties and could be used as functional additives for various PLA materials [27]. Moreover, ozonated vegetable oils are even more promising in this respect, since they could combine apparent antimicrobial activity with good plasticizing properties [28,29]. The most promising product of ozonation of the natural vegetable oils is 2lycerol-(9,10-trioxolane)-trioleate (ozonide of oleic acid triglyceride, OTOA) [30]. It is non-toxic, biocompatible, and has good biodegradability and antimicrobial activity. In a recent work [30] it was shown that introduction of OTOA into the nonwoven fibrous PLA materials has a significant effect on their physicochemical and functional properties and could provide additional antimicrobial functionality [31]. Thus, OTOA could be a promising additive for introduction into the PLA film matrix in order to provide additional functionality for various packaging applications.

In this work, we obtained novel film materials based on PLA with the addition of various amounts of OTOA (0–70 wt.%) as a modifier. DSC, DMA, X-ray diffraction analysis, FTIR spectroscopy, and optical and polarization microscopy were used to study the morphological, physicochemical, mechanical, and water absorption properties of PLA films with various OTOA contents. The obtained results provided a deeper understanding of the specific interactions of PLA with OTOA and made it possible to establish the effect of the OTOA content on the physicochemical and functional properties of the obtained film materials.

2. Materials and Methods

2.1. Materials

PLA was purchased from NatureWorks® Ingeo™ 3801X Injection Grade from Shenzhen Bright China Inc. (Shenzhen, China) with a viscosity average molecular weight of

1.9×10^5 g/mol; dry-cleaned chloroform, ≥99.5%, Sigma-Aldrich Inc. (St. Louis, MO, USA) was used to prepare solutions; Glycero-(9,10-trioxolane)-trioleate (ozonide of oleic acid triglyceride (OTOA)) (Scheme 1) was obtained from Medozon (Moscow, Russia). The chemical structure of OTOA has been described previously [30,32]. All reagents were used as received.

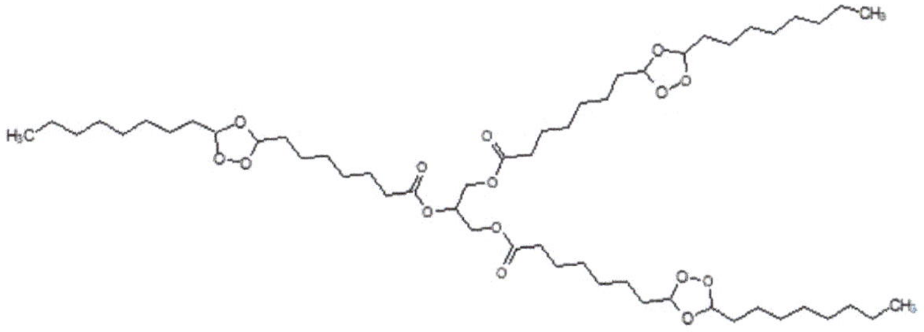

Scheme 1. Glycero-(9,10-trioxolane)-trioleate (ozonide of oleic acid triglyceride (OTOA)).

2.2. Preparation of Films

All films used in this work were prepared using solvent evaporation. PLA (2 g) was dissolved in chloroform (50 mL). Then, a certain amount of OTOA (5, 10, 30, 50, and 70 wt.%) was added to the PLA solution in chloroform and stirred for 12 h with a magnetic stirrer. After mixing, the solution was poured onto a glass plate and the film formed was dried to constant weight at ambient temperature. A film of the pristine PLA without additive was used as a control sample.

2.3. IR Spectroscopy

PLA films were analyzed using IR Fourier analysis with a NETZSCH TG 209 F1 (NETZSCH-Gerätebau GmbH, Selb, Germany) thermoanalytical balance and a Bruker Tensor 27 IR (Billerica, MA, USA) Fourier spectrometer with PIKE MIRacle™ accessory (PIKE Technologies, Madison, WI, USA) equipped with a germanium (Ge) crystal and an ATR attachment with a Teflon cell and cesium antimony electrode, which allows the measurements of solid samples. The sample was placed on the surface of the crystal and tightly clamped to ensure optical contact. IR spectra were recorded in the range of 4000–400 cm^{-1} with a resolution of 4 cm^{-1} and averaging over 16 successive scans.

2.4. DSC

The thermophysical properties of the film materials were determined on a NETZSCH DSC 204F1 Phoenix differential scanning calorimeter (NETZSCH-Gerätebau GmbH, Selb, Germany) in an inert atmosphere at an argon flow rate of 100 mL/min. PLA film samples (about 5 mg) were placed in aluminum sample crucibles and heated from 20 °C to 200 °C at a rate of 10 °C/min. The instrument was calibrated against indium, tin, and lead. After the first heating cycle, the samples were kept at 200 °C for 5 min, cooled to room temperature, and then reheated to 200 °C at a rate of 10 °C/min to record the second DSC curve. All samples were tested in triplicate.

The deconvolution of DSC peaks was carried out by means of NETZSCH (NETZSCH-Gerätebau GmbH, Selb, Germany) Peak Separation 2006.01 software employing the nonlinear regression method for the asymmetric DSC curves (Fraser-Suzuki algorithm) [33]. In calculation, the least squares (SLS) reduction was achieved using a hybrid procedure in which the Levenberg/Marquardt method was combined with step length optimization [34].

2.5. X-ray Diffraction Analysis

The structure of the PLA and PLA + OTOA films was studied by X-ray diffraction (XRD) using a DRON-3M X-ray diffractometer (Bourevestnik, St. Petersburg, Russia) with Cu Kα radiation ($\lambda = 1.5405$ Å) as an X-ray source. Scanning was carried out in the 2θ range from 10° to 50° with a scanning step of 0.1° and a data accumulation time of 5 s/step. The relative crystallinity of the polymer films was estimated using the following formula.

$$\chi = I_C / (I_C + I_A) \qquad (1)$$

where I_C and I_A are the integral intensities corresponding to the crystalline and amorphous phases, respectively [35].

2.6. Morphology

2.6.1. Optical and Polarized Light Microscopy

The surface morphology of PLA films with different OTOA contents was studied using an Olympus CX21 microscope (Olympus Corp., Tokyo, Japan) with a digital camera. The morphology of PLA crystalline spherulites was studied using polarized light microscopy on the Olympus BX51 microscope (Olympus Corp., Tokyo, Japan).

2.6.2. Sorption Capacity

To determine water sorption [36], PLA film samples were cut into uniform strips of 4×4 cm and placed under standard atmospheric conditions ($44 \pm 2\%$ RH and 20 ± 2 °C) for 24 h. After conditioning, the sample weight (m_1) was measured. Then the samples were immersed in distilled water for 24 h at a temperature of 20 ± 2 °C to ensure uniform water sorption. Thereafter, the wet samples were hung in open air at 20 ± 2 °C for 30 min to remove excess water on the surface of the samples. Then, the weight (m_2) of the wet samples was measured. The moisture content of the samples was calculated by the formula:

$$\%Q = (m_2 - m_1) / m_1 \times 100 \qquad (2)$$

with m_1 and m_2 being the sample weight before and after immersion in water. The degree of hydrophilicity of the samples was determined by the water contact angle measurements in the semi-angle modification [37] for a drop of water placed on the film surface. The droplet size measurements were carried out using an OLIMPUS CX21 optical microscope (Olympus Corp., Tokyo, Japan). The results were processed using the MICAM 3.02 software (https://micam.software.informer.com, Marien van Westen).

2.7. Mechanical Properties

The mechanical characteristics of the samples were studied on a Zwick Z010 testing machine (ZwickRoell GmbH & Co., Ulm, Germany) at room temperature. The layout of the tested samples is shown in Figure 1.

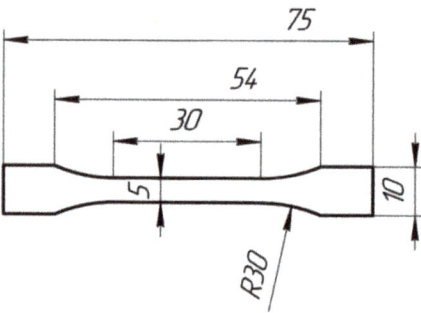

Figure 1. Scheme of the PLA film samples used for tests of mechanical characteristics.

The loading rate of the samples was 1 mm/min. During the tests, loading diagrams of the samples were recorded, namely the dependence of the load F on the deformation ε. From the diagrams obtained, tensile strength σ, elastic modulus E and elongation at break were determined. Mechanical tests were provided for five samples for each OTOA concentration, and the data were presented as mean value ± standard deviation at a significance level of $p < 0.05$.

2.8. Dynamic Mechanical Analysis

PLA films samples were studied by DMA under tension on a dynamic mechanical analyzer NETZSCH DMA 242 E Artemis (NETZSCH-Gerätebau GmbH, Selb, Germany). The films were cut into strips with a width of 5 mm and a length of the working part of 10 mm. The film samples were heated in the temperature range from 30 to 130 °C at a rate of 1 °C/min. During the measurement, the changes in the elastic modulus E' and the mechanical loss tangent tgδ were recorded with increasing temperature T according to the method described in the previous works [38,39]. The glass transition temperature of the PLA films was determined from the maximum of the tgδ versus temperature dependence.

3. Results and Discussions

3.1. Film Morphology

The introduction of OTOA significantly changed the appearance and morphology of the PLA films, their structure and surface properties, as could be seen in Figure 2. The thickness of films with different OTOA concentrations did not vary significantly and was 340 ± 50 μm. It should be noted that films with high concentrations of the OTOA additive above 30% show uniform morphology and do not undergo phase separation, as is also the case with PLA/PEG blends [40]. Only the PLA film with 70% OTOA showed the signs of non-uniform morphology. Like most of the flexible-chain polymers crystallizing isothermally from the solution, PLA is characterized by a spherulitic morphology, where a spherulite is an aggregate of crystals oriented relative to a common center. A feature of spherulites is such an arrangement of primary lamellas in them that the main macromolecular chains are always perpendicular to the radius of the spherulite. The optical images of PLA films (Figure 2) clearly show the increase in the size of spherulites when OTOA is added, which could affect their physicochemical and mechanical properties. The polarized light microscopy images of PLA films are displayed in Figure 2g,h, where spherulitic morphology with a typical Maltese-cross birefringence pattern could be seen. Additionally, the increase in the size of spherulites is obvious from these images. On the other hand, optical microscopy shows that for PLA films with 50% and 70% OTOA, spherulitic morphology is almost absent. Observed increase in the size of PLA spherulites could be attributed to the plasticizing effect of OTOA, since it was shown previously that addition of plasticizer increases the spherulitic growth rate and decrease crystallization kinetics of PLA spherulites [41]. Disturbed spherulitic morphology for the PLA films with 50% and 70% OTOA could be explained by the slow crystallization kinetics and low nucleation density at high plasticizer contents, leading to hardly detectable separate spherulites.

To determine the hydrophilic-hydrophobic properties of the film surface, the water contact angle for PLA films with variable OTOA content was measured (Figure 3). In the case of pristine PLA film, good wetting of the film surface with a contact angle of 47° by water was observed (the contact angle is noticeably lower than 90°). With the introduction of OTOA in an amount of 5%, the contact angle increased to 58°, indicating additional hydrophobization of the film surface. With an increase in the amount of OTOA in the PLA matrix, a significant increase in the hydrophobicity of the surface of the PLA + OTOA films was observed. The most pronounced hydrophobic properties were observed for the PLA film with 50% OTOA content. The PLA film with 70% OTOA demonstrated a decrease in the contact angle to 68.2°, which could be explained by the presence of a large number of defects in the film.

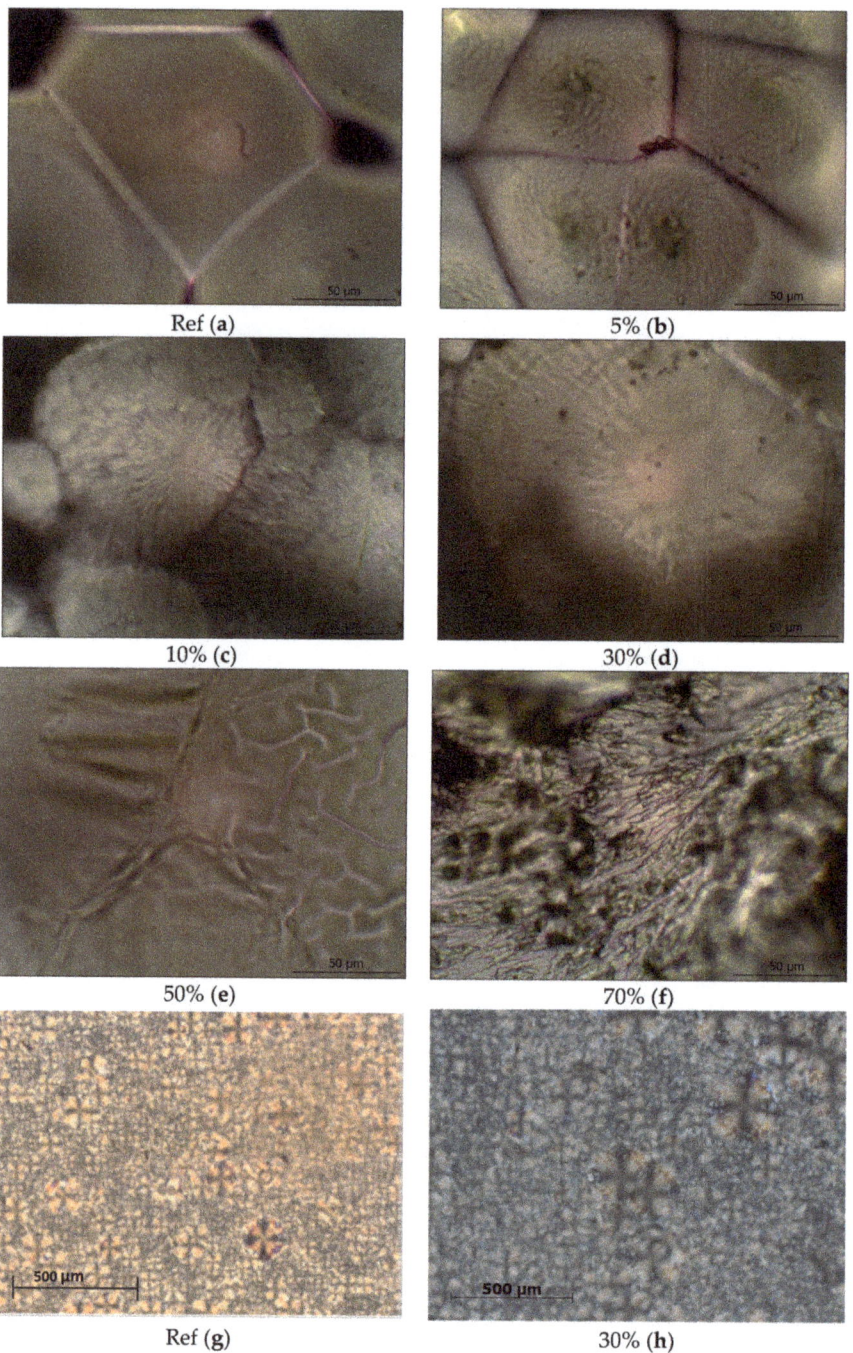

Figure 2. Microphotographs of the pristine PLA film and PLA + OTOA films obtained with an optical microscope (**a**–**f**) and photographs of the PLA film (**g**) and PLA +30% OTOA film (**h**) obtained with a polarizing microscope.

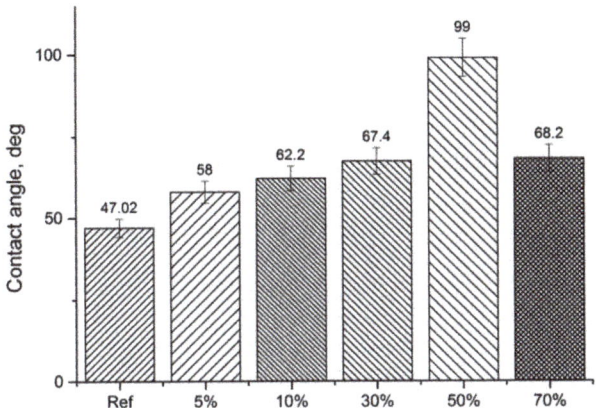

Figure 3. Water contact angle for PLA films with variable OTOA content.

Observed effects could be accounted for the distribution of long hydrophobic OTOA fragments in the hydrophobic regions of the PLA film and the weak interaction between them. Due to the chemical structure of PLA containing low-polarity ester groups, this biopolyester has low water solubility and belongs to moderately hydrophobic polymers [42]. At the same time, the structure of the OTOA molecule also consists mainly of hydrophobic fragments and includes, along with low-polarity ester groups, specific ozonide cycles [43–45]. Thus, the addition of OTOA to the PLA matrix in a wide range of concentrations (0–70%) leads to an increase in the hydrophobicity of the surface of the PLA + OTOA films.

In addition to hydrophobic properties of the film surface, water sorption capacity (Q) was studied for PLA films with variable OTOA content. As could be seen from the data presented in Figure 4, water sorption of films is about 1% and decreases with an increase in the amount of OTOA in the PLA matrix, which correlates with an increase in the water contact angle for the PLA + OTOA films (Figure 3). The PLA + OTOA film sample with the highest OTOA content (70%) demonstrates increased water sorption capacity up to 2.6%, most likely due to the non-uniform morphology and presence of defects in the film. The data obtained allows one to conclude that the addition of OTOA into the PLA matrix makes the film surface more hydrophobic, as compared to the surface of the pristine PLA film, which in turn leads to the decrease in the water sorption capacity of PLA + OTOA film materials.

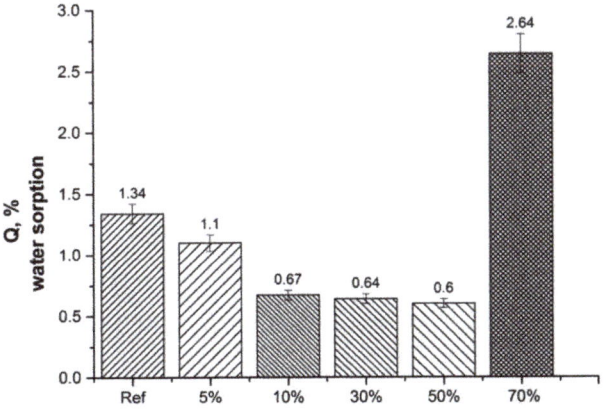

Figure 4. Sorption capacity (Q) of PLA films with variable OTOA content.

3.2. Mechanical Properties

Changes in the morphology of modified PLA materials could have a significant effect on their mechanical properties. Therefore, the elastic modulus (E), tensile strength (σ) and relative elongation at break for the PLA + OTOA films were estimated (Figure 5a,b). Generally, the introduction of OTOA into the PLA matrix led to a decrease in both tensile strength and elastic modulus of the PLA films. The addition of 5% OTOA sharply reduced the strength characteristics of the PLA film, whereas the PLA film with 10% OTOA showed the highest values of elastic modulus and tensile strength among PLA + OTOA samples. A further increase of the OTOA content in the PLA films led to a sharp decrease in the strength characteristics of film materials. It could be assumed that the tensile strength of the films decreases due to a decrease in the degree of crystallinity for PLA films and an increase in the proportion of the plasticizing agent (OTOA) in the PLA matrix [46,47].

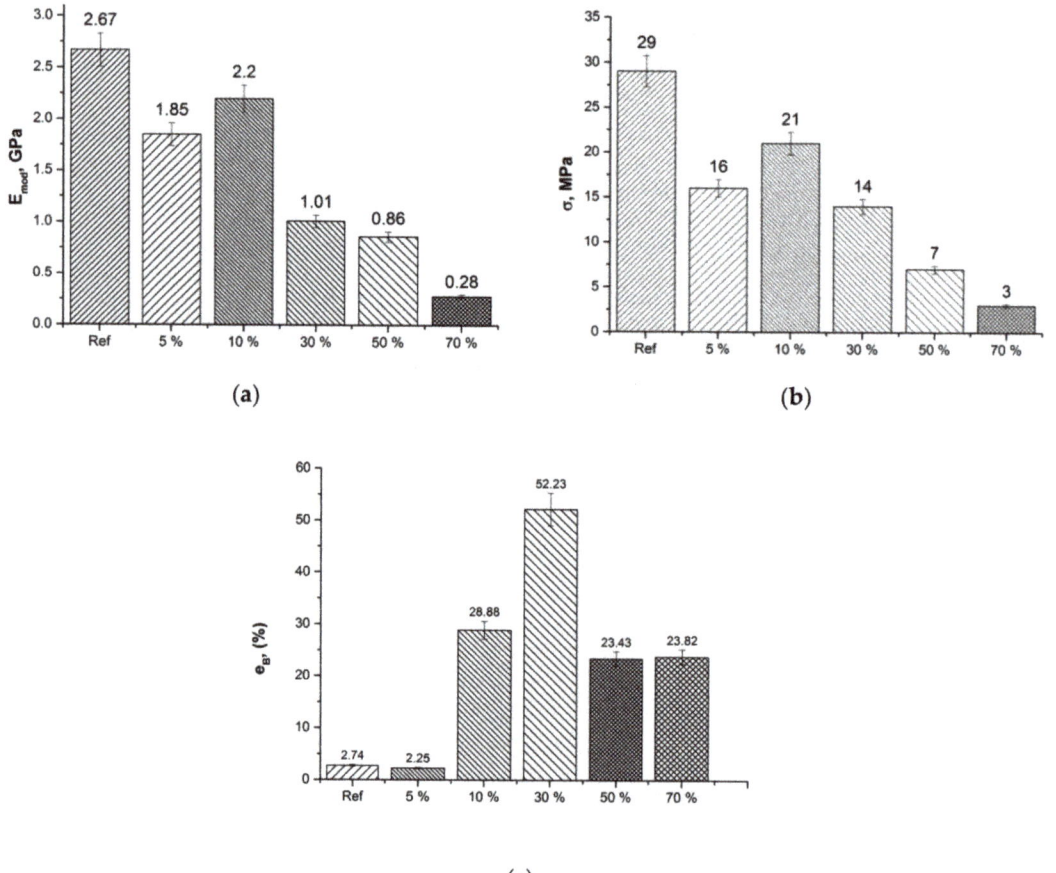

Figure 5. Elastic modulus (a), tensile strength (b), and relative elongation at break (c) for pristine PLA and PLA + OTOA film samples.

The effect of the OTOA addition on the relative elongation at break is somewhat different from the changes in the strength characteristics of the films described above (Figure 5c). This parameter for PLA film with 5% OTOA remains practically unchanged, as compared to the pristine PLA film, both showing poor ductility. Introduction of 10% OTOA into the PLA film led to a sharp increase in the relative elongation at the break (up to 29%), which is accompanied by a moderate decrease in the tensile strength of the material (from 29 to 21 MPa). The PLA film with 30% OTOA showed even greater value for the relative elongation at the break (52%). However, this is accompanied by a decrease by half in tensile strength. Further increase in the OTOA content above 30% was accompanied by a moderate decrease in relative elongation, while a sharp decrease in the tensile strength and elastic modulus was observed (σ = 3 MPa and E = 0.28 GPa for the PLA film containing 70% OTOA).

The observed behavior of the mechanical properties of PLA + OTOA films indicates the plasticizing effect of OTOA when it is introduced into the PLA matrix [48]. This effect could be associated with a change in the structural-dynamic state of the amorphous regions of PLA. In a recent work, it was shown that the dynamics of the amorphous phase of PLA changes significantly upon the addition of PHB [49]. The addition of PHB at concentrations of up to 30% leads to a decrease in the density of amorphous regions of PLA. A similar effect should be expected after the plasticization of PLA with OTOA molecules. The observed decrease in the relative elongation at break for the PLA films with the OTOA content above 30% could be attributed to the effect of steric hindrance, which the bulky OTOA molecule has on the segmental motion of PLA, provided that this effect exceeds the effect of plasticization. With an increase in OTOA content above 30%, the steric effect begins to prevail over the plasticization effect, increasing the PLA chain rigidity and thus reducing the relative deformation of the PLA films.

Summing up the obtained results on the mechanical properties of PLA films, it could be concluded that PLA film materials with 10% OTOA content maintain high tensile strength comparable to that of pristine PLA film and possess increased relative elongation at break. Observed combination of high elasticity and good strength characteristics obtained for PLA film materials with 10% and 30% OTOA content is important for potential applications of these materials [50].

3.3. Dynamic Mechanical Properties

In the DMA method, an oscillatory force is applied to a sample at a given temperature and frequency, and the material's response to this force is measured. For viscoelastic materials such as polymers, the magnitude of the material's response (i.e., strain amplitude) to an applied vibrational force is shifted by a phase angle δ, and the relationship between the applied stress and the strain occurring in the sample is calculated. The elastic modulus E' indicates the ability of the sample to store or return energy. Damping or mechanical losses are usually presented as tgδ [51,52].

Figure 5a shows the changes in the elastic modulus E' of the studied PLA films with temperature. At temperatures close to 30 °C, it could be seen that the value of the modulus increased when the OTOA content in PLA film is 10%. With an increase in the concentration of the OTOA in PLA films, a decrease in the elastic modulus is observed. The data obtained are in good agreement with the changes in the elastic modulus obtained from the mechanical tests (Figure 5a). With a further increase in temperature, a sharp decrease in the storage modulus is observed, which is associated with the transition of the material from the elastic to the highly elastic state, i.e., the glass transition of PLA film. The temperature interval of the transition for all studied PLA films is between 30 °C and 70 °C. It should be noted that PLA films with 5 and 10% OTOA are characterized by glass transition shifted to somewhat higher temperatures, as compared to glass transition for pristine PLA film. Further increase in OTOA concentration in PLA films shifts the glass transition interval to lower temperatures.

Characteristic temperatures associated with the glass transition (T_g) in studied PLA films could be estimated from the maximum of the tgδ versus temperature dependences (Figure 6b). All samples show a single T_g in the range of 61–65 °C, except for the PLA film with 30% OTOA, showing significantly lower T_g value. The glass transition temperature of the OTOA-modified PLA films with 5 and 10% OTOA (66.8 °C and 64.7 °C) showed somewhat higher values as compared to the T_g for the pristine PLA film (64.2 °C). The most striking difference was observed for the PLA film with 30% OTOA, showing the

T_g value being more than 10 °C lower than the corresponding value for the pristine PLA film. This result shows that the plasticizing effect of OTOA at concentration of 30% reaches its maximum.

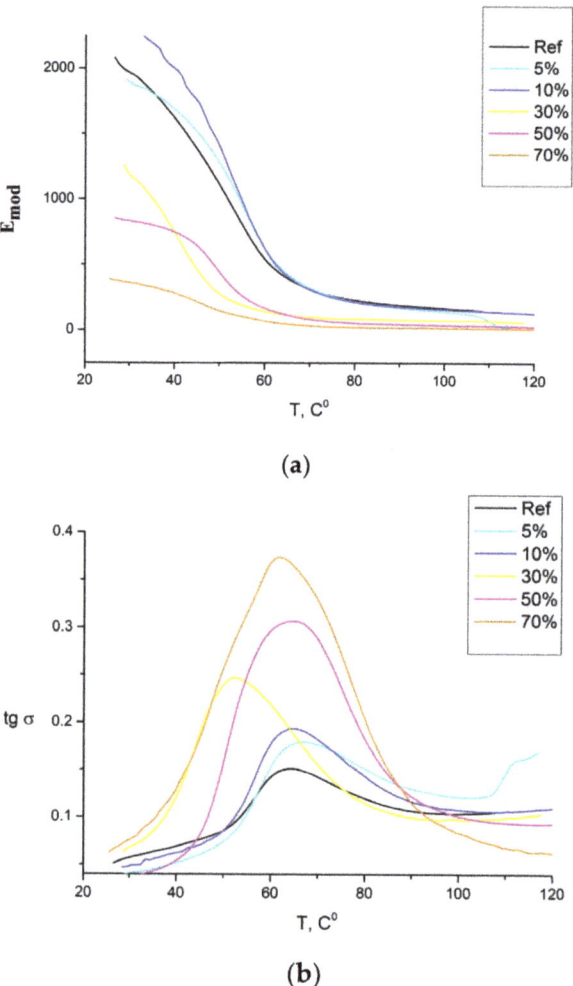

Figure 6. Viscoelastic behavior of pristine PLA and PLA + OTOA (5–70%) film samples: (**a**) elastic modulus, (**b**) mechanical loss tangent (tan δ).

As could be seen from the tgδ vs. temperature curves, the width of the temperature transition for the PLA films with 30, 50, and 70% OTOA is larger than that for the PLA films with lower OTOA concentrations. Increased width of the temperature transition evidence that PLA molecular chains exhibit a higher degree of mobility [53], which could be attributed to the plasticizing effect of OTOA.

3.4. FTIR Spectroscopy

To better understand the chemical interactions between PLA and OTOA, FTIR spectra of PLA + OTOA films were obtained (Figure 7a,b). The spectrum of the pristine PLA film shows characteristic bands at 1455 cm^{-1} and 1756 cm^{-1} resulting from -CH$_3$ bending vibrations and C=O group stretching vibrations [54]. The absorption bands at 2944 cm^{-1} and

2995 cm^{-1} refer to the asymmetric stretching vibration of the -CH group. As could be seen, the absorption bands of PLA and OTOA are partially overlapped, which significantly complicates the analysis of the effect of OTOA addition into the PLA matrix. Though, two additional bands appear at 2927 cm^{-1} and 2856 cm^{-1} in the spectra of PLA + OTOA films, which could be attributed to the symmetric and asymmetric stretching vibrations of the -CH$_2$ groups [55,56]. Since PLA does not contain -CH$_2$ groups in its chemical structure, in contrast to OTOA, in which -CH$_2$ group is in abundance, the presence of these bands in the FTIR spectra is the clear evidence for the inclusion of OTOA in the supramolecular structure of PLA films. As could be seen, the intensity of the absorption band at 2856 cm^{-1} shows good correlation with the OTOA content in PLA films.

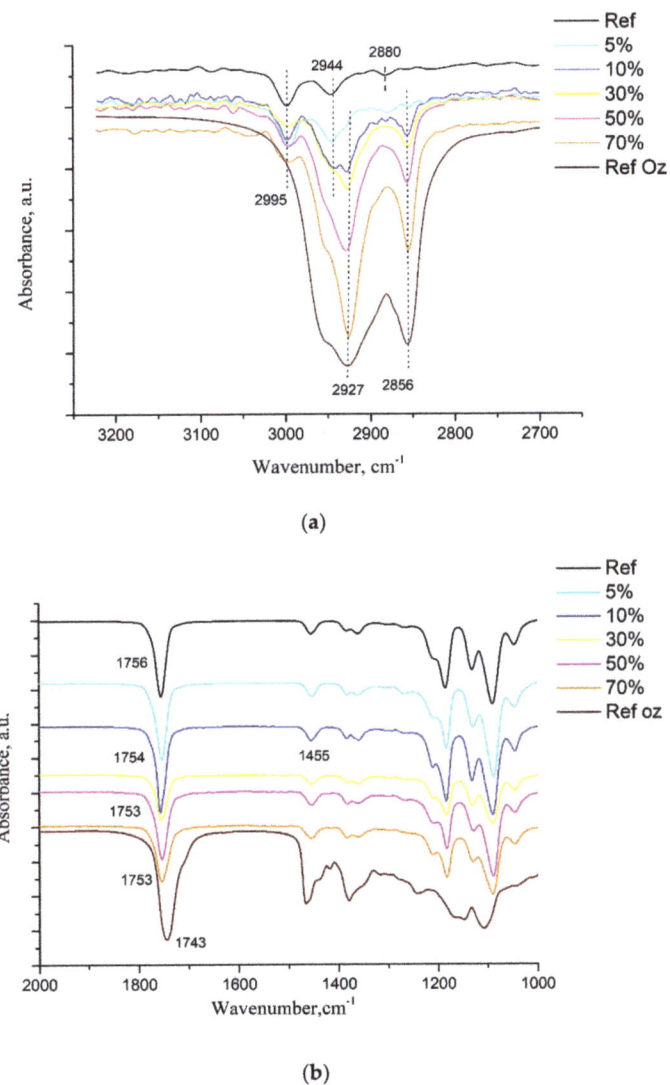

Figure 7. FTIR spectra of pristine PLA film, PLA + OTOA films and pure OTOA: (**a**) Close-up view of FTIR spectra at 2700–3000 cm^{-1} interval (**a**), FTIR spectra at 900–2000 cm^{-1} wavenumber region (**b**).

In addition, the band at 1756 cm^{-1} in the pristine PLA is shifted toward lower wavenumbers (1753 cm^{-1}) for PLA samples with the added OTOA. This is the evidence of weak interactions between C=O groups of PLA and polar ozonide cycles in the OTOA molecules (see below Table 1.) [57]. Based on the analysis of FTIR spectra, it could be assumed that observed interactions between polar groups in PLA and ozonide cycles in OTOA promote conformational changes associated with the reorientation of polar groups in PLA, which contributes to increased segmental mobility of PLA polymer chains [58]. Thus, low molecular weight OTOA could act as a plasticizer, affecting the mechanical and physicochemical properties of PLA film materials.

Table 1. Characteristic bands in the FTIR spectra for pristine PLA and PLA + OTOA films.

PLA Characteristic Bands, cm^{-1}	PLA + OTOA Characteristic Bands, cm^{-1}	Characteristic Band Assignment
2995	2995	-CH (asim)
2947	2947	-CH (sim)
2880	2880	-CH$_3$ stretching
1756	1753	C=O stretching
1455	1455	-CH$_3$ bending
	2928	-CH$_2$ (asim)
	2856	-CH$_2$ (sim)

3.5. Differential Scanning Calorimetry

Figure 8a shows the DSC curves of the first heating for the pristine PLA film and PLA + OTOA film samples, whereas Figure 8b shows the DSC thermograms of the subsequent cooling of the samples. The characteristic endothermic peak in the first heating DSC thermograms corresponds to melting PLA (T_m). The low-temperature step transition on the cooling thermograms in the interval of 40–60 °C could be attributed to the glass transition of PLA (T_g). The corresponding temperatures for both transitions (T_g and T_m) are shown in Table 2. The area of the melting peak was obtained, and the melting enthalpy (ΔH_m) was calculated, as well as the degree of crystallinity of the PLA films according to DSC (χ). The crystallinity degree (χ) for PLA films, assuming no cold crystallization taking place during heating, was calculated according to the equation:

$$\chi = \frac{\Delta H_m}{\Delta H_m^0} \times 100\% \qquad (3)$$

where ΔH_m is the experimental melting enthalpy, ΔH_m^0 is the melting enthalpy value for the 100% crystalline poly(L-lactide), being 93.6 J/g [59,60]. The values of χ for the studied PLA film samples are given in Table 2.

The DSC thermogram of the pristine PLA film shows an endothermic peak at 169.0 °C, which is a typical melting temperature of PLA [61]. With an increase in the mass fraction of OTOA in the film, a superposition of exo- and endothermic peaks is observed in the temperature range of 140–180 °C. The endothermic peak corresponds to the melting of PLA, whereas the exothermic peak is due to a complex reaction of thermal destruction of OTOA according to the mechanism of breaking of the C-O-O-C bonds and formation of C-OH groups [62]. Previously, it was shown that this exothermic reaction is irreversible. With an increase in the OTOA content in the film, the endothermic PLA melting peak demonstrates a decrease in its area (i.e., a decrease in crystallinity) and a shift in T_m toward lower temperatures. The deconvolution of the overlapping exo- and endothermic calorimetric peaks (Figure 9 and Table 3) made it possible to estimate more correctly the PLA melting enthalpy and thus determine the degree of crystallinity of the studied PLA films with high OTOA contents (Table 2).

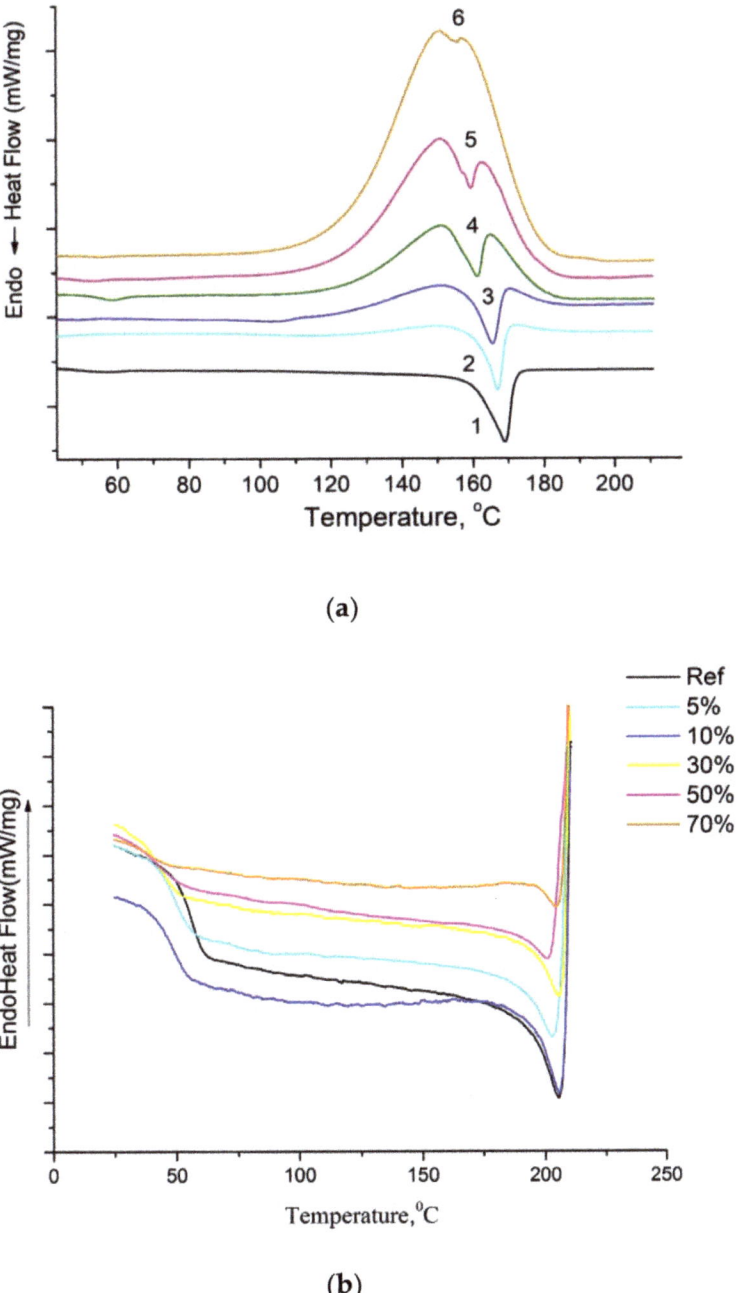

Figure 8. (**a**) DSC thermograms of the first heating for pristine PLA and PLA + OTOA (5–70%) film samples: PLA (1), PLA +5% OTOA (2), PLA +10% OTOA (3), PLA +30% OTOA (4), PLA +50% OTOA (5), PLA +70% OTOA (6). (**b**) DSC thermographs of colling the samples after the first scan.

Table 2. Thermodynamic and structural characteristics of PLA + OTOA films (0–70%) obtained using the DSC, DMA and XRD methods.

Sample	T_g (DSC) * (°C)	T_g (DMA) (°C)	T_m (DSC) * (°C)	ΔH_m (DSC) * (J/g)	χ (DSC) (%)	χ (XRD) (%)
PLA	57.8	64.2	169.0	35.7	38.1	35.8
PLA +5% OTOA	49.5	66.8	167.0	27.0	28.9	26.5
PLA +10% OTOA	48.0	64.7	165.0	26.6	28.5	26.1
PLA +30% OTOA	45.2	52.2	161.4	25.0	26.7	24.1
PLA +50% OTOA	43.4	64.6	159.5	17.6	18.8	18.9
PLA +70% OTOA	38.1	61.6	155.2	4.0	4.3	14.3

* T_g values were obtained from the DSC thermographs of cooling the PLA films after the first scan. T_m values were obtained as the result of deconvolution of the overlapping calorimetric peaks. The ΔH_m values are provided taking into account the mass fraction of the PLA in composition.

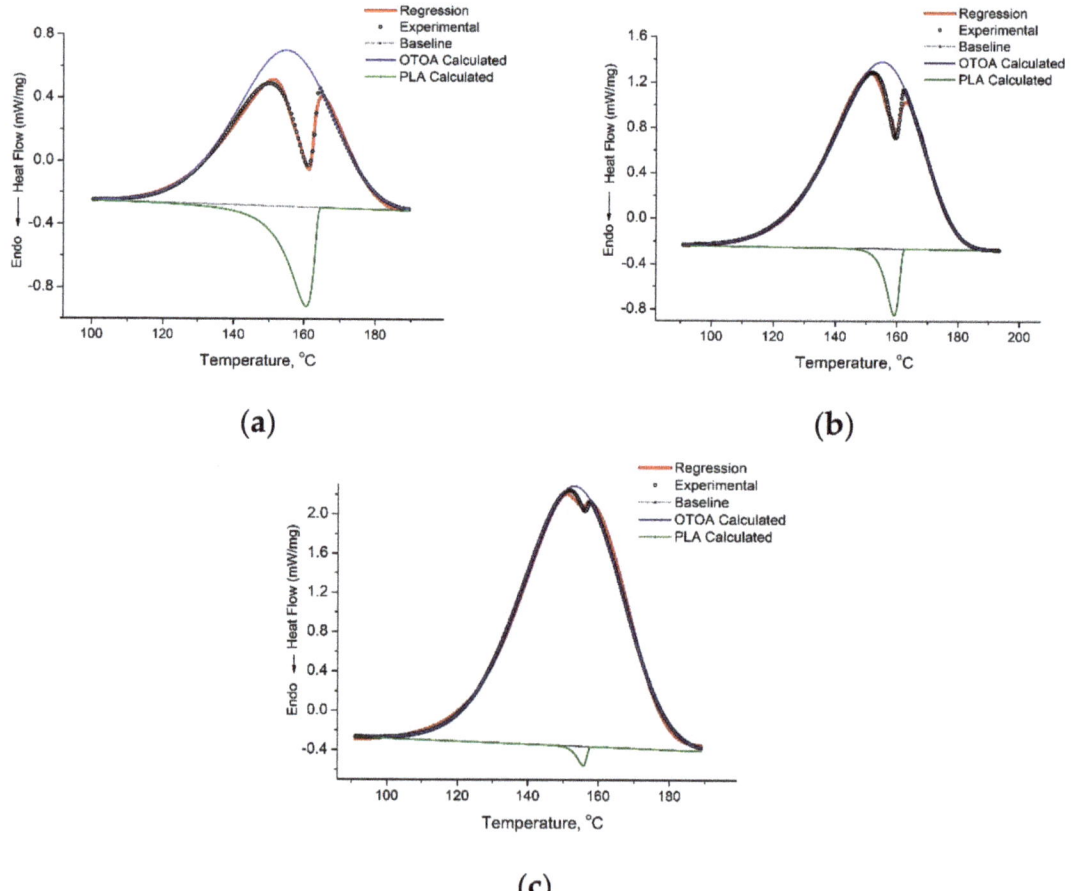

Figure 9. Deconvolution of the overlapping exo- and endothermic calorimetric peaks in the DSC thermograms for PLA + OTOA film samples with high OTOA content: (**a**) PLA +30% OTOA, (**b**) PLA +50% OTOA, (**c**) PLA +70% OTOA.

Table 3. Results of the deconvolution of the overlapping calorimetric peaks for PLA films with high OTOA contents.

	Deconvolution Parameters (Exothermic Peak)	Deconvolution Parameters (PLA Melting)
PLA + OTOA 30%	T = 155.0 °C; ΔH = 133.5 J/g	T = 160.7 °C; ΔH = -25.0 J/g
PLA + OTOA 50%	T = 155.0 °C; ΔH = 321.0 J/g	T = 159.8 °C; ΔH = -17.6 J/g
PLA + OTOA 70%	T = 153.0 °C; ΔH = 533.5 J/g	T = 155.8 °C; ΔH = -4.0 J/g

Neat PLA film was characterized by degree of crystallinity of 38.1%, as estimated by DSC, which is quite high for isothermally crystallized PLA films [63]. This could be explained by high mobility of PLA chains in the solution and high degree of supercooling, providing the thermodynamic driving force required for the growth of PLA spherulites. In addition, cold crystallization exothermic peak was not observed in DSC thermogram for the pristine PLA film and all PLA + OTOA film samples, being in contrast to the nonwoven fibrous materials based on PLA, for which the cold crystallization was previously observed [30]. This can be explained by the fact that under equilibrium crystallization conditions at ambient temperature, crystallization of neat PLA in solution proceeds with the formation of highly ordered semicrystalline structures with spherulitic morphology, which hinder further crystallization at temperatures above the glass transition due to a decrease in the segmental mobility of PLA chains. The absence of cold crystallization peaks for PLA + OTOA samples could be also attributed for the slower crystallization kinetics and decreased nucleation density due to the addition of OTOA [64].

The data shown in Table 2 show that the introduction of OTOA into PLA films led to the plasticizing effect, which was manifested in the decrease in the glass transition temperature from 57.8 to 38.1 °C and decrease in the melting temperature of PLA. The plasticizer occupies the intermolecular space between polymer chains, reducing the energy of molecular motion and the formation of hydrogen bonds between polymer chains, which in turn increases the free volume and molecular mobility [65–67]. As the content of the OTOA increases, the efficiency of the plasticizer in decreasing the T_g of PLA generally increases.

As can be seen in Table 2, both the melting temperature and the melting enthalpy decreased with the increase in the OTOA content in the PLA films. Decrease in T_m could be attributed to the plasticizing action of OTOA, as observed previously [30]. At the same time, a decrease in the melting enthalpy of PLA films observed upon OTOA addition evidence that along with the plasticizing effect, OTOA could impede PLA crystallization. Previously, it was shown that OTOA could hinder cold crystallization for the electrospun PLA fibers during the second heating cycle [30]. This effect could be attributed to the intermolecular interaction between the PLA terminal -OH groups and OTOA molecules observed by FTIR, leading to the decrease in the mobility of PLA polymer chains. This provides the physical hindrance for PLA crystallization and leads to the decrease in the crystallinity of PLA films after increase in OTOA content.

3.6. X-ray Diffraction Analysis

To independently confirm the effect of OTOA on the crystal characteristics of PLA + OTOA films, the samples were studied by X-ray diffraction (XRD) and the values of the degree of crystallinity of the films were obtained. The X-ray diffraction patterns of PLA films with different OTOA contents are shown in Figure 10. Pristine PLA film showed the main diffraction peaks at 2θ angles of 16.2° and 18.6°, confirming the presence of PLA crystal structures in the film. PLA has a strong diffraction at 16.2°, related to the crystalline α phase of PLA [68]. As could be seen, PLA and PLA + OTOA films have the same crystal structure. An increase in the mass fraction of OTOA does not lead to a significant change in the position of the diffraction peaks corresponding to the PLA crystalline phase. On the other hand, it leads to an increase in the contribution of the amorphous phase (amorphous halo), corresponding to the OTOA phase and/or amorphous regions of the semicrystalline PLA structure.

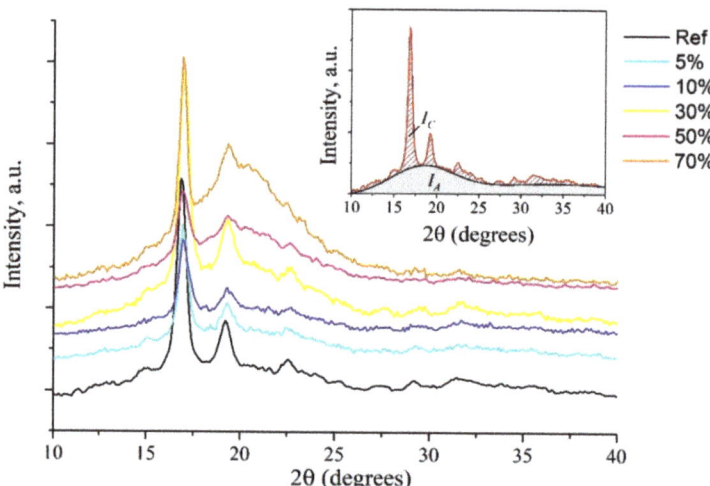

Figure 10. XRD patterns of PLA and PLA + OTOA samples (0–70%). The inset shows the determination of I_A and I_C areas used for the degree of crystallinity calculations.

The degree of crystallinity of PLA films was estimated by the formula given above (see materials and methods) and the corresponding values are given in Table 2. A noticeable decrease in the crystallinity of the system is observed with an increase in the mass fraction of OTOA in PLA films. For the film with the OTOA content of 70%, the crystallinity of the system drops up to 14.3%, being nonetheless quite high [41]. The crystallinity values estimated by the XRD method correlate well with the corresponding values obtained from the DSC data. The only exception is the PLA film with 70% OTOA content, for which estimation of the χ value from DSC data is complicated by the large overlapping calorimetric peaks.

Summing up the obtained results, it could be concluded that the morphology, sorption capacity, as well as thermodynamic and mechanical properties of PLA + OTOA films could be controlled by the OTOA content in the films. Thus, PLA film sample containing 10% OTOA is characterized by a combination of high elasticity and good strength characteristics (tensile strength 21 MPa, relative elongation at break 26%). The sample with the addition of 30% OTOA gives an even greater increase in relative elongation (up to 52%) with a significant loss of tensile strength. The combination of high elasticity and strength is of great importance for the various applications of PLA film materials. FTIR spectroscopy revealed an interaction between the OTOA molecules, which act as a plasticizer, and the PLA matrix. The plasticizing effect of OTOA, as shown by DSC, DMA, and mechanical tests, contributed to the changes in the thermal and mechanical properties of PLA films modified with OTOA. At the same time, at high OTOA concentrations (>30%), the effects related to the interaction of OTOA with PLA polymer chains and the hindrance of the segmental motion of PLA polymer ends could prevail over the plasticizing effect of OTOA, leading to the decrease in crystallinity of PLA films, deterioration of the morphology and mechanical properties of PLA films. It could be concluded that the PLA films with 10% and 30% OTOA possess optimal physicochemical properties, such as improved morphology, mechanical properties, and water sorption characteristics. Since OTOA exhibits a pronounced antimicrobial activity [31,62,69], it could be used not only as an efficient plasticizer, but also as a functional antimicrobial additive. Thus, developed PLA + OTOA films have a high potential to be used as biodegradable materials with antibacterial activity, for example, as antimicrobial packaging materials, which could provide protection from physical, chemical, and microbiological factors and reduce the negative impact on the environment.

4. Conclusions

In this article, the morphological, physicochemical, mechanical, and thermal properties of PLA films after the addition of different concentrations of oleic acid triglyceride ozonide (OTOA) were studied. Analysis of the film surface showed changes in the film morphology as a result of the addition of OTOA, which were manifested in an increase in the size of spherulites. PLA films containing 10% and 30% OTOA exhibited optimal mechanical and elastic properties, combining high elasticity and good tensile strength. The obtained results are important for potential packaging applications of the studied PLA film materials. Contact angle measurements showed that OTOA addition leads to significant hydrophobization of PLA films, whereas FTIR spectroscopy revealed weak interactions between OTOA and the PLA matrix. It was found that OTOA acts as a plasticizer and leads to an increase in PLA segmental mobility, which in turn contributes to changes in the thermodynamic and mechanical properties of PLA films. The results of DSC and XRD showed that OTOA promotes the process of PLA amorphization, therefore reducing the crystallinity of the resulting PLA + OTOA film materials. Eventually, the obtained results evidence that the morphological, thermodynamic, and mechanical properties of PLA + OTOA films could be controlled by the OTOA content in the films. Since OTOA possess pronounced antimicrobial activity, studied composite PLA + OTOA films could provide additional functionality as compared to biodegradable packaging materials based on pristine PLA. Developed PLA film materials with the optimal OTOA content of 10% and 30% could be used in various packaging and biomedical applications.

Author Contributions: Conceptualization, A.L.I., A.O., V.S. and O.A.; methodology, M.K., V.P., T.P., O.A. and O.K.; formal analysis, M.K., V.P., I.T. and O.K.; investigation, M.K., V.P., T.P., O.K., O.A. and S.L.; resources, T.P., M.K., V.P. and V.S.; data curation, I.T., O.K. and S.L.; writing—original draft preparation, O.A. and A.L.I.; writing—review and editing, O.A., A.L.I. and V.S.; supervision, A.L.I. and V.S.; project administration, A.O. and V.S. All authors have read and agreed to the published version of the manuscript.

Funding: This work was financially supported by the Ministry of Science and Higher Education of the Russian Federation, topic number 12041300207-2.

Institutional Review Board Statement: Not applicable.

Informed Consent Statement: Not applicable.

Data Availability Statement: Not applicable.

Acknowledgments: To S. S. Kozlov for his help in the analysis and discussion of the obtained data.

Conflicts of Interest: The authors declare no conflict of interest.

References

1. Ncube, L.K.; Ude, A.U.; Ogunmuyiwa, E.N.; Zulkifli, R.; Beas, I.N. Environmental Impact of Food Packaging Materials: A Review of Contemporary Development from Conventional Plastics to Polylactic Acid Based Materials. *Materials* **2020**, *13*, 4994. [CrossRef] [PubMed]
2. Sukumar, M.; Sudharsan, K.; Radha Krishnan, K. Polylactic Acid (PLA)-Based Composites in Food Packaging. In *Biopolymer-Based Food Packaging*; Kumar, S., Mukherjee, A., Dutta, J., Eds.; John Wiley & Sons, Inc.: Hoboken, NJ, USA, 2022; Chapter 8. [CrossRef]
3. Gutiérrez, T.J.; Mendieta, J.R.; Ortega-Toro, R. In-depth study from gluten/PCL-based food packaging films obtained under reactive extrusion conditions using chrome octanoate as a potential food grade catalyst. *Food Hydrocoll.* **2021**, *111*, 106255. [CrossRef]
4. DeStefano, V.; Khan, S.; Tabada, A. Applications of PLA in modern medicine. *Eng. Regen.* **2020**, *1*, 76–87. [CrossRef]
5. Donate, R.; Monzón, M.; Alemán-Domínguez, M. Additive manufacturing of PLA-based scaffolds intended for bone regeneration and strategies to improve their biological properties. *e-Polymers* **2020**, *20*, 571–599. [CrossRef]
6. Dybka-Stępień, K.; Antolak, H.; Kmiotek, M.; Piechota, D.; Koziróg, A. Disposable Food Packaging and Serving Materials—Trends and Biodegradability. *Polymers* **2021**, *13*, 3606. [CrossRef]
7. Masutani, K.; Kimura, Y. PLA Synthesis. From the Monomer to the Polymer. In *Poly(lactic acid) Science and Technology: Processing, Properties, Additives and Applications*; Jiménez, A., Peltzer, M., Ruseckaite, R., Eds.; Royal Society of Chemistry: London, UK, 2014; Chapter 1; pp. 1–36. [CrossRef]
8. Reneker, D.H.; Yarin, A.L. Electrospinning jets and polymer nanofibers. *Polymer* **2008**, *49*, 2387–2425. [CrossRef]

9. Siccardi, M.; Garcia-Fonte, X.X.; Simon, A.; Pettarin, V.; Abad, M.J.; Bernal, C. Effect of the Processing-Induced Morphology on the Mechanical Properties of Biodegradable Extruded Films Based on Poly(lactic acid) (PLA) Blends. *J. Polym. Environ.* **2019**, *27*, 2325–2333. [CrossRef]
10. Masmoudi, F.; Bessadok, A.; Dammak, M.; Jaziri, M.; Ammar, E. Biodegradable packaging materials conception based on starch and polylactic acid (PLA) reinforced with cellulose. *Environ. Sci. Pollut. Res.* **2016**, *23*, 20904–20914. [CrossRef]
11. Teixeira, S.; Eblagon, K.M.; Miranda, F.; Pereira, M.F.R.; Figueiredo, J.L. Towards Controlled Degradation of Poly(lactic) Acid in Technical Applications. *C* **2021**, *7*, 42. [CrossRef]
12. Zaaba, N.F.; Jaafar, M. A review on degradation mechanisms of polylactic acid: Hydrolytic, photodegradative, microbial, and enzymatic degradation. *Polym. Eng. Sci.* **2020**, *60*, 2061–2075. [CrossRef]
13. Seth, M.; Jana, S. Nanomaterials based superhydrophobic and antimicrobial coatings. *Nano World J.* **2020**, *6*, 26–28. [CrossRef]
14. Wu, F.; Misra, M.; Mohanty, A.K. Challenges and new opportunities on barrier performance of biodegradable polymers for sustainable packaging. *Prog. Polym. Sci.* **2021**, *117*, 101395. [CrossRef]
15. Zhou, L.; Xu, P.P.; Ni, S.H.; Xu, L.; Lin, H.; Zhong, G.-J.; Huang, H.-D.; Li, Z.-M. Superior Ductile and High-barrier Poly(lactic acid) Films by Constructing Oriented Nanocrystals as Efficient Reinforcement of Chain Entanglement Network and Promising Barrier Wall. *Chin. J. Polym. Sci.* **2022**. [CrossRef]
16. Risyon, N.P.; Othman, S.H.; Basha, R.K.; Talib, R.A. Characterization of polylactic acid/halloysite nanotubes bionanocomposite films for food packaging. *Food Packag. Shelf Life* **2020**, *23*, 100450. [CrossRef]
17. Thomas, M.S.; Pillai, P.K.S.; Faria, M.; Cordeiro, N.; Barud, H.; Thomas, S.; Pothen, L.A. Electrospun polylactic acid-chitosan composite: A bio-based alternative for inorganic composites for advanced application. *J. Mater. Sci. Mater. Med.* **2018**, *29*, 1–12. [CrossRef]
18. Siracusa, V.; Karpova, S.; Olkhov, A.; Zhulkina, A.; Kosenko, R.; Iordanskii, A. Gas Transport Phenomena and Polymer Dynamics in PHB/PLA Blend Films as Potential Packaging Materials. *Polymers* **2020**, *12*, 647. [CrossRef]
19. Wagner, A.; Poursorkhabi, V.; Mohanty, A.K.; Misra, M. Analysis of Porous Electrospun Fibers from Poly(L-lactic acid)/Poly(3-hydroxybutyrate-co-3-hydroxyvalerate) Blends. *ACS Sustain. Chem. Eng.* **2014**, *2*, 1976–1982. [CrossRef]
20. Arrieta, M.P.; López, J.; Hernández, A.; Rayón, E. Ternary PLA–PHB–Limonene blends intended for biodegradable food packaging applications. *Eur. Polym. J.* **2014**, *50*, 255–270. [CrossRef]
21. Arrieta, M.P.; Samper, M.D.; López, J.; Jiménez, A. Combined Effect of Poly(hydroxybutyrate) and Plasticizers on Polylactic acid Properties for Film Intended for Food Packaging. *J. Polym. Environ.* **2014**, *22*, 460–470. [CrossRef]
22. Malinconico, M.; Vink, E.T.H.; Cain, A. Applications of Poly(lactic Acid) in Commodities and Specialties. In *Industrial Applications of Poly(Lactic Acid)*; Di Lorenzo, M., Androsch, R., Eds.; Springer: Cham, Switzerland, 2018; pp. 35–50. [CrossRef]
23. Karakurt, I.; Ozaltin, K.; Pištěková, H.; Vesela, D.; Michael-Lindhard, J.; Humpolícek, P.; Mozetič, M.; Lehocky, M. Effect of Saccharides Coating on Antibacterial Potential and Drug Loading and Releasing Capability of Plasma Treated Polylactic Acid Films. *Int. J. Mol. Sci.* **2022**, *23*, 8821. [CrossRef]
24. Han, J.H. Antimicrobial packaging systems. In *Food Science and Technology, Innovations in Food Packaging*; Han, J.H., Ed.; Academic Press: Cambridge, MA, USA, 2005; pp. 80–107. [CrossRef]
25. Sung, S.-Y.; Sin, L.T.; Tee, T.-T.; Bee, S.-T.; Rahmat, A.; Rahman, W.; Tan, A.-C.; Vikhraman, M. Antimicrobial agents for food packaging applications. *Trends Food Sci. Technol.* **2013**, *33*, 110–123. [CrossRef]
26. Radzimierska-Kaźmierczak, M.; Śmigielski, K.; Sikora, M.; Nowak, A.; Plucińska, A.; Kunicka-Styczyńska, A.; Czarnecka-Chrebelska, K. Olive Oil with Ozone-Modified Properties and Its Application. *Molecules* **2021**, *26*, 3074. [CrossRef]
27. Malek, N.S.A.; Khuhairil, M.; Khusaimi, Z.; Bonnia, N.N.; Mahmood, M.R.; Asli, N.A. Polylactic acid (PLA) incorporated with antimicrobial substances for fruit packaging—A review. *AIP Conf. Proc.* **2021**, *2368*, 030002. [CrossRef]
28. Bétron, C.; Cassagnau, P.; Bounor-Legaré, V. Control of diffusion and exudation of vegetable oils in EPDM copolymers. *Eur. Polym. J.* **2016**, *82*, 102–113. [CrossRef]
29. Lim, K.M.; Ching, Y.C.; Gan, S.N. Effect of Palm Oil Bio-Based Plasticizer on the Morphological, Thermal and Mechanical Properties of Poly(Vinyl Chloride). *Polymers* **2015**, *7*, 2031–2043. [CrossRef]
30. Olkhov, A.; Alexeeva, O.; Konstantinova, M.; Podmasterev, V.; Tyubaeva, P.; Borunova, A.; Siracusa, V.; Iordanskii, A. Effect of Glycero-(9,10-trioxolane)-trialeate on the Physicochemical Properties of Non-Woven Polylactic Acid Fiber Materials. *Polymers* **2021**, *13*, 2517. [CrossRef] [PubMed]
31. Ugazio, E.; Tullio, V.; Binello, A.; Tagliapietra, S.; Dosio, F. Ozonated Oils as Antimicrobial Systems in Topical Applications. Their Characterization, Current Applications, and Advances in Improved Delivery Techniques. *Molecules* **2020**, *25*, 334–347. [CrossRef]
32. Miura, T.; Suzuki, S.; Sakurai, S.; Matsumoto, A.; Shinriki, N. Structure Elucidation of Ozonated Olive Oil. In Proceedings of the 15th Ozone World Congress: Medical Therapy Conference, London, UK, 11–15 September 2001.
33. Fraser, R.D.B.; Suzuki, E. Resolution of overlapping bands. Functions for simulating band shapes. *Anal. Chem.* **1969**, *41*, 37–39. [CrossRef]
34. Opfermann, J. Rechentechnik. *Datenverarbeitung* **1985**, *23*, 26.
35. Hsieh, Y.-T.; Nozaki, S.; Kido, M.; Kamitani, K.; Kojio, K.; Takahara, A. Crystal polymorphism of polylactide and its composites by X-ray diffraction study. *Polym. J.* **2020**, *52*, 755–763. [CrossRef]
36. Debnath, S.; Madhusoothanan, M. Water Absorbency of Jute—Polypropylene Blended Needle-punched Nonwoven. *J. Ind. Text.* **2010**, *39*, 215–231. [CrossRef]

37. Heib, F.; Schmitt, M. Statistical Contact Angle Analyses with the High-Precision Drop Shape Analysis (HPDSA) Approach: Basic Principles and Applications. *Coatings* **2016**, *6*, 57. [CrossRef]
38. Korokhin, R.A.; Shapagin, A.V.; Solodilov, V.I.; Zvereva, U.G.; Solomatin, D.V.; Gorbatkina, Y.A. Epoxy polymers modified with polyetherimide. Part I: Rheological and thermomechanical characteristics. *Polym. Bull.* **2020**, *78*, 1573–1584. [CrossRef]
39. ASTM D4473-08; Standard Test Method for Plastics: Dynamic Mechanical Properties: Cure Behavior. ASTM International: West Conshohocken, PA, USA, 2008. Available online: https://www.en-standard.eu (accessed on 2 August 2022).
40. Hu, Y.; Topolkaraev, V.; Hiltner, A.; Baer, E. Aging of poly(lactide)/poly(ethylene glycol) blends. Part 2. Poly(lactide) with high stereoregularity. *Polymer* **2003**, *44*, 5711–5720. [CrossRef]
41. Xiao, H.; Lu, W.; Yeh, J.-T. Effect of plasticizer on the crystallization behavior of poly(lactic acid). *J. Appl. Polym. Sci.* **2009**, *113*, 112–121. [CrossRef]
42. Pankova, Y.; Shchegolikhin, A.; Iordanskii, A.; Zhulkina, A.; Ol'Khov, A.; Zaikov, G. The characterization of novel biodegradable blends based on polyhydroxybutyrate: The role of water transport. *J. Mol. Liq.* **2010**, *156*, 65–69. [CrossRef]
43. Kiss, E.; Bertóti, I.; Vargha-Butler, E.I. XPS and wettability characterization of modified poly(lactic acid) and poly(lactic/glycolic acid) films. *J. Coll. Int. Sci.* **2002**, *245*, 91–98. [CrossRef]
44. Trifol, J.; Plackett, D.; Szabo, P.; Daugaard, A.E.; Baschetti, M.G. Effect of Crystallinity on Water Vapor Sorption, Diffusion, and Permeation of PLA-Based Nanocomposites. *ACS Omega* **2020**, *5*, 15362–15369. [CrossRef]
45. Pantani, R.; De Santis, F.; Auriemma, F.; De Rosa, C.; Di Girolamo, R. Effects of water sorption on poly(lactic acid). *Polymer* **2016**, *99*, 130–139. [CrossRef]
46. Ljungberg, N.; Wesslén, B. The effects of plasticizers on the dynamic mechanical and thermal properties of poly(lactic acid). *J. Appl. Polym. Sci.* **2002**, *86*, 1227–1234. [CrossRef]
47. Inácio, E.M.; Lima, M.C.P.; Souza, D.H.S.; Sirelli, L.; Dias, M.L. Crystallization, thermal and mechanical behavior of oligosebacate plasticized poly(lactic acid) films. *Polymer* **2018**, *28*, 381–388. [CrossRef]
48. Li, D.; Jiang, Y.; Lv, S.; Liu, X.; Gu, J.; Chen, Q.; Zhang, Y. Preparation of plasticized poly (lactic acid) and its influence on the properties of composite materials. *PLoS ONE* **2018**, *13*, e0193520. [CrossRef]
49. Iordanskii, A.L.; Samoilov, N.A.; Olkhov, A.A.; Markin, V.S.; Rogovina, S.Z.; Kildeeva, N.R.; Berlin, A.A. New Fibrillar Composites Based on Biodegradable Poly(3-hydroxybutyrate) and Polylactide Polyesters with High Selective Absorption of Oil from Water Medium. *Dokl. Phys. Chem.* **2019**, *487*, 106–108. [CrossRef]
50. Siracusa, V.; Rocculi, P.; Romani, S.; Rosa, M.D. Biodegradable polymers for food packaging: A review. *Trends Food Sci. Technol.* **2008**, *19*, 634–643. [CrossRef]
51. Van Krevelen, D.W. *Properties of Polymers*; Elsevier Science: Amsterdam, The Netherlands, 1997. [CrossRef]
52. Huda, M.; Yasui, M.; Mohri, N.; Fujimura, T.; Kimura, Y. Dynamic mechanical properties of solution-cast poly(l-lactide) films. *Mater. Sci. Eng. A* **2002**, *333*, 98–105. [CrossRef]
53. Müller, P.; Imre, B.; Bere, J.; Móczó, J.; Pukánszky, B. Physical ageing and molecular mobility in PLA blends and composites. *J. Therm. Anal.* **2015**, *122*, 1423–1433. [CrossRef]
54. Chieng, B.W.; Ibrahim, N.A.B.; Yunus, W.M.Z.W.; Hussein, M.Z. Poly(lactic acid)/Poly(ethylene glycol) Polymer Nanocomposites: Effects of Graphene Nanoplatelets. *Polymers* **2014**, *6*, 93–104. [CrossRef]
55. Arjmandi, R.; Hassan, A.; Eichhorn, S.J.; Haafiz, M.K.M.; Zakaria, Z.; Tanjung, F. Enhanced ductility and tensile properties of hybrid montmorillonite/cellulose nanowhiskers reinforced polylactic acid nanocomposites. *J. Mater. Sci.* **2015**, *50*, 3118–3130. [CrossRef]
56. Qu, P.; Gao, Y.; Wu, G.; Zhang, L. Nanocomposite of poly(lactid acid) reinforced with cellulose nanofibrils. *BioResources* **2010**, *5*, 1811–1823.
57. Chieng, B.W.; Ibrahim, N.A.; Then, Y.Y.; Loo, Y.Y. Epoxidized Vegetable Oils Plasticized Poly(lactic acid) Biocomposites: Mechanical, Thermal and Morphology Properties. *Molecules* **2014**, *19*, 16024–16038. [CrossRef] [PubMed]
58. Li, H.; Huneault, M.A. Effect of nucleation and plasticization on the crystallization of poly(lactic acid). *Polymer* **2007**, *48*, 6855–6866. [CrossRef]
59. Jia, S.; Yu, D.; Zhu, Y.; Wang, Z.; Chen, L.; Fu, L. Morphology, Crystallization and Thermal Behaviors of PLA-Based Composites: Wonderful Effects of Hybrid GO/PEG via Dynamic Impregnating. *Polymers* **2017**, *9*, 528. [CrossRef]
60. Davachi, S.M.; Kaffashi, B. Preparation and Characterization of Poly L-Lactide/Triclosan Nanoparticles for Specific Antibacterial and Medical Applications. *Int. J. Polym. Mater. Polym. Biomater.* **2015**, *64*, 497–508. [CrossRef]
61. Kolstad, J. Crystallization kinetics of poly(L-lactideco-meso-lactide). *J. Appl. Polym. Sci.* **1996**, *62*, 1079–1091. [CrossRef]
62. Kogawa, N.R.D.A.; de Arruda, E.J.; Micheletti, A.C.; Matos, M.D.F.C.; de Oliveira, L.C.S.; de Lima, D.P.; Carvalho, N.C.P.; de Oliveira, P.D.; Cunha, M.D.C.; Ojeda, M.; et al. Synthesis, characterization, thermal behavior, and biological activity of ozonides from vegetable oils. *RSC Adv.* **2015**, *5*, 65427–65436. [CrossRef]
63. Mngomezulu, M.E.; Luyt, A.S.; John, M.J. Morphology, thermal and dynamic mechanical properties of poly(lactic acid)/expandable graphite (PLA/EG) flame retardant composites. *J. Thermoplast. Compos. Mater.* **2019**, *32*, 89–107. [CrossRef]
64. Xiao, H.; Liu, F.; Jiang, T.; Yeh, J.-T. Kinetics and Crystal Structure of Isothermal Crystallization of Poly(lactic acid) Plasticized with Triphenyl Phosphate. *J. Appl. Polym. Sci.* **2010**, *117*, 2980–2992. [CrossRef]
65. Mohamed, H.F.M.; Ito, Y.; El-Sayed, A.M.A.; Abdel-Hady, E.E. Positron annihilation in polyvinylalcohol doped with CuCl2. *J. Radioanal. Nucl. Chem. Artic.* **1996**, *210*, 469–477. [CrossRef]

66. Abdel-Hady, E.E.; Mohamed, H.F.M.; Fareed, S.S. Temperature dependence of the free volume holes in polyhydroxybutyrate biopolymer: A positron lifetime study. *Phys. Status Solidi c* **2007**, *4*, 3907–3911. [CrossRef]
67. Mohamed, H.F.M.; Kobayashi, Y.; Kuroda, C.S.; Ohira, A. Impact of Heating on the Structure of Perfluorinated Polymer Electrolyte Membranes: A Positron Annihilation Study. *Macromol. Chem. Phys.* **2011**, *212*, 708–714. [CrossRef]
68. Farid, T.; Herrera, V.N.; Kristiina, O. Investigation of crystalline structure of plasticized poly (lactic acid)/Banana nanofibers composites. *IOP Conf. Series Mater. Sci. Eng.* **2018**, *369*, 012031. [CrossRef]
69. Moureu, S.; Violleau, F.; Haimoud-Lekhal, D.A.; Calmon, A. Ozonation of sunflower oils: Impact of experimental conditions on the composition and the antibacterial activity of ozonized oils. *Chem. Phys. Lipids* **2015**, *186*, 79–85. [CrossRef]

Article

The Distribution and Polymerization Mechanism of Polyfurfuryl Alcohol (PFA) with Lignin in Furfurylated Wood

Jindi Xu, Dongying Hu, Qi Zheng, Qiulu Meng and Ning Li *

School of Resources, Environment and Materials, Guangxi University, Nanning 530004, China; kittie7@163.com (J.X.); hdygxu@163.com (D.H.); zhengqist@163.com (Q.Z.); menglib06@163.com (Q.M.)
* Correspondence: ln33cn@gxu.edu.cn

Abstract: There is increasing interest in furfurylated wood, but the polymerization mechanism between its internal polyfurfuryl alcohol (PFA) and lignin is still uncertain. This paper investigated the distribution of PFA and the feasibility of the polymerization of PFA with lignin in furfurylated balsa wood. The wood first immersed in the furfuryl alcohol (FA) solution followed by in situ polymerization and the distribution of PFA was characterized by Raman, fluorescence microscopy, SEM, and CLSM. Then, the mill wood lignin (MWL) of balsa wood and lignin model molecules were catalytically polymerized with PFA, respectively, studying the mechanism of interaction between PFA and lignin. It was concluded that PFA was mainly deposited in cell corner with high lignin concentration, and additionally partly deposited in wood cell cavity due to high concentration of FA and partial delignification. TGA, FTIR, and NMR analysis showed that the cross-linked network structure generated by the substitution of MWL aromatic ring free position by PFA hydroxymethyl enhanced the thermal stability. New chemical shifts were established between PFA and C_5/C_6 of lignin model A and C_2/C_6 of model B, respectively. The above results illustrated that lignin-CH_2-PFA linkage was created between PFA and lignin in the wood cell wall.

Keywords: distribution; polymerization; polyfurfuryl alcohol; lignin; furfurylated wood

1. Introduction

Growing wood is net-carbon negative [1] and widely used in construction materials because of its environmentally friendly and technical advantages [2]. However, the major drawback of fast-growing wood is its poor durability [3], which can be effectively improved by impregnation modification. Generally, impregnation with low-molecular-weight resins, such as phenolic resin [4], urea-formaldehyde resin [5], and melamine formaldehyde resin [6], etc., is beneficial to improve the properties of wood, but these resins suffer from the problem of releasing free phenols and free aldehydes. In contrast, furfuryl alcohol is a green modification agent derived from pentose-rich agricultural residues and releases fewer volatile organic compounds or polycyclic aromatic hydrocarbons during the combustion and degradation of FA-modified wood [7,8]. In addition, the distribution of the modifier in the wood and the polymerization with the cell wall material are essential to improve the properties of the impregnated material [9]. These two aspects have been discussed by many scholars.

Numerous studies have shown that the FA monomer can penetrate into the cell wall and that FA resins are more evenly distributed in wood after partial removal of hemicellulose and lignin [10]. Theygesen et al. [11] analyzed by confocal laser scanning microscopy (CLSM) that conjugated PFA was formed in the wood cell wall, which significantly swells the wood cell wall. Li et al. [7] indirectly demonstrated chemically cross-linked FA resin in the wood cell wall by nanoindentation, as both reduced modulus and hardness of the furfurylated wood cell walls were significantly improved. Nevertheless, PFA polymerization with the cell wall in wood is extremely complicated, and it is unclear whether the FA monomer or oligomers are cross-linked with wood cell wall components. Some researchers believe that FA only polymerizes in the cell wall. For example, Yang et al. [12] confirmed that furfuryl alcohol had been polymerized in the wood by NMR, whereas chemical bonds between the FA resins and polymers within cell walls had not formed. Other studies have directly or indirectly demonstrated chemical reactions between lignin and FA monomer or oligomers. Lande et al. [8] have suggested that FA monomers incline to deposit in the cell wall and polymerize in areas with high lignin content, there may be a grafting reaction between lignin and furfuryl alcohol. Ehmcke et al. [13] used cellular ultraviolet micro spectrophotometry (UMSP) images of individual wood cell wall layers to support the hypothesis regarding the coagulation reaction between lignin and FA. The cell walls, especially in the regions with the highest lignin content, showed a significant increase in ultraviolet (UV) absorbance, which indicates a strong polymerization of aromatic compounds. The two studies have shown that the deposition of FA in wood is related to the distribution of lignin in wood, which indirectly indicates the chemical cross-linking of FA and lignin. Furthermore, Nordstierna et al. [14] demonstrated the formation of chemical bonds between cresol, a model lignin, and furfuryl alcohol using NMR. The free vacancies in the benzene ring of lignin generated methylene linkage with FA, which directly proved the chemical combination between lignin and FA. In this study we are illustrating in detail the types of chemical bonding between lignin, FA, and PFA, in addition to testing the previously unknown linkages between PFA and other cell wall polymers, namely cellulose and hemicellulose.

In this study, the balsa wood was first immersed in an FA solution, followed by in situ polymerization to obtain the furfurylated wood. Raman, fluorescence microscopy and SEM were performed to get the distribution of PFA information on different concentration FA and delignification-modified wood. To further discuss the mechanism of interaction between PFA and lignin, the MWL of balsa wood and lignin model molecules were catalytically polymerized with PFA, respectively. TGA, FTIR, and NMR measurements were done to analyze the crosslinking of lignin and FA monomer or PFA.

2. Experimental Sections

2.1. Materials

Balsa wood (*Ochroma pyramidale*) was obtained from the market (Guayaquil, Ecuador, South America), the average age of the tree was 6 years, the moisture content was about 10%, and the oven-dried density was 120 kg/m^3. We took a 20 mm × 20 mm × 20 mm (radial × tangential × longitudinal) sample treated with 1wt% NaClO$_2$ solution to obtain delignified wood (DW) [15]; The balsa wood was crushed and ball-milled, then dispersed in dioxane aqueous solution (96:4 volume ratio) to extract the milled wood lignin [16].

Furfuryl alcohol (FA), maleic anhydride (MA), 2-methoxy-4-methylphenol, and 2,6-dimethoxy-4-methylphenol were purchased from Shanghai Macklin Biochemical Co., Ltd. (Shanghai, China) Ethyl alcohol (EtOH), Borax (Na$_2$B$_4$O$_7$·10H$_2$O), sodium chlorite (NaClO$_2$), benzene, dioxane, dichloroethane, ethyl ether, and acetic acid were purchased from Damao Chemical Reagent Factory (Tianjin, China). All chemicals used in this study were as received and without further purification.

2.2. Preparation of Furfurylated Wood

The FA impregnation solution was prepared with 2 wt% MA (C$_4$H$_2$O$_3$), 2 wt% Borax (Na$_2$B$_4$O$_7$·10H$_2$O), 10 wt% EtOH (C$_2$H$_5$OH), (10, 20, 30, 40, and 50 wt%) FA (C$_5$H$_6$O$_2$), and DI water. Prior to the impregnation, the wood samples were divided into native wood (CW) and delignified wood (DW), with 10 samples in each group. All samples were oven-dried, and the mass was recorded as m$_0$. As-prepared solutions were impregnated into the wood samples under vacuum at −0.1 MPa for 24 h and then in the atmosphere for 48 h. After impregnation, the samples were wiped with tissue paper to remove excess solutions on the surface. To avoid solution evaporation during the curing stage, the samples were wrapped in aluminum foil, then cured in an oven at 103 °C for 3 h, allowing full polymerization of

FA within the wood matrix. Afterward, the aluminum foil was removed. The samples were heated from 40 to 103 °C at a rate of 10 °C/h until oven-dry (m_1) to obtain furfurylated wood (CFW) and furfurylated delignified wood (DFW). The weight percentage gain (WPG) of furfurylated wood was calculated accordingly,

$$WPG = \frac{m_1 - m_0}{m_0} \times 100\% \qquad (1)$$

2.3. Preparation of Furfurylated Lignin

MWL and FA were evenly mixed at 1:6 (wt%:wt%) and solidified after 24 h of reaction at 103 °C to obtain the PFA–lignin complex, LPFA, and the chemical bond between wood and furfuryl alcohol was investigated by an NMR test.

Two model molecules, 2-methoxy-4-methylphenol (A) and 2,6-dimethoxy-4-methylphenol (B), were homogeneously mixed 1:1 (wt%:wt%) with FA solution and completely cured after 24 h of reaction at 103 °C to derive PFA–lignin model molecule composites A-LPFA and B-LPFA, respectively, for further analysis of lignin and PFA chemical bonding.

2.4. Characterization

A Raman microscope (Renishaw, London, UK) equipped with a 50× microscope objective and a linear-polarized 633 nm laser was used to examine the distribution of lignin. The wavenumber range of 3000–2000 cm^{-1} at a resolution of 2 cm^{-1}. For mapping, 1 μm steps were chosen, and every pixel corresponded to one scan.

The morphology of the delignified wood and furfurylated wood was characterized by Sigma 300 field emission gun scanning electron microscope (ZEISS, Oberkochen, Germany).

The microscopic distribution of PFA resin within the furfurylated wood was examined using a Image Z2 fluorescent microscope (ZEISS, Oberkochen, Germany) with a 40× microscope objective. The small pieces of wood from a specimen with size of 3 mm × 5 mm × 0.2 mm were stained with 0.5% toluidine blue solution for 10 min to suppress the autofluorescence of lignin for acquiring images.

The microscopic distribution of PFA resin within the furfurylated wood was examined using LEICA-TCS-SP8MP confocal laser scanning microscopy (CLSM) (Leica, Wetzlar, Germany) with a 63× (oil immersion) objective lens. The excitation laser wavelength was 633 nm, and the detector range was 650–700 nm.

The thermal degradation behaviors of MWL, LPFA, and PFA were investigated using a DTG-60 (H) thermogravimetric analyzer (Shimadzu, Kyoto, Japan) from room temperature (25 °C) to 800 °C at a heating rate of 10 °C/min in a flowing nitrogen atmosphere of 100 mL/min.

FTIR were recorded on a Thermo Scientifific Nicolet iS50 spectrophotometer (Waltham, MA, USA) over the wavenumber range of 4000–400 cm^{-1} at a resolution of 4 cm^{-1}, using the KBr pellet method.

The polymerization mechanism of PFA resin, LPFA, A-LPFA, and B-LPFA were investigated by ^{13}C NMR analysis. The Spectra were recorded on a Avance HD500 spectrometer (Bruker, Karlsruhe, Germany), and the DMSO-d6 cross-peak at δC13 39.52 ppm was used as an internal reference. Solid-state NMR spectra were measured at room temperature with an Bruker III 400 NMR spectrometer (Bruker, Karlsruhe, Germany) fitted with a 4 mm magic-angle spinning (MAS) probe head.

3. Results and Discussion

3.1. Distribution of PFA Resin in Wood

In order to study the distribution of PFA resin in wood, various characterization methods were used. Firstly, fluorescence imaging of furfurylated wood was conducted under a fluorescence microscope (Figure 1). To effectively suppress the autofluorescence of lignin in cell walls, staining treatment with toluidine blue dye was performed in advance. The bright red spots in the figure are FA resin filled in the cell wall of the wood, and the fluorescence intensity reflects the filling degree of FA resin. The increase in FA concentration improved the opportunity for PFA resin to adhere to the interior of the wood, and more PFA resin polymerized on the cross section of the wood, contributing to a WPG of 225% for the 50% FA CFW (Figure 2f). In addition, fluorescence intensity analysis performed on the fluorescence image by Image J (Table 1) showed that the penetration area of 50% FA CFW was 110,196.67 μm^2, accounting for 35.19%. The infiltration area increased by 58.58% compared to that of 10% FA CFW. The SEM micrographs shown in Figure 1 exhibit the distribution of PFA resin. The figure reveals that in the low concentration of CFW, only a small amount of the lumen was filled with PFA resin, and most of it may have been present in the cell wall. As the concentration increased, the FA resin inclined to deposit near the vessels and rays, as shown by the arrows in Figure 1. On the one hand, the vessels are conduits for the delivery of water and nutrients to the wood, pits in the inner wall of the vessels are connected to the radial tracheids, allowing FA to penetrate radially and deposit in the cell cavity [17]. On the other hand, FA monomers easily volatilize during heating and curing, causing fewer deposits of PFA resin in the vessels and cell cavity. Finally, CLSM fluorescence imaging was performed on the specimens in order to have a clearer observation of the distribution of PFA resin in the wood microstructure. As shown in Figure 1g, there was a faint fluorescence on the cell wall of CW, which was the autofluorescence of lignin, and the fluorescence disappeared after delignification treatment. In contrast, CFW fluoresced strongly in the cell wall and part of the cell cavity. This suggests that the PFA resin was not only deposited in the cell lumen, but may have also been present in the cell wall region.

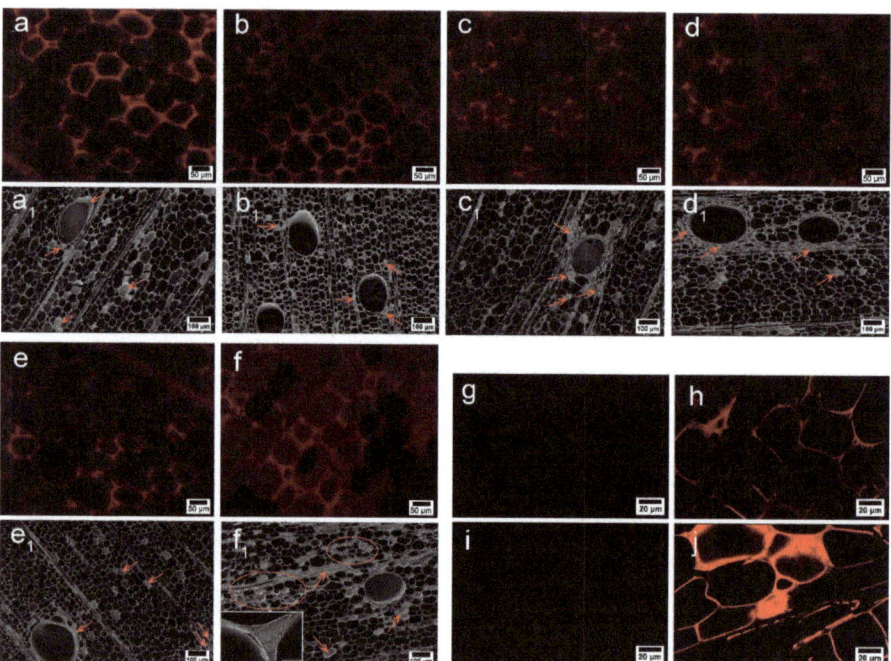

Figure 1. Surface fluorescence microscopy of 10% FA CFW (**a**), 20% FA CFW (**b**), 30% FA CFW (**c**), 40% FA CFW (**d**), 50% FA CFW (**e**), and 50% FA DFW (**f**); SEM image of 10% FA CFW (a_1), 20% FA CFW (b_1), 30% FA CFW (c_1), 40% FA CFW (d_1), 50% FA CFW (e_1), and 50% FA DFW (f_1); CLSM image of CW (**g**), CFW (**h**), DW (**i**), and DFW (**j**).

According to previous reports, FA monomers tend to be deposited on the cell walls in regions with high lignin content. To that end, in this study we partially delignified balsa wood to investigate whether there was a correlation between the distribution of lignin and

PFA in wood. In the Raman spectrum of CW and DW (Figure 2e), the bands belonging to lignin structures were annotated at 1270, 1330, 1598, 1660 and 2945 cm^{-1} [18]. The two band at 1270 and 1330 cm^{-1} were attributable to G and S units in balsa samples, where the G units were assigned to aromatic ethers stretch, and S units were attributed to bending vibration of phenolic hydroxyl lignin [19], respectively. Other lignin features were detected at 1598, 1600, and 2945 cm^{-1}. There was a strong band at 2945 cm^{-1} from the C–H stretch of the methoxyl groups of the lignin. The peak around 1598 cm^{-1} pointed to the lignin stretching vibrations of the aromatic ring, and that at 1660 cm^{-1}—to ring-conjugated C=C bonds in coniferyl alcohol units [20,21]. It follows that the characteristic peak absorption intensity of lignin decreased with delignification treatment. The distribution of lignin can be visualized in Figure 2a,b by integral calculation in the range of 1540–1700 cm^{-1}. Raman imaging showed that high fluorescence intensity of the cell corner (CC) and compound middle lamella (CML) of fibers in untreated samples. Chlorine dioxide generated by acidic sodium chlorite oxidizes phenolic lignin to o-quinone and p-quinone structures [22,23] while cutting the lignin and cellulose, hemicellulose into water-soluble small molecules, thus removing lignin and dropping the Raman intensity, which is more reactive CML than in CC. It can be seen that the CC was highly lignified, and lignin was difficult to remove.

DW was impregnated with PFA resin and observed by a fluorescence microscope and CLSM. It was found that the fluorescence intensity at CC in the fiber cells without PFA in the cell cavity was stronger, suggesting a possible correlation between PFA and lignin. Moreover, a large amount of FA resin filled the cell cavity of DFW compared to that of CFW (Figure 1f), resulting in a WPG of 363% for DFW with 50% FA, which was 61% higher compared to that of CFW, due to formation of a nanonetwork structure after acidic sodium chlorite treatment, exposing more active sites and increasing the permeability of the cell wall [24]. Previous studies have also attested that delignification promotes the impregnation of the modifier and its interaction with cellulose [25]. The analysis of fluorescence intensity calculation also indicated that the delignification treatment significantly increased the glue spots on the DFW cross section. The permeable area of 50% FA DWF was 210,576.16 μm^2, accounting for 67.24%, which increased by 91.08% compared to that of 50% FA CFW. In addition, SEM images displayed a macroscopic shape, and the microscopic honeycomb-like cell wall architecture of native wood was preserved in delignified samples (Figure 2d). The DW cell wall changed into thin, and deformation came into being. After lignin removal, there was a phenomenon where cracks presented in the cell corner, and the cell wall was stratified (Figure 2d). However, these cracks were filled with PFA resin after impregnation with FA (Figure 1f$_1$). Moreover, PFA resin was also found in the intercellular layer, and the cell walls of furfurylated wood were thickened.

Table 1. Distribution of PFA at different concentrations in the cross section of wood.

Samples	Infiltration Area (μm^2)	Ratio of Permeable Area (%)
CW	4.10	6.07
dyed CW	2.96	4.39
10% FA CFW	69,484.04	22.19
20% FA CFW	84,619.40	27.02
30% FA CFW	93,675.93	29.91
40% FA CFW	106,763.16	34.09
50% FA CFW	110,196.67	35.19
50% FA DFW	210,576.16	67.24

CW was native wood, CFW and DFW were FA modified native wood and delignified wood, respectively.

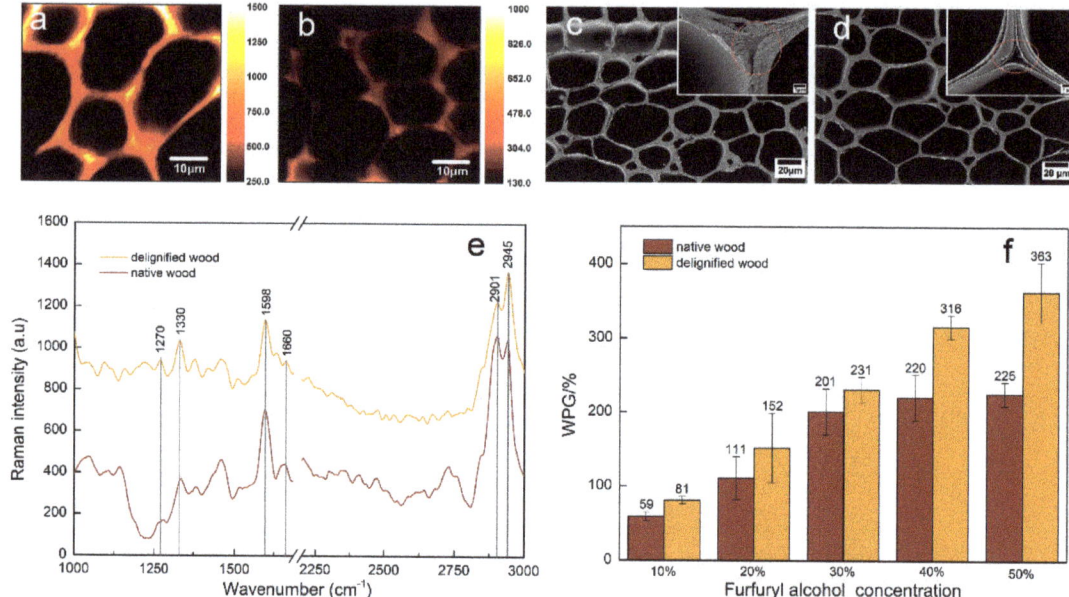

Figure 2. (**a**,**b**)—Raman images of lignin distribution of CW and DW, respectively; (**c**,**d**)—SEM analysis of CW and DW, respectively; (**e**)—average Raman spectra extracted from the cell walls of CW and DW; (**f**)—WPG for different concentrations of furfuryl alcohol in CW and DW.

3.2. The Polymerization Mechanism of PFA and CW

In order to investigate the possible chemical functional groups changes and chemical reactions between PFA and lignin, FTIR spectra of CW, CFW, and PFA were used to obtain more information (Figure 3a). There were new bands at 1562 and 789 cm^{-1} in CFW compared to those of CW, assigned to skeletal vibration of 2,5-disubstituted furan rings and conjugated C=C species [8], respectively. There was also a little peak at 1712 cm^{-1} of furfurylated wood attributed to the C=O stretch of γ-diketones formed from hydrolytic ring opening of the furan rings, covered up by 1741 cm^{-1} assigned to C=O stretching vibrations in unconjugated ketones, carbonyls, and esters. In addition, the intensity of the small peak at 897 cm^{-1} assigned to the β-glycosidic linkages between the sugar units decreased compared with that of control wood, suggesting that acid hydrolysis of hemicelluloses occurred during the furfurylation of wood [20]. This demonstrated that the polymerization of FA was successfully completed within wood, but there was insufficient evidence of chemical reactions between PFA and wood molecules. Based on previous reports and the distribution of PFA resin in wood in this paper, it is known that there may be a relationship between PFA and lignin. Therefore, MWL and lignin model molecules were used for catalytic polymerization with PFA, respectively, to further explore the polymerization between PFA and lignin.

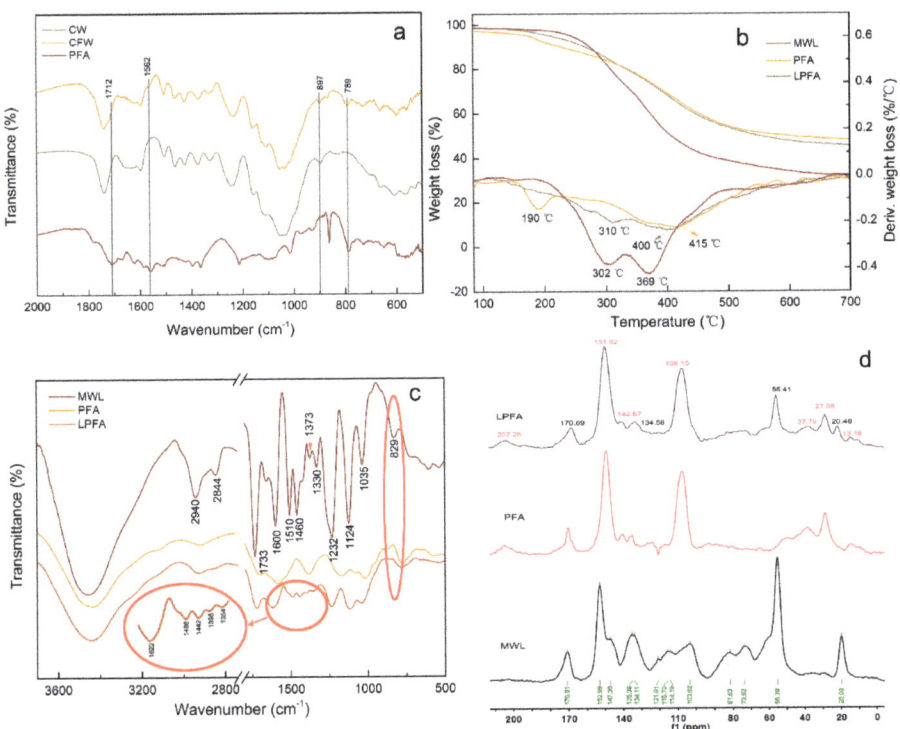

Figure 3. (**a**)—FTIR spectra of CW, CFW, and PFA; (**b**)—TGA and DGA of MWL, PFA and LPFA; (**c**)—FTIR spectra of MWL, PFA and LPFA; (**d**)—^{13}C NMR spectra of MWL, LPFA, and PFA.

3.3. The Polymerization Mechanism of PFA and MWL

Firstly, the thermal degradation stability of MWL, PFA, and LPFA was obtained by thermogravimetric analysis. The thermograms (TGA) and derivatives of thermograms (DTA) are gathered in Figure 3b. The DTA curves showed that the main peak around 415 °C for PFA was attributed to the degradation of methylene and the furan ring in the PFA chain; the peak at 190 °C was owing to the volatiles that evolved during further condensation polymerization of the PFA resin [26]. MWL lost a little weight before 200 °C, mainly water molecules and some small molecule impurities in the sample, and its decomposition temperature range was mainly between 200 and 500 °C. The weight loss was faster at 302 and 369 °C, and the final carbon residue was 31%. LPFA started to decompose slowly at about 120 °C, and the decomposition rate was faster at about 310 and 400 °C. The carbon residue rate was 44%, which was 42% higher than that of MWL, with a slower degradation rate and higher thermal oxidation degradation temperature. This denoted that PFA with a relatively stable furan ring structure polymerized with MWL to form a more complex cross-linked network structure, resulting in better thermal stability of LPFA.

Next, FTIR was used to qualitatively analyze the functional groups on the structure of MWL, PFA, and LPFA, and the spectra are shown in Figure 3c. The characteristic absorption of each functional group in MWL was mainly concentrated in the range of 3800–800 cm^{-1} [27], with typical infrared spectral characteristics of lignin, that is, benzene C=C skeleton vibration occurred at 1600 and 1510 cm^{-1}. The absorption at 2940 and 2844 cm^{-1} arose from the C-H stretching vibrations in the methoxyl group and methyl or methylene groups, respectively [28]. The strong band at 1232 cm^{-1} was indicative of the absorption peak of p-hydroxyphenyl. The band at 1035 cm^{-1} originated from the C-O stretching vibration in primary alcohol and C-H vibration on G unit, and 1733 cm^{-1} due to C=O stretching vibration in unconjugated ketones and carboxyl groups. Additionally, the sharp absorption band at 1124 cm^{-1} and the absorption bands at 829 and 1330 cm^{-1} in these spectra indicated that balsa MWL had typical wood quality characteristics of hardwood [29]. Obviously, the spectra of LPFA were offset between 1600 and 1300 cm^{-1}, such as the C=C vibration at the benzene ring and the characteristic peak shift of the S unit aromatic ring, marked in Figure 3c. Moreover, the O-H stretching vibration of phenolic hydroxyl groups was also affected. Consequently, condensation with PFA hydroxyl may have existed at the free position on the aromatic ring of MWL, leading the peak position of the aromatic ring being affected. Furthermore, the absorption of the Syringyl unit C-H (829 cm^{-1}) disappeared in LPFA, indicating that the C_2/C_6 position of the S unit was occupied by the other groups. In summary, the MWL of Balsa was successfully extracted with a large number of Syringyl units and a small amount of Guaiacyl units, and the hydroxyl group of PFA and the free position of MWL aromatic ring were condensed.

The structures were further examined, and ^{13}C NMR was utilized to characterize MWL and LPFA, as shown in Figure 3d. The attribution of related signals refers to Holtman [30] and Pang [31], as shown in Table 2, and the main unit structures of MWL are shown in Figure 4. The intensities of the signal belonging to methyl (δ20 ppm) in the aliphatic region (10–50 ppm) were significantly high. In the side chain region (50–90 ppm), the intensities of the methoxy group (δ55.38 ppm) presented prominently, and the signals of C_α (δ73.92 ppm) and C_β (δ81.63 ppm) of β-O-4′ aryl ether linkages (III), which are typical bonds of natural lignin, were obvious. In the aromatic region (100–160 ppm), Syringyl (S) and Guaiacyl (G) lignin units, main characteristic units of hardwood lignin, could be detected. The S unit presented a C_2/C_6 (δ103.62 ppm) chemical shift signals as well as a strong signal of C_3/C_5 (δ152.99 ppm). The chemical shift signals of C_1 (δ135.36 ppm) were strong, and C_3, C_5, and C_6 were weak on the G unit. In general, the S unit signal was more intense than that of the G unit, which was in line with the results of the FTIR spectrum. Comparative analysis of LPFA NMR spectra of MWL after furfurylation suggested that it had more PFA polymer signals, and the attribution of the relevant signals is given in Table 3. The signals at δ151.62 and δ108.15 ppm in the PFA polymer were stronger than MWL signals, representing C_1 and C_3 on the furan ring in the PFA chained structure, respectively. The chemical shifts of -CH_2- (δ27.38 ppm) between the furan rings were also remarkable. The chemical shift signals of the methoxy group (δ55.41 ppm) and p-coumaric acid (δ170.69 ppm) were also obtained in LPFA NMR spectra, along with a small amount of methyl (δ20.46 ppm). In brief, no new chemical shift signals were found in the LPFA NMR, except for those of PFA and MWL. The possible reasons were as follows: Firstly, it is proposed that the C_5 position of the G unit was more likely to attract the attack of carbon atoms on FA and participate in electrophilic aromatic substitution reactions [14,32], while the MWL of Balsa was dominated by the Syringyl unit, and the content of Guaiacyl unit was less, thus, no cross-linking reaction between Balsa lignin and PFA was detected. Secondly, no new chemical shift signal was found due to the complex structure unit of lignin as well as the low resolution of solid-state NMR resulting in broad peak positions, leading to overwriting of new chemical shifts generated by the PFA and MWL reactions.

Table 2. Assignment of the ^{13}C CP/MAS NMR spectra of MWL.

Peak	δ	Assignment	Peak	δ	Assignment
1	170.91	C=O in $_\alpha$	8	114.19	C_3/C_5 in G′
2	152.99	C_3/C_5 in S	9	103.62	C_2/C_6 in S
3	147.30	C_3/C_5 in S′	10	81.63	C_β in III, C_α in I
4	135.36	C_1 in G	11	73.92	C_α in III
5	134.11	C_1 in G′	12	55.38	-OCH$_3$
6	121.01	C_6 in G	13	20.00	-CH$_3$
7	115.70	C_3/C_5 in G	——	——	——

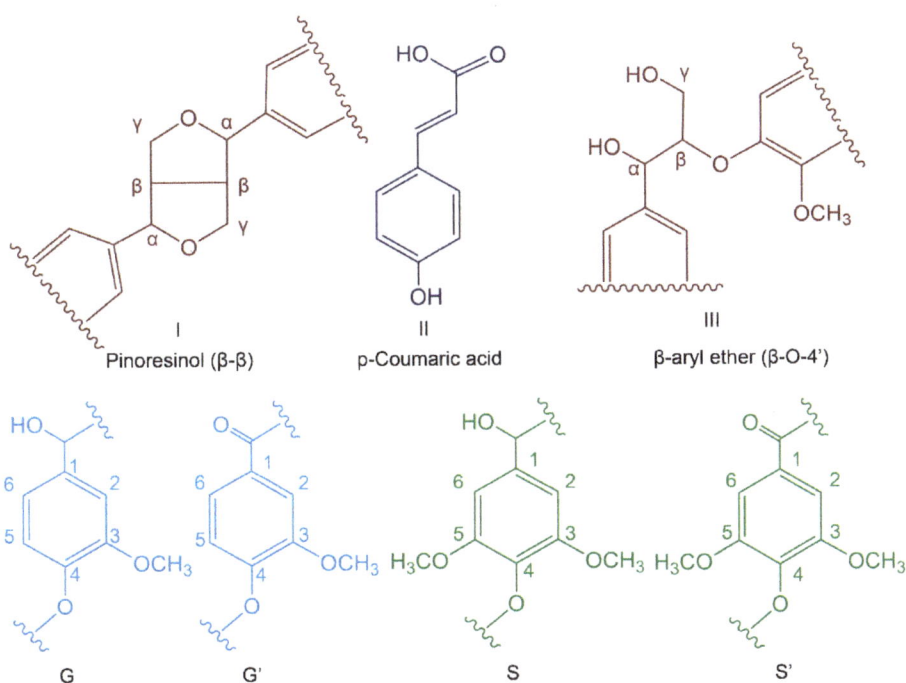

Figure 4. Main structures present in balsa wood MWL.

Table 3. Assignment of the ^{13}C CP/MAS NMR spectra of LPFA.

Peak	δ	Assignment
1	207.28	C=O in Furan ring opening
2	170.69	C=O in p-Coumaric acid
3	151.62	C_1 in chain structure of PFA, C_3/C_5 in S
4	142.67	C_4 in chain structure of PFA
5	134.58	C_1 in G/G'
6	108.15	C_3 in chain structure of PFA
7	55.41	-OCH$_3$ in MWL
8	37.79	-CH$_2$ in reticular conformation of PFA
9	27.38	-CH$_2$ in chain structure of PFA
10	20.46	-CH$_3$ in MWL
11	13.18	-CH$_3$ in chain structure of PFA

3.4. The Polymerization Mechanism of PFA and Lignin Model Molecules

The main MWL unit of Balsa wood is the S unit, and the structure of pure lignin is too complex, resulting in an obscure mechanism of MWL and PFA. Therefore, the mechanism of interaction with PFA was explored by simplifying lignin, for example, by using a lignin model molecule. In this paper, two lignin major unit model molecules, 2-methoxy-4-methylphenol (A) and 2,6-dimethoxy-4-methylphenol (B), were mixed with an FA impregnating solution at 1:1 (wt%:wt%) and thoroughly reacted at 103 °C for 24 h to obtain the furfurylated lignin models A-LPFA and B-LPFA, respectively. A-LPFA and B-LPFA were dissolved in DMSO-D$_6$, ^{13}C NMR detection (Figure 5) was performed, and the chemical structure changes in the furfurylated model were analyzed (Figure 6).

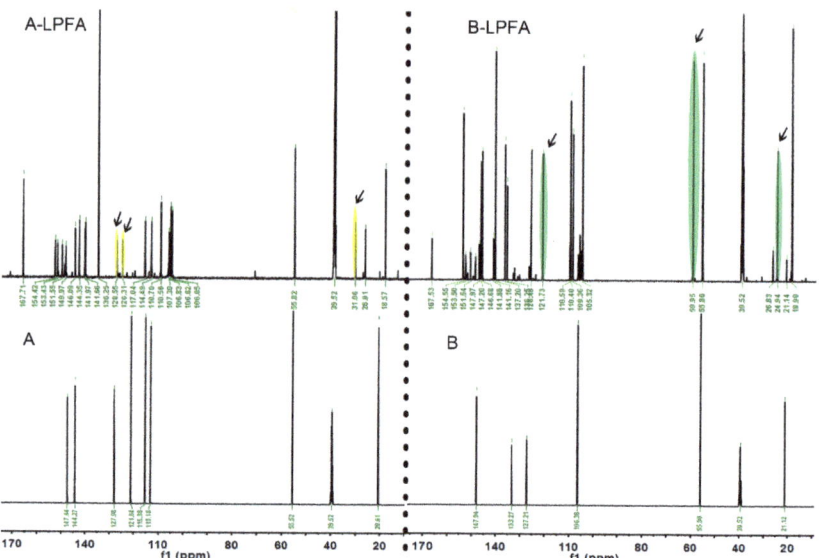

Figure 5. ^{13}C NMR spectra of lignin model (**A,B**) and (**A-LPFA,B-LPFA**).

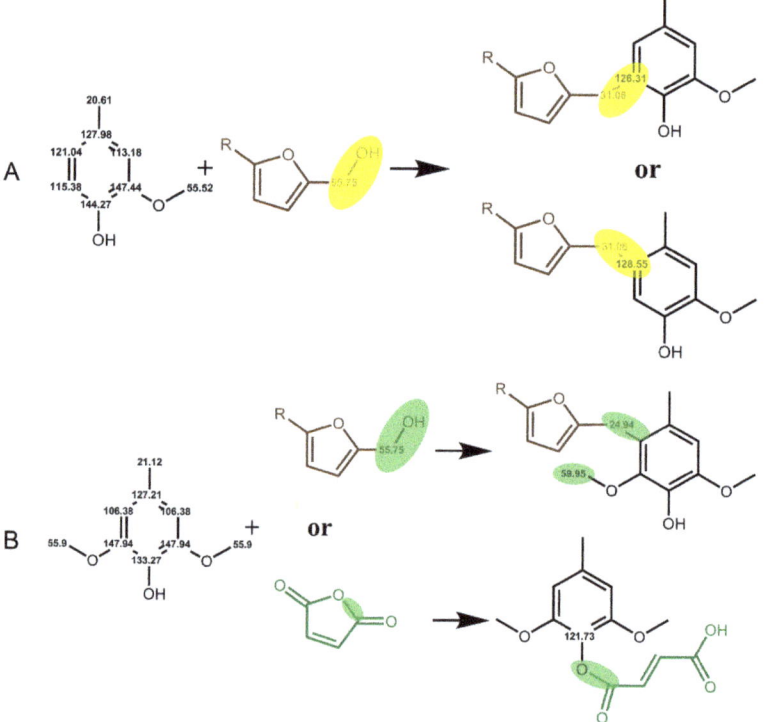

Figure 6. Hypothesized reaction between FA and lignin models (**A,B**). Substituent R is either a poly(furfuryl alcohol) chain or a hydrogen.

Model A was catalytically polymerized with FA to obtain A-LPFA. In addition to chemical shift signals of model A and PFA, a new polymer ^{13}C signal was detected in modified model A (Figure 5A-LPFA), which was not observed for any other combinations or treatments of the starting materials. Significantly, excluding the -CH$_2$- links (δ26.91 ppm) between the furan rings, one of the new methylene signals (δ31.06 ppm) appeared as indicated by the arrow, which belonged to a covalent bridge between model A and PFA resin. Based on the NMR spectra, it was speculated that the covalent bridges between model A and PFA resin might be at the position of C$_5$ and C$_6$ of model A (Figure 6). As an electrophilic aromatic substitution reaction, both the hydroxyl group and the methoxy group on the benzene ring of model A could be utilized as the positioning group. According to the density functional theory (DFT) model calculated by Barsberg [33] et al., the enthalpy of C$_5$ and C$_6$ positions of cresols in ethanol and aqueous solution do not differ appreciably, indicating that both positions on the benzene ring may be connected to PFA. Moreover, the NMR spectra showed the chemical shifts of C$_5$ (δ126.31ppm) and C6 (δ128.55 ppm) shifted toward lower magnetic by the PFA connection. Nordstierna [14] et al. also indicated that -CH$_2$-, as a link between cresol and FA, can be observed in the chemical shift signal at δ31.7 ppm.

Model B was catalytically polymerized with FA to obtain B-LPFA, and a new polymer ^{13}C signal were detected (Figure 5B-LPFA), one for new methylene signals (δ24.94 ppm) appeared as indicated by the arrow, which belonged to the covalent bridge between model B and PFA resin. The second was δ121.73 ppm, which was inferred to be the chemical shift of an esterification reaction of maleic anhydride with the hydroxyl group on the benzene ring, and not FA (Figure 6). Guigo [34] et al. suggested that the connection of FA with the hydroxyl group on C$_4$ produces a new chemical shift around δ66 ppm, but no such signal was detected in this experiment. Therefore, it can be inferred that the entry of maleic anhydride shifts the C$_4$ chemical shift on the benzene ring of model B to the high magnetic field. Another new chemical shift signal was δ59.95 ppm, which was influenced by the electrophilic substitution reaction between FA and model B at C$_2$/C$_6$, causing the methoxyl carbon on model B move to a lower magnetic field. According to the directing effect and reactivity principle of electrophilic substitution reaction on the aromatic ring, the activity of the hydroxyl group was higher than that of the methoxyl group, but since the para position of hydroxyl group was occupied, the methoxyl group acted as the directing group and developed -CH$_2$- bridging with PFA at its adjacent and para positions. The results of FTIR spectra also testified that the disappearance of C-H absorption of the Syringyl unit was associated with the access of PFA hydroxymethyl. It is thus proposed that the hydroxymethyl group of PFA had a new chemical bond after condensation at the free position of the aromatic ring of model B, and maleic anhydride catalyst may have also reacted with model B.

In conclusion, model A and B, representing the Guaiac unit (G) and the Syringyl unit (S), respectively, could basically explain the formation of lignin in the cell wall, both of which could have an electrophilic aromatic substitution reaction with PFA, indicating the presence of lignin-CH$_2$-PFA and the chemical cross-linking between PFA and wood cell wall polymers.

4. Conclusions

In this work, a variety of characterization methods were utilized to display the distribution of PFA in furfurylated wood, and a new chemical link was established between PFA and lignin, demonstrated by the reaction of two types of lignin with it. Raman imaging, fluorescence microscopy, SEM, and CLSM observations revealed that PFA presented more in the intercellular layer and the cell corner where the concentration of lignin was higher than that in the cell cavity. In addition, we found that delignified wood allowed for better infiltration by PFA compared to that in the control, but the mechanism of action was not investigated, and we will continue to study it in the future. MWL and lignin model molecules were catalytically polymerized with PFA, respectively, and TGA and FTIR measurements showed that the cross-linked network structure generated by the substitution of the MWL aromatic ring-free position by PFA hydroxymethyl enhanced the thermal stability. NMR analysis confirmed that there was no obvious connection signal between FA and MWL in LPFA, nevertheless, both methoxy and hydroxyl groups on the lignin model molecules could be as directing groups to activate the benzene ring and attract carbocation on PFA to attack, and -CH$_2$- covalent connection with PFA came into being. In addition, the catalyst maleic anhydride may have also esterified with the phenolic hydroxyl group.

Author Contributions: Conceptualization: J.X. and N.L.; methodology: Q.M.; validation: J.X. and D.H.; formal analysis: Q.Z.; investigation: Q.Z.; resources: N.L.; data curation: Q.M.; writing—original draft preparation: J.X.; writing—review and editing: J.X., D.H. and Q.Z.; project administration: N.L.; funding acquisition: D.H. All authors have read and agreed to the published version of the manuscript.

Funding: This work was supported by the National Natural Science Foundation of China (31960293).

Institutional Review Board Statement: Not applicable.

Informed Consent Statement: Not applicable.

Data Availability Statement: The data presented in this study are available on request from the corresponding author.

Acknowledgments: The authors acknowledge the instrument support by State key Laboratory for Conservation and Utilization of Subtropical Agro-Bioresources of Guangxi University China.

Conflicts of Interest: The authors declare no conflict of interests.

References

1. Chheda, J.N.; Huber, G.W.; Dumesic, J.A. Liquid-phase catalytic processing of biomass-derived oxygenated hydrocarbons to fuels and chemicals. *Angew. Chem. Int. Ed.* **2007**, *46*, 7164–7183. [CrossRef] [PubMed]
2. Dong, Y.; Altgen, M.; Mäkelä, M.; Rautkari, L.; Hughes, M.; Li, J.; Zhang, S. Improvement of interfacial interaction in impregnated wood via grafting methyl methacrylate onto wood cell walls. *Holzforschung* **2020**, *74*, 967–977. [CrossRef]
3. Hill, C.; Hughes, M.; Gudsell, D. Environmental Impact of Wood Modification. *Coatings* **2021**, *11*, 366. [CrossRef]
4. Furuno, T.; Imamura, Y.; Kajita, H. The modification of wood by treatment with low molecular weight phenol-formaldehyde resin: A properties enhancement with neutralized phenolic-resin and resin penetration into wood cell walls. *Wood Sci. Technol.* **2004**, *37*, 349–361. [CrossRef]
5. Jiang, Y.F.; Wu, G.F.; Chen, H.Y.; Song, S.P.; Pu, J.W. Preparation of nano-SiO2 modified urea-formaldehyde performed polymer to enhance wood properties. *Rev. Adv. Mater. Sci.* **2013**, *33*, 46–50.
6. Deka, M.; Gindl, W.; Wimmer, R.; Christian, H. Chemical modification of Norway spruce (Picea abies (L) Karst) wood with melamine formaldehyde resin. *Indian J. Chem. Technol.* **2007**, *14*, 134–138.
7. Li, W.J.; Ren, D.; Zhang, X.X.; Wang, H.K.; Yu, Y. The Furfurylation of Wood: A Nanomechanical Study of Modified Wood Cells. *Bioresources* **2016**, *11*, 3614–3625. [CrossRef]
8. Lande, S.; Eikenes, M.; Westin, M. Chemistry and ecotoxicology of furfurylated wood. *Scand. J. For. Res.* **2004**, *19*, 14–21. [CrossRef]
9. Keplinger, T.; Cabane, E.; Chanana, M.; Hass, P.; Merk, V.; Gierlinger, N.; Burgert, I. A versatile strategy for grafting polymers to wood cell walls. *Acta. Biomater.* **2015**, *11*, 256–263. [CrossRef]
10. Yang, T.; Ma, E.; Cao, J. Synergistic effects of partial hemicellulose removal and furfurylation on improving the dimensional stability of poplar wood tested under dynamic condition. *Ind. Crops Prod.* **2019**, *139*, 111550. [CrossRef]
11. Thygesen, L.G.; Barsberg, S.; Venås, T.M. The fluorescence characteristics of furfurylated wood studied by fluorescence spectroscopy and confocal laser scanning microscopy. *Wood Sci. Technol.* **2009**, *44*, 51–65. [CrossRef]
12. Yang, T.; Cao, J.; Ma, E. How does delignification influence the furfurylation of wood? *Ind. Crops Prod.* **2019**, *135*, 91–98. [CrossRef]
13. Ehmcke, G.; Pilgård, A.; Koch, G.; Richter, K. Topochemical analyses of furfuryl alcohol-modified radiata pine (Pinus radiata) by UMSP, light microscopy and SEM. *Holzforschung* **2017**, *71*, 821–831. [CrossRef]
14. Nordstierna, L.; Lande, S.; Westin, M.; Karlsson, O.; Furo, I. Towards novel wood-based materials: Chemical bonds between lignin-like model molecules and poly(furfuryl alcohol) studied by NMR. *Holzforschung* **2008**, *62*, 709–713. [CrossRef]
15. Fu, Q.L.; Ansari, F.; Zhou, Q.; Berglund, L.A. Wood Nanotechnology for Strong, Mesoporous, and Hydrophobic Biocomposites for Selective Separation of Oil/Water Mixtures. *ACS Nano* **2018**, *12*, 2222–2230. [CrossRef]
16. BjÖrkman, A. Isolation of Lignin from Finely Divided Wood with Neutral Solvents. *Nature* **1954**, *174*, 1057–1058. [CrossRef]
17. Tarmian, A.; Zahedi Tajrishi, I.; Oladi, R.; Efhamisisi, D. Treatability of wood for pressure treatment processes: A literature review. *Eur. J. Wood Wood Prod.* **2020**, *78*, 635–660. [CrossRef]

18. Jin, K.; Liu, X.; Wang, K.; Jiang, Z.; Tian, G.; Yang, S.; Shang, L.; Ma, J. Imaging the dynamic deposition of cell wall polymer in xylem and phloem in Populus x euramericana. *Planta* **2018**, *248*, 849–858. [CrossRef]
19. Saariaho, A.M.; Jaaskelainen, A.S.; Nuopponen, M.; Vuorinen, T. Ultra violet resonance Raman spectroscopy in lignin analysis: Determination of characteristic vibrations of p-hydroxyphenyl, guaiacyl, and syringyl lignin structures. *Appl. Spectrosc.* **2003**, *57*, 58–66. [CrossRef]
20. Agarwal, U.P. Raman imaging to investigate ultrastructure and composition of plant cell walls: Distribution of lignin and cellulose in black spruce wood (Picea mariana). *Planta* **2006**, *224*, 1141–1153. [CrossRef]
21. Larsen, K.L.; Barsberg, S. Theoretical and Raman spectroscopic studies of phenolic lignin model monomers. *J. Phys. Chem. B* **2010**, *114*, 8009–8021. [CrossRef] [PubMed]
22. Dolk, M.; Yan, J.F.; Mccarthy, J.L. Lignin-25—Kinetics of Delignification of Western Hemlock in Flow-through Reactors under Alkaline Conditions. *Holzforschung* **1989**, *43*, 91–98. [CrossRef]
23. Brogdon, B.N.; Mancosky, D.G.; Lucia, L.A. New Insights into Lignin Modification During Chlorine Dioxide Bleaching Sequences (II): Modifications in Extraction (E) and Chlorine Dioxide Bleaching (D1). *J. Wood Chem. Technol.* **2005**, *24*, 221–237. [CrossRef]
24. Meng, Y.; Majoinen, J.; Zhao, B.; Rojas, O.J. Form-stable phase change materials from mesoporous balsa after selective removal of lignin. *Compos. Part B* **2020**, *199*, 108296. [CrossRef]
25. Korotkova, E.; Pranovich, A.; Wärnå, J.; Salmi, T.; Murzin, D.Y.; Willför, S. Lignin isolation from spruce wood with low concentration aqueous alkali at high temperature and pressure: Influence of hot-water pre-extraction. *Green Chem.* **2015**, *17*, 5058–5068. [CrossRef]
26. Lems, E.-M.; Winklehner, S.; Hansmann, C.; Gindl-Altmutter, W.; Veigel, S. Reinforcing effect of poly(furfuryl alcohol) in cellulose-based porous materials. *Cellulose* **2019**, *26*, 4431–4444. [CrossRef]
27. Jahan, M.S.; Mun, S.P. Characteristics of milled wood lignins isolated from different ages of nalita wood (Trema orientalis). *Cell Chem. Technol.* **2006**, *40*, 457–467.
28. Alves, A.; Santos, S.; Simoes, R.; Rodrigues, J. Characterization of residual lignin in cellulose isolated by the diglyme method from three Pinus species by IR spectroscopy and analytical pyrolysis. *Holzforschung* **2018**, *72*, 91–96. [CrossRef]
29. Faix, O. Classification of Lignins from Different Botanical Origins by FT-IR Spectroscopy. *Holzforschung* **1991**, *45*, 21–28. [CrossRef]
30. Holtman, K.M.; Chang, H.M.; Kadla, J.F. Solution-state nuclear magnetic resonance study of the similarities between milled wood lignin and cellulolytic enzyme lignin. *J. Agric. Food Chem.* **2004**, *52*, 720–726. [CrossRef]
31. Pang, B.; Yang, S.; Fang, W.; Yuan, T.-Q.; Argyropoulos, D.S.; Sun, R.-C. Structure-property relationships for technical lignins for the production of lignin-phenol-formaldehyde resins. *Ind. Crops Prod.* **2017**, *108*, 316–326. [CrossRef]
32. Jacobs, A. *Understanding Organic Reaction Mechanisms*; Cambridge University Press: Cambridge, UK, 1997. [CrossRef]
33. Barsberg, S.T.; Thygesen, L.G. A Combined Theoretical and FT-IR Spectroscopy Study of a Hybrid poly(furfuryl alcohol)—Lignin Material: Basic Chemistry of a Sustainable Wood Protection Method. *ChemistrySelect* **2017**, *2*, 10818–10827. [CrossRef]
34. Guigo, N.; Mija, A.; Vincent, L.; Sbirrazzuoli, N. Eco-friendly composite resins based on renewable biomass resources: Polyfurfuryl alcohol/lignin thermosets. *Eur. Polym. J.* **2010**, *46*, 1016–1023. [CrossRef]

Article

Particleboard Production from *Paulownia tomentosa* (Thunb.) Steud. Grown in Portugal

Bruno Esteves [1,2,*], Pedro Aires [1], Umut Sen [3], Maria da Glória Gomes [4], Raquel P. F. Guiné [2,5], Idalina Domingos [1,2], José Ferreira [1,2], Hélder Viana [6,7] and Luísa P. Cruz-Lopes [2,8]

1. Department of Wood Engineering, Polytechnic Institute of Viseu, Av. Cor. José Maria Vale de Andrade, 3504-510 Viseu, Portugal
2. Centre for Natural Resources, Environment and Society-CERNAS-IPV, Av. Cor. José Maria Vale de Andrade, 3504-510 Viseu, Portugal
3. Forest Research Centre (CEF), The School of Agriculture, Tapada da Ajuda, 1349-017 Lisbon, Portugal
4. CERIS, Department of Civil Engineering, Architecture and Georesources, Instituto Superior Técnico, Universidade de Lisboa, Av. Rovisco Pais, 1049-001 Lisbon, Portugal
5. Department of Food Engineering, Agrarian School of Viseu, Quinta da Alagoa, Estrada de Nelas, Ranhados, 3500-606 Viseu, Portugal
6. Department of Ecology and Sustainable Agriculture, Polytechnic Institute of Viseu, Av. Cor. José Maria Vale de Andrade, 3504-510 Viseu, Portugal
7. Centre for the Research and Technology of Agro-Environmental and Biological Sciences (CITAB), University of Trás-os-Montes and Alto Douro (UTAD), Quinta de Prados, 5000-801 Vila Real, Portugal
8. Department of Environmental Engineering, Polytechnic Institute of Viseu, Av. Cor. José Maria Vale de Andrade, 3504-510 Viseu, Portugal
* Correspondence: bruno@estgv.ipv.pt

Abstract: Paulownia wood has raised high attention due to its rapid growth and fire resistance. The number of plantations in Portugal has been growing, and new exploitation methods are needed. This study intends to determine the properties of particleboards made with very young Paulownia trees from Portuguese plantations. Single layer particleboards were produced with 3-year-old Paulownia trees using different processing parameters and different board composition in order to determine the best properties for use in dry environments. The standard particleboard was produced at 180 °C and a 36.3 kg/cm^2 pressure for 6 min using 40 g of raw material with 10% urea-formaldehyde resin. Higher particle size lead to lower-density particleboards, while higher resin contents lead to higher density of the boards. Density has a major effect on board properties with higher densities improving mechanical properties such as bending strength, modulus of elasticity (MOE) and internal bond, lower water absorption but higher thickness swelling and thermal conductivity. Particleboards meeting the requirements for dry environment according to NP EN 312 standard, could be produced with young Paulownia wood with acceptable mechanical and thermal conductivity properties with density around 0.65 g/cm^3 and a thermal conductivity of 0.115 W/mK.

Keywords: young Paulownia; particleboard; density; thermal conductivity; mechanical properties; physical properties; water absorption

1. Introduction

Wood is a scarce commodity, and neither natural or coppice forests are unlimited reserves, so technological and industrial centers have begun the development of alternative products capable of responding to the growing demand for wood and wood-based panels [1–3]. In Portugal, the particleboard industry works almost exclusively with pine wood (*Pinus pinaster*), which is one of the most abundant resources in the Portuguese forest. In recent years, there have been some studies on new species that may have a higher production yield in the short term, as is the case of the Paulownia species [2,4–6]. Also several forest and agricultural wastes have been tested for particleboard production; nevertheless, most of these materials have high heterogeneity and seasonal availability [1,4].

The genus Paulownia, belonging to the Family Paulowniaceae, consists of nine species: *P. albiphloea, P. australis, P. catalpifolia, P. elongata, P. fargesii, P. fortunei, P. kawakamii, P. taiwaniana,* and *P. tomentosa* and several hybrids such as cotevisa 2, produced in Spain, which is the result of the crossing *Paulownia elongata × Paulownia fortunei* [7,8].

Paulownia wood is easily air-dried without severe drying defects. It is more resistant to fires than other fast-growing species due to its high ignition temperature, high water content in the fire season and large leaves. The vessel structure of Paulownia wood is very large and independent, making it difficult for oxygen to be adequately supplied and making this wood difficult to ignite [9]. It has a high resistance-to-weight ratio, a low shrinkage coefficient and does not deform or crack easily [8,10]. The workability and finishing properties of the wood are excellent, but it is considered an underused species [11]. Nevertheless, Paulownia wood has been used for plywood, paper production, veneers, and some objects such as, for example, rice pots, bowls or spoons [12,13]. *Paulownia tomentosa* has shown a very fast growing rate, which permits its utilization in short rotation periods. Fast growing has the downsides of producing low-density wood with poor mechanical properties and, therefore, not being suitable for structural applications. On the other hand, low density also leads to low thermal conductivity, which might be important for thermal insulation boards. This is a great advantage in relation to pine wood since it can allow the elaboration of particleboards with lower density and thermal conductivity. According to Akyildiz and Kol [10], this species has a specific mass of approximately 35 g/cm^3, but the density can vary according to the age of the tree and plantation location. According to Kim et al. [14], wood grown in Korea has a lower density, around 0.27 g/cm^3, while wood growth in Portugal has a higher density, around 0.44 g/cm^3, possibly due to the slower growth [15]. Even in the same plantation, there are significant differences seen in Paulownia trees planted in Hungary that had a density of 0.28 g/cm^3 on average but ranged from 0.24 g/cm^3 to 0.33 g/cm^3 [16]. On the other hand, *Paulownia tomentosa × elongata* planted in Spain, Bulgaria and Serbia presented a similar density of around 0.26 g/cm^3 [17]. The thermal conductivity of *Paulownia tomentosa* wood has been reported to be between 0.073–0.100 W/mK [18].

Several parameters affect the final properties of particleboards, such as properties of the raw material, like for instance, species, density or particle size and shape; the properties of the resin, such as kind, quality and quantity; and those of the pressing system like pressure, temperature, press-closing time and pressing time [19–23]. The density of the raw material influences the density of the panel and, consequently, its mechanical and physical properties. This relationship is called compaction ratio. Using raw materials with low density is better for the production of medium density panels since they have an adequate compaction ratio (1.3) which allows good contact area between particles during pressing, leading to a good bonding [24,25]. Generally, higher particle sizes have been reported to lead to lower density due to the lower compaction ratio of coarse particles in relation to finer particles, as mentioned by Cosereanu et al. [20] for particleboard made with sunflower seed husk. Ferrandez-Garcia [26], with boards made from Washingtonia Palm rachis with citric acid, Hegazy and Ahmed [27], with particleboards manufactured from Date Palm fronds or Osarenmwinda, and Nwachukwu [28], with rice husk particleboard, all obtained a similar decrease for bigger particles, while Farrokh Payam et al. [29] obtained an increase followed by a decrease. However, in accordance with Bazetto et al. [24], who tested the particle size effect on bamboo particleboard properties, no significant influence was found on board density, but bending strength and modulus of elasticity (MOE) decreased. These results are possibly because of the narrow range of particle size used (between 0.210 and 0.420 mm) in the study. The shape of the particles also influences the particleboard properties. For instance, the concave particles of sunflower seed husk decrease the compaction of the particleboard, negatively affecting the board properties [20]. The spreading or mat density is related to particle dimensions. As the particle thickness and width increase, the mat density also increases while the particle length has the opposite effect [21]. Another study showed that the longer the particles, the higher the mechanical properties (IB, and MOR, respectively) and the lower the water adsorption and thickness swelling [30]. The type of resin has an obvious influence on board properties. For example, for particleboards made with Asian bamboo and three different resins, melamine formaldehyde (MF), melamine urea phenol formaldehyde (MUPF) and phenol formaldehyde (PF), Malanit et al. [31] reported that melamine formaldehyde obtained the best results. Also, particleboards produced using isocyanate resins result in better mechanical properties and dimensional stability than those produced using urea formaldehyde resins [32]. Higher amounts of resin generally improve the board properties, mainly thickness swelling and mechanical properties as stated before by Rathke et al. [33] in a study of particleboards made from willow, poplar, and locust, or by Nemli et al. [23] in a study of particleboard panels consisting of 45% beech (*Fagus orientalis*), 35% pine (*Pinus nigra*) and 20% poplar (*Populus nigra*), or even by Arabi et al. [34] who reported that higher resin content lead to increased mechanical properties for single-layer particleboards made from poplar wood. Increased hot pressing time and temperature has often been mentioned to improve the mechanical properties of the particleboards, like, for instance, for Asian bamboo [31] or Jatropha Fruit Hulls treated in acidic conditions, particleboard made from sorghum bagasse as reported by Iswanto et al. [35,36] or particleboard made from recycled particles bonded with a new natural adhesive composed of tannin and sucrose [37].

The density of the panel is probably the most important factor since it considerably affects both physical and mechanical properties of the panels [5,27,34,38]. Several studies have shown that boards with higher density have higher thickness swelling but lower water absorption and better mechanical properties such as MOR, MOE and Internal bond strength [11,19,38–40].

This work intended to determine for the first time, the feasibility of using very young trees from *Paulownia tomentosa* wood with only 3 years of growth, grown in Portugal, for single-layer particleboard production. Its fast-growing rate would allow a sustainable forest management since the wood can be harvested sooner than traditional wood species.

2. Materials and Methods

2.1. Sampling and Material Preparation

Young age Paulownia wood (*Paulownia tomentosa* (Thunb.) Steud.) used in the work was harvested in an experimental plantation field at the Agriculture High School of Polytechnic Institute of Viseu, cultivated under the Carbo Energy and Biomass coppice Project (PROJ/CI&DETS/CGD/0008).

In this study, 3-year-old Paulownia wood was cut into small logs (Figure 1) and air dried until it reached a moisture content of around 12%. Afterwards, the logs were debarked, turned into chips with a chisel and milled into particles in a knife Retsch SMI mill (Haan, Germany), followed by sifting in a Retsch AS200 (Haan, Germany) sifter for 20 min at 50 rpm. Four distinct fractions (<0.25 mm; 0.25–0.4 mm; 0.4–1.18 mm and >1.7 mm) were obtained. After this separation, the particles were dried until they reached a final moisture content of between 3 and 4%.

Figure 1. *Paulownia tomentosa* samples after being cut.

2.2. Preparation of Particleboard

For the formation of single-layer particleboards, the samples were mixed with a UF resin with 64% resin solids content (EuroResinas—Indústrias Químicas S. A., Sines, Portugal) in an Ika Ost Basic mixer at 750 rpm. In the mixing process, the resin was slowly added so that it was evenly distributed. The mattress was formed in a square stainless-steel mold of 100 mm × 100 mm (Figure 2) lined with aluminum foil to ensure that the mixture did not adhere to the mold during pressing.

Figure 2. Stainless steel mold with ready-to-press mixture.

The samples were pressed in a Carver press 3889CE (Wabash, IN, USA) with a standard temperature of 180 °C, 10% resin content and a 36.3 kg/cm^2 pressure for 6 min. After this step, the mold was removed from the press, and the board cooled until it was possible to remove the foil. The final weight and dimensions of the board were determined in an analytical scale and with the aid of a caliper. In each board, an average of 40 g of raw material was used with 10% urea-formaldehyde resin.

In order to test the influence on particleboard properties, the percentage of resin used varied between 8–12% while particle size varied between 0.25–0.4 mm; 0.4–1.18 mm and >1.7 mm. These variations allowed us to more accurately evaluate their influence on the quality of particleboard boards, which are visible in Figure 3.

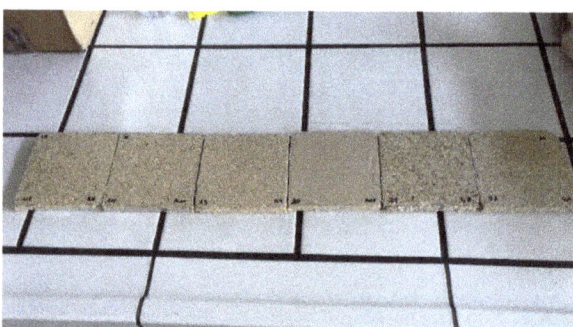

Figure 3. *Paulownia tomentosa* particleboard boards.

2.3. Particleboard Testing

Density was determined for particleboard conditioned at 20 °C and 65% relative humidity by weighing and measuring the board dimensions in the three directions (approximately 100 mm × 100 mm × 8 mm).

Before all the tests, particleboard was conditioned at 20 °C and 65% humidity.

Bending strength and modulus of elasticity (MOE) were determined in a universal test machine Servosis I—405/5 using a 3-point bending test according to NP EN 310 standard [41] with some modifications. Each board was cut into 25 mm × 100 mm long samples.

The samples were placed on the machine with an 80 mm span and subjected to growing tension with an applied load at a constant rate of cross-head movement so that the maximum load was reached within 60 ± 30 s. Tests were made in triplicate. Bending strength (BS) and modulus of elasticity (MOE) were determined in accordance to:

$$BS\ (MPa) = \frac{3 * F_{max} * l_1}{2 * b * t^2} \quad (1)$$

l_1 is the span in millimeters;
b is the width of the test piece, in millimeters;
t is the thickness of the test piece, in millimeters;
F_{max} is the maximum load, in Newtons.

$$MOE(MPa) = \frac{(F_2 - F_1) \times l_1}{(a_2 - a_1) \times 4 \times b \times t^3} \quad (2)$$

l_1 is the span in millimeters;
b is the width of the test piece, in millimeters;
t is the thickness of the test piece, in millimeters;
$F_2 - F_1$ is the increment of load on the straight-line portion of the load-deflection curve, in N;
$a_2 - a_1$ is the increment of deflection at the mid-length of the test piece (corresponding to $F_2 - F_1$) in millimeters.

Internal bond Measurement was done in accordance with NP EN 319 [42]. Samples with 50 × 50 mm were glued in aluminum blocks with the same dimensions. A tension load with a loading speed of approximately 2 mm = min, adjusted so that the maximum load was reached within 60 ± 30 s, was applied vertically to the board face. The maximum load (P) supported by the sample with dimensions ($b \times L$) was recorded, and internal bond was calculated according to the following equation:

$$\text{Internal bond (N/mm}^2\text{)} = \frac{P}{b \times L} \quad (3)$$

Water absorption (WA) and thickness swelling (TS) were determined in accordance to NP EN 317 [42] by conditioning the samples at 65% relative humidity and 20 °C prior to water immersion. The samples were kept vertically in water at 20 °C and pH 7. After 24 h, the samples were removed from the water bath, wiped with an absorbing paper, weighed and measured with a digital caliper on the middle of the sample. Water absorption and thickness swelling were determined in accordance with Equations (3) and (4):

$$\text{Water absorption} = \frac{m_f - m_i}{m_i} \times 100 \quad (4)$$

$$\text{Thickness swelling} = \frac{t_f - t_i}{t_i} \quad (5)$$

where f means final after the soaking period, and i means initial conditions (65% RH and 20 °C).

The thermal conductivity (λ, W/mK) was determined by means of a modified transient pulse method (MTPS), following the ASTM-D-5334 [43], ASTM D 5930 [44] and EN 22007-2 [45] testing procedures, using an ISOMET 2114 portable hand-held heat transfer analyzer, from Applied Precision Enterprise. Each measurement took approximately 25 min and was performed with a surface measurement probe placed on top of each sample surface at three different points. The surface probe has a constantly powered resistor heater that imposes a heat impulse on the sample in thermal equilibrium with the surrounding environment [43,44].

2.4. Statistical Analysis

All the tests with the particle size, resin content and pressing temperature were made in triplicate. Error bars represent ±σ (standard deviation). A linear regression model was used to test the relation between the independent variables. The coefficient of variation was also determined to test the dispersion of data points around the mean.

3. Results and Discussion

There are many factors influencing the board properties, such as board constitution (material, size and shape of particles, resin kind and content) or board production parameters (temperature, pressure or pressing time) [21–23,34,36]. All these parameters influence the board density, which in turn influences mechanical properties like bending strength and modulus of elasticity, internal bond, and water absorption and thickness swelling. Density has also proven to be the most important property in relation to the thermal conductivity of particleboards. Therefore, several parameters were studied in order to obtain the lowest conductivity possible while fulfilling the mechanical and physical properties requirements of NP EN 312 [46].

3.1. Influence of Constitution on Board Properties

The results showed that the initial wood density was around 0.42 g/cm^3 and that the final densities of the boards varied between 0.39 g/cm^3 and 0.92 g/cm^3, depending on pressing parameters or board factors such as particle size or percentage of resin used. These variations will be discussed in detail in this section. Earlier studies suggested that a board density lower than the wood density is not advisable; therefore, the boards should have the lowest density compatible with the minimum requirements for particleboard [40], and the board density has to be higher than the wood species density so that there is a good inter-particle contact or else the resin would polymerize in the void spaces, leading to poor bonding [40].

The size of the wood fractions has proven to be one of the most important parameters in the final density of the particleboards. According to Figure 4, it is possible to verify that, at similar conditions, when larger particles are used, lower-density boards are produced. The smaller fraction (0.25–0.4 mm) has an average density of 0.75 g/cm^3, while in the largest fraction (1.7–2 mm), it is around 0.55 g/cm^3. The coefficient of determination (R^2 = 0.92) is considered good, considering that the linear regression model explains 92% of the variance obtained. The standard deviation between samples is not very small, but the coefficient of variation ranges from 13% to 18%, which means that there is a low-to-medium dispersion of the results. This shows that, since the final objective is the production of lower-density boards to obtain lower thermal conductivity, particles of larger dimensions should be used if they do not significantly impair the panel properties.

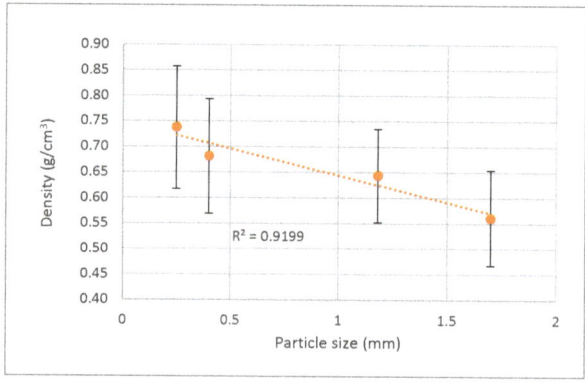

Figure 4. Correlation between board density variation and raw wood particle size.

The increase in board density with decreasing particle size is probably due to higher compaction when using smaller particles. Similar results have been reported before for single-layer particleboard made from sunflower seed husks [20]. Particle size has been mentioned before as one of the most important factors that can be used to influence the physical and mechanical properties of particleboards [47]. These authors stated that, by increasing the length of the particles, the bending strength and MOE increases, but, on the other hand, internal bond decreases [34,48,49].

It is not only the particle size that can affect the board density. Figure 5 presents the effect of resin content on board density. Results show that a higher percentage of resin leads to a higher density of the board, which was expected since wood density was around 0.42 g/cm^3, and the UF resin density is much higher 1.2–1.3 g/cm^3. The relationship between resin content and density appears to be linear, at least between 8–12% resin content. Similar results were presented before by Rathke et al. [33] that tested the effects of alternative raw materials and varying resin content on the mechanical properties of particleboards. Nevertheless, higher resin contents increase the price of the boards and the formaldehyde release during use. Therefore, a high resin content is to be avoided. Once again, the determination coefficient is very high, R^2 = 0.999, and the dispersion of the results is low with a maximum coefficient of variation of 7%.

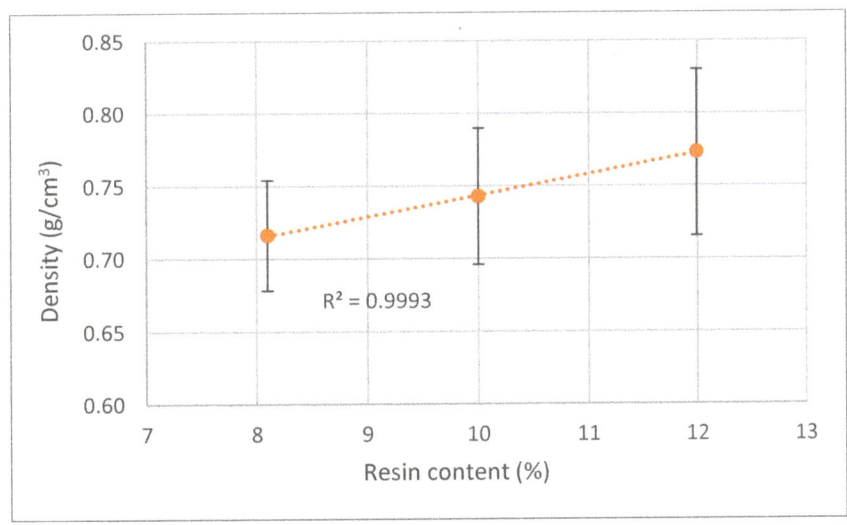

Figure 5. Correlation between board density variation and resin content.

In order to study the resin content effect on board properties with the same density, a target density of 0.50 g/cm^3 was used. At similar density, a higher resin content leads to better mechanical properties, with a higher bending strength and modulus of elasticity, as can be seen in Figure 6. Bending strength increased from around 19 MPa to 23 MPa for 8–12% resin content which represents a 21% increase, while MOE changed from 2000 MPa to 3170 MPa (58% increase).

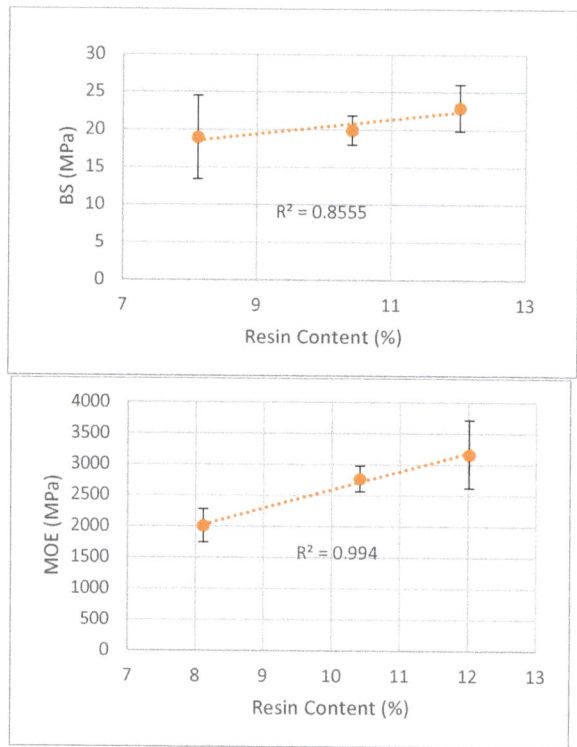

Figure 6. Correlation between bending strength and MOE variation and resin content.

Similar results were presented before by, for instance, Lehmann [50] with 2, 4 and 8 percent resin solids of urea-formaldehyde resin that obtained increases in both bending strength and modulus of elasticity, by Nemli et al. [51] with kiwi (*Actinidia sinensis* Planch.) pruning particleboard or by Rathke et al. [33] with different wood species. Arabi et al. [34] stated that MOR and MOE increased with resin content, and that an exponential function can better describe the simultaneous effect of slenderness and resin content than the linear equation. On the other hand, Kimoto et al. [52] tested resin contents from 8–15% and stated that only minor improvements were found in strength with 15% resin content compared to 10%. In relation to the determination coefficient for the relation between both bending strength and modulus of elasticity with resin content, the first was only 0.856 and the second was 0.994. On the other hand, the standard deviations that seems very small due to the scale, range from 5% to 25% which is in the limit of medium dispersion.

Internal bond (Figure 7) increases with the increase in resin content from around 0.7 MPa for 6% resin content to more than 1.5 MPa for 17% resin content. Similarly, Nemli et al. [23] reported that internal bond increased from 0.483 to 0.648 MPa for commercially produced particleboard panels with 45% beech (*Fagus orientalis*), 35% pine (*Pinus nigra*) and 20% poplar (*Populus nigra*) with 8–10% and 10–12% resin content. The same was reported by Dai et al. [53] for strand boards. In relation to internal bond, the coefficient of determination is 0.986, and the coefficient of variation reaches a maximum of 20% for the highest resin content; therefore, the dispersion of the results can be considered between small to medium.

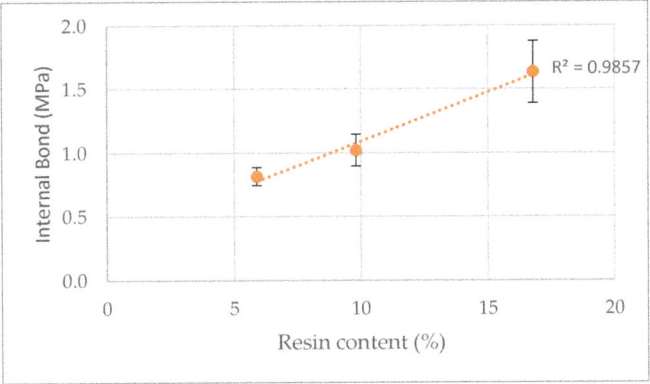

Figure 7. Correlation between internal bond variation and resin content.

Water absorption variation and thickness swelling with resin content are presented in Figure 8. Results show that higher resin content leads to lower water absorption and lower thickness swelling. Similar results were reported before by Sekaluvu et al. [54] with particleboards made with Maize Cob. These authors observed an increase in MOE and MOR and a decrease in water absorption and thickness swelling for higher resin contents. Similarly, Ashori and Nourbakhshb [55], who studied the effect of press cycle time and resin content on the physical and mechanical properties of particleboard panels made from Date palm, Eucalyptus, Mesquite and Saltcedar, concluded that higher resin content decreased the thickness swelling for all the studied materials. Nemli et al. [23] also reported a lower thickness swelling for higher resin content.

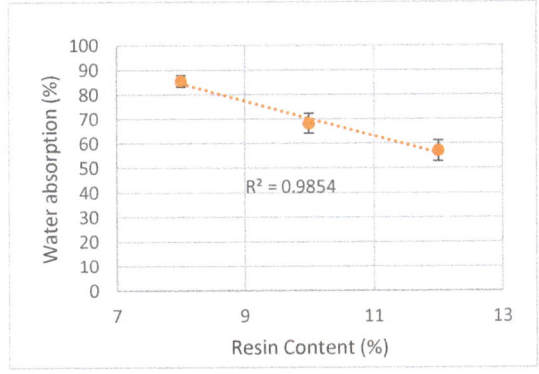
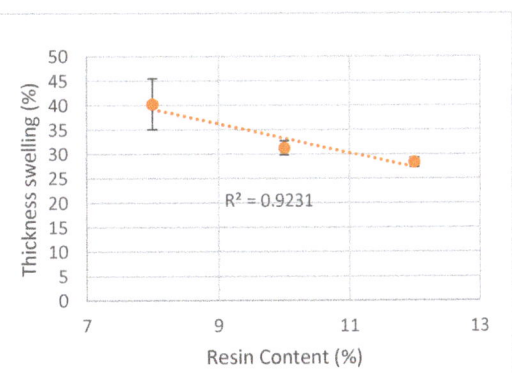

Figure 8. Correlation between water absorption and thickness swelling variation and resin content.

The pressing temperature presented in Figure 9 was tested for a target density of 0.50 g/cm^3 with 10% UF resin content. The pressing time used was 6 min at 36.3 kg/cm^2 pressure. Generally, mechanical properties increased with the pressing temperature. Results showed that a temperature of at least 160 °C was needed in order to obtain good mechanical properties, which could be due to the low pressing time. A higher temperature is needed for the core temperature to reach the degree required to cure the resin. Bending strength presented higher values for 160 °C than for 180 °C, although values presented high standard deviations, as seen in Figure 6. Similar results were presented before for particleboard made from sorghum bagasse [36], where mechanical properties increased for higher pressing temperature, with the exception of 180 °C in relation to 170 °C. This could be due to some thermal degradation at higher temperatures. Similarly, Iswanto et al. [35,36]

reported increased mechanical properties with a higher pressing temperature of particleboard made from Jatropha Fruit Hulls treated in acidic conditions and particleboard made from sorghum bagasse. However, the temperatures tested were lower, ranging from 110 °C to 130 °C. Malanit et al. [31] reported that higher pressing temperature resulted in higher mechanical strength of UF bonded Asian Bamboo particleboards, for pressing temperatures ranging from 150 °C to 210 °C but using three different resins (melamine formaldehyde, melamine urea-formaldehyde and phenol formaldehyde). The variation between pressing temperature and BS and MOE does not seem to be linear with determination coefficients of 0.766 and 0.938. The dispersion of the results goes from very small for 100 °C to medium for 160 °C.

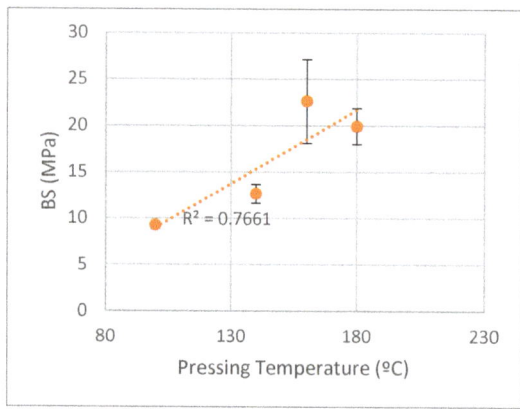 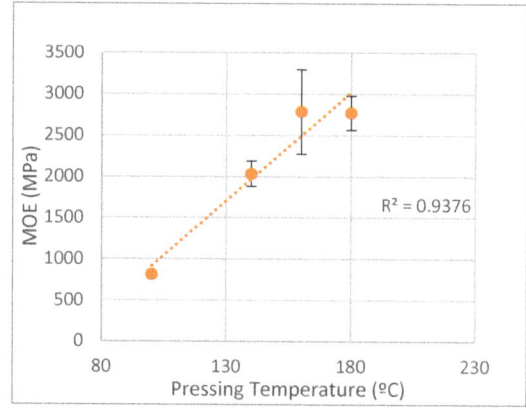

Figure 9. Correlation between bending strength and MOE variation with pressing temperature.

3.2. Influence of Density on Board Properties

Density is most likely the major factor when addressing the board properties. Figures 10–13 present the variation of mechanical properties (bending strength, MOE and internal bond), water absorption, TS and thermal conductivity with density. The mechanical properties of particleboards increased with the board density. Even with several different parameters, it is clear that bending strength increases with density and that this increase has a linear trend. Bending strength varied between under 5 MPa for the lowest density boards to over 340 MPa for the ones with the highest density. MOE behavior followed a similar trend to bending strength ranging from 200 MPa to 5000 MPa. Similar results were presented before by several authors with different materials. For example, De Melo et al. [19] studied the board density effect on the physical and mechanical properties of particleboards made from *Eucalyptus grandis* W. Hill ex Maiden, and they concluded that when density increased linearly from 0.6 g/cm^3 to 0.8 g/cm^3, both MOR and MOE increased. Kalaycioglu et al. [11] produced three-layer boards from Paulownia wood with 0.35 g/cm^3 manufactured with densities of 0.55 g/cm^3 and 0.65 g/cm^3 and reported an increase in MOR, MOE and IB with the board density. Mechanical properties like bending strength and MOE seem to be directly proportional to density with high determination coefficients.

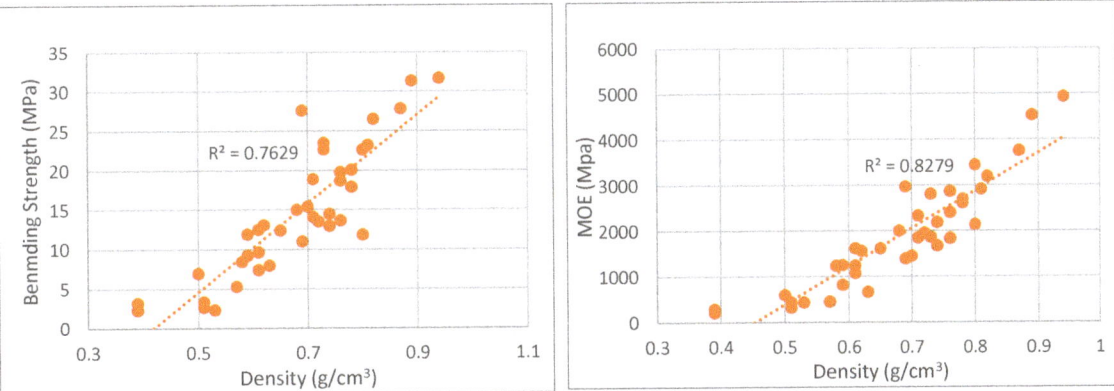

Figure 10. Correlation between bending strength and MOE variation with particleboard density.

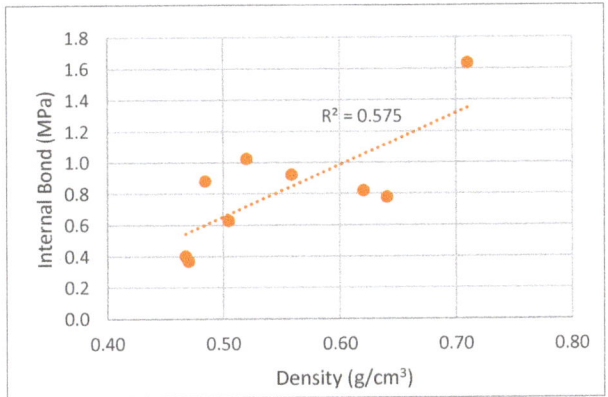

Figure 11. Correlation between internal bond variation with particleboard density.

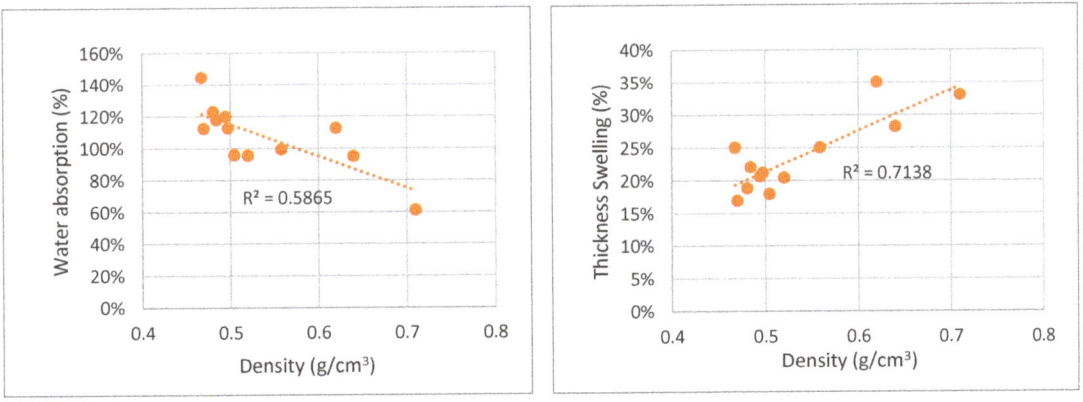

Figure 12. Correlation between water absorption and thickness swelling variation with particleboard density.

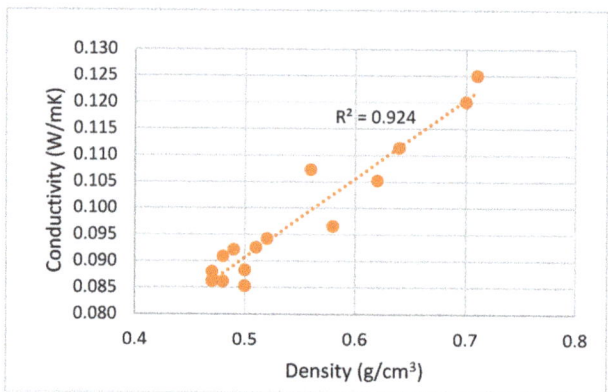

Figure 13. Correlation between conductivity variation and particleboard density.

Internal bond increased with the density of the boards, similarly to bending strength and modulus of elasticity. Values ranged between 0.4 MPa and 0.7 MPa. Likewise, De Melo and Del Menezzi [19] obtained internal bond strength varying from around 0.25–0.4 MPa for densities of 0.6–0.8 g/cm³ for particleboards made from *Eucalyptus grandis*.

Contrary to the mechanical properties, water absorption decreased for higher-density boards. Water absorption ranged between 150% and 50%. Similar results were presented before by several author, such as, for instance, De Melo and Del Menezzi [19] for particleboards made from *Eucalyptus grandis* W. Hill ex Maiden and Cravo et al. [38] for particleboards based on cement packaging.

Thickness swelling has been reported to increase for higher-density boards. Khedari et al. [39], who studied a new insulating particleboard from durian peel and coconut coir, concluded that a higher density resulted in an increase of thickness swelling higher than the standard specification requirements for particleboards reaching about 29% and 35% for board density around 0.590 g/cm³ for Durian peel and coconut coir, respectively [39]. Similar results were presented by De Melo and Del Menezzi [19]. The same was observed in the present study, with 24 h thickness swelling increasing with density, as can be seen in Figure 12.

Moreover, the thermal conductivity of Paulownia particleboards increased with the board density, similar to what was observed before [56]. This has been attributed to the lower space and void for higher density particleboards, since the air in voids has a low thermal conductivity. Thermal conductivity ranged between 0.085–0.125 W/mK. The thermal conductivity was lower than was obtained before for a particleboard made with a mixture of durian peel and coconut coir (10:90) which was around 0.134 W/mK [56]. Nevertheless, this conductivity was obtained for boards with 0.856 g/cm³ density. If the comparison is made at a similar density, for instance around 0.600 g/cm³, durian peel and coconut coir have a lower thermal conductivity of 0.085 W/mK against 0.105 W/mK [56].

In accordance to NP EN 312 [46] the requirements for general purpose boards used in a dry environment (Type P1) are bending strength higher than 10.5 MPa for 8 mm thickness boards and 0.28 MPa for the internal bond. If these boards are to be used in furniture (Type P2), the requirements are 11 MPa for bending strength, 1800 MPa for MOE and 0.4 MPa for internal bond. Thickness swelling is only a requirement for boards used in a humid environment. For example, for non-structural boards, thickness swelling should be lower than 17% swelling in 24 h. Therefore, in order to fulfill the requirements for a dry environment, a density of around 0.65 g/cm³ is needed to achieve the necessary mechanical strength, and this would lead to a thermal conductivity of around 0.115 W/mK. If these boards were to be used in a humid environment, they would not fulfill the requirements because, to achieve a swelling under 17%, the maximum density would be around 50 g/cm³, and, for this density, mechanical properties would be lower than the minimum requirements. Bending strength would be around 5 MPa and MOE around 1000 Mpa. However, if no mechanical strength is needed, a lower thermal conductivity of around 0.085 W/mK could be obtained.

4. Conclusions

The main objective of this study was to produce particleboards from very young Paulownia trees with acceptable mechanical properties and lower thermal conductivity. The objectives were accomplished.

The specific conclusions of the study are:

1. Particleboards can be produced with density ranging from 0.39 g/cm^3 to 0.92 g/cm^3, bending strength 2–32 MPa and MOE 200–4900 MPa, Internal bond 0.4–1.6 MPa, water absorption 60–140%, thickness swelling 16–44% and thermal conductivity 0.085–0.125 W/mK.
2. In order to meet the requirements for a dry environment according to NP EN 312 standard, Paulownia particleboards need to have a density higher than 0.65 g/cm^3 which leads to a thermal conductivity of around 0.115 W/mK.

Author Contributions: Conceptualization, B.E.; methodology, B.E., M.d.G.G., U.S., I.D. and L.P.C.-L.; formal analysis, P.A., B.E., I.D., L.P.C.-L., M.d.G.G. and U.S.; investigation, H.V., B.E., I.D., M.d.G.G. and L.P.C.-L.; resources, H.V. and J.F.; writing—original draft preparation, B.E.; writing—review and editing, B.E., M.d.G.G., U.S., I.D., H.V., R.P.F.G., L.P.C.-L. and J.F.; project administration, B.E.; funding acquisition, B.E., M.d.G.G., U.S., I.D., H.V., R.P.F.G., L.P.C.-L. and J.F. All authors have read and agreed to the published version of the manuscript.

Funding: This work was conducted in the framework of Project VALPT (PROJ/IPV/ID&I/003) financed by Caixa Geral de Depósitos and financed by national funds through FCT—Fundação para a Ciência e Tecnologia, through the CERNAS Research Centre—project UIDB/00681/2020 (I.D.), CITAB, Centre for the Research and Technology of Agro-Environmental and Biological Sciences—Project UIDB/04033/2020 and CERIS, Civil Engineering Research and Innovation for Sustainability—Project UIDB/04625/2020.

Institutional Review Board Statement: Not applicable.

Informed Consent Statement: Not applicable.

Data Availability Statement: Data are available on request from the corresponding author.

Conflicts of Interest: The authors declare no conflict of interest.

References

1. Neitzel, N.; Hosseinpourpia, R.; Walther, T.; Adamopoulos, S. Alternative Materials from Agro-Industry for Wood Panel Manufacturing—A Review. *Materials* **2022**, *15*, 4542. [CrossRef]
2. Pędzik, M.; Janiszewska, D.; Rogoziński, T. Alternative Lignocellulosic Raw Materials in Particleboard Production: A Review. *Ind. Crops Prod.* **2021**, *174*, 114162. [CrossRef]
3. Nunes, L.J.R.; Pimentel, C.; Garrido Azevedo, S.; Matias, J.C.O. (Eds.) *New Trends for Biomass Energy Development: From Wood to Circular Economy*; Energy science, engineering and technology; Nova Science Publishers, Inc.: New York, NY, USA, 2020; ISBN 978-1-5361-7190-7.
4. Lee, S.H.; Lum, W.C.; Boon, J.G.; Kristak, L.; Antov, P.; Pędzik, M.; Rogoziński, T.; Taghiyari, H.R.; Lubis, M.A.R.; Fatriasari, W.; et al. Particleboard from Agricultural Biomass and Recycled Wood Waste: A Review. *J. Mater. Res. Technol.* **2022**, *20*, 4630–4658. [CrossRef]
5. Nelis, P.A.; Mai, C. The Influence of Low-Density (*Paulownia* Spp.) and High-Density (*Fagus sylvatica* L.) Wood Species on Various Characteristics of Light and Medium-Density Three-Layered Particleboards. *Wood Mater. Sci. Eng.* **2021**, *16*, 21–26. [CrossRef]
6. Kozakiewicz, P.; Laskowska, A.; Ciolek, S. A Study of Selected Features of Shan Tong Variety of Plantation Paulownia and Its Wood Properties. *Ann. Wars. Univ. Life Sci. SGGW. For. Wood Technol.* **2020**, *111*, 116–123. [CrossRef]
7. He, T.; Vaidya, B.; Perry, Z.; Parajuli, P.; Joshee, N. Paulownia as a Medicinal Tree: Traditional Uses and Current Advances. *Eur. J. Med. Plants* **2016**, *14*, 1–15. [CrossRef]
8. Yadav, N.K.; Vaidya, B.N.; Henderson, K.; Lee, J.F.; Stewart, W.M.; Dhekney, S.A.; Joshee, N. A Review of *Paulownia* Biotechnology: A Short Rotation, Fast Growing Multipurpose Bioenergy Tree. *Am. J. Plant Sci.* **2013**, *4*, 2070. [CrossRef]
9. Li, P.; Oda, J. Flame Retardancy of Paulownia Wood and Its Mechanism. *J. Mater. Sci.* **2007**, *42*, 8544–8550. [CrossRef]

10. Akyildiz, M.H.; Kol, H.S. Some Technological Properties and Uses of Paulownia (*Paulownia tomentosa* Steud.) Wood. *J. Environ. Biol.* **2010**, *31*, 351–355.
11. Kalaycioglu, H.; Deniz, I.; Hiziroglu, S. Some of the Properties of Particleboard Made from Paulownia. *J. Wood Sci.* **2005**, *51*, 410–414. [CrossRef]
12. Jakubowski, M. Cultivation Potential and Uses of Paulownia Wood: A Review. *Forests* **2022**, *13*, 668. [CrossRef]
13. Jakubowski, M.; Tomczak, A.; Jelonek, T.; Grzywiński, W. The Use of Wood and the Possibility of Planting Trees of the Paulownia Genus. *Acta Sci. Pol. Silvarum Colendarum Ratio Et Ind. Lignaria* **2018**, *17*, 291–297.
14. Kim, Y.K.; Kwon, G.J.; Kim, A.R.; Lee, H.S.; Purusatama, B.; Lee, S.H.; Kang, C.W.; Kim, N.H. Effects of Heat Treatment on the Characteristics of Royal Paulownia (*Paulownia tomentosa* (Thunb.) Steud.) Wood Grown in Korea. *J. Korean Wood Sci. Technol.* **2018**, *46*, 511–526. [CrossRef]
15. Esteves, B.; Ferreira, H.; Viana, H.; Ferreira, J.; Domingos, I.; Cruz-Lopes, L.; Jones, D.; Nunes, L. Termite Resistance, Chemical and Mechanical Characterization of *Paulownia tomentosa* Wood before and after Heat Treatment. *Forests* **2021**, *12*, 1114. [CrossRef]
16. Koman, S.; Feher, S.; Vityi, A. Physical and Mechanical Properties of *Paulownia Tomentosa* Wood Planted in Hungary. *Wood. Res.* **2017**, *62*, 335–340.
17. Barbu, M.C.; Buresova, K.; Tudor, E.M.; Petutschnigg, A. Physical and Mechanical Properties of *Paulownia Tomentosa x elongata* Sawn Wood from Spanish, Bulgarian and Serbian Plantations. *Forests* **2022**, *13*, 1543. [CrossRef]
18. Sedeer, E.-S.; Nabil, E.-S. The Paulownia Tree an Alternative for Sustainable Forestry. 2003. Available online: http://www.cropdevelopment.org/docs/PaulowniaBrochure_print.pdf (accessed on 24 January 2023).
19. De Melo, R.R.; Del Menezzi, C.H.S. Influência Da Massa Específica Nas Propriedades Físico-Mecânicas de Painéis Aglomerados. *Silva Lusit.* **2010**, *18*, 59–73.
20. Cosereanu, C.N.; Brenci, L.-M.N.; Zeleniuc, O.I.; Fotin, A.N. Effect of Particle Size and Geometry on the Performance of Single-Layer and Three-Layer Particleboard Made from Sunflower Seed Husks. *BioResources* **2015**, *10*, 1127–1136. [CrossRef]
21. Niemz, P.; Sandberg, D. Critical Wood-Particle Properties in the Production of Particleboard. *Wood Mater. Sci. Eng.* **2022**, *17*, 386–387. [CrossRef]
22. Kelly, M.W. *Critical Literature Review of Relationships between Processing Parameters and Physical Properties of Particleboard*; U.S. Department of Agriculture, Forest Service, Forest Products Laboratory: Madison, WI, USA, 1977.
23. Nemli, G.; Aydın, I.; Zekoviç, E. Evaluation of Some of the Properties of Particleboard as Function of Manufacturing Parameters. *Mater. Des.* **2007**, *28*, 1169–1176. [CrossRef]
24. De Bazzetto, J.T.L.; Bortoletto, G.; Brito, F.M.S. Effect of Particle Size on Bamboo Particle Board Properties. *Floresta E Ambiente* **2019**, *26*. [CrossRef]
25. Maloney, T.M. Modern Particleboard & Dry-Process Fiberboard Manufacturing, Forest Prod. *Soc. Madison Wis.* **1993**, 681.
26. Ferrandez-Garcia, M.T.; Ferrandez-Garcia, A.; Garcia-Ortuño, T.; Ferrandez-Garcia, C.E.; Ferrandez-Villena, M. Influence of Particle Size on the Properties of Boards Made from Washingtonia Palm Rachis with Citric Acid. *Sustainability* **2020**, *12*, 4841. [CrossRef]
27. Hegazy, S.S.; Ahmed, K. Effect of Date Palm Cultivar, Particle Size, Panel Density and Hot Water Extraction on Particleboards Manufactured from Date Palm Fronds. *Agriculture* **2015**, *5*, 267–285. [CrossRef]
28. Osarenmwinda, J.O.; Nwachukwu, J.C. Effect of Particle Size on Some Properties of Rice Husk Particleboard. In Proceedings of the Advanced Materials Research. *Trans. Tech. Publ.* **2007**, *18*, 43–48.
29. Farrokhpayam, S.R.; Valadbeygi, T.; Sanei, E. Thin Particleboard Quality: Effect of Particle Size on the Properties of the Panel. *J. Indian Acad. Wood Sci.* **2016**, *13*, 38–43. [CrossRef]
30. Hashim, R.; Saari, N.; Sulaiman, O.; Sugimoto, T.; Hiziroglu, S.; Sato, M.; Tanaka, R. Effect of Particle Geometry on the Properties of Binderless Particleboard Manufactured from Oil Palm Trunk. *Mater. Des.* **2010**, *31*, 4251–4257. [CrossRef]
31. Malanit, P.; Barbu, M.C.; Frühwald, A. The Gluability and Bonding Quality of an Asian Bamboo (" Dendrocalamus Asper") for the Production of Composite Lumber. *J. Trop. For. Sci.* **2009**, *21*, 361–368.
32. Papadopoulos, A.N. Property Comparisons and Bonding Efficiency of UF and PMDI Bonded Particleboards as Affected by Key Process Variables. *BioResources* **2006**, *1*, 201–208. [CrossRef]
33. Rathke, J.; Sinn, G.; Harm, M.; Teischinger, A.; Weigl, M.; Müller, U. Effects of Alternative Raw Materials and Varying Resin Content on Mechanical and Fracture Mechanical Properties of Particle Board. *BioResources* **2012**, *7*, 2970–2985.
34. Arabi, M.; Faezipour, M.; Gholizadeh, H. Reducing Resin Content and Board Density without Adversely Affecting the Mechanical Properties of Particleboard through Controlling Particle Size. *J. For. Res.* **2011**, *22*, 659–664. [CrossRef]
35. Iswanto, A.H.; Febrianto, F.; Hadi, Y.S.; Ruhendi, S.; Hermawan, D. The Effect of Pressing Temperature and Time on the Quality of Particle Board Made from Jatropha Fruit Hulls Treated in Acidic Condition. *Makara J. Technol.* **2013**, *17*, 8. [CrossRef]
36. Iswanto, A.H.; Azhar, I.; Susilowati, A. Effect of Resin Type, Pressing Temperature and Time on Particleboard Properties Made from Sorghum Bagasse. *Agric. For. Fish.* **2014**, *3*, 62–66. [CrossRef]
37. Zhao, Z.; Umemura, K. Investigation of a New Natural Particleboard Adhesive Composed of Tannin and Sucrose. 2. Effect of Pressing Temperature and Time on Board Properties, and Characterization of Adhesive. *Bioresources* **2015**, *10*, 2444–2460. [CrossRef]
38. Cravo, J.C.M.; de Lucca Sartori, D.; Mármol, G.; Schmidt, G.M.; de Carvalho Balieiro, J.C.; Fiorelli, J. Effect of Density and Resin on the Mechanical, Physical and Thermal Performance of Particleboards Based on Cement Packaging. *Constr. Build. Mater.* **2017**, *151*, 414–421. [CrossRef]

39. Khedari, J.; Charoenvai, S.; Hirunlabh, J. New Insulating Particleboards from Durian Peel and Coconut Coir. *Build. Environ.* **2003**, *38*, 435–441. [CrossRef]
40. Wong, K.K. Optimising Resin Consumption, Pressing Time and Density of Particleboard Made of Mixes of Hardwood Sawmill Residue and Custom Flaked Softwood. Ph.D. Thesis, RMIT University, Melbourne, Australia, 2012.
41. *NP EN 310:2002*; Placas de Derivados de Madeira—Determinação Do Módulo de Elasticidade Em Flexão e Da Resistência à Flexão. Instituto Português Da Qualidade: Caparica, Portugal, 2002.
42. *NP EN 317:2002*; Aglomerado de Partículas de Madeira e Aglomerado de Fibras de Madeira—Determinação Do Inchamento Em Espessura Após Imersão Em Água. Instituto Português Da Qualidade: Caparica, Portugal, 2002.
43. *ASTM D5334*; Standard Test Method for Determination of Thermal Conductivity of Soil and Soft Rock by Thermal Needle Probe Procedure. ASTM: West Conshohocken, PA, USA, 2014.
44. *ASTM D5930*; Standard Test Method for Thermal Conductivity of Plastics by Means of a Transient Line-Source Technique. ASTM: West Conshohocken, PA, USA, 2009.
45. *ISO 22007-2:2015*; Plastics—Determination of Thermal Conductivity and Thermal Diffusivity—Part 2: Transient Plane Heat Source (Hot Disc) Method. ISO: Geneva, Switzerland, 2015.
46. *NP EN 312:2017*; Aglomerado de Partículas de Madeira—Especificações. Instituto Português Da Qualidade: Caparica, Portugal, 2017.
47. Widyorini, R.; Umemura, K.; Isnan, R.; Putra, D.R.; Awaludin, A.; Prayitno, T.A. Manufacture and Properties of Citric Acid-Bonded Particleboard Made from Bamboo Materials. *Eur. J. Wood Wood Prod.* **2016**, *74*, 57–65. [CrossRef]
48. Barnes, D. A Model of the effect of strand length and strand thickness on the strength properties of oriented wood composites. *For. Prod. J.* **2001**, *51*, 36–47.
49. Miyamoto, K.; Nakahara, S.; Suzuki, S. Effect of Particle Shape on Linear Expansion of Particleboard. *J. Wood Sci.* **2002**, *48*, 185–190. [CrossRef]
50. Lehmann, W.F. Resin Efficiency in Particleboard as Influenced by Density, Atomization, and Resin Content. *For. Prod. J.* **1970**, *20*, 48–54.
51. Nemli, G.; Kırcı, H.; Serdar, B.; Ay, N. Suitability of Kiwi (Actinidia Sinensis Planch.) Prunings for Particleboard Manufacturing. *Ind. Crops Prod.* **2003**, *17*, 39–46. [CrossRef]
52. Kimoto, K.; Ishimori, E.; Sasaki, H.; Maku, T. Studies on the Particle Boards: Report 6: Effects of Resin Content and Particle Dimension on the Physical and Mechanical Properties of Low-Density Particle Boards. *Wood Res.* **1964**, *32*, 1–14.
53. Dai, C.; Yu, C.; Jin, J. Theoretical Modeling of Bonding Characteristics and Performance of Wood Composites. Part IV. Internal Bond Strength. *Wood Fiber Sci.* **2008**, *40*, 146–160.
54. Sekaluvu, L.; Tumutegyereize, P.; Kiggundu, N. Investigation of Factors Affecting the Production and Properties of Maize Cob-Particleboards. *Waste Biomass Valoriz.* **2014**, *5*, 27–32. [CrossRef]
55. Ashori, A.; Nourbakhsh, A. Effect of Press Cycle Time and Resin Content on Physical and Mechanical Properties of Particleboard Panels Made from the Underutilized Low-Quality Raw Materials. *Ind. Crops Prod.* **2008**, *28*, 225–230. [CrossRef]
56. Khedari, J.; Nankongnab, N.; Hirunlabh, J.; Teekasap, S. New Low-Cost Insulation Particleboards from Mixture of Durian Peel and Coconut Coir. *Build. Environ.* **2004**, *39*, 59–65. [CrossRef]

Disclaimer/Publisher's Note: The statements, opinions and data contained in all publications are solely those of the individual author(s) and contributor(s) and not of MDPI and/or the editor(s). MDPI and/or the editor(s) disclaim responsibility for any injury to people or property resulting from any ideas, methods, instructions or products referred to in the content.

Article

Plastic/Natural Fiber Composite Based on Recycled Expanded Polystyrene Foam Waste

Wilasinee Sriprom *, Adilah Sirivallop, Aree Choodum, Wadcharawadee Limsakul and Worawit Wongniramaikul

Integrated Science and Technology Research Center, Faculty of Technology and Environment, Phuket Campus, Prince of Songkla University, Kathu, Phuket 83120, Thailand; adey.siriwallop@gmail.com (A.S.); aree.c@phuket.psu.ac.th (A.C.); wadcharawadee.n@phuket.psu.ac.th (W.L.); worawit.won@phuket.psu.ac.th (W.W.)
* Correspondence: wilasinee.s@phuket.psu.ac.th; Tel.: +66-(0)-7627-6486

Abstract: A novel reinforced recycled expanded polystyrene (r-EPS) foam/natural fiber composite was successfully developed. EPS was recycled by means of the dissolution method using an accessible commercial mixed organic solvent, while natural fibers, i.e., coconut husk fiber (coir) and banana stem fiber (BSF) were used as reinforcement materials. The treatment of natural fibers with 5% (w/v) sodium hydroxide solution reduces the number of –OH groups and non-cellulose components in the fibers, more so with longer treatments. The natural fibers treated for 6 h showed rough surfaces that provided good adhesion and interlocking with the polymer matrix for mechanical reinforcement. The tensile strength and impact strength of r-EPS foam composites with treated fibers were higher than for non-filled r-EPS foam, whereas their flexural strengths were lower. Thus, this study has demonstrated an alternative way to produce recycled polymer/natural fiber composites via the dissolution method, with promising enhanced mechanical properties.

Keywords: natural fiber; recycled expanded polystyrene foam; natural fiber-recycled plastic composites; mechanical properties; dissolution

1. Introduction

Expanded polystyrene (EPS) foam is a thermoplastic that has been used to make a wide variety of consumer products. Due to its low thermal conductivity, high compressive strength, extremely light weight, versatility, durability, and moisture resistance, it is often used as an insulator, in lightweight protective packaging, and in food packaging. With the increased use of EPS foam, its waste has also increasingly accumulated around the world. However, discarded foam waste is non-biodegradable and resistant to photolysis [1]. It can break down when exposed to sunlight, rain, and ocean water (especially in tropical waters) into its constituents, including styrene monomers [2] that are classified as possible human carcinogens [3]. Styrene trimers may also increase thyroid hormone levels [4].

Fortunately, EPS foam waste has been reported as an excellent material for recycling [5]. To recycle polystyrene (PS) foams, they are typically cleaned first and then densified to shippable logs using a thermal treatment [6]. The densified or compressed PS obtained is then simply chopped up, heated, and recast into plastic pellets that can be used as raw materials for plastic products [7]. Dissolution of PS foam with a suitable solvent has become an alternative method for waste volume reduction or recycling, as it is one of the least costly alternatives and uses less energy than melting or compressing the waste [8]. Various organic solvents, such as toluene, acetone, limonene, and other liquid hydrocarbons have been reported for PS foam dissolution [5,8–11]. The dissolved EPS can be used for packaging, similar to the brand new polymer [12], and various products can be produced, e.g., nanofibers [13,14], or polymer–cement composites [5].

The loss of mechanical properties of the polymer during recycling has been reported, and reinforcement with glass fibers or natural fibers is applied to conquer this problem

and extend the potential applications of recycled plastic materials [15–17]. There are recent studies on the characterization of wood-recycled plastic composites, showing potential as alternative materials [18,19]. Several kinds of natural fibers [20] have been reported as reinforcing fillers, e.g., banana/jute/flax fiber [21,22], kenaf/coir [23–25], and bagasse/Napier grass fiber–polyester composites [26,27]. Nevertheless, most of the previous studies have reinforce virgin plastic (polymer pellets) instead of recycled plastic, which may influence mechanical properties of the filled composite.

The aim of this work was to investigate the feasibility to produce a novel composite using EPS recycled via dissolution (r-EPS) and reinforced with natural fiber. A commercial grade thinner containing mixed organic solvents was used, along with acetone, as a solvent to recycle the EPS. Coconut husk fiber (coir) and banana stem fiber (BSF) were incorporated into r-EPS to enhance the mechanical properties of the composite. The coir and BSF were first treated with an alkaline solution and mixed with r-EPS to various filler loadings in the composite sheets. The mechanical properties of both r-EPS and the composite sheets were then evaluated.

2. Materials and Methods

2.1. Materials

The used and clean expanded polystyrene (EPS) foam boards were collected from Prince of Songkla University, Phuket Campus. It was analyzed using gel permeation chromatography (GPC) (Shodex Standard SM-105) equipped with Shodex GPC KF-806 M and KF-803 L (300 mm Length × 8.0 mm ID) using RI-Detector, obtaining the number average molecular weight ($\overline{M_n}$) (120,491 g/mol) and the polydispersity index (PDI) (2.17). THF was used as the eluent at a flow rate of 1.0 mL/min at 40 °C. The GPC system was calibrated with polystyrene (PS) standards, with molecular weights ranging from 3790 to 3,053,000 g/mol. Acetone (99.98%) was purchased at the highest purity available from Fisher Chemical (Loughborough, England). Commercial thinner containing mixed organic solvents, including toluene (70%), acetone (15.4%), ethyl acetate (4.9%), 2-butoxyethanol (3.9%), 2-propanol (2.9%), and 2-methyl-1-propanol (2.9%), was supplied by TOA Paint (Thailand) Co., Ltd. (Samutprakan, Thailand) [28]. Coconut and banana fibers were prepared from coconut husks and banana stems collected from a fruit garden in Phuket, Thailand.

2.2. Dissolution of EPS Foam

An EPS foam board was broken into small pieces (30 g) before dissolving in 200 mL of the mixed organic solvents (thinner: acetone in a 3:1 volume ratio). The mixture was constantly stirred at 750 rpm for 4 h at room temperature to obtain a homogenous solution. The dissolved EPS solution (r-EPS) was used for the preparation of natural fiber-reinforced recycled EPS foam composites.

2.3. Preparation and Characterization of Natural Fiber

Coconut husk fiber (coir) and banana stem fiber (BSF) were cleaned by washing with tap water and then dried in an oven at 100 °C for 24 h. They were chopped into small pieces and sieved to a length of 1 to 3 mm. An alkali treatment was applied on both coir and BSF separately by immersing the chopped fibers in 5% aqueous sodium hydroxide (NaOH) solution for 6, 12, or 24 h at room temperature. The treated fibers were vacuum filtered and washed with distilled water until the water became neutral (pH = 7). Then, the treated fibers were dried in an oven at 80 °C for 24 h.

The chemical functionality of treated and untreated natural fibers was evaluated by Fourier transform infrared spectroscopy (FTIR, Perkin-Elmer Frontier). Transmittance was measured over a range from 4000 to 600 cm^{-1}. The surface topography and compositions of treated and untreated natural fibers were observed by using a scanning electron microscope, SEM Quanta 400, operated at 20 kV at 1000×.

2.4. Preparation of Natural Fiber-Reinforced Recycled EPS Foam Composites

Both untreated and treated coir and BSF were mixed with r-EPS at 2%, 5%, and 10% by total weight to obtain natural fiber-reinforced recycled EPS foam composites. Next, the composites were slowly poured to fill a 25 × 150 mm Petri dishes and then set aside to dry at room temperature in the fume hood. The mass of samples was measured and recorded until the constant mass was obtained within 72 h. With the ease of preparing the composites using these conditions, the solvent blend could be recovered for further use as a solvent for EPS recycling.

2.5. Mechanical Properties of Natural Fiber-Reinforced Recycled EPS Foam Composites

The tensile strength, flexural strength, and impact strength of recycled EPS foam and recycled EPS foam natural fiber composites were investigated, as summarized in Table 1. Each sample was cut into three specimens, and the same test was performed on these replicates. The average test results with standard deviations (SD) are reported.

Table 1. Materials prepared from recycled EPS to evaluate the mechanical properties.

Material *	Tensile Strength	Flexural Strength	Impact Strength
r-EPS	✓	✓	✓
r-EPS/u-coir (2, 5, and 10%)	✓	-	-
r-EPS/u-BSF (2, 5, and 10%)	✓	-	-
r-EPS/t-coir (2, 5, and 10%)	✓	✓	✓
r-EPS/t-BSF (2, 5, and 10%)	✓	✓	✓

* r-EPS = recycled expanded polystyrene foam; u-coir = untreated coir; t-coir = treated coir; u-BSF = untreated banana stem fiber; t-BSF = treated banana stem fiber.

Tensile testing: Each sample was cut into three dumbbell-shaped specimens (gauge length, L_0 of 60 mm). The tensile test was conducted using the Instron (Model 5566) universal testing machine with a 1kN load cell at 1 mm/min crosshead speed. The test was continued until tensile failure occurred.

Flexural testing: Three-point bending flexural testing was carried out with an Instron universal testing machine (Model 55R4502). The load cell and the crosshead speed were 1 kN and 1.13 mm/min, respectively, while the support span was 42 mm. The rectangular specimens had 80 mm (L) × 12.7 mm (W) × 2.3 mm (T) dimensions.

Impact testing: Notched Izod impact testing was performed according to the ASTM D256 standard method for determining the impact resistance of plastic samples. The testing specimen was 65 mm (L) × 13 mm (W) × 2.5 mm (T). The depth under the notch of the specimen was 10 mm. A pendulum energy of 1 joule was employed in the testing.

3. Results and Discussion

3.1. Characterization of Natural Fibers

The treated natural fibers are shown in Figure 1, and their chemical functionalities are shown in Figure 2. The FTIR of both untreated BSF and coir showed characteristic peaks of cellulose, hemicellulose, and lignin. The absorption bands at 3300 and 2910 cm^{-1} indicate hydroxyl groups (–OH) and –CH stretching vibrations, respectively, from the chemical structures of cellulose and hemicellulose. The peak at 1733 cm^{-1} was attributed to C=O stretching vibrations of the acetyl group in the hemicelluloses. Adsorption at 1507, 1436, and 1250 cm^{-1} was attributed to C=C aromatic symmetrical stretching, HCH and OCH in plane bending vibrations, and C-O stretching vibrations of the acetyl groups, respectively, and these are typical absorption peaks of lignin [29]. After alkali treatment of BSF and coir, all absorption peaks decreased with treatment time. The absorption peaks around 1733–1250 cm^{-1} of both fibers disappeared after 24 h of treatment, indicating that lignin and hemicellulose might be removed, matching some previous reports [20,30–32].

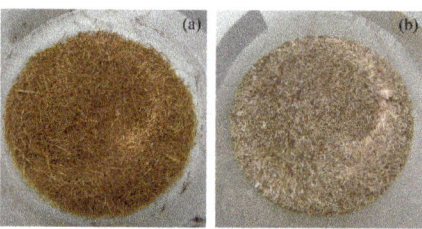

Figure 1. Treated fibers. (a) coir, and (b) BSF.

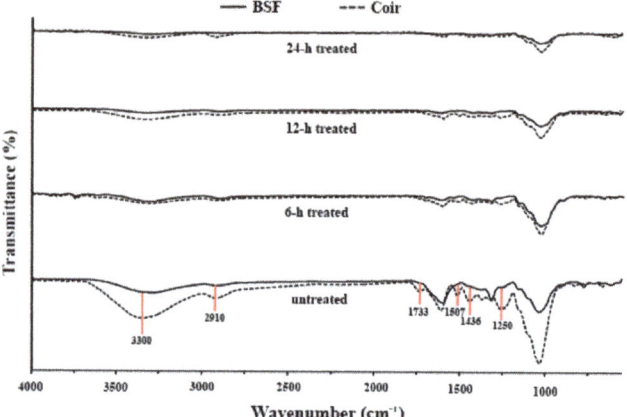

Figure 2. FTIR spectra of untreated and alkali-treated BSF and coir.

The surface topography of BSF and coir, both untreated and treated with 5% NaOH solution, was investigated, and the SEM micrographs are shown in Figure 3. The surfaces of untreated BSF and coir were not smooth, spread with nodes, and covered with irregular strips (Figure 3a) that may represent lignin, hemicellulose, or impurities [33]. After alkali treatment of both fibers for 6 h, the layer of substances on the fiber surface seemed to be removed (Figure 3b), matching the FTIR results indicating that lignin and hemicellulose contents were decreased. Some holes and rough surfaces were obviously observed, especially for coir fiber, that could improve the mechanical interlocking of the fiber and polymer matrix [33–35]. However, after alkali treatments for 12 or 24 h, the surfaces of the fibers had become smoother (Figure 3c,d), so that potentially, the interfacial bonding with the polymer matrix would be weaker than with the 6 h treatment of the fibers. It can be expected that the alkali treatment of fibers for longer than 6 h may cause damage by removing hemicelluloses, lignin, and bound cellulose from the fibers, which weakens the fiber strength [32]. Therefore, 6 h treated fiber was considered the most suitable for producing fiber-reinforced polymer composites, possibly having a good interfacial bonding with the polymer matrix and a desirable amount of cellulose exposed on the fiber surfaces.

The mechanical and physical properties of coir and banana fiber have been reported in the literature, as summarized in Table 2 [36–39]. However, Yue et al. [40] reported that the mechanical properties of many natural plant fibers may vary, to a large extent due to inappropriate measurement.

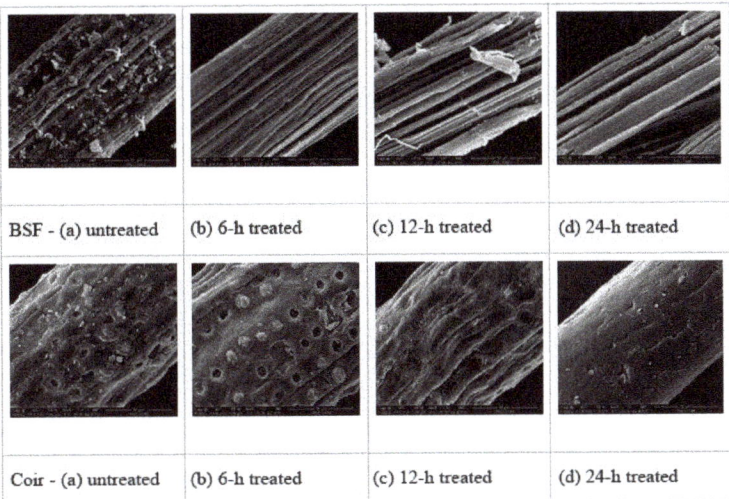

Figure 3. SEM micrographs (at 1000×) of BSF and coir: (**a**) untreated, (**b**) 6 h treated, (**c**) 12 h treated, and (**d**) 24 h treated.

Table 2. Mechanical and physical properties of the fibers.

Properties	Coir	Banana
Diameter (μm)	150–250 [36]	100–250 [36]
Density (g/cm^3)	1.2 [36,37]	0.8 [36]
Tensile Strength (MPa)	175 [36], 131–220 [37]	161.8 [36]
Young's modulus (GPa)	4–6 [36,37]	8.5 [36]
Elongation at break (%)	30 [36], 15–30 [37]	2.0 [36]
Surface energy (mJ/m^2)	35.1 ± 1.3 [38]	39.49 [39]

3.2. Characterization of Recycled EPS Foam/Natural Fiber Composites

Composite materials of r-EPS foam with BSF and coir at 2%, 5%, and 10% by weight were prepared using both the untreated (u) and the treated (t) fibers (treatment with 5% NaOH aqueous solution for 6 h). Composite sheets with 3 mm thickness and 130 mm diameter were obtained. Their weights were increased from the total mass of EPS and fiber by ~6%, which might be attributed to organic solvent trapping during the curing process. The distribution of fibers in the EPS matrix and the mechanical properties tensile strength, flexural strength, and impact strength of the composites were then investigated.

3.2.1. The Distribution of Fibers in r-EPS

Images of both treated and untreated BSF and coir in r-EPS at different fiber loadings are shown in Figure 4. The untreated fibers were gathered mostly at the center of the composite sheet, whereas the treated fibers were distributed more evenly in the matrix. This is because of the improved fiber-polymer matrix adhesion [33–35]. At the same mass of the fibers mixed in the composite, the BSF spreads more thoroughly within the composite sheet than does the coir, due to it being a greater density fiber (BSF has the smaller volume) [41]. Increasing the fiber loading caused the fibers to be more evenly distributed all over the composite sheet, and the fibers were aligned in the polymer matrix. However, it was found that the composite of EPS with 10% w/w untreated coir had some excess fibers appearing on the polymer surfaces. The results confirmed that the wettability of fibers by polymer solution was enhanced by the alkali treatment, as the –OH groups were modified to –O$^-$Na$^+$ groups [32,42], resulting in the reduction in the polarity of the fibers [43]. Consequently, the treated fibers more readily form stronger interfacial adhesion with the polymer matrix [36,43,44], so they dispersed throughout the composite sheet better than the untreated fibers. Maximizing the interfacial adhesion between the fibers and the polymer matrix would provide the final composite material with the highest strength [34,45–48].

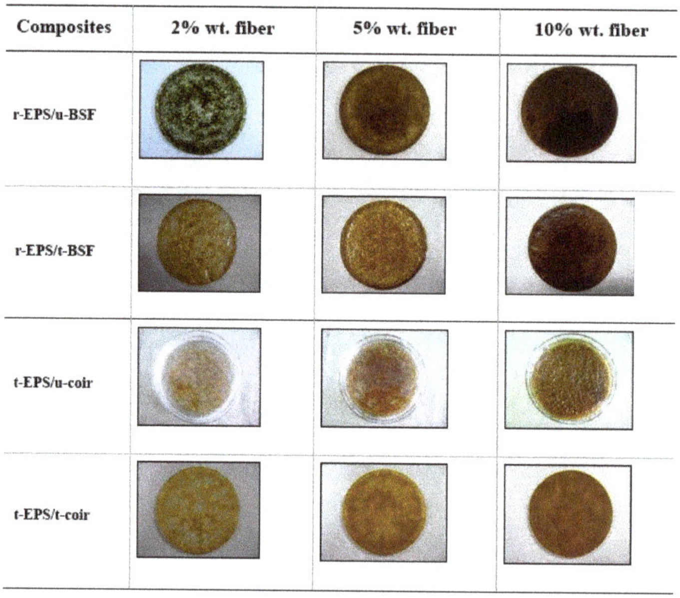

Composites	2% wt. fiber	5% wt. fiber	10% wt. fiber
r-EPS/u-BSF			
r-EPS/t-BSF			
t-EPS/u-coir			
t-EPS/t-coir			

Figure 4. Digital photographs showing the distribution characteristics of fibers in the recycled EPS (r-EPS) foam/natural fiber composite sheets prepared at different fiber loadings.

3.2.2. Mechanical Properties

The tensile strength, flexural strength, and impact strength of r-EPS/natural fiber composites and r-EPS (without fiber) were evaluated and compared. The composites with untreated coir (r-EPS/u-coir) had lower tensile strength than the r-EPS, whereas composites with treated coir (r-EPS/t-coir) provided similar tensile strength. On the other hand, the tensile strength of all composites with treated BSF (r-EPS/t-BSF) was higher than that of r-EPS, whereas the composites with untreated BSF (r-EPS/u-BSF) showed reduced tensile strength (Figure 5). As described in the previous section, the alkali treatment of fiber not only caused the fiber surfaces to be less polar, but also increased surface roughness, enabling a stronger interlocking of fibers with the polymer matrix. Therefore, the alkali treated fibers should provide better interfacial adhesion with the polymer matrix than untreated fibers [27,34,36,44,48]. On the other hand, the addition of untreated fibers resulted in lower tensile strength, due to the unevenness of fiber distribution in the matrix and the weak adhesion of fibers. Furthermore, it was found that the higher content of treated BSF distributed in the polymer matrix decreased the tensile strength of the composites because of fiber agglomeration by fiber–fiber interactions, more so at higher fiber loadings. Similar results were observed by Ibrahim et al. [45] and Ramesh et al. [49].

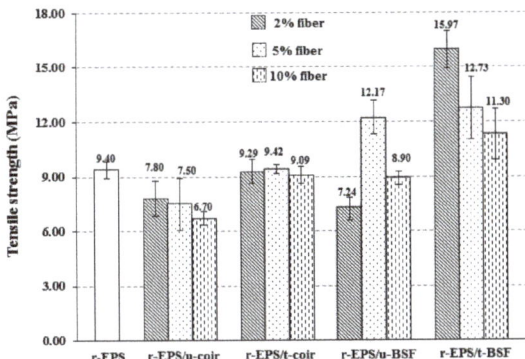

Figure 5. Tensile strengths of r-EPS composites with untreated and treated BSF or coir fibers at 2%, 5%, or 10% wt. loadings.

In addition, it can be observed that the tensile strength of the treated BSF composite (r-EPS/t-BSF) was higher than that of the coir composite (r-EPS/t-coir). This indicates that the BSF provides better reinforcement in the composite than coir, potentially because the BSF contains more crystalline cellulose [20]. The BSF fibers are also less thick, which provides better wettability (fewer gaps at the interface) between the fibers and the polymer matrix. This was confirmed by the tear surfaces of dumbbell-shaped specimens after tensile testing of both r-EPS/5% t-coir and r-EPS/5% t-BSF (Figure 6). The composite of treated coir showed fibers slipping out from the polymer matrix more frequently than what was observed in the composite filled with treated BSF.

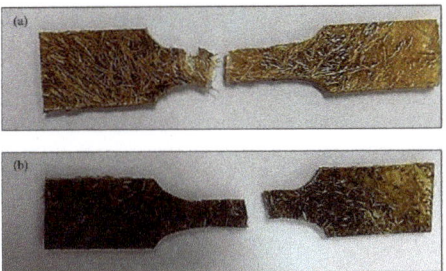

Figure 6. Specimens after tensile testing: (**a**) r-EPS/5% t-coir, and (**b**) r-EPS/5% t-BSF.

The flexural strength of treated BSF and coir reinforced r-EPS composites with different fiber loadings was evaluated by three-point bending flexural testing. The flexural strength of both the r-EPS/t-coir and r-EPS/t-BSF composites was smaller than that of the r-EPS (Figure 7), indicating that the natural fibers diminished bending resistance. This is possibly because the orientation of fibers may be at right angles to the direction of the force acting on them [49,50]. When compared with the tensile test discussed previously, the treated fibers added in the r-EPS enhanced the tensile strength of the material, as the pulling force is at right angles to the bending force, confirming the parallel arrangement of the fibers to the pulling direction, or the perpendicular arrangement to the bending direction. In addition, the flexural strength of r-EPS/t-coir was found to be greater than that of r-EPS/t-BSF at the same fiber loadings. This may be attributed to the fact that most of the treated coir in the composite was aligned in the direction of the bending force [47,51], more so than in the treated BSF. The flexural strength of the composites also increased as the fiber content increased from 2% to 5%, but it decreased to the lowest level for 10% of either treated fiber.

This is likely due to the less uniform fiber distribution with greater fiber loading in the polymer matrix. Similar results were obtained for the tensile test, as described previously.

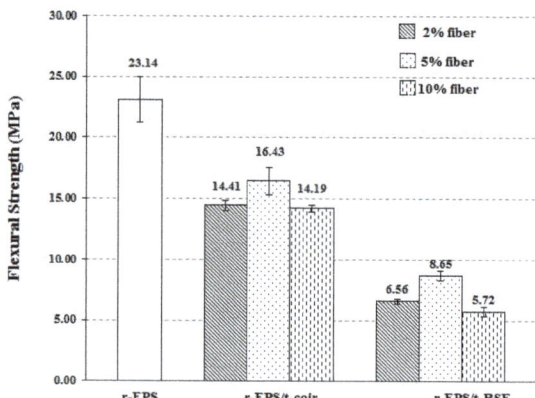

Figure 7. Flexural strength of r-EPS composites with treated BSF or coir at 2%, 5%, or 10% wt. fiber loadings.

The impact strength of the composites was found to be increased from that of r-EPS (Figure 8), and it increased with the loading of treated fibers. Moreover, the r-EPS/t-coir composite showed higher impact strength than the r-EPS/t-BSF composite at the same loading. This indicates that the coir may absorb greater force during the impact test than the BSF. Since the coir is larger in size or lower in density than the BSF, the greater volume of the coir presents in the composite at the same mass added [41].

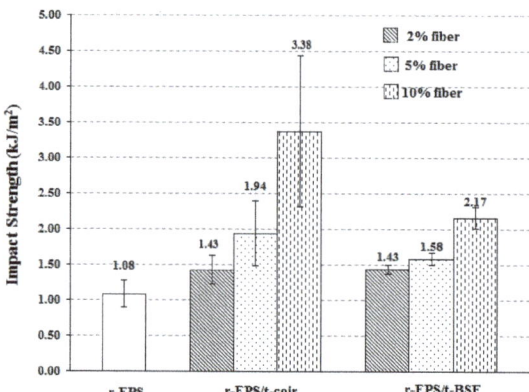

Figure 8. Impact strength of r-EPS composites with treated BSF or coir at 2%, 5%, or 10% wt. fiber loadings.

4. Conclusions

This study has demonstrated the production of r-EPS foam composites reinforced with coir and BSF by the dissolution process. The natural fibers were treated with 5% (w/v) NaOH for 6, 12,, or 24 h. The FTIR analyses revealed that the number of hydroxyl groups and non-cellulose components in the fibers decreased with treatment time, and SEM imaging showed that alkali treatment modified the fiber structure. The 6 h treated BSF (2% wt.) reinforced r-EPS foam composite provided the greatest increase in tensile strength by about 70%, whereas r-EPS foam composite reinforced with 10% wt. of 6 h treated coir showed the maximum increase in impact strength, by about 210% compared to r-EPS. This was due to the good adhesion and interlocking of treated fibers with the polymer matrix. This study presented an alternative method to produce recycled polymer/natural fiber composites via the dissolution method, with promising enhanced mechanical properties.

Author Contributions: Conceptualization, W.S.; data curation, W.S., A.C., W.L. and W.W.; formal analysis, W.S. and A.C.; funding acquisition, W.S. and W.L.; investigation, A.S.; methodology, W.S.; project administration, W.S.; resources, W.S.; supervision, W.S.; validation, W.S. and W.L.; visualization, A.S.; writing—original draft, W.S.; writing—review and editing, W.S., A.C., W.L. and W.W. All authors have read and agreed to the published version of the manuscript.

Funding: This research was funded by the Faculty of Technology and Environment, Prince of Songkla University, Phuket Campus, and the Graduate School, Prince of Songkla University.

Institutional Review Board Statement: Not applicable.

Informed Consent Statement: Not applicable.

Data Availability Statement: All data are available from the corresponding author on reasonable request.

Acknowledgments: The authors would like to thank the English correction service of the Research and Development Office (RDO), Prince of Songkla University (Seppo Karrila).

Conflicts of Interest: The authors declare no conflict of interest.

References

1. Bandyopadhyay, A.; Basak, G.C. Studies on photocatalytic degradation of polystyrene. *Mater. Sci. Technol.* **2007**, *23*, 307–314. [CrossRef]
2. Breaking Down Ocean Polystyrene–Fauna & Flora International. Available online: https://www.fauna-flora.org/app/uploads/2020/07/FFI_2020_Breaking-Down-Ocean-Polystyrene_Scoping-Report.pdf (accessed on 7 April 2022).
3. International Agency for Research on Cancer; World Health Organization. *IARC Monographs on the Evaluation of Carcinogenic Risks to Humans (Some Industrial Chemicals)*; IARC: Lyon, France, 1994; Volume 60, pp. 233–320.
4. Yanagiba, Y.; Ito, Y.; Yamanoshita, O.; Zhang, S.; Watanabe, G.; Taya, K.; Li, C.M.; Inotsume, Y.; Kamijima, M.; Gonzalez, F.J.; et al. Styrene trimer may increase thyroid hormone levels via down-regulation of the aryl hydrocarbon receptor (AhR) target gene UDP-glucuronosyltransferase. *Environ. Health. Perspect.* **2008**, *116*, 740–745. [CrossRef] [PubMed]
5. Eskander, S.B.; Tawfik, M.E. Polymer–cement composite based on recycled expanded polystyrene foam waste. *Polym. Compos.* **2011**, *32*, 1430–1438. [CrossRef]
6. Kan, A.; Demirboğa, R. A new technique of processing for waste-expanded polystyrene foams as aggregates. *J. Mater. Process. Technol.* **2009**, *209*, 2994–3000. [CrossRef]
7. Singh, N.; Hui, D.; Singh, R.; Ahuja, I.P.S.; Feo, L.; Fraternali, F. Recycling of plastic solid waste: A state of art review and future applications. *Compos. Part B Eng.* **2017**, *115*, 409–422. [CrossRef]
8. García, M.T.; Duque, G.; Gracia, I.; de Lucas, A.; Rodríguez, J.F. Recycling extruded polystyrene by dissolution with suitable solvents. *J. Mater. Cycles Waste Manag.* **2009**, *11*, 2–5. [CrossRef]
9. Gutiérrez, C.; García, M.T.; Gracia, I.; de Lucas, A.; Rodríguez, J.F. Recycling of extruded polystyrene wastes by dissolution and supercritical CO_2 technology. *J. Mater. Cycles Waste Manag.* **2012**, *14*, 308–316. [CrossRef]
10. Miller-Chou, B.A.; Koenig, J.L. A review of polymer dissolution. *Prog. Polym. Sci.* **2003**, *28*, 1223–1270. [CrossRef]
11. Shin, C.; Chase, G.G. Nanofibers from recycle waste expanded polystyrene using natural solvent. *Polym. Bull.* **2005**, *55*, 209–215. [CrossRef]
12. Noguchi, T.; Inagaki, Y.; Miyashita, M.; Watanabe, H. A new recycling system for expanded polystyrene using a natural solvent. Part 2. Development of a prototype production system. *Packag. Technol. Sci.* **1998**, *11*, 29–37. [CrossRef]
13. Shin, C. Filtration application from recycled expanded polystyrene. *J. Colloid Interface Sci.* **2006**, *302*, 267–271. [CrossRef] [PubMed]
14. Shin, C.; Chase, G.G.; Reneker, D.H. Recycled expanded polystyrene nanofibers applied in filter media. *Colloids Surf. A Physicochem. Eng. Asp.* **2005**, *262*, 211–215. [CrossRef]
15. Homkhiew, C.; Ratanawilai, T.; Thongruang, W. Effects of natural weathering on the properties of recycled polypropylene composites reinforced with rubberwood flour. *Ind. Crops Prod.* **2014**, *56*, 52–59. [CrossRef]
16. Khanam, N.P.; AlMaadeed, M.A. Improvement of ternary recycled polymer blend reinforced with date palm fibre. *Mater. Des.* **2014**, *60*, 532–539. [CrossRef]
17. Zadeh, K.M.; Ponnamma, D.; Al-Maadeed, M.A.A. Date palm fibre filled recycled ternary polymer blend composites with enhanced flame retardancy. *Polym. Test.* **2017**, *61*, 341–348. [CrossRef]

18. Arnandha, Y.; Satyarno, I.; Awaludin, A.; Irawati, I.S.; Prasetya, Y.; Prayitno, D.A.; Winata, D.C.; Satrio, M.H.; Amalia, A. Physical and mechanical properties of WPC board from sengon sawdust and recycled HDPE plastic. *Procedia Eng.* **2017**, *171*, 695–704. [CrossRef]
19. Turku, I.; Keskisaari, A.; Kärki, T.; Puurtinen, A.; Marttila, P. Characterization of wood plastic composites manufactured from recycled plastic blends. *Compos. Struct.* **2017**, *161*, 469–476. [CrossRef]
20. Faruk, O.; Bledzki, A.K.; Fink, H.P.; Sain, M. Biocomposites reinforced with natural fibers: 2000–2010. *Prog. Polym. Sci.* **2012**, *37*, 1552–1596. [CrossRef]
21. Bledzki, A.K.; Mamun, A.A.; Faruk, O. Abaca fibre reinforced PP composites and comparison with jute and flax fibre PP composites. *Express Polym. Lett.* **2007**, *1*, 755–762. [CrossRef]
22. Bledzki, A.K.; Faruk, O.; Mamun, A.A. Influence of compounding processes and fibre length on the mechanical properties of abaca fibre-polypropylene composites. *Polimery* **2008**, *53*, 120–125. [CrossRef]
23. Abu Bakar, M.A.; Ahmad, S.; Kuntjoro, W. The mechanical properties of treated and untreated kenaf fibre reinforced epoxy composite. *J. Biobased Mater. Bioenergy* **2010**, *4*, 159–163. [CrossRef]
24. Akil, H.M.; Omar, M.F.; Mazuki, A.A.M.; Safiee, S.; Ishak, Z.A.M.; Abu Bakar, A. Kenaf fiber reinforced composites: A review. *Mater. Des.* **2011**, *32*, 4107–4121. [CrossRef]
25. Chandra Rao, C.H.; Madhusudan, S.; Raghavendra, G.; Venkateswara Rao, E. Investigation in to wear behavior of coir fiber reinforced epoxy composites with the Taguchi method. *Int. J. Eng. Res. Appl.* **2012**, *2*, 371–374. Available online: http://www.ijera.com/papers/Vol2_issue5/BK25371374.pdf (accessed on 7 April 2022).
26. Haameem, J.A.M.; Abdul Majid, M.S.; Afendi, M.; Marzuki, H.F.A.; Hilmi, E.A.; Fahmi, I.; Gibson, A.G. Effects of water absorption on Napier grass fibre/polyester composites. *Compos. Struct.* **2016**, *144*, 138–146. [CrossRef]
27. Naguib, H.M.; Kandil, U.F.; Hashem, A.I.; Boghdadi, Y.M. Effect of fiber loading on the mechanical and physical properties of "green" bagasse–polyester composite. *J. Radiat. Res. Appl. Sci.* **2015**, *8*, 544–548. [CrossRef]
28. Barco Thinner AAA MSDS. Available online: https://04a77950-65bb-464b-99ba-845b033effcb.usrfiles.com/ugd/04a779_6d97802ffb23440d82d3d2ea114cf854.pdf (accessed on 20 May 2022).
29. Punyamurthy, R.; Sampathkumar, D.; Ranganagowda, R.P.G.; Bennehalli, B.; Srinivasa, C.V. Mechanical properties of abaca fiber reinforced polypropylene composites: Effect of chemical treatment by benzenediazonium chloride. *J. King Saud. Univ. Eng. Sci.* **2017**, *29*, 289–294. [CrossRef]
30. Samal, R.K.; Panda, B.B.; Rout, S.K.; Mohanty, M. Effect of chemical modification on FTIR spectra. I. Physical and chemical behavior of coir. *J. Appl. Polym. Sci.* **1995**, *58*, 745–752. [CrossRef]
31. Sgriccia, N.; Hawley, M.C.; Misra, M. Characterization of natural fiber surfaces and natural fiber composites. *Compos. Part A Appl. Sci. Manuf.* **2008**, *39*, 1632–1637. [CrossRef]
32. Williams, T.; Hosur, M.; Theodore, M.; Netravali, A.; Rangari, V.; Jeelani, S. Time effects on morphology and bonding ability in mercerized natural fibers for composite reinforcement. *Int. J. Polym. Sci.* **2011**, *2011*, 192865. [CrossRef]
33. Karthikeyan, A.; Balamurugan, K. Effect of alkali treatment and fiber length on impact behavior of coir fiber reinforced epoxy composites. *J. Sci. Ind. Res.* **2012**, *71*, 627–631. Available online: http://nopr.niscair.res.in/handle/123456789/14634 (accessed on 7 April 2022).
34. Gopinath, S.; Vadivu, K.S. Mechanical behavior of alkali treated coir fiber and rice husk reinforced epoxy composites. *IJIRSET* **2014**, *3*, 1268–1271. Available online: http://www.ijirset.com/upload/2014/icets/265_ME517.pdf (accessed on 7 April 2022).
35. Mulinari, D.R.; Baptista, C.A.R.P.; Souza, J.V.C.; Voorwald, H.J.C. Mechanical properties of coconut fibers reinforced polyester composites. *Procedia Eng.* **2011**, *10*, 2074–2079. [CrossRef]
36. Narayana, V.L.; Rao, L.B. A brief review on the effect of alkali treatment on mechanical properties of various natural fiber reinforced polymer composites. *Mater. Today Proc.* **2021**, *4*, 1988–1994. [CrossRef]
37. Zhang, Z.; Cai, S.; Li, Y.; Wang, Z.; Long, Y.; Yu, T.; Shen, Y. High performances of plant fiber reinforced composites—A new insight from hierarchical microstructures. *Compos. Sci. Technol.* **2020**, *194*, 108151. [CrossRef]
38. Tran, L.Q.N.; Fuentes, C.A.; Dupont-Gillain, C.; Van Vuure, A.W.; Verpoest, I. Understanding the interfacial compatibility and adhesion of natural coir fibre thermoplastic composites. *Compos. Sci. Technol.* **2013**, *80*, 23–30. [CrossRef]
39. Alonso, E.; Pothan, L.A.; Ferreira, A.; Cordeiro, N. Surface modification of banana fibers using organosilanes: An IGC insight. *Cellulose* **2019**, *26*, 3643–3654. [CrossRef]
40. Yue, H.; Rubalcaba, J.C.; Cui, Y.; Fernández-Blázquez, J.P.; Yang, C.; Shuttleworth, P.S. Determination of cross-sectional area of natural plant fibres and fibre failure analysis by in situ SEM observation during microtensile tests. *Cellulose* **2019**, *26*, 4693–4706. [CrossRef]
41. Gurunathan, T.; Mohanty, S.; Nayak, S.K. A review of the recent developments in biocomposites based on natural fibres and their application perspectives. *Compos. Part A Appl. Sci. Manuf.* **2015**, *77*, 1–25. [CrossRef]
42. John, M.J.; Anandjiwala, R.D. Recent developments in chemical modification and characterization of natural fiber-reinforced composites. *Polym. Compos.* **2008**, *29*, 187–207. [CrossRef]
43. Paul, S.A.; Joseph, K.; Mathew, G.; Pothen, L.A.; Thomas, S. Influence of polarity parameters on the mechanical properties of composites from polypropylene fiber and short banana fiber. *Compos. Part A Appl. Sci. Manuf.* **2010**, *41*, 1380–1387. [CrossRef]

44. Magagula, S.I.; Sefadi, J.S.; Mochane, M.J.; Mokhothu, T.H.; Mokhena, T.C.; Lenetha, G.G. 2-The effect of alkaline treatment on natural fibers/biopolymer composites. In *Surface Treatment Methods of Natural Fibres and Their Effects on Biocomposites*; Shahzad, A., Tanasa, F., Teaca, C., Eds.; Woodhead Publishing: Sawston, UK, 2022; pp. 19–45. [CrossRef]
45. Ibrahim, M.M.; Dufresne, A.; El-Zawawy, W.K.; Agblevor, F.A. Banana fibers and microfibrils as lignocellulosic reinforcements in polymer composites. *Carbohydr. Polym.* **2010**, *81*, 811–819. [CrossRef]
46. Ku, H.; Wang, H.; Pattarachaiyakoop, N.; Trada, M. A review on the tensile properties of natural fiber reinforced polymer composites. *Compos. B Eng.* **2011**, *42*, 856–873. [CrossRef]
47. Masuelli, M.A. Introduction of fibre-reinforced polymers−polymers and composites: Concepts, properties and processes. In *Fiber Reinforced Polymers—The Technology Applied for Concrete Repair*; Masuelli, M.A., Ed.; IntechOpen: London, UK, 2013; pp. 3–40. [CrossRef]
48. Merlini, C.; Soldi, V.; Barra, G.M.O. Influence of fiber surface treatment and length on physico-chemical properties of short random banana fiber-reinforced castor oil polyurethane composites. *Polym. Test.* **2011**, *30*, 833–840. [CrossRef]
49. Ramesh, M.; Atreya, T.S.A.; Aswin, U.S.; Eashwar, H.; Deepa, C. Processing and mechanical property evaluation of banana fiber reinforced polymer composites. *Procedia Eng.* **2014**, *97*, 563–572. [CrossRef]
50. Bagherpour, S. Fibre reinforced polyester composites. In *Polyester*; Saleh, H., Ed.; IntechOpen: London, UK, 2012; pp. 135–166. [CrossRef]
51. Pickering, K.L.; Efendy, M.G.A.; Le, T.M. A review of recent developments in natural fibre composites and their mechanical performance. *Compos. Part A Appl. Sci. Manuf.* **2016**, *83*, 98–112. [CrossRef]

Article

Chiral Polymers from Norbornenes Based on Renewable Chemical Feedstocks

Ivan V. Nazarov [1], Danil P. Zarezin [1], Ivan A. Solomatov [1], Anastasya A. Danshina [2,3], Yulia V. Nelyubina [2], Igor R. Ilyasov [4] and Maxim V. Bermeshev [1,*]

1. A.V. Topchiev Institute of Petrochemical Synthesis, RAS, 29 Leninskiy Pr., 119991 Moscow, Russia
2. A.N. Nesmeyanov Institute of Organoelement Compounds, Russian Academy of Sciences, Vavilov Street 28, 119991 Moscow, Russia
3. Moscow Institute of Physics and Technology, National Research University, Institutskiy Per., 9, 141700 Dolgoprudny, Russia
4. Nelubin Institute of Pharmacy, Sechenov First Moscow State Medical University, Trubetskaya Str. 8/2, 119991 Moscow, Russia
* Correspondence: bmv@ips.ac.ru; Tel.: +7-495-647-59-27 (ext. 379)

Abstract: Optically active polymers are of great interest as materials for dense enantioselective membranes, as well as chiral stationary phases for gas and liquid chromatography. Combining the versatility of norbornene chemistry and the advantages of chiral natural terpenes in one molecule will open up a facile route toward the synthesis of diverse optically active polymers. Herein, we prepared a set of new chiral monomers from *cis*-5-norbornene-2,3-dicarboxylic anhydride and chiral alcohols of various natures. Alcohols based on cyclic terpenes ((-)-menthol, (-)-borneol and pinanol), as well as commercially available alcohols (S-(-)-2-methylbutanol-1, S-(+)-3-octanol), were used. All the synthesized monomers were successfully involved in ring-opening metathesis polymerization, affording polymers in high yields (up to 96%) and with molecular weights in the range of 1.9×10^5–5.8×10^5 (M_w). The properties of the metathesis polymers obtained were studied by TGA and DSC analysis, WAXD, and circular dichroism spectroscopy. The polymers exhibited high thermal stability and good film-forming properties. Glass transition temperatures for the prepared polymers varied from -30 °C to $+139$ °C and, therefore, the state of the polymers changed from rubbery to glassy. The prepared polymers represent a new attractive platform of chiral polymeric materials for enantioselective membrane separation and chiral stationary phases for chromatography.

Keywords: chiral polymer; renewable feedstock; pinene; menthol; borneol; norbornene; polynorbornene; ring-opening metathesis polymerization

1. Introduction

Synthetic optically active polymers are of great interest both in academic and industrial fields. On the one hand, such polymers are widely studied and used in asymmetric catalysis [1–3] and as materials for dense enantioselective membranes [4–6], chiral stationary phases for gas and liquid chromatography [7–9], and also for mimicking biological properties [10–12]. On the other hand, these polymers are suitable objects for systematic structure–property study. There are three main approaches to achieving the optical activity of polymers. The first way is the formation of chiral centers in main chains [13,14]. The second one is the introduction of chiral moieties in side chains of polymers [15,16] and in the case of the third way, optical activity appears due to a supramolecular structure [17,18].

Metathesis polymerization of various monomers is a powerful tool for obtaining all three types of polymers. At the same time, ring-opening metathesis polymerization (ROMP) of functionalized norbornenes is the most versatile methodology in the design and preparation of various polymers with the required architecture [19–21]. There are several significant benefits of this tool. Norbornene-type monomers can be readily prepared by using well-known cycloaddition (e.g., the Diels–Alder and [2 + 2 + 2]-cycloaddition reactions) and other reactions [22–25], and these monomers show high reactivity in ROMP polymerization. Moreover, such monomers can be polymerized according to three different mechanisms, affording polymers with different structures of main chains (Figure 1) [19,26–29]. In turn, well-defined Grubbs catalysts exhibit good tolerance towards various functional groups and allow the control of tacticity, configuration of backbones, the composition of copolymers, molecular weights, and molecular weight distributions.

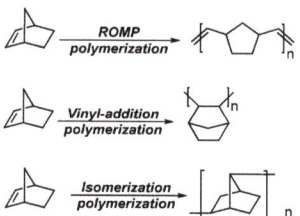

Figure 1. The ways of norbornene-type monomers polymerization.

Some chiral polymers from optically pure isomers of norbornene derivatives were earlier described. The most investigated among these monomers are N-substituted norbornene-2,3-dicarboximides with a chiral group at nitrogen atom [15,30] and norbornenes bearing ester [31–33] or ether [34,35] moiety between norbornene skeleton and a chiral fragment. At the same time, monoterpenes, such as α- and β-pinenes, borneol, and menthol are attractive synthetic building blocks and the source of available chiral moieties that are obtained from renewable feedstock. They are widely represented in nature. Therefore, it seems an attractive strategy to combine a norbornene moiety as a polymerizable part with a fragment of a terpene as the source of chirality in the same molecule to develop new optically active polymers for the structure–property study.

Here we report the preparation and ROMP polymerization of a new set of norbornene-type monomers based on alcohols from cyclic terpenes ((-)-menthol, (-)-borneol, and pinanol) as well as commercially available alcohols (S-(-)-2-methylbutanol-1, S-(+)-3-octanol). As a result, a series of high-molecular-weight polymers with good film-forming properties was successfully obtained. The prepared polymers represent a new attractive platform of chiral polymeric materials for enantioselective membrane separation and chiral stationary phases for chromatography. The properties of the metathesis polymers obtained were studied by TGA and DSC analysis, WAXD, and circular dichroism spectroscopy.

2. Materials and Methods

2.1. Materials

(-)-Menthol, (-)-borneol, S-(+)-3-octanol, S-(-)-2-methylbutanol-1, the 2nd generation Grubbs catalysts, dicyclopentadiene, 1,2,4-trichlorbenzene, maleic anhydride, Et$_3$N, SOCl$_2$, and cis-5-norbornene-endo-2,3-dicarboxylic anhydride (endo-NDA) were purchased from commercial suppliers (Sigma-Aldrich, ABCR GmbH, and TCI) and used as received unless otherwise noted. Methylene chloride and 1,2-dichloroethane were distilled over CaH$_2$ under argon atmosphere and stored over molecular sieves (4 Å). Toluene was distilled in argon atmosphere and stored over molecular sieves (4 Å). ((1R,2S,5R)-6,6-Dimethylbicyclo[3.1.1]heptan-2-yl)methanol (pinanol) was prepared according to the procedure described earlier [36].

2.2. Physico-Chemical Characterization

NMR spectra were recorded on a Bruker Ascend 400 spectrometer at 400.1 MHz (^{1}H), 100.6 MHz (^{13}C) and on a Bruker MSL-300 spectrometer at 300 MHz (^{1}H). Chemical shifts (δ) are reported in parts per million (ppm) relative to the reference (residual CHCl$_3$ signal) for ^{1}H and ^{13}C NMR spectra. Each sample of a polymer for NMR analysis was dissolved in CDCl$_3$ up to a concentration of 10%.

Gel-permeation chromatography (GPC) analysis of the polymers was performed on Agilent 1280 Infinity II system with a differential refractometer (THF as the eluent, flow rate 0.3 mL/min). Molecular mass and polydispersity were calculated by standard procedure relative to monodispersed polystyrene standards.

Calorimetric measurements were conducted using a "Mettler" TA-4000 differential scanning calorimeter (Mettler Toledo, Giesen, Germany) at a heating rate of 20 °C/min under argon. TGA measurements were carried out on "TGA/DSC 1" (Mettler Toledo, Polaris Parkway, OH, USA) in argon and in air at the heating rate of 10 °C/min from 30 to 1000 °C.

Wide-angle X-ray diffraction (WAXD) data were obtained using a two-coordinate AXS detector (Bruker, Bremen, Germany) and Cu Kα emission (wavelength of 0.154 nm).

Specific rotation was measured using KRÜSS P3000 polarimeter in CHCl$_3$ or THF (HPLC grade). Circular dichroism (CD) spectra of the monomers and polymers were measured in chloroform or THF at the concentration of 1 mg/mL. Spectra were recorded with spectropolarimeter J810 (Jasco, Hachioji city, Tokio, Japan) in the 220–450 nm range (50 nm/min, 1 nm slit width) in 0.1 cm path-length quartz cells with a detachable window (Hellma, Jena, Germany). The baseline spectrum was recorded from pure chloroform or THF.

The density of polymer films was calculated by the hydrostatic weighing method according to the following procedure. A sample of the film was weighed on analytical balance (m_{dry}). Thin copper wire was hung on a beam over the balance. The sample was fixed on the lower end of the wire, immersed in the beaker with methanol placed on the balance, and weighed (m_1). Then, the sample was withdrawn and the free wire end immersed in methanol was weighed (m_2). The density of the film sample was calculated by the formula: $d = d_S m_{dry}/(m_{dry} - (m_1 - m_2))$, where $d_S = 0.791$ g/mL is the density of methanol.

X-ray diffraction data for **M2** were collected at 120 K with a Bruker APEXII DUO CCD diffractometer, using graphite monochromated Cu-Kα radiation (λ = 1.54178 Å, ω-scans), while those for **M1**, at 100 K with a Bruker Quest D8 CMOS diffractometer, using graphite monochromated Mo-Kα radiation (λ = 0.71073 Å, ω-scans). Structures were solved using Intrinsic Phasing with the ShelXT [37] structure solution program in Olex2 [38] and then refined with the XL [39] refinement package using Least-Squares minimization against F^2 in the anisotropic approximation for non-hydrogen atoms. Positions of hydrogen atoms were calculated, and they were refined in the isotropic approximation within the riding model. Crystal data and structure refinement parameters are given in Table S1. CCDC 2213611 (**M2**) and 2213612 (**M1**) contain supplementary crystallographic information for this paper.

High-resolution mass spectra were recorded on Bruker maXis QTOF (tandem quarrupole/time-of-flight mass analyzer) mass spectrometers equipped with an ESI source. The m/z scanning range was 50–1600. External calibration of the mass scale was carried out using a low-concentration calibration solution "Tuning mix" (Agilent Technologies, Santa Clara, California, USA). Samples were injected using a 500 μL Hamilton RN 1750 syringe (Hamilton, Biel, Switzerland). The measurements were carried out in the positive ion mode (+) (grounded spray needle, −4500 V high-voltage capillary; HV End Plate Offset: −500 V). Nitrogen was used as a nebulizer gas (1.0 bar) and dry gas (4.0 L/min, 200 °C). The data were processed using the Bruker Data Analysis 4.0 software.

2.3. Film Preparation

The polymer films were prepared by casting from 5 wt.% chloroform solution of a polymer. The solution was poured into a steel cylinder with a diameter of 7 cm and a stretched cellophane bottom. The solvent was allowed to evaporate slowly at room temperature to yield the desired polymer film. After the formation of the films, the cellophane was detached, and the films were dried under vacuum at room temperature to a constant weight. A thermal treatment was not applied. The thickness of the films formed was in the range of 40–60 µm.

2.4. Synthetic Part

Cis-5-norbornene-*exo*-2,3-dicarboxylic anhydride (*exo*-NDA). In a 1000 mL two-necked round bottom flask fitted with reflux, dropping funnel with dicyclopentadiene (DCPD, 122.4 mL, 0.909 mol) and a magnetic stirring bar, was placed 300 mL of 1,2,4-trichlorbenzene and 178 g of maleic anhydride (1.818 mol). The solution was heated to 200 °C and DCPD added dropwise over 20 min. The solution acquired a yellow color, which changed to dark brown upon further heating. The resulting solution was stirred for 4 h at 200 °C. After that, it was cooled down and left overnight after the addition of *n*-hexane. The next day, the crystal precipitate was formed in the flask. The precipitate was separated and recrystallized 4 times from the minimum amount of benzene. The white crystalline substance was obtained (54.6 g, yield 18%). M.p.—145–149 °C

^{1}H NMR (300 MHz, CDCl$_3$, δ, ppm): 1.50 (dd, 2H, RCH$_2$, J = 10.20); 2.97 (s, 2H); 3.42 (s, 2H); 6.30 (s, 2H, R-CH=CH-R).

^{13}C NMR (100 MHz, CDCl3, δ, ppm): 44.07; 46.82; 48.71; 137.91; 171.59.

2.5. General Procedure for the Synthesis of Monomers M1–M3

The synthesis of **M1**. The first step was performed similarly to the previously described procedure with minor differences [40]. A 250 mL round-bottom flask was equipped with a magnetic stirring bar, and then was filled with 13.4 g (82 mmol) of *exo*-NDA, 200 mL of dry toluene, 29.5 mL of triethylamine (210 mmol) and 57.56 g (370 mmol) of (-)-menthol. The mixture was stirred for 96 h (24 for borneol ester) at 55 °C with the conversion control by ^{1}H NMR spectroscopy. The remaining solvent and triethylamine were evaporated at a rotary evaporator. The resulting residue was redissolved in 125 mL of diethyl ether. The solution was washed with 10% hydrochloric acid (3 × 120 mL). The aqueous layer was extracted with diethyl ether (5 × 200 mL). The organic layers were combined and extracted with saturated K$_2$CO$_3$ solution (4 × 250 mL). The resulting aqueous phase was washed once with diethyl ether (150 mL). Then, the aqueous phase was acidified with hydrochloric acid and extracted with methylene chloride (5 × 200 mL). The combined organic layers were dried with anhydrous magnesium sulfate filtered and the remaining solvents were evaporated on a rotary evaporator. As a result, 26.24 g (96% yield) of crude (-)-menthyl ester of *exo*-norbornene dicarboxylic acid was obtained.

The second and third steps. In a 100 mL round-bottom flask equipped with a magnetic stirring bar, 11.34 g (35 mmol) of (-)-menthyl *exo*-norbornene dicarboxylic acid monoester, 25.5 mL (351 mmol) of thionyl chloride, and 80 mL of absolute toluene were placed with a couple of drops of DMF. The mixture was stirred for 4 h at 90 °C. The remaining excess of thionyl chloride was removed under vacuum, and the resulting product was dissolved in 150 mL of dry toluene and 3.66 mL of pyridine (1.3 eq.). (-)-Menthol (4.66 g, 30 mmol) was added to the prepared solution. The resulting mixture was heated to 90 °C and stirred for two hours. Then, toluene was evaporated on a rotary evaporator. The resulting product was recrystallized from ethanol. The yield of **M1** was 13.82 g (86% at this stage). The crystals of **M1** suitable for X-ray analysis were prepared by the slow evaporation of methanol solution of **M1**. $[\alpha]_D^{25} = -75$ deg·mL·g^{-1}·dm^{-1} in CHCl$_3$ (C = 1). ESI-(+)HRMS, m/z: 459.3472, calcd for C$_{29}$H$_{47}$O$_4$: 459.3469 [M + H]$^+$.

^{1}H NMR (400 MHz, CDCl$_3$, δ, ppm): 6.26–6.24 (m, 2H, R-HC=CH-R), 4.61–4.76 (m, 2H), 3.15–3.14 (m, 1H), 3.05–3.04 (m, 1H), 2.63–2.62 (m, 1H), 2.55–2.54 (m, 1H), 2.13–2.04 (m, 3H), 2.03–1.86 (m, 2H), 1.76–1.60 (m, 4H), 1.56–1.30 (m, 5H), 1.62–0.76 (m, 24H).

^{13}C NMR (100 MHz, CDCl$_3$, δ, ppm): 172.93–172.78, 138.15–137.77, 74.52–74.24, 47.53, 47.34, 46.95, 46.88, 46.78, 45.68, 45.19, 40.68, 40.23, 34.33, 31.49–31.35, 26.19, 25.96, 23.42, 23.35, 22.07, 20.95, 16.57, 16.25.

Monomer **M2**

Endo-NDA was used instead of *exo*-NDA for **M2** synthesis. The yield of **M2** was 76%. The crystals of **M2** suitable for X-ray analysis were prepared by the slow evaporation of the methanol solution of **M2**. $[\alpha]_D^{20} = -125$ deg·mL·g^{-1}·dm^{-1} in CHCl$_3$ (C = 1). ESI-(+)HRMS, m/z: 459.3461, calcd for C$_{29}$H$_{47}$O$_4$: 459.3469 [M + H]$^+$.

^{1}H NMR (400 MHz, CDCl$_3$, δ, ppm): 6.28 (br. S., 1H), 6.20 (br.s., 1H), 4.68–4.53 (m, 2H), 3.25–3.15 (m, 4H), 2.05–1.96 (m, 2H), 1.95–1.82 (m, 2H), 1.68–1.62 (m, 4H), 1.47–1.23 (m, 7H), 1.10–0.96 (m, 2H), 0.94–0.79 (m, 17H), 0.78–0.70 (m, 6H).
^{13}C NMR (100 MHz, CDCl$_3$, δ, ppm): 171.98, 171.79, 135.32, 134.52, 74.36, 74.00, 48.62, 48.18, 48.14, 47.48, 47.04, 46.78, 40.95, 40.84, 34.50, 34.45, 31.59, 31.45, 26.25, 25.96, 23.42, 23.37, 22.19, 21.09, 21.06, 16.46, 16.36.

Monomer **M3**

The yield of **M3** was 72%. $[\alpha]_D^{20} = -42$ deg·mL·g^{-1}·dm^{-1} in CHCl$_3$ (C = 1). ESI-(+)HRMS, m/z: 455.3156, calcd for C$_{29}$H$_{43}$O$_4$: 455.3156 [M + H]$^+$.

^{1}H NMR (400 MHz, CDCl$_3$, δ, ppm): 6.24 (br. S., 2H), 3.11–3.07 (m, 2H), 2.68–2.63 (m, 2H), 2.43–2.24 (m, 2H), 2.21–2.14 (m, 1H), 2.00–1.85 (m, 2H), 1.81–1.63 (m, 4H), 1.55–1.46 (m, 1H), 1.36–1.12 (m, 5H), 1.12–0.96 (m, 2H), 0.92–0.82 (m, 20H).
^{13}C NMR (100 MHz, CDCl$_3$, δ, ppm): 173.62, 173.61, 137.99, 137.94, 80.40, 80.15, 48.96, 48.58, 47.82, 47.70, 47.38, 45.91, 45.35, 44.86, 44.84, 36.87, 36.14, 28.03, 27.24, 27.22, 19.71, 18.85, 13.65, 13.51.

2.6. General Procedure for the Synthesis of Monomers **M4–M6**

Synthesis of **M5**. A 50 mL one-neck flask was equipped with a reflux condenser, 0.93 g (0.57 mmol) *exo*-NDA, (S)-(-)-2-methylbutanol 1.1 g (1.25 mmol), 0.06 g (0.61 mmol) *para*-toluenesulfonic acid, and 30 mL toluene(abs) as a solvent. Molecular sieves (4 Å) were used as a drying agent. The reaction mixture was refluxed for 14 h and the conversion was monitored by ^{1}H NMR spectroscopy. The mixture was cooled and washed with saturated solution of K$_2$CO$_3$ (2 × 15 mL). After that, the organic layer was washed with water (15 mL), brine (15 mL) and dried over MgSO$_4$. Then, it was filtered and the toluene was evaporated on a rotary evaporator. The product was purified by silica gel column chromatography using hexanes/ethyl acetate (40/1) as an eluent. The product was obtained as a colorless viscous oil (1.3 g, 71% yield). $[\alpha]_D^{20} = +7$ deg·mL·g^{-1}·dm^{-1} in THF (C = 1). ESI-(+)HRMS, m/z: 321.2038, calcd for C$_{29}$H$_{29}$O$_4$: 321.2060 [M + H]$^+$.

^{1}H NMR (300 MHz, CDCl$_3$, δ, ppm): 6.15 (br. s, 2H, R-CH=CH-R), 3.96–3.67 (M, 4H), 3.02 (s, 2H), 2.57 (s, 2H), 2.07 (m, 2H), 1.67–1.55 (m, 2H), 1.45–1.29 (m, 3H), 1.23–1.03 (m, 2H), 0.92–0.77 (m, 12H).
^{13}C NMR (100 MHz, CDCl$_3$, δ, ppm): 172.99, 137.36, 68.82, 46.77, 45.13, 44.80, 33.50, 33.46, 25.48, 15.87, 15.83, 10.66, 10.61.

Monomer **M4**

The monomer yield was 68%. $[\alpha]_D^{20} = -11$ deg·mL·g^{-1}·dm^{-1} in CHCl$_3$ (C = 1). ESI-(+)HRMS, m/z: 455.3153, calcd for C$_{29}$H$_{43}$O$_4$: 455.3156 [M + H]$^+$.

^{1}H NMR (400 MHz, CDCl$_3$, δ, ppm): 6.20 (s, 2H, R-CH=CH-R), 4.08–4.02 (m, 2H), 3.94–3.89 (m, 2H), 3.08–3.05 (m, 2H), 2.60–2.57 (m, 2H), 2.38–2.30 (m, 4H), 2.13–2.08 (m, 1H), 1.95–1.84 (m, 10H), 1.52–1.41 (m, 3H), 1.19–1.17 (m, 6H), 1.00–0.96 (m, 6H), 0.94–0.90 (m, 2H).
^{13}C NMR (100 MHz, CDCl$_3$, δ, ppm): 173.72, 173.67, 138.09, 138.07, 65.51, 69.30, 47.49, 47.47, 45.90, 45.82, 45.73, 45.54, 46.40, 43.18, 41.39, 41.37, 40.41, 40.30, 38.64, 33.12, 33.07, 27.99, 27.97, 25.98, 23.39, 23.34, 18.94, 18.81.

Monomer **M6**

The monomer yield was 54%. $[\alpha]_D^{20} = +4$ deg·mL·g^{-1}·dm^{-1} in CHCl$_3$ (C = 1). ESI-(+)HRMS, m/z: 407.3155, calcd for C$_{25}$H$_{43}$O$_4$: 407.3156 [M + H]$^+$.

^{1}H NMR (400 MHz, CDCl$_3$, δ, ppm): 6.17 (s, 2H, R-CH=CH-R), 4.74–4.71 (m, 2H), 3.03–3.02 (m, 2H), 2.54–2.53 (m, 2H), 2.11–2.09 (m, 1H), 1.52–1.42 (m, 9H), 1.37–1.17 (m, 12H), 0.87–0.83 (m, 12H).

^{13}C NMR (100 MHz, CDCl$_3$, δ, ppm): 173.15, 138.03, 137.99, 75.84, 75.79, 47.45, 47.37, 46.19, 46.08, 45.33, 33.21, 33.13, 31.92, 31.82, 26.55, 26.54, 25.05, 24.93, 22.62, 14.09, 14.06, 9.63, 9.50.
Epoxide based on **M5**
^{1}H NMR (400 MHz, CDCl$_3$, δ, ppm): 4.03–3.72 (M, 4H), 3.18 (s, 2H), 2.82 (s, 2H), 2.78 (s, 2H), 1.76–0.84 (m, 20H).
^{13}C NMR (100 MHz, CDCl$_3$, δ, ppm): 172.11, 69.68, 50.68, 47.08, 40.97, 34.02, 33.98, 26.03, 23.27, 16.43, 16.40, 11.23, 11.19.
MS (EI, m/z (intensity, %)): 338 (2%, M$^+$), 181 (60%, C$_9$H$_9$O$_4^+$).

*2.7. General Procedure of ROMP Polymerization (**MP1**–**MP6**)*

The synthesis of **MP5**. The monomer **M5** (3.12 g, 9.7 mmol) was dissolved in 15.9 mL of absolute 1,2-dichloroethane and the solution of the 2-st generation Grubbs catalyst 3.52 mL (0.0129 mmol, 3.36 × 10^{-3} M) was added to the monomer solution at stirring. The stirring was continued for 2 h at 45 °C. The polymerization was terminated by the addition of ethyl vinyl ether with following stirring for 10 min. Then, the polymer solution was precipitated by methanol containing the inhibitor (2,2'-methylenebis(6-tert-butyl-4-methylphenol)). The polymer was separated, washed with several portions of methanol, and dried in vacuum. The polymer was twice reprecipitated by methanol from the chloroform solution and dried in vacuum to a constant weight (3.00 g, 96% yield). $M_w = 4.57 \times 10^5$, $M_w/M_n = 2.2$. $[α]_D^{22} = 9$ deg·mL·g^{-1}·dm^{-1} in THF (C = 1).

^{1}H NMR (400 MHz, CDCl$_3$, δ, ppm): 5.51–5.23 (m, 2H, HRC=CRH), 3.94–0.72 (m, 31H).
^{13}C NMR (100 MHz, CDCl$_3$, δ, ppm): 178.14–178.10 (C=O), 133.37–131.80, 69.74–69.30, 53.91–52.62, 45.71–44.80, 41.25–39.45, 34.27–33.96, 26.31–26.05, 16.72–16.33, 11.41–11.23.
Polymer **MP1**

The polymer yield was 96%. $M_w = 4.88 \times 10^5$, $M_w/M_n = 2.1$. $[α]_D^{22} = -59$ deg·mL·g^{-1}·dm^{-1} in CHCl$_3$ (C = 1).

^{1}H NMR (400 MHz, CDCl$_3$, δ, ppm): 5.55–4.51 (m, 4H), 3.56–0.56 (m, 42H).
^{13}C NMR (100 MHz, CDCl$_3$, δ, ppm): 172.87–171.44, 134.54–131.65, 74.50–74.00, 54.31–52.07, 47.55–44.51, 41.26–39.18, 34.72–34.13, 31.90–31.32, 26.62–25.70, 23.86–23.18, 22.61–21.82, 24.41–20.85, 16.97–16.10.
Polymer **MP2**

The polymer yield was 74%. $M_w = 1.9 \times 10^5$, $M_w/M_n = 1.6$. $[α]_D^{22} = -64$ deg·mL·g^{-1}·dm^{-1} in CHCl$_3$ (C = 1).

^{1}H NMR (400 MHz, CDCl$_3$, δ, ppm): 5.74–5.28 (m, 2H), 4.81–4.55 (m, 2H), 3.31–0.45 (m, 42H).
^{13}C NMR (100 MHz, CDCl$_3$, δ, ppm): 171.94–170.91, 132.14–130.53, 75.12–73.81, 52.21–15.62.
Polymer **MP3**

The polymer yield was 94%. $M_w = 4.69 \times 10^5$, $M_w/M_n = 2.1$. $[α]_D^{22} = -34$ deg·mL·g^{-1}·dm^{-1} in CHCl$_3$ (C = 1).

^{1}H NMR (400 MHz, CDCl$_3$, δ, ppm): 5.66–5.12 (m, 2H), 5.01–4.71 (m, 2H), 3.31–0.45 (m, 38H).
^{13}C NMR (100 MHz, CDCl$_3$, δ, ppm): 173.15–171.92, 133.42–131.48, 80.68–79.12, 53.62–13.52.
Polymer **MP4**

The polymer yield was 95%. $M_w = 5.8 \times 10^5$, $M_w/M_n = 2.9$. $[α]_D^{22} = -11$ deg·mL·g^{-1}·dm^{-1} in CHCl$_3$ (C = 1).

^{1}H NMR (400 MHz, CDCl$_3$, δ, ppm): 5.50–5.11 (m, 2H, HRC=CRH), 4.12–0.77 (m, 40H).
^{13}C NMR (100 MHz, CDCl$_3$, δ, ppm): 172.94–172.52, 133.20–132.26, 69.29–69.27, 53.58–52.59, 45.25–38.60, 33.20–33.00, 28.12–27.95, 25.98–25.96, 23.42–23.41, 19.05–18.95.
Polymer **MP6**

The polymer yield was 92%. $M_w = 2.69 \times 10^5$, $M_w/M_n = 2.2$. $[α]_D^{22} = +4$ deg·mL·g^{-1}·dm^{-1} in CHCl$_3$ (C = 1).

^{1}H NMR (400 MHz, CDCl$_3$, δ, ppm): 5.54–5.13 (m, 2H), 4.81–4.64 (m, 2H), 3.53–0.76 (m, 38H).
^{13}C NMR (100 MHz, CDCl$_3$, δ, ppm): 172.93–171.85, 134.12–131.61, 76.16–75.43, 54.30–52.16, 46.07–44.27, 41.03–39.23, 33.73–32.86, 32.16–31.58, 27.12–26.25, 25.50–24.51, 22.95–22.43, 14.55–13.79, 9.97–9.39.

2.8. Synthesis of Polymer MP7

The monomer **M1** (0.30 g, 0.66 mmol) was dissolved in 0.80 mL of absolute toluene. RuCl$_3$·3H$_2$O (3.5 mg, 0.011 mmol) was dissolved in 1.0 mL of EtOH (abs) and the solution was stirred for 15 min. After that, the solution of **M1** in toluene was added to the solution of RuCl$_3$ in ethanol. The resulting mixture was heated for 48 h at 75 °C. Then, the reaction mixture was precipitated by methanol containing the inhibitor (2,2′-methylenebis(6-tert-butyl-4-methylphenol)). The polymer was separated, washed with several portions of methanol, and dried in vacuum. The polymer was twice reprecipitated by methanol from the chloroform solution and dried in vacuum to a constant weight. The polymer yield was 18% (52.5 mg). $M_w = 2.9 \times 10^5$, $M_w/M_n = 2.6$. $[α]_D^{22} = -46$ deg·mL·g^{-1}·dm^{-1} in CHCl$_3$ (C = 1). $T_g = 81$ °C.

^{1}H NMR (400 MHz, CDCl$_3$, δ, ppm): 5.54–5.37 (m, 2H), 4.75–458 (m, 2H), 3.10–0.55 (m, 42H).

3. Results and Discussion

3.1. Synthesis of Monomers

Herein we designed and synthesized two sets of new chiral di-substituted norbornene-type monomers (Schemes 1 and 2). The chiral groups in these monomers are attached to norbornene moiety via ester linkages. As the source of norbornene moiety, *cis*-5-norbornene-*exo*-2,3-dicarboxylic anhydride (*exo*-NDA) was chosen due to its preparative availability and high reactivity. As starting materials for introducing chiral fragments, renewable optically active alcohols ((-)-menthol, (-)-borneol, and pinanol), as well as commercial chiral primary and secondary alcohols, were used.

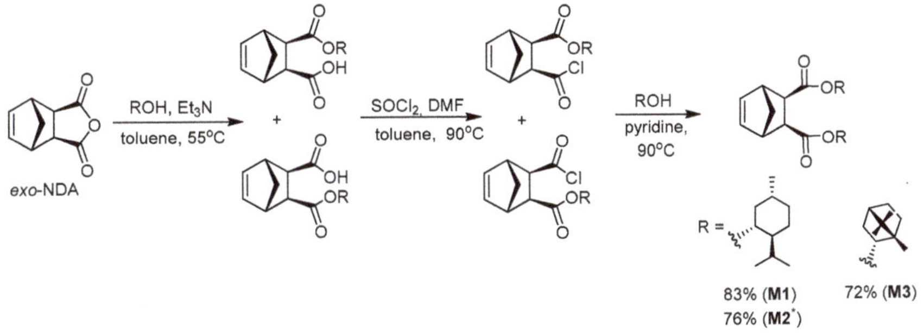

Scheme 1. The synthesis of monomers **M1–M3** (*M2 is *endo*-isomer of **M1**).

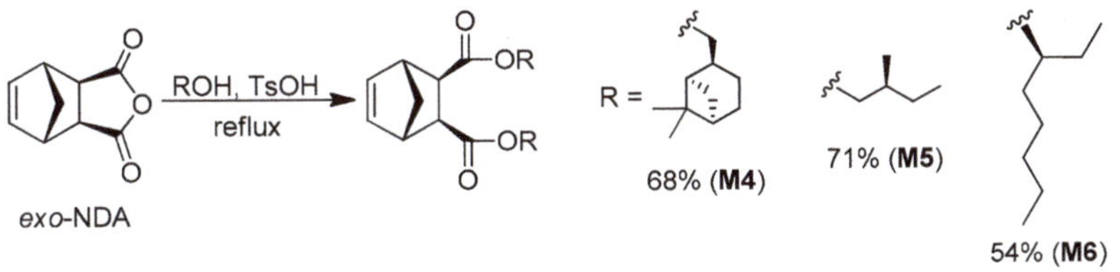

Scheme 2. The synthesis of monomers **M4–M6**.

The required monomers were readily prepared in moderate or good yields according to one- and three-step procedures (Table 1). Primary and less sterically hindered alcohols reacted directly with *exo*-NDA with the formation of the corresponding diesters (Scheme 2).

This scheme was found to be inefficient for the preparation of diesters of *cis*-5-norbornene-*exo*-2,3-dicarboxylic acid based on (-)-menthol and (-)-borneol. Therefore, an alternative way for the synthesis of the corresponding diesters from (-)-menthol and (-)-borneol was realized (Scheme 1). By the usage of this scheme, a diester **M2**, which is an isomer of **M1**, was also obtained from *endo*-NDA and (-)-menthol to evaluate the influence of *exo-/endo*-isomerism on the properties of resulting polymers. It is worth noting that intermediates formed in the first step of the three-step procedure can be isolated as mixtures of the diastereomers (Scheme 1).

Table 1. The data on the synthesis and some properties of the monomers.

Monomer	Reaction Conditions [a]	Yield, %	$[\alpha]_D$ [b], deg·mL·g^{-1}·dm^{-1}	T_m, °C
M1	i	83	−75 (T = 25 °C, CHCl$_3$)	86–89
M2	i	76	−125 (T = 20 °C, CHCl$_3$)	97–100
M3	i	72	−42 (T = 25 °C, CHCl$_3$)	96–98
M4	ii	68	−11 (T = 20 °C, CHCl$_3$)	Colorless oil
M5	ii	71	+7 (T = 20 °C, THF)	Colorless oil
M6	ii	54	+4 (T = 20 °C, CHCl$_3$)	Colorless oil

[a]: *i*—three-step procedure, (1) alcohol (4.5 eq.), Et$_3$N, toluene, 55 °C. (2) SOCl$_2$, DMF (few drops), toluene, 90 °C. (3) alcohol (0.9 eq.), pyridine, 90 °C; *ii*—one-step procedure, alcohol (1.05 eq.), toluene, TsOH, reflux. [b]: the specific optical rotation was measured at C = 1.

All the synthesized monomers were isolated as individual compounds. Their structure and purity were analyzed by ^{1}H, ^{13}C NMR spectroscopy (Figures S1–S11), and HRMS (Figures S13–S18). In the case of solid monomers **M1** and **M2**, X-ray analysis was additionally performed (Figure 2, Table S1), which confirmed the expected structures of the studied monomers. The specific optical rotations for the monomers were nonzero (Table 1). The optical activity of these compounds was also confirmed by circular dichroism (CD) spectroscopy, showing a Cotton effect (Figures S31–S35).

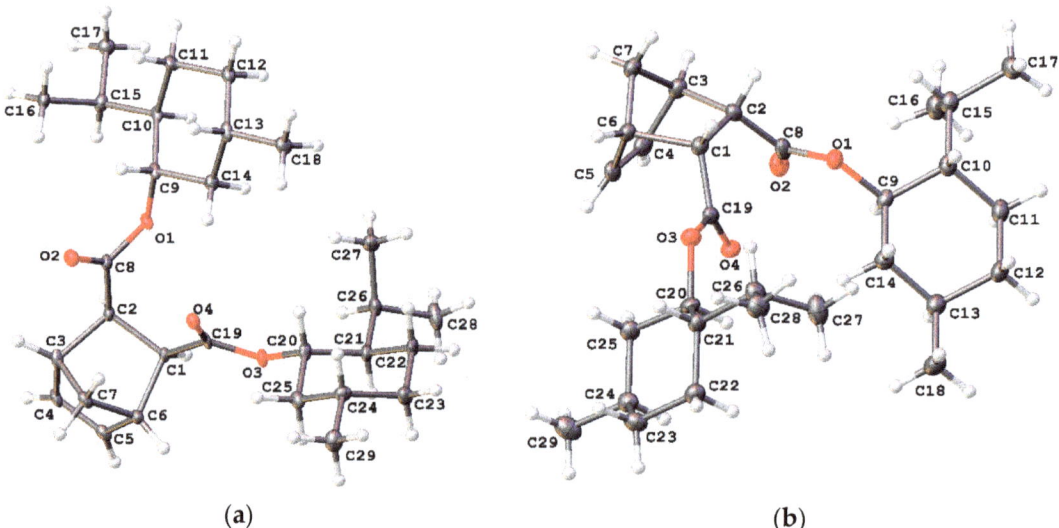

Figure 2. The structure of **M1** (**a**) and **M2** (**b**) according to X-ray analysis.

Interestingly, the synthesized diesters showed a tendency to be slowly oxidized with atmospheric oxygen. We found that **M5** was oxidized with atmospheric oxygen for several months, affording the corresponding epoxide in the quantitative yield (Scheme 3, Figures 3 and S12). This is different from the related N-substituted imides of *cis*-5-norbornene-

exo-2,3-dicarboxylic acid which are stable in the air [41]. Therefore, it is desirable to store such diesters in an inert atmosphere.

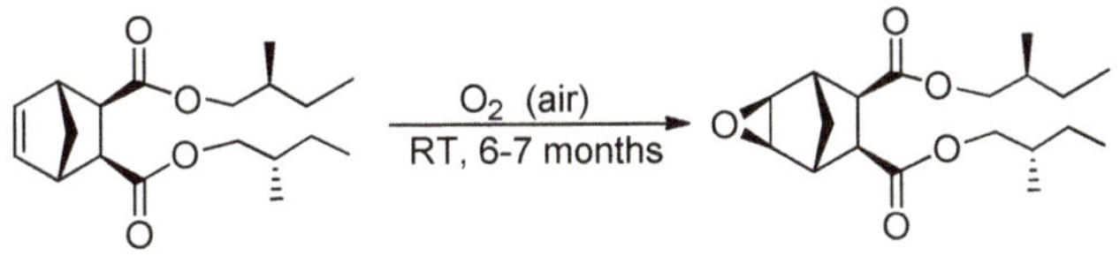

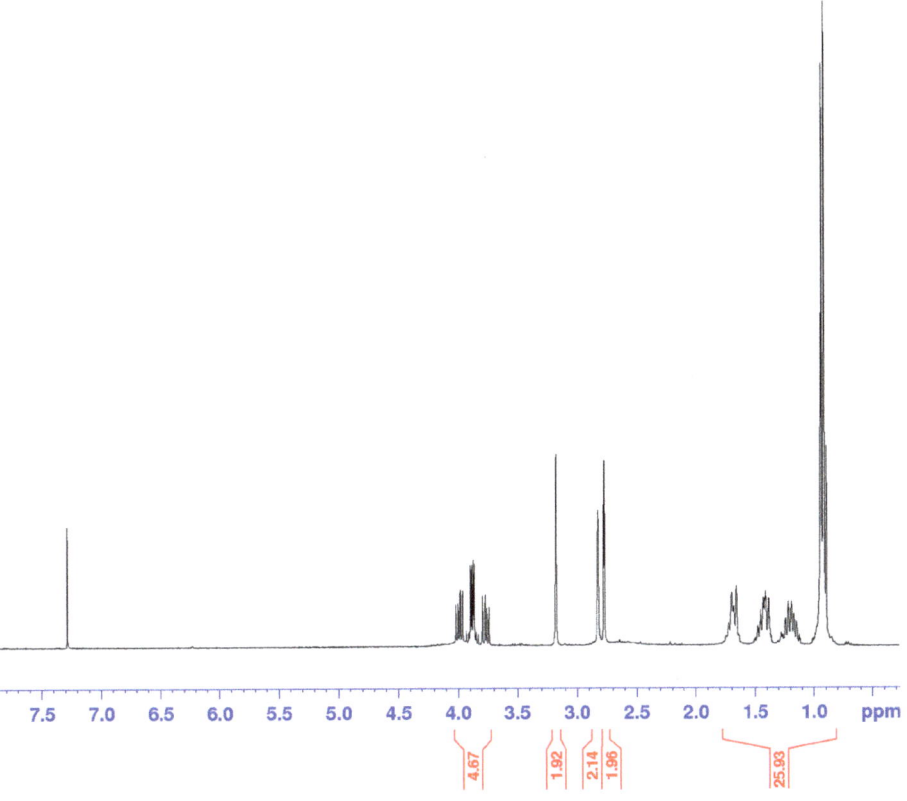

Scheme 3. The oxidation of **M5** during storage in air.

Figure 3. ^{1}H NMR spectrum of the epoxide based on **M5** (CDCl$_3$).

3.2. Ring-Opening Metathesis Polymerization

Ring-opening metathesis polymerization of the synthesized monomers was studied in the presence of the second-generation Grubbs catalyst (Scheme 4). All the monomers exhibited high reactivity. ROMP polymerization of diesters of *cis*-5-norbornene-*exo*-2,3-dicarboxylic acid proceeded for a couple of hours at catalyst loadings of 0.1–0.2 mol.% with the formation of soluble high-molecular weight products in good and high yields (Table 2). For the synthesized polymers, relatively narrow values of molecular weight distributions were found.

Table 2. ROMP polymerization of diesters of *cis*-5-norbornene-2,3-dicarboxylic acid over the 2nd generation Grubbs catalyst [a].

Polymer	Monomer/[Ru] Molar Ratio	C, M [b]	Yield, %	$M_w \times 10^{-3}$	M_w/M_n [c]	T_g, °C [d]	$[\alpha]_D$ [e], deg·mL·g^{-1}·dm^{-1}
MP1	1000/1	0.5	96	488	2.1	80	−59
MP2	750/1	0.5	74	190	1.6	136	−64
MP3	500/1	0.5	94	469	2.1	139	−34
MP4	1000/1	0.5	95	582	2.9	68	−11
MP5	500/1	0.5	96	457	2.2	−9	+10 [f]
MP6	500/1	0.5	92	269	2.2	−30	+4

[a]—reaction temperature—+45 °C, solvent—1,2-dichloroethane, [Ru]—the 2nd generation Grubbs catalyst; [b]—the initial concentration of a monomer; [c]—the molecular weights of polymers were estimated by means of gel permeation chromatography in THF; [d]—glass transition temperature according to DSC; [e]—the specific optical rotation was measured at 22 °C in chloroform solution at C = 1; [f]—the specific optical rotation was measured in THF solution.

The reactivity of the *endo*-isomer was noticeably lower than that of the *exo*-isomer (**M2** vs. **M1**, Table 2). Nevertheless, we succeeded to obtain ROMP polymer from **M2** with high M_w (above 1×10^5) in a good yield. The analysis of the prepared ROMP polymers with NMR-spectroscopy confirmed the expected structure (Figures 4 and S19–S30). The *cis*-/*trans*-ratio of double bonds in monomer units of these polymers (Figures 5 and 6) was analyzed with ^{1}H NMR spectroscopy. The signals of protons at *cis*- and *trans*-double bonds for these polymers are at around 5.2 ppm and 5.35 ppm, correspondingly. This was confirmed by the synthesis of the isomeric ROMP polymer from **M1** (polymer **MP7**, Figure 6) in the presence of RuCl$_3$·3H$_2$O/ethanol catalytic system, which gives ROMP polymers with predominantly *trans*-double bonds (>90%). The studied ROMP polymers obtained over the second generation Grubbs catalyst contained a little more *cis-* than *trans*-double bonds (53–60%, e.g., the *cis-/trans*-double bond ratio was 54/46 for **MP1**).

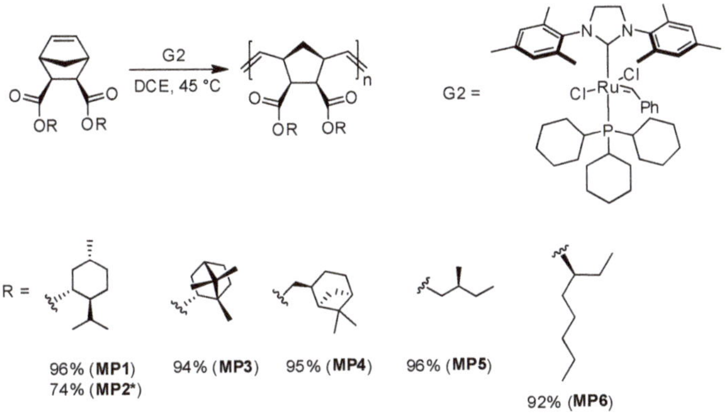

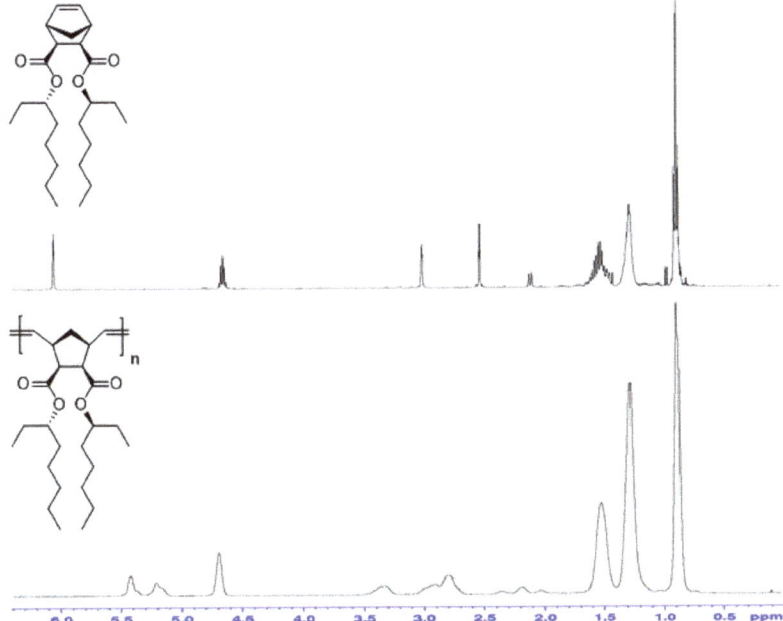

Scheme 4. Ring-opening metathesis polymerization of the prepared monomers (***MP2** was obtained from **M2** (*endo*-isomer of **M1**)).

Figure 4. ^{1}H NMR spectra of **M6** and the corresponding ROMP polymer from **M6** (CDCl$_3$).

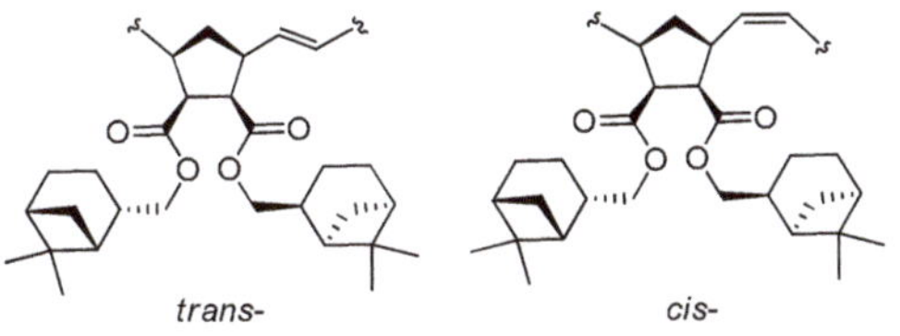

Figure 5. The structure of monomer units of **MP4** with *trans-* and *cis*-double bonds.

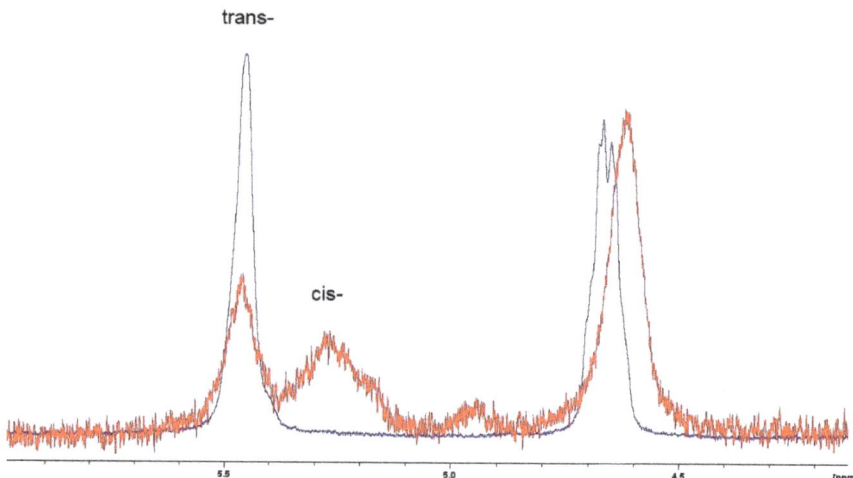

Figure 6. The parts of ^{1}H NMR spectra of the isomeric ROMP polymers obtained from **M1**: **MP1** (red curve) and **MP7** (blue curve).

These ROMP polymers are optically active as it was established by the measurements of the specific optical rotations (Table 2) and by CD spectroscopy (Figures S31–S35). The *cis/trans*-ratio of double bonds has an effect on optical activity. For instance, **MP1** and **MP7** are isomeric ROMP polymers differing in the ratio of *cis/trans*-double bonds (>98% *trans*-double bonds in **MP7** and 46% of *trans*-double bonds in **MP1**). The specific optical rotations are -59 deg·mL·g^{-1}·dm^{-1} and -46 deg·mL·g^{-1}·dm^{-1} for **MP1** and **MP7**, correspondingly. It should be noted that the *cis/trans*-ratio of double bonds can affect the secondary structure of the obtained polymers, forming, for example, a helix in the case of polymers containing predominantly one type of double bond. This study is in progress now.

The resulting ROMP polymers showed good solubility in THF and chloroform while swelling in many other common organic solvents (Table 3). The good solubility and high molecular weights allowed preparing free-standing and robust thin films from these polymers (Figure 7).

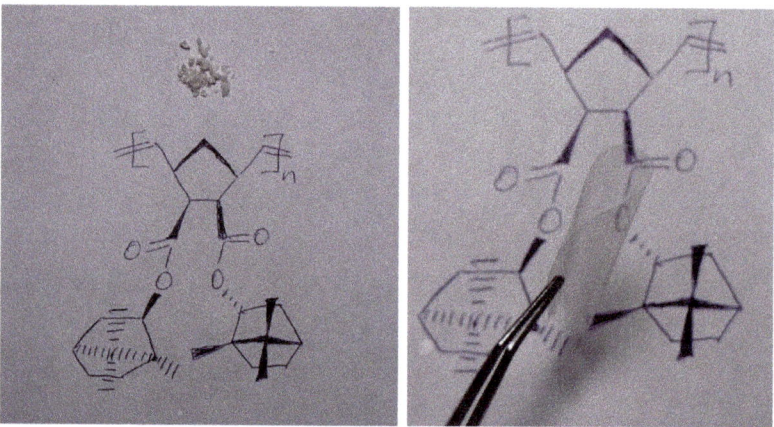

Figure 7. The photographs of the powder and thin films of **MP3**.

Table 3. Solubility of the ROMP polymers synthesized from diesters of *cis*-5-norbornene-2,3-dicarboxylic acid [a].

Polymer	CH$_3$CN	Acetone	DMA	DMSO	*n*-Hexane	PhCH$_3$	THF	CHCl$_3$
MP1	±	±	±	−	+	±	+	+
MP3	−	±	±	−	±	+	+	+
MP4	±	±	±	±	±	+	+	+
MP5	±	±	±	±	±	+	+	+
MP6	−	+	±	±	±	+	+	+

[a] —"+"—soluble, "−"—insoluble, and "±"—swelling.

The density of the obtained polymers exceeded 1.0 g/cm^3 (Table 4), which is unusual because, for related halogen-free polymers, the density is typically less than 1 g/cm^3. Despite the high density, the fraction free volume (FFV) for these polymers is in the range of 16–17%, which is higher than the values typical for polymers with low free volume [42]. It is possible that the obtained FFV values can be explained by the presence of bulky and rigid side-substituents in such polymers, which create free-volume elements near them.

The synthesized polymers showed glass transition temperatures (T_g) in the range of −30 to +139 °C (Table 2), depending on the substituents' nature. At the same time, there were no melting peaks in the DSC curves (Figure S36), which indicates the amorphous nature of the prepared polymers. The ROMP polymers with long flexible alkyl groups had the lowest values of T_g, while the highest T_g was observed for the ROMP polymer with bulky and rigid borneol moieties in side chains. Interestingly, isomeric polymers based on *exo*-NDA and *endo*-NDA diesters (**MP1** vs. **MP2**) showed very different T_g values, and **MP2** derived from *endo*-NDA had a higher T_g value, evidencing that **MP2** possesses more rigid polymeric backbones.

TGA analysis of the synthesized ROMP polymers showed their high thermal stability (Figures 8 and S37–S42). The decomposition temperatures (5 wt.% loss) in argon and in air were close and exceeded 300 °C. For some of the resulting polymers, a small weight loss (1–2%) was observed around 180–200 °C, probably due to traces of sorbed water

(Figure 8b). Char yields at 1000 °C were close to zero, which indicates the almost complete decomposition of the polymers at high temperatures.

Table 4. Density and fractional free-volume (FFV) for ROMP polymers synthesized from diesters of cis-5-norbornene-2,3-dicarboxylic acid [a].

Polymer	Density, g/cm^3	FFV, %
MP1	1.033	16
MP3	1.064	17
MP4	1.110	13

[a]—the density of polymer films was determined by the hydrostatic weighing; fractional free-volume was calculated by Bondi's method [43] from the density of polymer films.

The state of polymers is another important characteristic that determines the performance of a polymer in many applications. According to the WAXD study, all the synthesized polymers are amorphous (Figure 9). WAXD patterns for **MP1**, **MP2**, **MP5** and **MP6** are represented by two broad signals. At the same time, for **MP3** and **MP4** the second peak was not observed and WAXD patterns are represented by one intensive broad peak at larger 2θ angles. The WAXD patterns of **MP1** were the same before and after heating the polymer sample, and the intensity of the peak in WAXD patterns only slightly changed after heating (Figure S43), confirming the amorphous nature of the polymer. It is worth noting that usually there is a single broad peak for conventional amorphous glassy polymers in their WAXD patterns. Two signals in WAXD patterns of **MP1**, **MP2**, **MP5** and **MP6** are not usual for amorphous glassy polymers. It was shown that the signals at higher 2θ angles are related to intrasegmental interactions and the signals at lower 2θ angles to intersegmental interactions [44,45]. Therefore, it seems that both intra- and inter-segmental interactions are significant in the case of **MP1**, **MP2**, **MP5** and **MP6**. Within this group of polymers, d_1- and d_2-distances are practically the same (Table 5). In turn, **MP3** and **MP4** had d_1-distances close to d_2-distances for **MP1**, **MP2**, **MP5** and **MP6** polymers. It should be noted, although there is no clear trend between WAXD data and real porosity, the polymers with larger d-spacings often have a more porous structure [46]. Thus, a more porous structure in the case of **MP1**, **MP2**, **MP5** and **MP6** can be expected.

Table 5. The intersegmental and intrasegmental distances (d, Å) evaluated with Bragg-formula for the prepared ROMP polymers.

Polymer	$2\theta_1$,°	d_1, Å	$2\theta_2$,°	d_2, Å
MP1	9.2	9.6	17.6	5.0
MP2	9.1	9.7	17.8	5.0
MP3	15.2	5.8	-	-
MP4	15.9	5.6	-	-
MP5	9.3	9.5	17.6	5.0
MP6	9.1	9.7	19.0	4.7

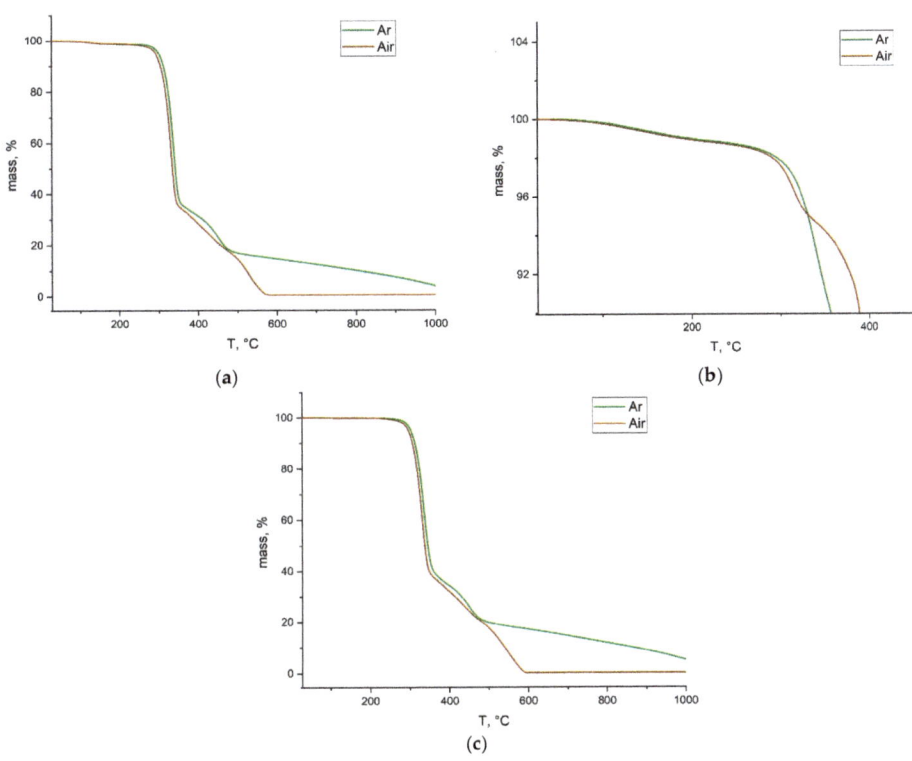

Figure 8. TGA curves for **MP1** (**a**), the enlarged TGA curves for **MP1** (**b**), and TGA curves for **MP6** (**c**).

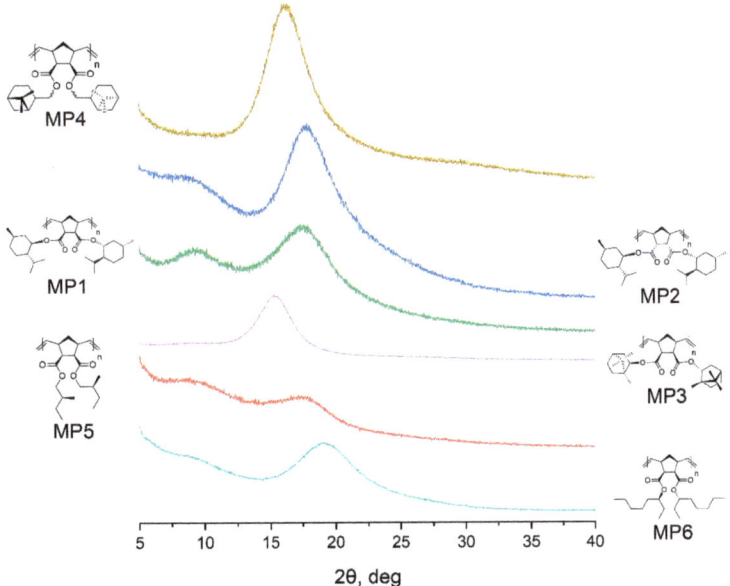

Figure 9. Wide-angle X-ray diffraction (WAXD) patterns for the prepared ROMP polymers.

229

The amorphous nature of the synthesized polymers, good film-forming properties, and high thermal stability, combined with the facile method of the synthesis, make these polymers attractive for the detailed study of their properties in the future as promising materials for dense enantioselective membranes and chiral stationary phases for chromatography.

4. Conclusions

A set of new optically active polymers was developed based on chiral norbornene-type monomers. The desired monomers were readily prepared in good yields by the esterification of *cis*-5-norbornene-2,3-dicarboxylic anhydride with alcohols from the renewable sources ((-)-menthol, (-)-borneol, and pinanol) and with commercially available alcohols (S-(-)-2-methylbutanol-1, S-(+)-3-octanol). ROMP polymerization of these monomers over the second-generation Grubbs catalyst afforded the metathesis polymers in good yields with high-molecular weights. The synthesized ROMP polymers showed high thermal stability. They had good film-forming properties, giving free-standing thin films from solutions suitable for studying membrane properties. The properties of the polymers strongly depended both on the nature of side groups and the type of their bonding with monomer units (*exo-/endo*-isomerism). Glass transition temperatures for the prepared polymers can be varied in a wide range (from $-30\ °C$ to $+139\ °C$) and, therefore, the state of the polymers can be changed from rubbery to glassy. Thus, it is possible to fine-tune the properties of these polymers to achieve the desired characteristics by changing the nature of the side chains. The prepared polymers represent a new attractive platform of chiral polymeric materials for enantioselective membrane separation and chiral stationary phases for chromatography.

Supplementary Materials: The following supporting information can be downloaded at: https://www.mdpi.com/article/10.3390/polym14245453/s1, Figures S1–S12: NMR spectra of monomers; Figures S13–S18: HRMS spectra of monomers; Figures S19–S30: NMR spectra of polymers; Figures S31–S35: CD spectra of monomers and polymers; Figures S36–S42: DSC and TGA curves for polymers; Figure S43: WAXD patterns of MP1; Table S1: Crystal data and structure refinement details for M1 and M2.

Author Contributions: Visualization, formal analysis, investigation, I.V.N.; investigation, D.P.Z.; investigation, I.A.S.; investigation, A.A.D.; investigation, Y.V.N.; investigation, I.R.I.; conceptualization, validation, methodology, and writing, M.V.B. All authors have read and agreed to the published version of the manuscript.

Funding: The design and synthesis of ROMP polymers were supported by the Russian Science Foundation (grant no. 20-13-00428). The preparation of pinanol was carried out within the State Program of TIPS RAS. X-ray diffraction data were collected using the equipment of Center for molecular composition studies of INEOS RAS with financial support from the Ministry of Science and Higher Education of the Russian Federation (Contract/agreement No. 075-00697-22-00).

Institutional Review Board Statement: Not applicable.

Informed Consent Statement: Not applicable.

Data Availability Statement: NMR, CD spectra, and TGA curves of the monomers and the polymers are available free of charge as a file of Supplementary Materials.

Acknowledgments: This work was performed using the equipment of the Shared Research Center «Analytical center of deep oil processing and petrochemistry of TIPS RAS».

Conflicts of Interest: The authors declare no conflict of interest.

References

1. Bolm, C.; Tanyeli, C.; Grenz, A.; Dinter, C.L. ROMP-Polymers in Asymmetric Catalysis: The Role of the Polymer Backbone. *Adv. Synth. Catal.* **2002**, *344*, 649–656. [CrossRef]
2. Abdelkawy, M.A.; Davin, C.; Aly, E.-S.A.; El-Badawi, M.A.; Itsuno, S. Chiral Polyureas Derived Cinchona Alkaloids: Highly Efficient Bifunctional Organocatalysts for the Asymmetric Michael Addition Reaction. *ChemistrySelect* **2021**, *6*, 11971–11979. [CrossRef]

3. Wali Ullah, M.; Haraguchi, N. Asymmetric Diels-Alder Reaction Catalyzed by Facile Recoverable Ionically Core-Corona Polymer Microsphere-Immobilized MacMillan Catalyst. *ChemistrySelect* **2022**, *7*, e202202568. [CrossRef]
4. Han, H.; Liu, W.; Xiao, Y.; Ma, X.; Wang, Y. Advances of enantioselective solid membranes. *New J. Chem.* **2021**, *45*, 6586–6599. [CrossRef]
5. Vedovello, P.; Marcio Paranhos, C.; Fernandes, C.; Elizabeth Tiritan, M. Chiral polymeric membranes: Recent applications and trends. *Sep. Purif. Technol.* **2022**, *280*, 119800. [CrossRef]
6. Choi, H.-J.; Ahn, Y.-H.; Koh, D.-Y. Enantioselective Mixed Matrix Membranes for Chiral Resolution. *Membranes* **2021**, *11*, 279. [CrossRef]
7. Guntari, S.N.; Nam, E.; Pranata, N.N.; Chia, K.; Wong, E.H.H.; Blencowe, A.; Goh, T.K.; Caruso, F.; Qiao, G.G. Fabrication of Chiral Stationary Phases via Continuous Assembly of Polymers for Resolution of Enantiomers by Liquid Chromatography. *Macromol. Mater. Eng.* **2014**, *299*, 1285–1291. [CrossRef]
8. Teixeira, J.; Tiritan, M.E.; Pinto, M.M.M.; Fernandes, C. Chiral Stationary Phases for Liquid Chromatography: Recent Developments. *Molecules* **2019**, *24*, 865. [CrossRef]
9. Betzenbichler, G.; Huber, L.; Kräh, S.; Morkos, M.-L.K.; Siegle, A.F.; Trapp, O. Chiral stationary phases and applications in gas chromatography. *Chirality* **2022**, *34*, 732–759. [CrossRef]
10. Wang, L.; Lu, N.; Huang, S.; Wang, M.; Chen, X.-M.; Yang, H. Optically Active Nucleobase-Functionalized Polynorbornenes Mimicking Double-Helix DNA. *CCS Chem.* **2021**, *3*, 1787–1796. [CrossRef]
11. Dana, D.M.; Yitzhak, M. Biomimetic Polymers for Chiral Resolution and Antifreeze Applications. In *On Biomimetics*; Lilyana, D.P., Ed.; IntechOpen: Rijeka, Croatia, 2011.
12. Green, J.J.; Elisseeff, J.H. Mimicking biological functionality with polymers for biomedical applications. *Nature* **2016**, *540*, 386–394. [CrossRef] [PubMed]
13. De Rosa, C.; Di Girolamo, R.; Talarico, G. Expanding the Origin of Stereocontrol in Propene Polymerization Catalysis. *ACS Catal.* **2016**, *6*, 3767–3770. [CrossRef]
14. Emmerich, A.; Daniliuc, C.G.; Studer, A. Synthesis of Polymers Bearing a Chiral Backbone via Stereospecific Ionic Ring-Opening Polymerization of Chiral Donor-Acceptor Cyclopropanes. *Macromol. Rapid Commun.* **2021**, *42*, 2100030. [CrossRef]
15. Mizuta, K.; Fukutomi, S.; Yamabuki, K.; Onimura, K.; Oishi, T. Ring-opening metathesis polymerization of N-substituted-5-norbornene-2,3-dicarboximides in the presence of chiral additives. *Polym. J.* **2010**, *42*, 534–539. [CrossRef]
16. Zang, Y.; Aoki, T.; Teraguchi, M.; Kaneko, T.; Ma, L.; Jia, H.; Miao, F. Synthesis of Well-Defined Chiral Oligopinanylsiloxane Graft Copoly(phenylacetylene)s Using the Macromonomer Method and Their Enantioselective Permeability. *ACS Appl. Polym. Mater.* **2020**, *2*, 853–861. [CrossRef]
17. Zhou, Y.; Zhang, C.; Zhou, Z.; Zhu, R.; Liu, L.; Bai, J.; Dong, H.; Satoh, T.; Okamoto, Y. Influence of different sequences of l-proline dipeptide derivatives in the pendants on the helix of poly(phenylacetylene)s and their enantioseparation properties. *Polym. Chem.* **2019**, *10*, 4810–4817. [CrossRef]
18. Daeffler, C.S.; Miyake, G.M.; Li, J.; Grubbs, R.H. Partial Kinetic Resolution of Oxanorbornenes by Ring-Opening Metathesis Polymerization with a Chiral Ruthenium Initiator. *ACS Macro Lett.* **2014**, *3*, 102–104. [CrossRef]
19. Ogba, O.M.; Warner, N.C.; O'Leary, D.J.; Grubbs, R.H. Recent advances in ruthenium-based olefin metathesis. *Chem. Soc. Rev.* **2018**, *47*, 4510–4544. [CrossRef]
20. Alentiev, D.A.; Bermeshev, M.V. Design and Synthesis of Porous Organic Polymeric Materials from Norbornene Derivatives. *Polym. Rev.* **2021**, *62*, 400–437. [CrossRef]
21. Ponkratov, D.O.; Shaplov, A.S.; Vygodskii, Y.S. Metathesis Polymerization in Ionic Media. *Polym. Sci. Ser. C* **2019**, *61*, 2–16. [CrossRef]
22. Petrov, V.A.; Vasil'ev, N.V. Synthetic chemistry of quadricyclane. *Curr. Org. Synth.* **2006**, *3*, 215–259. [CrossRef]
23. Houk, K.N.; Liu, F.; Yang, Z.; Seeman, J.I. Evolution of the Diels–Alder Reaction Mechanism since the 1930s: Woodward, Houk with Woodward, and the Influence of Computational Chemistry on Understanding Cycloadditions. *Angew. Chem. Int. Ed.* **2021**, *60*, 12660–12681. [CrossRef] [PubMed]
24. Guseva, M.A.; Alentiev, D.A.; Bermesheva, E.V.; Zamilatskov, I.A.; Bermeshev, M.V. The selective hydrosilylation of norbornadiene-2,5 by monohydrosiloxanes. *RSC Adv.* **2019**, *9*, 33029–33037. [CrossRef] [PubMed]
25. Sundell, B.J.; Lawrence, J.A.; Harrigan, D.J.; Lin, S.; Headrick, T.P.; O'Brien, J.T.; Penniman, W.F.; Sandler, N. Exo-selective, Reductive Heck Derived Polynorbornenes with Enhanced Molecular Weights, Yields, and Hydrocarbon Gas Transport Properties. *ACS Macro Lett.* **2020**, *9*, 1363–1368. [CrossRef]
26. Bermeshev, M.V.; Chapala, P.P. Addition polymerization of functionalized norbornenes as a powerful tool for assembling molecular moieties of new polymers with versatile properties. *Prog. Polym. Sci.* **2018**, *84*, 1–46. [CrossRef]
27. Bermesheva, E.V.; Wozniak, A.I.; Andreyanov, F.A.; Karpov, G.O.; Nechaev, M.S.; Asachenko, A.F.; Topchiy, M.A.; Melnikova, E.K.; Nelyubina, Y.V.; Gribanov, P.S.; et al. Polymerization of 5-Alkylidene-2-norbornenes with Highly Active Pd–N-Heterocyclic Carbene Complex Catalysts: Catalyst Structure–Activity Relationships. *ACS Catal.* **2020**, *10*, 1663–1678. [CrossRef]
28. Bermeshev, M.V.; Bulgakov, B.A.; Genaev, A.M.; Kostina, J.V.; Bondarenko, G.N.; Finkelshtein, E.S. Cationic Polymerization of Norbornene Derivatives in the Presence of Boranes. *Macromolecules* **2014**, *47*, 5470–5483. [CrossRef]
29. García-Loma, R.; Albéniz, A.C. Vinylic Addition Polynorbornene in Catalysis. *Asian J. Org. Chem.* **2019**, *8*, 304–315. [CrossRef]

30. Coles, M.P.; Gibson, V.C.; Mazzariol, L.; North, M.; Teasdale, W.G.; Williams, C.M.; Zamuner, D. Amino acid derived homochiral polymers via ring-opening metathesis polymerisation. *J. Chem. Soc. Chem. Commun.* **1994**, *21*, 2505–2506. [CrossRef]
31. Sutthasupa, S.; Terada, K.; Sanda, F.; Masuda, T. Ring-opening metathesis polymerization of amino acid-functionalized norbornene diester monomers. *Polymer* **2007**, *48*, 3026–3032. [CrossRef]
32. Rosebrugh, L.E.; Marx, V.M.; Keitz, B.K.; Grubbs, R.H. Synthesis of Highly Cis, Syndiotactic Polymers via Ring-Opening Metathesis Polymerization Using Ruthenium Metathesis Catalysts. *J. Am. Chem. Soc.* **2013**, *135*, 10032–10035. [CrossRef] [PubMed]
33. Choinopoulos, I.; Koinis, S.; Pitsikalis, M. Synthesis and characterization of chiral poly(l-lactide-b-hexyl isocyanate) macromonomers with norbornenyl end groups and their homopolymerization through ring opening metathesis polymerization to afford polymer brushes. *J. Polym. Sci. Part A Polym. Chem.* **2017**, *55*, 1102–1112. [CrossRef]
34. Fadlallah, S.; Peru, A.A.M.; Flourat, A.L.; Allais, F. A straightforward access to functionalizable polymers through ring-opening metathesis polymerization of levoglucosenone-derived monomers. *Eur. Polym. J.* **2020**, *138*, 109980. [CrossRef]
35. Fadlallah, S.; Peru, A.A.M.; Longé, L.; Allais, F. Chemo-enzymatic synthesis of a levoglucosenone-derived bi-functional monomer and its ring-opening metathesis polymerization in the green solvent Cyrene™. *Polym. Chem.* **2020**, *11*, 7471–7475. [CrossRef]
36. Souto, J.A.; Stockman, R.A.; Ley, S.V. Development of a flow method for the hydroboration/oxidation of olefins. *Org. Biomol. Chem.* **2015**, *13*, 3871–3877. [CrossRef]
37. Sheldrick, G. Crystal structure refinement with SHELXL. *Acta Crystallogr. Sect. C* **2015**, *71*, 3–8. [CrossRef]
38. Dolomanov, O.V.; Bourhis, L.J.; Gildea, R.J.; Howard, J.A.K.; Puschmann, H. OLEX2: A complete structure solution, refinement and analysis program. *J. Appl. Crystallogr.* **2009**, *42*, 339–341. [CrossRef]
39. Sheldrick, G. A short history of SHELX. *Acta Crystallogr. Sect. A* **2008**, *64*, 112–122. [CrossRef]
40. Bolm, C.; Atodiresei, I.; Schiffers, I. Asymmetric alcoholysis of meso-anhydrides mediated by alkaloids. *Org. Synth.* **2005**, *82*, 120–125. [CrossRef]
41. Bermesheva, E.V.; Nazarov, I.V.; Kataranova, K.D.; Khrychikova, A.P.; Zarezin, D.P.; Melnikova, E.K.; Asachenko, A.F.; Topchiy, M.A.; Rzhevskiy, S.A.; Bermeshev, M.V. Cocatalyst versus precatalyst impact on the vinyl-addition polymerization of norbornenes with polar groups: Looking at the other side of the coin. *Polym. Chem.* **2021**, *12*, 6355–6362. [CrossRef]
42. Yushkin, A.; Grekhov, A.; Matson, S.; Bermeshev, M.; Khotimsky, V.; Finkelstein, E.; Budd, P.M.; Volkov, V.; Vlugt, T.J.H.; Volkov, A. Study of glassy polymers fractional accessible volume (FAV) by extended method of hydrostatic weighing: Effect of porous structure on liquid transport. *React. Funct. Polym.* **2015**, *86*, 269–281. [CrossRef]
43. Askadskii, A.A. *Computational Materials Science of Polymers*; Cambridge International Science Publishing Ltd.: Cambridge, UK, 2003; p. 16.
44. Wilks, B.R.; Chung, W.J.; Ludovice, P.J.; Rezac, M.R.; Meakin, P.; Hill, A.J. Impact of average free-volume element size on transport in stereoisomers of polynorbornene. I. Properties at 35 °C. *J. Polym. Sci. Part B Polym. Phys.* **2003**, *41*, 2185–2199. [CrossRef]
45. Wilks, B.R.; Chung, W.J.; Ludovice, P.J.; Rezac, M.E.; Meakin, P.; Hill, A.J. Structural and free-volume analysis for alkyl-substituted palladium-catalyzed poly(norbornene): A combined experimental and Monte Carlo investigation. *J. Polym. Sci. Part B Polym. Phys.* **2006**, *44*, 215–233. [CrossRef]
46. Gringolts, M.; Bermeshev, M.; Yampolskii, Y.; Starannikova, L.; Shantarovich, V.; Finkelshtein, E. New High Permeable Addition Poly(tricyclononenes) with Si(CH3)3 Side Groups. Synthesis, Gas Permeation Parameters, and Free Volume. *Macromolecules* **2010**, *43*, 7165–7172. [CrossRef]

Article

Self-Healable, Strong, and Tough Polyurethane Elastomer Enabled by Carbamate-Containing Chain Extenders Derived from Ethyl Carbonate

Pengcheng Yi, Jingrong Chen, Junyao Chang, Junbo Wang, Ying Lei, Ruobing Jing, Xingjiang Liu, Ailing Sun, Liuhe Wei * and Yuhan Li *

Zhengzhou Key Laboratory of Elastic Sealing Materials, College of Chemistry and Green Catalysis Center, Zhengzhou University, Zhengzhou 450001, China; yipengcheng@gs.zzu.edu.cn (P.Y.); cjrkate@stu.zzu.edu.cn (J.C.); cjyailyh1212@stu.zzu.edu.cn (J.C.); 202023000516@stu.zzu.edu.cn (J.W.); 202023000403@stu.zzu.edu.cn (Y.L.); 202023000105@stu.zzu.edu.cn (R.J.); xingjiangliu@zzu.edu.cn (X.L.); ailingsun@zzu.edu.cn (A.S.)
* Correspondence: weiliuhe@zzu.edu.cn (L.W.); liyuhan@zzu.edu.cn (Y.L.)

Citation: Yi, P.; Chen, J.; Chang, J.; Wang, J.; Lei, Y.; Jing, R.; Liu, X.; Sun, A.; Wei, L.; Li, Y. Self-Healable, Strong, and Tough Polyurethane Elastomer Enabled by Carbamate-Containing Chain Extenders Derived from Ethyl Carbonate. *Polymers* 2022, 14, 1673. https://doi.org/10.3390/polym14091673

Academic Editors: Valentina Siracusa, Nadia Lotti, Michelina Soccio and Alexey Iordanskii

Received: 17 March 2022
Accepted: 15 April 2022
Published: 20 April 2022

Publisher's Note: MDPI stays neutral with regard to jurisdictional claims in published maps and institutional affiliations.

Copyright: © 2022 by the authors. Licensee MDPI, Basel, Switzerland. This article is an open access article distributed under the terms and conditions of the Creative Commons Attribution (CC BY) license (https://creativecommons.org/licenses/by/4.0/).

Abstract: Commercial diol chain extenders generally could only form two urethane bonds, while abundant hydrogen bonds were required to construct self-healing thermoplastic polyurethane elastomers (TPU). Herein, two diol chain extenders bis(2-hydroxyethyl) (1,3-pheny-lene-bis-(methylene)) dicarbamate (BDM) and bis(2-hydroxyethyl) (methylenebis(cyclohexane-4,1-diy-l)) dicarbamate (BDH), containing two carbamate groups were successfully synthesized through the ring-opening reaction of ethylene carbonate (EC) with 1,3-benzenedimetha-namine (MX-DA) and 4, 4′-diaminodicyclohexylmethane (HMDA). The two chain extenders were applied to successfully achieve both high strength and high self-healing ability. The BDM-1.7 and BDH-1.7 elastomers had high comprehensive self-healing efficiency (100%, 95%) after heated treatment at 60 °C, and exhibited exceptional comprehensive mechanical performances in tensile strength (20.6 ± 1.3 MPa, 37.1 ± 1.7 MPa), toughness (83.5 ± 2.0 MJ/m^3, 118.8 ± 5.1 MJ/m^3), puncture resistance (196.0 mJ, 626.0 mJ), and adhesion (4.6 MPa, 4.8 MPa). The peculiar mechanical and self-healing properties of TPUs originated from the coexisting short and long hard segments, strain-induced crystallization (SIC). The two elastomers with excellent properties could be applied to engineering-grade fields such as commercial sealants, adhesives, and so on.

Keywords: carbamate-containing chain extenders; high strength; high self-healing ability; adhesive properties

1. Introduction

Artificial polymer materials are often unusable or have their service life greatly reduced after physical damage, resulting in environmental pollution and a great waste of resources [1–3]. Studies have found that almost all natural organisms can spontaneously heal themselves after physical injury, greatly increasing their survival capacity and life span [4,5]. The salamander, for example, has an amazing ability of self-healing and can heal almost any organ [6]. Based on the inspiration that living things can heal themselves, people have been devoted efforts to the research of self-healing polymer materials in recent decades, which can increase the service life of materials, preserve their stability, reduce the maintenance cost, and further relieve the pressure of resource waste and environmental pollution [7,8].

In terms of the healing mechanism, the healing modes of polymer materials can be divided into external healing and internal healing. The latter has attracted more attention given its repeatable self-healing and low influence on material properties. Internal healing introduces dynamic reversible covalent bond [9–15] or non-covalent bond [16–19] into the polymer chain to achieve self-healing more efficiently. Internally healed polymer materials are widely used considering their high-level applicability and mass production [20–22] while polyurethane elastomer has become one of the most widely used polymer materials by virtue of its excellent properties [23–25]. Self-healing polyurethane materials are widely applied in electronic skin, intelligent sensors, biomedical materials, and many other areas, where the mechanical properties of self-healing polyurethane materials are generally poor. However, in some sealants and commercial coatings [26], polyurethane materials are required to be self-healing and have high requirements for mechanical properties such as strength and toughness. Therefore, it is of great significance to develop self-healing polyurethane elastomers with a strong self-healing ability, as well as excellent strength and toughness performance in these fields.

It is undeniable that the mobility of chain segments is the most critical factor in promoting self-healing in the internal healing mode of polyurethane elastomers [2,27]. High self-healing efficiency of elastomers has poor mechanical properties in most cases [28–31]. The strength rise of polyurethane elastomer is generally accompanied by the content rise of hard segments, forming more physical cross-linking points between soft segments. It is not conducive to self-healing when the movement of the chain segments is limited. It composes a paradox that improvement in mechanical properties is followed by a decline in self-healing efficiency. Therefore, it is still challenging to find rational molecular chain design and innovative strategies to resolve such a contradictory relation [4,32].

An innovative strategy was proposed to develop polyurethane elastomers with high strength self-healing ability, and solve the contradiction between mechanical properties and the self-healing ability of TPU. Polytetramethylene ether glycol (PTMEG) was selected as the polymeric diol because its SIC would produce a mechano-response self-strengthening effect, which will further synchronously strengthen and toughen the elastomer [33–36]. Carbamate-containing chain extenders, BDM and BDH, were successfully synthesized from inexpensive raw materials. BDM and BDH were incorporated into the polyurethane backbone, with high steric hindrance benzene and cyclohexane. These large steric groups made the whole hard segments loosely-stacked, which rendered the chain segments strong fluidity, and endowed the elastomers with high self-healing performance. The special design of the long and short hard segments in BDM and BDH elastomers facilitated the formation of SIC and introduced suitable amounts of hydrogen bonds, which granted TPU high mechanical properties.

2. Experimental

2.1. Materials

PTMEG (M_n = 1000 g/mol, f = 2), dibutyltin dilaurate (DBTDL), and ethylene glycol (EG) were purchased from Aladdin (Shanghai, China). EC, MXDA, HMDA and 4,4′-dicyclohexylmethane diisocyanate (HMDI) were from Macklin (Shanghai, China). None of the above chemicals were further purified. Both toluene and tetrahydrofuran (THF) were redistilled by CaH_2 for use.

2.2. Synthesis of BDM, BDH

BDM and BDH were synthesized from EC, MXDA, and HMDA via ring-opening reactions. At room temperature, EC and MXDA or HMDA were mixed thoroughly in a 250-mL flask in a molar ratio of 2.05/1 (MXDA strongly reacting with EC and should be added in batches). The temperature of the reaction system was raised to 120 °C, when the reaction would last for 12 h. To ensure the full reaction of MXDA and HMDA, the amount of EC was slightly excessive during such a process. After the reaction, the product was correspondingly dissolved, recrystallized, filtered and dried to get BDM and BDH powder (Figures S1 and S2 from Supplementary Materials). FTIR (Figures S3a and S4a, Tables S1 and S2) and ^{1}HNMR (Figures S3b and S4b) proved the successful synthesis of BDM and BDH.

2.3. Synthesis of Polyurethane Elastomers Using BDM and BDH as Chain Extenders

Two types of linear polyurethane elastomers with different R values (defined as the molar ratio of isocyanate group to hydroxyl group in PTMEG[NCO]/[OH]) were synthesized via the same process. Take the synthesis process of BDM-2.0 elastomer as an example: firstly, PTMEG (20.0 g, 20.0 mmol) was added into a 250-mL four-neck flask with a mechanical stirrer and a mercury thermometer, then heated at 120 °C with an electric heating jacket for 1 h under vacuum to remove the residual moisture in PTMEG, and cooled to 60 °C after that. HMDI (10.5 g, 40.0 mmol) was added into the flask and stirred at 80 °C for 1 h under argon gas protection. Then, DBTDL (about 2.5 mg, 4.0×10^{-3} mmol) was added and reacted at 80 °C for 2 h to obtain the prepolymer. After that, BDM (6.2 g, 20.0 mmol) was added, and then 70 mL of THF was added to the system and stirred at 60 °C for about 0.5 h (to adjust the viscosity and dissolve BDM powder). After BDM was completely dissolved, 100 mL of toluene solution and DBTDL (25.3 mg, 4.0×10^{-2} mmol) were added to react at 80 °C for 2 h, so that all NCO groups could be completely consumed. The product was poured into the prepared rectangular mold, dried at 80 °C for 48 h, and then dried in a vacuum drying oven at 80 °C for 24 h to remove the residual solvent. The obtained polyurethane film (1 ± 0.2 mm) was labeled as BDM-2.0. With EG as chain extender and R = 1.7, the polymer was synthesized and labeled as EG-1.7.

2.4. Tensile Tests

Mechanical properties including puncture, tensile strength, elongation at break and toughness were tested by a 500 N load sensor tension tester (TH-8203A, TOPHUNG Inc., Suzhou, China): while dumbbell (DIN 53504, Type S1) samples were cut from the prepared elastomer film (1 ± 0.2 mm) and tensile tests were also performed at a constant rate of 100 mm/min at room temperature. A steel needle was mounted on the same tensile testing machine, and the puncture test was carried out at a constant speed of 50 mm/min until about 0.5 mm of the film was penetrated. Puncture energy was calculated by integral of force–displacement curve. At room temperature, the spline was stretched to 300% strain at a constant rate of 50 mm/min. After five cycles of continuous loading and unloading, the cyclic stretching curve could be obtained after staying at room temperature for 12 h and after one recirculating. Self-healing performance was evaluated by tensile test according to the following methods: at least five dumbbell samples were cut with a razor in the middle and splice immediately. The tensile test was carried out after a certain period of treatment at different temperatures. Self-healing efficiency (S.E.) was calculated by tensile strength, elongation and toughness: $(P_{healed}/P_{original}) \times 100\%$, where P is tensile strength, elongation at break or toughness.

2.5. Adhesion Tests

Tensile shear strength tests were conducted by a 100 KN load sensor tension tester (TH-8100A, TOPHUNG Inc., Suzhou, China) with a 100 mm × 25 mm × 0.8 mm steel plate lap whose, lap area was 25 mm × 12.5 mm. Not less than 5 samples were heated in the oven at 80 °C and 150 °C for a certain time, then adjusted 24 h to the equilibrium state at room temperature and tested at a constant speed of 50 mm/min. Tensile shear strength $\tau = F/B$ (unit: MPa), where F is the maximal shear failure load (N) and B is the lap area (mm^2).

2.6. Characterizations

The film samples were under FTIR test in the attenuated total reflection (ATR) mode using a Bruker ALPHA II (Baden, Württemberg, Germany) Fourier transform infrared spectrometer. The resolution was 4 cm^{-1}, the amount of scanning, 32; and the spectral scanning, ranging from 4000 to 500 cm^{-1}. Nicolet 6700 produced by Thermo Electron Company (Waltham, MA, USA) in the United States was used for variable-temperature FTIR experiments. The test temperature was from 25 to 105 °C while the interval temperature was 10 °C, and the residence time was 3 min. The ^{1}HNMR test was performed by Bruker Avance 400 MHz (Baden, Württemberg, Germany), DMSO and CDCl$_3$ as solvent. Gel permeation chromatography (GPC, Agilent LC1200, Agilent Technologies, Santa Clara, CA, USA) evaluated molecular weight and polydispersion index with THF as a mobile phase. DSC measurements were carried out using TA-DSC250 (TA Instruments, New Castle, DE, USA) at a heating rate of 10 °C/min in a nitrogen atmosphere. During the first operation, the temperature dropped from to −50 °C for 3 min, and then increased to 150 °C and kept for 3 min. For the second run, the temperature was lowered from 150 to −50 °C and finally to 150 °C. Each sample was scanned at the rate of 10 °C/min. Polarization optical microscope (POM, Olympus BX61, Olympus, Tokyo, Japan) was used to observe the surface self-healing of the samples. Small Angle X-ray scattering (SAXS) was performed by Bruker-Anaostar (Bruker, Karlsruhe, Germany) at 50 KV and 0.6 mA. The radiation source was Cu Kα, λ= 1.54056 Å, and the distance between the sample and the detector was 450 mm. Rheological properties were tested using a rotational rheometer (Malvern, London, UK) with a 20 mm circular membrane. Frequency sweeping was at a strain amplitude of 0.1%, in the range of 0.1–100 Hz (90 °C). Temperature sweeping was in the range of 25 to 150 °C.

3. Results and Discussion

3.1. Molecular Design

The two chain extenders with carbamate groups, BDM and BDH, were synthesized by ring-opening reaction (Figure 1a): given that their benzene ring and two cyclohexyl groups, which possessed high steric resistance and loosely-stacked hard segments structure, could be formed in the elastomer, their carbamate groups within the synthesized chain extenders also introduce further more hydrogen bonds. The special structure of these two chain extenders laid a foundation for the preparation of high mechanical properties and high self-healing elastomers. BDM and BDH were skillfully introduced into TPU by the two-step prepolymer method. In the first step, HMDI reacted with PTMEG to generate prepolymers and form short hard segments. The specific feed ratio of HMDI/PTMEG was showed in Tables S3 and S4. In the second step, BDM and BDH, as chain extenders, formed loosely-stacked short hard segments and long hard segments (Figure 1b,c). FTIR (Figure S5a,b and Tables S5 and S6), ^{1}HNMR (Figure S6a,b) and GPC (Figure S7a,b and Tables S7 and S8) proved that the above methods could successfully synthesize BDM-based, BDH-based and EG-1.7 thermoplastic polyurethane elastomers.

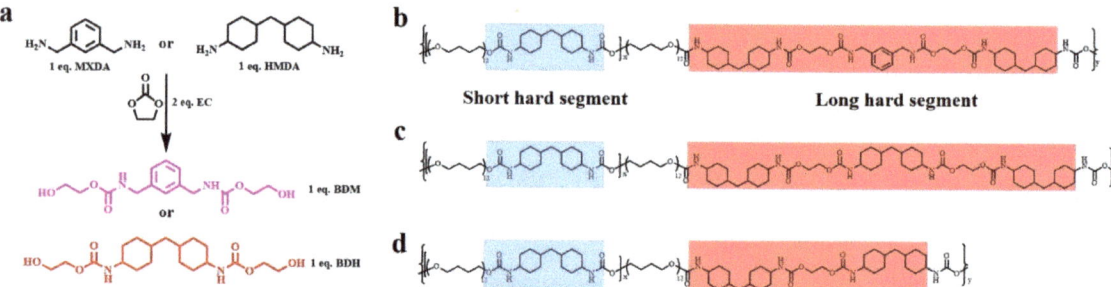

Figure 1. (**a**) Synthetic routes of BDM and BDH; (**b**) Molecular structure of BDM-1.7 elastomers embedded with short and long hard segments; (**c**) Molecular structure of BDH-1.7 elastomers embedded with short and long hard segments; (**d**) Molecular structure of EG-1.7 elastomers embedded with short and long hard segments.

Polyurethane elastomer contained polar groups such as carbamate. Due to its large cohesion, hydrogen bonds could be formed between molecules and spontaneously aggregate into a hard phase region, while the chain segments of polyesters or polyether diols with weak polarity aggregated into a soft phase region. The hard phase and soft phase were thermodynamically incompatible, resulting in a special phase morphology-microscopic phase separation. The special phase morphology of BDM-1.7 and BDH-1.7 elastomers was characterized by SAXS. As showed in Figure 2a, an obvious scattering peak existed inevitably due to different electron densities of the soft phase and the hard phase. The results showed that there was a microphase separation structure between these two phases, when the hard phase acted as physical cross-linking points within the soft phase, which enhances BDM-1.7 and BDH-1.7 elastomer strength.

BDM-1.7 and BDH-1.7 elastomers were tested by variable-temperature FTIR, and the test results were shown in Figure 2b,c. The characteristic peaks of the C=O group in BDM-1.7 and BDH-1.7 were subjected to the same trend with the temperature. BDM-1.7 spectrum was taken as an example to explain that, from 25 to 105 °C, the intensity of C=O characteristic peak (1716 cm^{-1}) in the free state of amide I gradually increased, while the trend of C=O characteristic peak (1705 cm^{-1}) in the hydrogen bond association state of amide I performed opposite, indicating the dynamic reversibility of hydrogen bonds in BDM-based and BDH-based elastomers with the increase of temperature.

3.2. Mechanical Properties

Next, the mechanical properties of BDM-based and BDH-based elastomers were discussed from the aspects of tensile, puncture, and cyclic tensile experiment, showing that the hard phase possessed the advantage of both short and long hard segment structure. Stress-strain curves of the BDM-based and BDH-based elastomers were shown in Figure 2d,e, and specific data were shown in Tables S9 and S10. It was obvious that the elongation at break decreased with increasing R values while the ultimate tensile strength augmented with increasing R values. It also showed that the content of the long hard segment greatly influenced strain hardening. The tensile strength of BDM-1.7 and BDH-1.7 elastomers was found to be significantly increased, due to the SIC effect of soft phase PTMEG chain in synthetic TPU elastomers, which played a positive role in the strengthening and toughening. The tensile strength of BDM-1.4 and BDH-1.1 was only 0.61 ± 0.01 MPa and 0.19 ± 0.03 MPa, respectively. Although the molecular weight of synthesized TPU was high, the low content of long hard segments made the amounts of physical cross-linking points too small (Tables S3 and S4), and the density was not enough to maintain the integrity of polyurethane network structure during the tensile process. The chain segment would be broken when the strain was low. The tensile strength of BDH-1.7 and BDH-1.7 elastomers was 20.55 ± 1.3 MPa and 37.07 ± 1.7 MPa, and the elongation at break was 1667 ± 56% and 1128 ± 53%, respectively. The special design of the long and short hard segments in BDM-1.7 and BDH-1.7 elastomers facilitates the formation of SIC and introduced suitable amounts of hydrogen bonds. At this time, the elastomer possessed a high content of long hard segments, forming a large number of physical cross-linking points, so that the whole polymer network could restrict chain migration in the tensile process, resulting in higher mechanical properties.

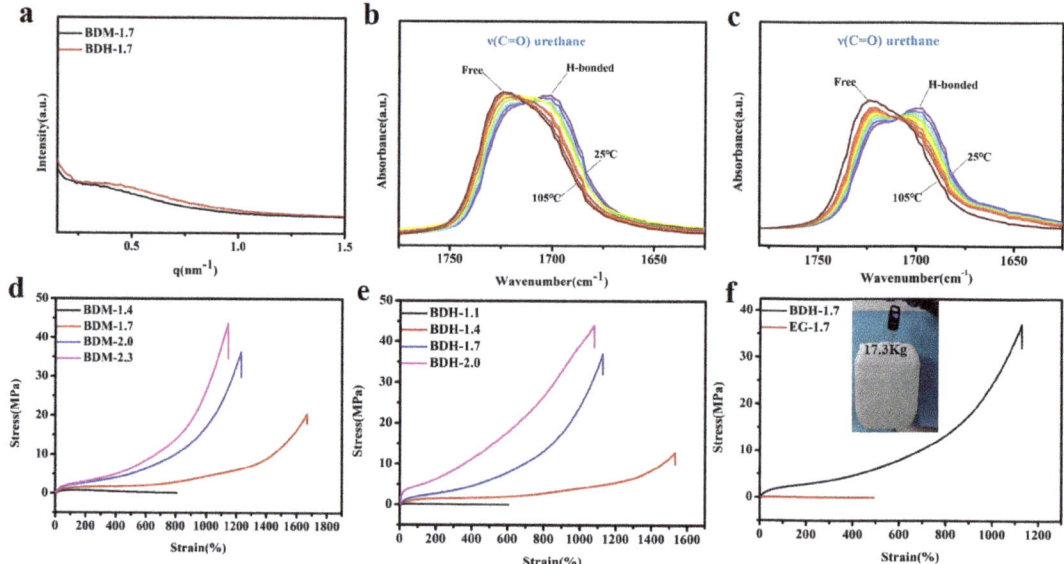

Figure 2. (**a**) SAXS results of BDM-1.7, BDH-1.7 elastomers; (**b**) Variable-temperature FTIR spectra of BDM-1.7; (**c**) Variable-temperature FTIR spectra of BDH-1.7; (**d**) Stress-strain curves of BDM-based elastomers; (**e**) Stress-strain curves of BDH-based elastomers; (**f**) Stress-strain curves of BDH-1.7, EG-1.7 elastomers and a tape of BDH-1.7 elastomer lifting a heavy bucket.

To intuitively highlight the advantages of the special structures of BDM and BDH in enhancing the mechanical properties of polyurethane elastomers, EG-1.7 elastomer was synthesized, with EG, which had no carbamates and a low steric resistance, as a chain extender. Figure 2f showed the stress-strain curves of BDH-1.7 and EG-1.7 elastomers. EG-1.7 had a high molecular weight (Table S6), but its tensile strength was only 0.13 ± 0.02 MPa, since the content of the hard segments of EG-1.7 elastomer was low, among which, the content of long hard segment (Figure 1d) was particularly low (Table S4). At this time, the amounts of physical cross-linking points were too small, and the density of physical cross-linking points was not enough to maintain the integrity of the polyurethane network structure during the tensile process. It would be broken when the strain was low, since BDH-1.7 elastomer possessed a high content of long hard segments that provided more physical cross-linking points and contained rigid cyclohexyl structure and carbamate groups which mattered considerably in enhancing the mechanical properties, the tensile strength of BDH-1.7 elastomer was 37.07 ± 1.7 MPa, much higher than EG-1.7. It was found that the special structure of BDM and BDH contributed for improving the mechanical properties of polyurethane elastomers. Benefited from strain hardening, BDH-1.7 elastomer is provided with a high strength and BDH-1.7 tape (thickness of 1.3 mm) that could lift a bucket weighing 17.3 kg (Figure 2f).

Puncture resistance is defined as material ability to avoid the puncture of a sharp needle. BDM-based and BDH-based elastomers possess excellent puncture resistance. Their puncture experiments on the tensile testing machines could obtain force-displacement curves (Figure 3a,c). It could be seen from the above data that with the increased of R value, the maximum puncture force of BDM-based and BDH-based elastomers increased, while the maximum puncture displacement decreased. When the R value increased, the number of hydrogen bonds in the elastomer increased correspondingly, making it difficult for the molecular chain to move. At this time, when the steel needle penetrated the film, it presented a higher penetration force and higher penetration resistance. The maximal puncture force of BDH-2.0 was 39.2 N, slightly lower than that of BDH-1.7. This might result from the fact that the hard segment content was too high when the R value of BDH elastomer was 2.0, which gave rise to too many hydrogen bonds that limited the chain movement. This puncture resistance of BDH-2.0 elastomer was weakened, resulting in a short maximal puncture displacement and failure to achieve the desired puncture force. The results showed that with the increase of R value, the rise of BDM and BDH content was the key to improving puncture resistance. Puncture energy could be obtained by integrating the force-displacement curves (Figure 3b,d). It could intuitively reflect the energy dissipation capacity during puncture deformation. The maximal puncture energy of the BDM-1.7 elastomer was 196.0 mJ, and that of the BDH-1.7 elastomer was 627.0 mJ. BDM-1.7 and BDH-1.7 elastomers were subjected to cyclic tensile test (Figure 3e,f) for their resilience. After staying at room temperature for 12 h, the dissociated hydrogen bonds were re-associated, and the last cycle curve was close to the first one, indicating that BDM-1.7 and BDH-1.7 elastomers have high resilience which could also prove the dynamic reversibility of hydrogen bonds. It was also found that BDM-1.7 and BDH-1.7 elastomers had a high strength, high puncture resistance and high resilience, laying a good foundation for their practical application.

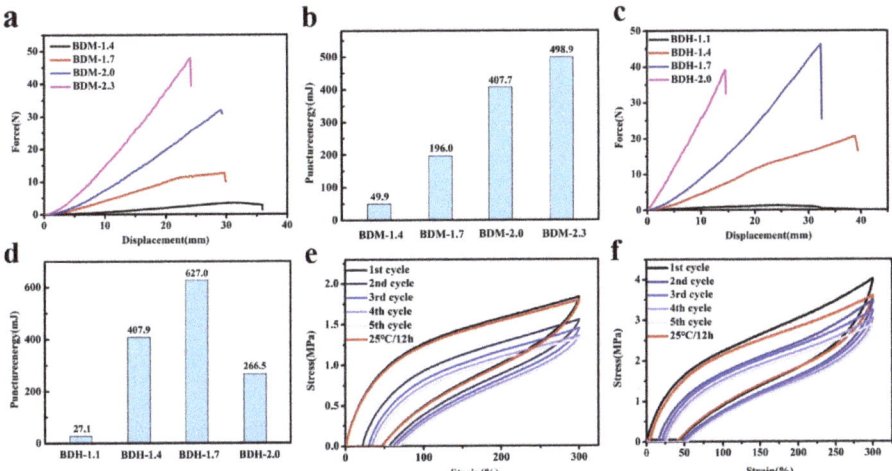

Figure 3. (**a**) Force-displacement curves for BDM-based elastomer; (**b**) Puncture energies of BDM-based elastomers calculated from force-displacement curves; (**c**) Force-displacement curves for BDH-based elastomer; (**d**) Puncture energies of BDH-based elastomers calculated from force-displacement curves; (**e,f**) Cyclic test curves of BDM-1.7, BDH-1.7 elastomer.

3.3. Self-Healing Properties

According to the DSC results (Figure 4a), the T_g of hard segment of BDM-1.7 and BDH-1.7 elastomers were 30.4 and 29.8 °C, respectively. Theoretically, when the temperature was higher than the T_g temperature, these two kinds of elastomers were in high elastic state and had certain self-healing ability. Both DSC and variable-temperature FTIR results indicate that BDM-based and BDH-based elastomers may have self-healing ability after heating. To confirm the self-healing ability of BDM-based and BDH-based elastomers after heating, the elastomer specimens of BDM-1.7 and BDH-1.7 were cut and spliced together immediately. After heating treatment, the BDM-1.7 and BDH-1.7 elastomers were subjected to suspension experiments (Figure S8a,b). It was found that these two types of elastomers could successfully withstand a weight of 500 g after the healing with a short period of heat treatment. The results showed that BDM-1.7, BDH-1.7 elastomer did have a strong self-healing ability after heat treatment. The scratches before and after the self-healing of BDM-1.7 and BDH-1.7 elastomers were observed with an optical microscope (Figure 4b,c), and the self-healing ability of BDM-1.7 and BDH-1.7 elastomers could be more clearly marked. It could be seen that scratches would almost disappear completely after heating at 60 °C for 24 h. It further indicated that BDM-1.7 and BDH-1.7 elastomers possessed a strong self-healing ability.

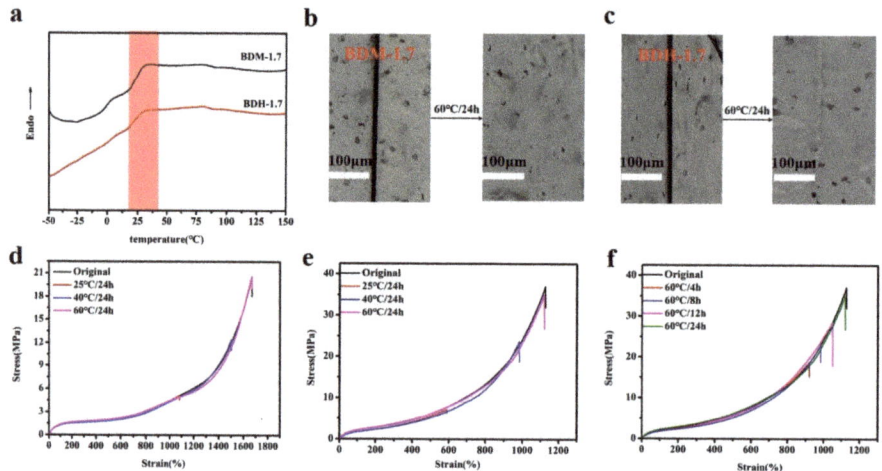

Figure 4. (**a**) DSC curves of BDM-1.7, BDH-1.7 elastomers; (**b**) and (**c**) BDM-1.7, BDH-1.7 elastomers self-healing at 60 °C for 24 h, change of scratches; (**d**) and (**e**) Stress-strain curves of BDM-1.7, BDH-1.7 elastomers self-healing at 25, 40, and 60 °C for 24 h; (**f**) Stress-strain curves for BDH-1.7 elastomers self-healing at 60 °C for 4, 8, 12, and 24 h.

To further quantify the self-healing efficiency of BDM-based and BDH-based elastomers, specimens of BDM-1.7 and BDH-1.7 were cut in the middle and quickly spliced together. They were heat-treated in ovens at 25, 40, and 60 °C, and tensile tests were conducted to obtain the corresponding stress-strain curves (Figure 4d,e). Herein, considering the strain hardening of the SIC effect, it should be that self-healing efficiency was synthetically evaluated by the repairing of tensile strength, elongation and toughness instead of an individual one. According to their self-healing data, the self-healing efficiency of the elastomer increased with the increase of temperature. When BDM-1.7 and BDH-1.7 elastomers were treated at 60 °C for 24 h, the self-healing efficiency (the tensile strength, elongation at break and toughness) of BDM-1.7 elastomers reached nearly 100%, and the self-healing efficiency(the tensile strength, elongation at break and toughness) of BDH-1.7 elastomers reached more than 95% (Tables S11 and S12). The results showed that under these conditions, the special phase morphology of long and short hard segment in elastomer and the design of BDM and BDH structure endowed the elastomer with high mechanical properties and the chain segment highly fluidity. Under the heating condition, the fluidity of the chain segment of the elastomer increased, and new hydrogen bonds were rapidly formed at the scratches, making the scratches remerge quickly. When healing time increased, the self-healing ability was further enhanced, and the mechanical properties were basically restored. BDM-1.7 and BDH-1.7 elastomers also possessed a high toughness, reaching 85.5 ± 2.0 MJ/m^3 and 118.8 ± 5.1 MJ/m^3, respectively. The strength and toughness of BDM-2.0, BDM-2.3, and BDH-2.0 with a higher R value were greatly improved, but the self-healing ability was decreased instead, indicating that the mechanical properties and self-healing efficiency of BDM-based and BDH-based elastomers could be regulated by changing the R value. BDH-1.7 was heated at 60 °C for 4, 8, and 12 h to obtain the stress-strain curve (Figure 4f). The results showed that the self-healing ability of BDH-1.7 elastomer was significantly improved with the prolongation of the healing time.

To further analyze the self-healing mechanism of BDM-1.7 and BDH-1.7 elastomers, rheological tests were therefore carried out, and their rheological curves were plotted at 90 °C (Figure 5a,c). The relaxation time (τ_f) of the whole chain flow was calculated from the reciprocal of G'=G" used to indicate the size of chain mobility. The relaxation time of BDM-1.7 and BDH-1.7 was 34 and 18 s, respectively, extremely short, which means that BDM-1.7 and BDH-1.7 both possessed an extremely high chain mobility at 90 °C. According to the results of the self-healing experiment, BDM-1.7 and BDH-1.7 elastomers had a higher self-healing efficiency at 60 °C after the heat treatment of 24 h. These results indicated that the hard segments of BDM-1.7 and BDH-1.7 elastomers were loosely-stacked and had a high chain mobility at 60 °C. It also represented a rapid rearrangement of their hard segments after heating. The increase of self-healing time and temperature improved the corresponding efficiency, which could be obviously explained by the increase of segment mobility of BDM-based and BDH-based elastomers.

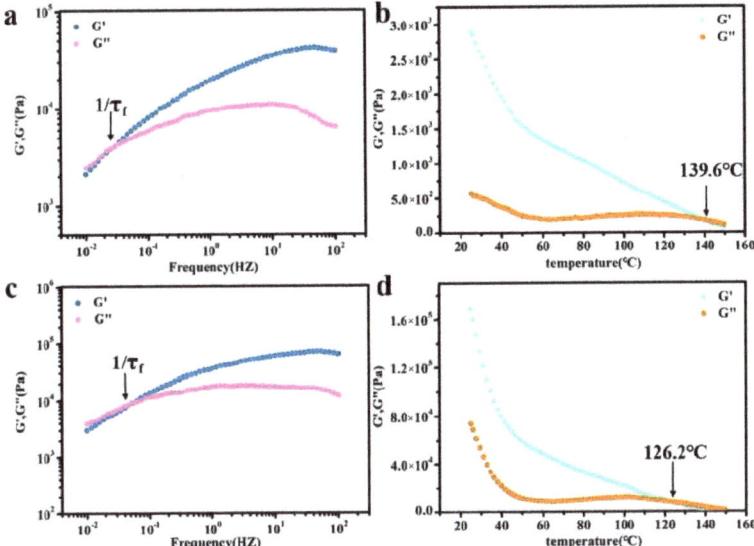

Figure 5. (a) Rheological curves of BDM-1.7 elastomers (90 °C) for deducing relaxation time of chain flow transition τ_f; (b) Temperature sweep of rheological testing for the BDM-1.7 elastomer; (c) Rheological curves of BDH-1.7, elastomers (90 °C) for deducing relaxation time of chain flow transition τ_f; (d) Temperature sweep of rheological testing for the BDH-1.7 elastomer.

3.4. Adhesion Properties

In the above tests, BDM-based and BDH-based elastomers were found to be soft and viscous when heated, similar to adhesives. The adhesion properties of these two types of elastomers were tested. They were heated in an oven at 80 °C for 8 h and 150 °C for 0.5 h, respectively, and then adjusted at room temperature for 24 h for the tensile shear test. The shear strength-displacement curves (Figure 6a,c) and shear strength (Figure 6b,d) of BDM elastomer at different temperatures were obtained, and the shear strength-displacement curves (Figure 6g,i) and shear strength (Figure 6h,j) of BDH elastomer at different temperatures were obtained. The adhesion performance of BDM-based and BDH-based elastomers at 80 °C was much weaker than that at 150 °C, since their melting temperature was higher. It could be observed in the temperature scanning curves (Figure 5b,d) that the melting temperatures were 139.6 and 126.2 °C, respectively. The elastomers become soft when treated at 80 °C and could not nicely adhere to the substrate, so the shear strength was low. When heated at 150 °C for 0.5 h, large amounts of hydrogen bonds could interact with the surface of the substrate, and the hydrogen bonds in the material could also achieve rapid dynamic exchange and recombination to obtain higher cohesive energy. It could nicely adhere with the substrate, so it had a higher shear strength. With the increase of R value, the adhesion properties of BDM-based and BDH-based elastomers increase. Because of the increase of R value, the content of hard segment increased continuously, increasing the number of hydrogen bonds, and the elastomer had a strong interaction with the substrate after heating. BDM-1.7 and BDH-1.7 elastomers possessed relatively excellent adhesion properties. The shear strength of BDM-1.7 and BDH-1.7 elastomers was 1.37 and 2.29 MPa, respectively, when heated at 80 °C for 8 h. When heated at 150 °C for 0.5 h, the shear strength was 4.59 and 4.82 MPa, respectively, which were comparable with some commercially available adhesives. As R value continues to increase, the shear strength of BDM-2.3 and BDH-2.0 elastomers became the largest, reaching an astonishing 12.91 and 11.32 MPa when heated at 150 °C for 0.5 h. The shear strength reached 5.70 and 4.55 MPa at 80 °C for 8 h, respectively. The adhesion properties of BDM-2.3 and BDH-2.0 elastomers with maximal shear strength were tested at 80 °C for different heating times, and the shear strength-displacement curves (Figure 6e,k) and shear strength (Figure 6f,l) of BDM-2.3 and BDH-2.0 elastomers were obtained at 80 °C for different heating times. It could be observed that the highest shear strength was achieved when BDH-2.0 elastomers were heated for 2 h at 80 °C. The shear strength of BDM-2.3 reached the maximal after heated for 8 h. These results indicated that elastomers with different adhesion properties could be obtained by adjusting the R value, temperature and heating time. The elastomer with high adhesion performance could be widely used in electronic assembly, aerospace, and other fields.

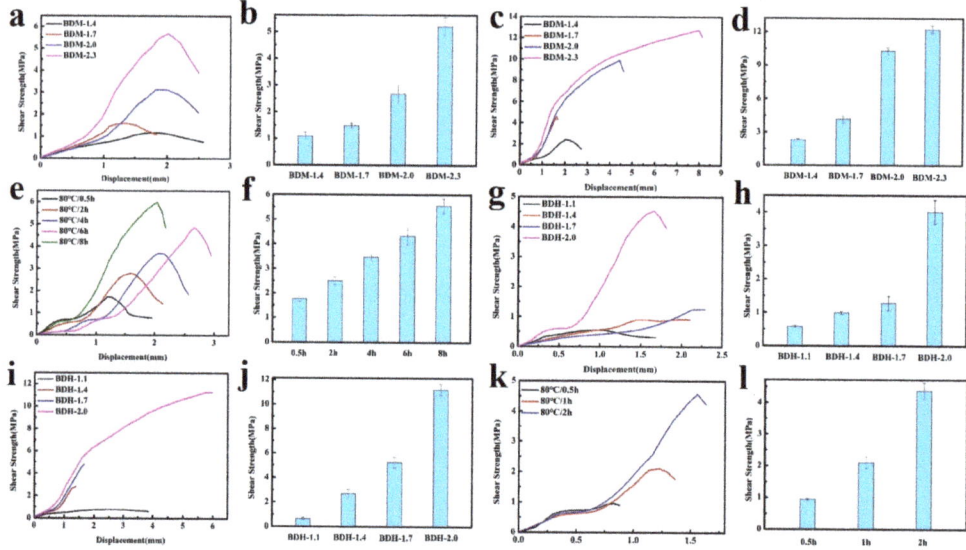

Figure 6. (**a**) Shear strength-displacement curves of BDM-based elastomer at 80 °C for 8 h; (**b**) Shear strength of BDM-based elastomers at 80 °C for 8 h; (**c**) Shear strength-displacement curves of BDM-based elastomer at 150 °C for 0.5 h; (**d**) Shear strength of BDM-based elastomers at 150 °C for 0.5 h; (**e**) Shear strength-displacement curve of BDM-2.3 elastomers at 80 °C for 0.5, 2, 4, 6, and 8 h; (**f**) Shear strength of BDM-2.3 elastomers at 80 °C for 0.5, 2, 4, 6, and 8 h; (**g**) Shear strength-displacement curves of BDH-based elastomer at 80 °C for 8 h; (**h**) Shear strength of BDH-based elastomers at 80 °C for 8 h; (**i**) Shear strength-displacement curves of BDH-based elastomer at 150 °C for 0.5 h; (**j**) Shear strength of BDH-based elastomers at 150 °C for 0.5 h; (**k**) Shear strength-displacement curve of BDH-2.0 elastomers at 80 °C for 0.5, 1, and 2 h; (**l**) Shear strength of BDH-2.0 elastomers at 80 °C for 0.5, 1, and 2 h.

4. Conclusions

In summary, two chain extenders, BDM and BDH, were synthesized by ring-opening reaction with the five-membered EC. BDM and BDH as chain extenders were adopted and the special structure of short hard and long hard segments was skillfully introduced by regulating the R value. This special structure facilitated the formation of SIC and introduced suitable amounts of hydrogen bonds, which contributed to the improvement of their mechanical properties. Due to the special structure of BDM and BDH, the synthesized polyurethane elastomer was proven to possess the loosely-stacked hard segment structure, beneficial for the improvement of their self-healing ability. Two polyurethane elastomers with excellent comprehensive properties were synthesized. The results showed that the tensile strength of the BDM-1.7 and BDH-1.7 elastomers was 20.6 ± 1.3 MPa and 37.1 ± 1.7 MPa, the toughness, 83.5 ± 2.0 MJ/m^3 and 118.8 ± 5.1 MJ/m^3, and the puncture energy, 196.0 and 627.0 mJ, respectively. Under the condition of 60 °C for 24 h, the comprehensive self-healing efficiency of BDM-1.7 elastomer reached almost 100%, and that of BDH-1.7 elastomer is above 95%. In terms of adhesive properties, the shear strength of BDM-1.7 and BDH-1.7 elastomers was 4.59 and 4.82 MPa, respectively, after being treated at 150 °C for 0.5 h. Based on all of these, this study provides an innovative strategy for further exploration on elastomers with high mechanical properties and excellent self-healing ability.

Supplementary Materials: The following are available online at https://www.mdpi.com/article/10.3390/polym14091673/s1, Figure S1: BDM powder. Figure S2: BDH powder. Figure S3: (a) FTIR spectra of BDM; (b) ^{1}HNMR spectrum of BDM. Figure S4: (a) FTIR spectra of BDH; (b) ^{1}HNMR spectrum of BDH. Figure S5: (a) FTIR spectra of BDM-based elastomers; (b) FTIR spectra of BDH-based and EG-1.7 elastomers. Figure S6: (a) 1HNMR spectrum of BDM-1.7 elastomer, chemical shift for peak 1, 2 and 3 is 7.51, 7.20 and 7.00 ppm, respectively; (b) ^{1}HNMR spectrum of BDH-1.7 elastomer, chemical shift for peak 1 and 2 is 7.55, 7.01 ppm, respectively. Figure S7: (a) GPC profiles for BDM-based elastomers; (b) GPC profiles for BDH-based and EG-1.7 elastomers. Figure S8: (a)and (b) The photos show that BDM-1.7 and BDH-1.7 elastomer tapes can be cut and processed at 60 °C for 0.5 h after splicing, which can increase the load of 500 g. Table S1: Characteristic peak assignments of BDM. Table S2: Characteristic peak assignments of BDH. Table S3: Ingredients of BDM-based elastomers. Table S4: Ingredients of BDH-based elastomers. Table S5: Characteristic peak assignments of BDM-based elastomers. Table S6: Characteristic peak assignments of BDH-based elastomers. Table S7: Information of molecular weight of BDM-based elastomers. Table S8: Information of molecular weight of BDH-based and EG-1.7 elastomers. Table S9: Mechanical properties of BDM-based elastomers obtained from tensile tests. Table S10: Mechanical properties of BDH-based and EG-1.7 elastomers obtained from tensile tests. Table S11: The Self-healing efficiencies of BDM-1.7 elastomers. Table S12: The Self-healing efficiencies of BDH-1.7 elastomers.

Author Contributions: Validation, P.Y.; data curation, J.C. (Jingrong Chen), Y.L. (Ying Lei) and R.J.; software, J.W. and A.S.; project administration, X.L.; writing—original draft preparation, J.C. (Junyao Chang); writing—review and editing, L.W. and Y.L. (Yuhan Li). All authors have read and agreed to the published version of the manuscript.

Funding: This work was funded by Zhengzhou Major Collaborative Innovation Project (Zhengzhou University, Grant No.18XTZX12001).

Institutional Review Board Statement: Not applicable.

Informed Consent Statement: Not applicable.

Data Availability Statement: Not applicable.

Conflicts of Interest: The authors declare no conflict of interest.

References

1. Fang, Y.; Du, X.; Du, Z.; Wang, H.; Cheng, X. Light and heat-triggered polyurethane based on dihydroxyl anthracene derivatives for self-healing applications. *J. Mater. Chem. A* **2017**, *5*, 8010–8017. [CrossRef]
2. Liang, Z.; Huang, D.; Zhao, L.; Nie, Y.; Zhou, Z.; Hao, T.; Li, S. Self-healing Polyurethane Elastomer Based on Molecular Design: Combination of Reversible Hydrogen Bonds and High Segment Mobility. *J. Inorg. Organomet. Polym. Mater.* **2021**, *31*, 683–694. [CrossRef]
3. Wu, H.; Xie, H.; Tian, X.; Sun, Y.; Shi, B.; Zhou, Y.; Sheng, D.; Liu, X. Hard, tough and fast self-healing thermoplastic polyurethane. *Prog. Org. Coat.* **2021**, *159*, 106409. [CrossRef]
4. Chen, Y.; Kushner, A.M.; Williams, G.A.; Guan, Z. Multiphase design of autonomic self-healing thermoplastic elastomers. *Nat. Chem.* **2012**, *4*, 467–472. [CrossRef]
5. Yuan, Y.; Yin, T.; Rong, M.; Zhang, M. Self healing in polymers and polymer composites. Concepts, realization and outlook: A review. *Express Polym. Lett.* **2008**, *2*, 238–250. [CrossRef]
6. Kragl, M.; Knapp, D.; Nacu, E.; Khattak, S.; Maden, M.; Epperlein, H.H.; Tanaka, E.M. Cells keep a memory of their tissue origin during axolotl limb regeneration. *Nature* **2009**, *460*, 60–65. [CrossRef]
7. Song, K.; Ye, W.; Gao, X.; Fang, H.; Zhang, Y.; Zhang, Q.; Li, X.; Yang, S. Synergy between dynamic covalent boronic ester and boron–nitrogen coordination: Strategy for self-healing polyurethane elastomers at room temperature with unprecedented mechanical properties. *Mater. Horiz.* **2021**, *8*, 216–223. [CrossRef]
8. Zhang, W.; Wang, M.; Zhou, J.; Sheng, Y.; Xu, M.; Jiang, X.; Ma, Y.; Lu, X. Preparation of room-temperature self-hea-ling elastomers with high strength based on multiple dynamic bonds. *Eur. Polym. J.* **2021**, *156*, 110614. [CrossRef]
9. Cash, J.J.; Kubo, T.; Bapat, A.P.; Sumerlin, B.S. Room-Temperature Self-Healing Polymers Based on Dynamic-Covalent Boronic Esters. *Macromolecules* **2015**, *48*, 2098–2106. [CrossRef]
10. Feng, Z.; Yu, B.; Hu, J.; Zuo, H.; Li, J.; Sun, H.; Ning, N.; Tian, M. Multifunctional Vitrimer-Like Polydimethylsiloxane (PDMS): Recyclable, Self-Healable, and Water-Driven Malleable Covalent Networks Based on Dynamic Imine Bond. *Ind. Eng. Chem. Res.* **2019**, *58*, 1212–1221. [CrossRef]
11. Ji, S.; Cao, W.; Yu, Y.; Xu, H. Visible-Light-Induced Self-Healing Diselenide-Containing Polyurethane Elastomer. *Adv. Mater.* **2015**, *27*, 7740–7745. [CrossRef] [PubMed]
12. Kuhl, N.; Bode, S.; Bose, R.K.; Vitz, J.; Seifert, A.; Hoeppener, S.; Garcia, S.J.; Spange, S.; van der Zwaag, S.; Hager, M.D.; et al. Acylhydrazones as Reversible Covalent Crosslinkers for Self-Healing Polymers. *Adv. Funct. Mater.* **2015**, *25*, 3295–3301. [CrossRef]
13. Peng, Y.; Yang, Y.; Wu, Q.; Wang, S.; Huang, G.; Wu, J. Strong and tough self-healing elastomers enabled by dual r-eversible networks formed by ionic interactions and dynamic covalent bonds. *Polymer* **2018**, *157*, 172–179. [CrossRef]
14. Rekondo, A.; Martin, R.; Ruiz de Luzuriaga, A.; Cabañero, G.; Grande, H.J.; Odriozola, I. Catalyst-free room-temperature self-healing elastomers based on aromatic disulfide metathesis. *Mater. Horiz.* **2014**, *1*, 237–240. [CrossRef]
15. Xu, C.; Cui, R.; Fu, L.; Lin, B. Recyclable and heat-healable epoxidized natural rubber/bentonite composites. *Compos. Sci. Technol.* **2018**, *167*, 421–430. [CrossRef]
16. Chen, S.; Mo, F.; Yang, Y.; Stadler, F.J.; Chen, S.; Yang, H.; Ge, Z. Development of zwitterionic polyurethanes with multi-shape memory effects and self-healing properties. *J. Mater. Chem. A* **2015**, *3*, 2924–2933. [CrossRef]
17. Li, C.H.; Zuo, J.L. Self-Healing Polymers Based on Coordination Bonds. *Adv. Mater.* **2020**, *32*, 1903762. [CrossRef]
18. Song, Y.; Liu, Y.; Qi, T.; Li, G. Towards Dynamic but Supertough Healable Polymers through Biomimetic Hierarchi-cal Hydrogen-Bonding Interactions. *Angew. Chem. Int. Ed.* **2018**, *57*, 13838–13842. [CrossRef]
19. Wang, Z.; An, G.; Zhu, Y.; Liu, X.; Chen, Y.; Wu, H.; Wang, Y.; Shi, X.; Mao, C. 3D-printable self-healing and mechanically reinforced hydrogels with host–guest non-covalent interactions integrated into covalently linked networks. *Mater. Horiz.* **2019**, *6*, 733–742. [CrossRef]
20. An, S.Y.; Noh, S.M.; Nam, J.H.; Oh, J.K. Dual Sulfide–Disulfide Crosslinked Networks with Rapid and Room Temperature Self-Healability. *Macromol. Rapid Commun.* **2015**, *36*, 1255–1260. [CrossRef]
21. Bekas, D.G.; Tsirka, K.; Baltzis, D.; Paipetis, A.S. Self-healing materials: A review of advances in materials, evaluation,characterization and monitoring techniques. *Compos. Part B Eng.* **2016**, *87*, 92–119. [CrossRef]
22. Priemel, T.; Degtyar, E.; Dean, M.N.; Harrington, M.J. Rapid self-assembly of complex biomolecular architectures during musselbyssus biofabrication. *Nat. Commun.* **2017**, *8*, 14539. [CrossRef] [PubMed]
23. Appel, A.K.; Thomann, R.; Mülhaupt, R. Hydroxyalkylation and Polyether Polyol Grafting of Graphene Tailored for Graphene/Polyurethane Nanocomposites. *Macromol. Rapid Commun.* **2013**, *34*, 1249–1255. [CrossRef] [PubMed]
24. Hunley, M.T.; Pötschke, P.; Long, T.E. Melt Dispersion and Electrospinning of Non-Functionalized Multiwalled Carbo-n Nanotubes in Thermoplastic Polyurethane. *Macromol. Rapid Commun.* **2009**, *30*, 2102–2106. [CrossRef] [PubMed]
25. Wang, X.; Pan, Y.; Shen, C.; Liu, C.; Liu, X. Facile Thermally Impacted Water-Induced Phase Separation Approach for the Fabrication of Skin-Free Thermoplastic Polyurethane Foam and Its Recyclable Counterpart for Oil–Water Separation. *Macromol. Rapid Commun.* **2018**, *39*, 1800635. [CrossRef] [PubMed]
26. Nevejans, S.; Ballard, N.; Fernández, M.; Reck, B.; Asua, J.M. Flexible aromatic disulfide monomers for high-perform-ance self-healable linear and cross-linked poly(urethane-urea) coatings. *Polymer* **2019**, *166*, 229–238. [CrossRef]
27. Yu, K.; Xin, A.; Du, H.; Li, Y.; Wang, Q. Additive manufacturing of self-healing elastomers. *NPG Asia Mater.* **2019**, *11*, 7. [CrossRef]

28. Ding, F.; Shi, X.; Wu, S.; Liu, X.; Deng, H.; Du, Y.; Li, H. Flexible Polysaccharide Hydrogel with pH-Regulated Reco-very of Self-Healing and Mechanical Properties. *Macromol. Mater. Eng.* **2017**, *302*, 1700221. [CrossRef]
29. Li, C.; Wang, C.; Keplinger, C.; Zuo, J.; Jin, L.; Sun, Y.; Zheng, P.; Cao, Y. A highly stretchable autonomous self-healing elastomer. *Nat. Chem.* **2016**, *8*, 618–624. [CrossRef]
30. Pettignano, A.; Häring, M.; Bernardi, L.; Tanchoux, N.; Quignard, F.; Díaz, D. Self-healing alginate–gelatin bioh-ydrogels based on dynamic covalent chemistry: Elucidation of key parameters. *Mater. Chem. Front.* **2017**, *1*, 73–79. [CrossRef]
31. Yabuki, A.; Shiraiwa, T.; Fathona, I.W. pH-controlled self-healing polymer coatings with cellulose nanofibers providing an effective release of corrosion inhibitor. *Corros. Sci.* **2016**, *103*, 117–123. [CrossRef]
32. Khan, A.; Huang, K.; Sarwar, M.G.; Rabnawaz, M. High modulus, fluorine-free self-healing anti-smudge coatings. *Prog. Org. Coat.* **2020**, *145*, 105703. [CrossRef]
33. Gent, A.N.; Kawahara, S.; Zhao, J. Crystallization and Strength of Natural Rubber and Synthetic cis-1,4-Polyisoprene. *Rubber Chem. Technol.* **1998**, *71*, 668–678. [CrossRef]
34. Li, W.; Liu, X.; Sun, A.; Wei, L.; Li, R.; He, H.; Yang, S.; Wang, S.; Niu, Y.; Li, Y. Exploring piperazine for intrinsic weather-proof, robust and self-healable poly(urethane urea) toward surface and tire protection. *Polymer* **2021**, *227*, 123829. [CrossRef]
35. Liu, X.; Liu, X.; Li, W.; Ru, Y.; Li, Y.; Sun, A.; Wei, L. Engineered self-healable elastomer with giant strength and toughness via phase regulation and mechano-responsive self-reinforcing. *Chem. Eng. J.* **2021**, *410*, 128300. [CrossRef]
36. Tang, M.; Zhang, R.; Li, S.; Zeng, J.; Luo, M.; Xu, Y.; Huang, G. Towards a Supertough Thermoplastic Polyisoprene Elastomer Based on a Biomimic Strategy. *Angew. Chem.* **2018**, *130*, 16062–16066. [CrossRef]

Article

Biodegradation of PBSA Films by Elite *Aspergillus* Isolates and Farmland Soil

Hsiao-Lin Chien [1,†], Yi-Ting Tsai [1,†], Wei-Sung Tseng [1], Jin-An Wu [2], Shin-Liang Kuo [2], Sheng-Lung Chang [2], Shu-Jiuan Huang [2] and Chi-Te Liu [1,3,4,*]

1. Institute of Biotechnology, National Taiwan University, Taipei 106, Taiwan; d99642014@ntu.edu.tw (H.-L.C.); r08642004@ntu.edu.tw (Y.-T.T.); t3885863412@gmail.com (W.-S.T.)
2. Material and Chemical Research Laboratories, Industrial Technology Research Institute, Hsinchu 300, Taiwan; itriA50018@itri.org.tw (J.-A.W.); slkuo@itri.org.tw (S.-L.K.); itria00534@itri.org.tw (S.-L.C.); amyhuang@itri.org.tw (S.-J.H.)
3. Department of Agricultural Chemistry, National Taiwan University, Taipei 106, Taiwan
4. Agricultural Biotechnology Research Center, Academia Sinica, Taipei 115, Taiwan
* Correspondence: chiteliu@ntu.edu.tw
† Co-first author.

Citation: Chien, H.-L.; Tsai, Y.-T.; Tseng, W.-S.; Wu, J.-A.; Kuo, S.-L.; Chang, S.-L.; Huang, S.-J.; Liu, C.-T. Biodegradation of PBSA Films by Elite *Aspergillus* Isolates and Farmland Soil. *Polymers* **2022**, *14*, 1320. https://doi.org/10.3390/polym14071320

Academic Editor: Valentina Siracusa

Received: 14 February 2022
Accepted: 21 March 2022
Published: 24 March 2022

Publisher's Note: MDPI stays neutral with regard to jurisdictional claims in published maps and institutional affiliations.

Copyright: © 2022 by the authors. Licensee MDPI, Basel, Switzerland. This article is an open access article distributed under the terms and conditions of the Creative Commons Attribution (CC BY) license (https://creativecommons.org/licenses/by/4.0/).

Abstract: Plastic films are widely used in current agricultural practices; however, most mulch films used are discarded and buried in the land after harvest, having adverse environmental impacts. To solve this environmental problem, the demand for biodegradable mulch has been increasing in recent years. Polybutylene succinate-co-adipate (PBSA) is a biodegradable polymer with good ductility and can be used for packaging and mulching. In this study, we isolated two elite fungal strains for PBSA degradation from farmlands, i.e., *Aspergillus fumigatus* L30 and *Aspergillus terreus* HC, and the latter showed better degradation ability than the former. It is noteworthy that biodegradation of PBSA by *A. terreus* is reported for the first time, which revealed unique characteristics. In the soil burial test, even the soil with relatively poor degradation ability could be improved by the addition of elite fungal mycelia. In substrate specificity analyses of soil samples, PBSA could induce the synthesis of lipolytic enzymes of indigenous microbes to degrade substrates with medium and long carbon chains in soil. Furthermore, PBSA residues or fungal mycelia supplementation in soils had no adverse effect on the seed germination rate, seedling growth, or mature plant weight of the test green leafy vegetable. Taken together, the results of this study not only advance our understanding of the biodegradation of PBSA films by filamentous fungi but also provide insight into improving the efficiency of biodegradation in soil environments.

Keywords: polybutylene succinate-co-adipate (PBSA); biodegradation; *Aspergillus*; phytotoxicity; ecotoxicity; lipolytic enzyme

1. Introduction

Plastics are widely used in our daily life and industries worldwide. The materials of conventional plastics are mostly derived from nonrenewable fossil fuels. The treatment of plastic wastes is mainly divided into three categories: incineration, recycling, and landfill burying [1]. However, the recycling rate of plastic is relatively low at approximately 9%, and almost 60% of plastic waste is discarded in landfills or natural environments [2]. Although marine plastic pollution is an environmental issue that has attracted much attention in recent years, the amount of plastic waste discarded and retained in the terrestrial environment is estimated to be 4–23 times that released into the ocean [3].

The main sources of terrestrial plastics include sewage sludge (i.e., application as fertilizer for agricultural lands), controlled-release fertilizers, and agricultural plastics (e.g., polytunnels and plastic mulches) [3]. In particular, the impact of agricultural mulch film on the environment is usually underestimated. Mulch film can effectively increase the surface temperature of farmland, maintain soil moisture, prevent soil loss, inhibit weed growth, promote seed germination, exceed seedling growth, and indirectly increase crop yields [4,5]. Currently, the majority of commercially available mulch film on global markets is prepared with polyethylene (PE) and low-density polyethylene (LDPE) (Transparency Market Research, 2019) [4]. More than 2 million tons of mulch film are used globally every year, most of which becomes mulch film waste within one to two years [6]. Mulch films are damaged to varying degrees due to weathering and photocatalytic degradation during crop cultivation [7]. After use, farmers need to pay additional expenses for mulch film waste disposal. Most of the mulch films used are incinerated or buried. In addition, conventional mulch films, such as LDPE and other raw materials buried in farmland, affect soil drainage and crop absorption of water because of their low degradation rate, and these films also pollute soil and water environments by the release of microplastics [3,8].

Based on the consideration of sustainable environmental development and reduction in the health hazards of plastics, in addition to imposing taxes on the production of traditional nondegradable plastic products or directly imposing use bans, many countries also actively encourage the development and use of biodegradable plastic materials [4]. Taking biodegradable plastic mulch films as an example, in recent years, the demand for these films has continued to increase. The estimated output value of plastic mulch films will reach USD 5 billion in 2024 (Zion Market Research, 2018, New York, NY, USA), and these films will have a compound annual growth rate (CAGR) of 9.4% (2017–2023) (Prescient & Strategic Intelligence Private Limited, 2018, USA). Common biodegradable mulch film materials can be divided into three categories: polysaccharides, aliphatic polyesters, and aromatic polyesters, including polylactic acid (PLA), polyhydroxybutyrate (PHB), starch blends, cellulose, polybutylene adipate terephthalate (PBAT), and polybutylene succinate (PBS) [9]. PBAT mulch films are commonly used in agriculture because of their strong mechanical properties, and PLA is often blended with other polymers because of its strength and low cost, but these compounds need to be decomposed by microorganisms in a relatively high-temperature or high-humidity environment [4]. PHB is produced by microorganisms, but its use is limited because of its brittleness, crystalline characteristics, and high production cost [4,9]. PBS is often copolymerized with other polymers or forms copolymers such as polybutylene succinate-co-adipate (PBSA) to reduce its crystallinity and improve decomposition efficiency [4].

PBSA is a biodegradable, aliphatic copolymer synthesized by 1,4-butanediol (BDO) with succinic acid (SA) and adipic acid [10,11]. PBS is produced by SA and BDO, and PBA is produced by BDO and AA [12]. PBSA possesses high flexibility, processability, excellent impact strength, and good thermal and chemical resistance [13,14]. The semicrystalline characteristic of PBSA makes this material relatively easy to biodegrade and is applied to packaging and mulch film production [12,15]. In a previous study, Yamamoto-Tamura and colleagues collected cultivated field soil samples from 11 sites in Japan, where PBSA films (2 × 2 cm^2, thickness: 20 µm) were buried on a laboratory scale. According to their results, the degradation rates of PBSA films ranged from 1.4% to 95.9% after four weeks of incubation [16].

In the depolymerization steps of biodegradation, microbes secrete a variety of extracellular enzymes to turn polymer chains into oligomers or monomers [17]. The hydrolytic enzymes working on carboxyl ester bonds in lipids are called lipolytic enzymes, which include carboxylesterases and lipases [18]. Lipases (EC 3.1.1.3, triacylglycerol hydrolases) and esterases (EC3.1.1.1, carboxyl ester hydrolases) are the major enzymes for the degradation of polyester-based biodegradable plastics [19]. Lipase has been considered as one of the good candidate enzymes for PBSA degradation [20]. Several studies have reported that fungal lipases could degrade PBSA. For example, the lipases secreted from *Pseudozyma antarctica* JCM 10317, *Rhizopus niveus*, *Rhizopus oryzae*, *Candida cylindracea*, and *Rhizopus delemar* have been demonstrated [21–23]. Similarly, esterase also plays an important role in the degradation of bioplastics by hydrolyzing polymer materials [24]. The esterase activity of *Leptothrix* sp. was found to be increased along with PBSA degradation and microbial growth [25]. The number average molecular weight [26] of PBSA was decreased from 5.9×10^4 to 1.5×10^4 by treating *Pseudozyma antarctica* esterase [27]. In addition, the recombinant protein (FSC) derived from *Fusarium solani* showed strong PBSA degradation activity (179.6 U/mg), with relatively high esterase activity (287 U/mg) [28].

According to the literature, many fungal strains belonging to the genus *Aspergillus* possess the potential for bioplastic degradation. For example, Merugu and colleagues isolated seven *Aspergillus* species from soil (*A. niger*, *A. flavus*, *A. ochraceus*, *A. nidulans*, *A. terreus*, *A. fumigatus*, and *A. parasiticus*) showed PHB degradability [29]. *Aspergillus* is a versatile plastic degrader. *A. fumigatus* is reported to be able to degrade PHB, polysuccinic acid [30], PBS, polycaprolactone, PCL, and PLA [31].

The aim of this study is to select elite microorganisms that are able to be applied in the soil environments for biodegradation of PBSA mulch film. We reported two PBSA-degrading *Aspergillus* strains (*A. fumigatus* L30 and *A. terreus* HC) isolated from farmland soils. Their PBSA-degrading activity was assessed under medium culture conditions. Additionally, we evaluated the biodegradability of PBSA films buried in farmland soils. We also determined the activities of the lipolytic enzymes of the two isolates and those of the soil samples for PBSA degradation. After degradation of the PBSA films in soil, we conducted a phytotoxicity assay to evaluate any harmful effects of the soil containing PBSA residues on the growth of seedlings and adult plants of Chinese cabbage.

2. Materials and Methods

2.1. Plastic Material

PBSA (number average molecular weight, Mn = 51,899 Daltons; weight average molecular weight, Mw = 77,951 Daltons) particles and films (thickness: 50 μm) were provided by the Industrial Technology Research Institute (ITRI). The constituent ratios of SA, AA, and BDO for the PBSA film were 37%, 13%, and 49.9%, respectively. The film was cut into 2.5 cm × 5.0 cm or 5.0 cm × 5.0 cm fragments for degradation assessment under liquid conditions or for the soil burial degradation test, respectively.

2.2. Screening and Isolation of PBSA-Degrading Microorganisms from Soils

The test soils were collected from the following 8 sites (9 samples in total) in Taiwan: Four samples were collected from the bank of the Xindian River topsoil (XDR-A (24°59'09.7" N 121°31'37.4" E), XDR-B (24°59'21.3" N 121°31'51.0" E), XDR-C (24°59'30.5" N 121°31'56.5" E), and XDR-D (24°59'30.9" N 121°31'56.4" E)); two samples were collected from a composting yard at National Taiwan University (25°00'57.9" N 121°32'31.4" E) (NTUCS-S collected from the surface (0–20 cm depth), and NTUCS-D collected at a 50 cm depth in the compost soil); two samples were collected from a rice paddy field topsoil in Xinfeng Township, Hsinchu County (HCP-A (24°55'47.9" N 121°00'34.4" E) and HCP-B (24°55'45.2" N 121°00'33.8" E)); and the final sample was collected from a tomato farm topsoil located in Xiushui Township, Changhua County (CHT (24°01'08.8" N 120°31'02.5" E)). All the soil samples were sieved through 3 mm mesh before use.

We applied a rapid screening platform (i.e., clear zone method) to select elite microorganisms for PBSA degradation, as proposed by [32]. Ten grams of PBSA plastic particles were dissolved in 40 mL of chloroform (Sigma–Aldrich, Merck & Co., Taipei, Taiwan), and the solution was emulsified with 1 g/L commercial detergent (PAOS®, Nice Co., Chiayi County, Taiwan), mixed with 200–300 mL of carbon-free basal medium (0.7 g/L KH_2PO_4, 0.7 g/L $MgSO_4 \cdot 7 H_2O$, 1 g/L $NH_4 NO_3$, 0.005 g/L NaCl, 0.002 g/L $FeSO_4 \cdot 7 H_2O$, 0.002 g/L $ZnSO_4 \cdot 7 H_2O$, and 0.001 g/L $MnSO_4 \cdot H_2O$) [33,34]. The plastic solution was homogenized by a sonicator for at least 15 min, and then the volume was brought to 1 L with a carbon-free medium. Volatile chloroform was removed in a fume hood overnight, and the pH was adjusted to 7.0. Then, 1.2% agar was added and autoclaved (121 °C, 20 min) to prepare PBSA agar plates [34].

One gram of individual soil sample was dissolved in 5 mL of distilled water and shaken at 100 rpm for 15 min at room temperature. The respective soil suspension (100 μL) was spread over the surface of the plastic solid medium by glass beads and incubated at 25, 30, or 37 °C for 9 days. When clear zones were formed on the plates, a single colony of individual PBSA-degrading microorganisms was isolated by the streaking plate method. According to the morphological characteristics of the isolates with hyphae and spores, we speculated that these isolates were fungi.

2.3. Phylogenetic Analysis of ITS Gene Sequences

Genomic DNA from the PBSA-degrading fungal strains was isolated by a Presto™ Mini gDNA kit (Geneaid Biotech Ltd., Taipei, Taiwan). The primer sets ITS4 (5′-TCC TCC GCT TAT TGA TATGC-3′), and ITS5 (5′-GGA AGT AAA AGT CGT AAC AAG G-3′) were used to amplify the ITS region of fungi [35,36]. PCR products of approximately 400 bp were expected and sequenced through paired-end sequencing. Amplicons were Sanger sequenced at the Center of Biotechnology at National Taiwan University. The sequencing results were edited with BioEdit 7.2.6 software [37]. The obtained DNA sequences were identified against the GenBank database using the Basic Local Alignment Search Tool (BLAST) (https://blast.ncbi.nlm.nih.gov/Blast.cgi/, accessed on 20 July 2021) of the National Center for Biotechnology Information (NCBI). The phylogenetic tree was generated by the neighbor-joining method (1000 bootstrap repeats) on version X Molecular Evolutionary Genetics Analysis (MEGA) software (The Pennsylvania State University, State College, PA, USA).

2.4. Determination of the PBSA Film Degradation Ability of Isolated Fungal Strains

PBSA film (size: 2.5×5.0 cm^2; thickness: 50 μm) was sterilized with 6% sodium hypochlorite (Sigma–Aldrich Co., St. Louis, MO, USA) and 70% ethanol for 10 min, washed twice with sterile water, and then kept in the carbon-free medium before use. The spore suspension for inoculation was prepared by culturing the respective elite strain (*Aspergillus fumigatus* L30; *Aspergillus terreus* HC) and the reference strain for PBSA-degradation *Aspergillus oryzae* RIB40 (ATCC 42149) cultured on potato dextrose agar (PDA) for 7–10 days. The spores were collected with sterile distilled water containing 0.05% Tween 20 (Bioman Scientific Co., Ltd., Taipei, Taiwan). The final concentration of the inoculated spore suspension was 2.5×10^6 spores/mL, and the spores were cultured with 12 pieces of PBSA film in 100 mL of the carbon-free medium at 30 °C and 80 rpm. The plastic films were stirred gently every week to prevent uneven decomposition caused by film overlap. Three pieces of the films were removed from the flask after 14 and 30 days of incubation, and the attached mycelia were removed from the surface of the films. After washing the films carefully with distilled water to remove the attached hyphae, the dry weights were determined after drying in a 60 °C oven overnight. All tests were performed in at least three independent biological replicates.

The effect of the addition or replacement of fresh medium for long-term (60 days) degradation of PBSA film was also conducted with *A. fumigatus* L30 and *A. terreus* HC, respectively. The preparation for PBSA film degradation by the respective elite fungal culture was the same as mentioned above. After 30 days of incubation, 30 mL of fresh carbon-free basal medium was added to the culture fluid, or the total culture fluid (100 mL) was removed carefully by pipette and replaced with an equal amount of fresh basal medium. The attached mycelia were removed from the surface of PBSA films after 30 and 60 days of incubation, and their weights were determined.

The degree of degradation was evaluated by weight loss (WL) using the following modified equation [38]: Plastic remaining weight (%) = $100\% - (W_i - W_t)/W_i \times 100\%$, where W_i is the initial weight of the sample, and W_t is the weight after the incubation time.

2.5. Scanning Electron Microscopic Analysis

To observe the plastic surface erosion or decomposition after microbial degradation, we used scanning electron microscopy [39] (Jeol, JSM-6510, Tokyo, Japan). The SEM sample preparation protocol followed that of the Joint Center for Instruments and Researchers,

College of Bio-Resources and Agriculture, NTU. Plastic films cultured with *A. fumigatus* L30 or *A. terreus* HC for 30 days were removed from the flask, and each film was cut into 2 pieces and divided into washed and unwashed groups. For the washed group, the plastic film was washed with distilled water to remove the mycelia attached to the plastic surface. The plastic films were immediately soaked in 2.5% (w/v) glutaraldehyde (Sigma–Aldrich Co.) at 4 °C and then shaken overnight for cell fixation. The fixed films were washed with 0.1 M sodium phosphate buffer (pH = 7.3) and postfixed in 1% osmium tetroxide (w/v) in an ice bath for 1 h. The samples were dehydrated gradually with different concentrations of ethanol (30%, 50%, 70%, 85%, 90%, 95%, 100%) for 60 min at each concentration, except for 70% and 100% ethanol for which overnight dehydration was conducted. The dehydrated samples were placed in a critical point dryer for critical spot drying, and subsequently, the samples were coated with gold and observed under a scanning electron microscope (Jeol, JSM-6510) [40].

2.6. NMR Analysis

For NMR sample preparation, PBSA plastic films after 30 days of degradation (0.1~0.05 g) were dissolved in 3 mL of deuterated chloroform ($CDCl_3$). The 1 H NMR analysis was conducted by the Industrial Technology Research Institute/Material and Chemical Laboratories (ITRI/MCL) with a Varian INOVA 500 NMR spectrometer (Varian Medical Systems, Inc., Palo Alto, CA, USA). The 1 H NMR spectra were recorded at 25 °C (128 scans, 1 s relaxation delay), and used tetramethylsilane (TMS) was employed as an internal reference for the reported chemical shifts.

2.7. Soil Burial Tests of PBSA Plastic Films

To verify the degradation of the PBSA plastic films in the general soil environment, we conducted a soil burial test following a previous method with some modifications [10]. The soil was collected from the experimental farm of the College of Bioresources and Agriculture, National Taiwan University. The effect of temperature on biodegradation was also evaluated by sampling soil in September (26.3 °C~32.1 °C, avg. 27.8 °C) and December (15 °C~24.7 °C, avg. 18.1 °C). The soil samples were sieved in advance through a 3 mm stainless steel sifting screen and transferred to a sterilized plastic box (18.1 (L) × 12.8 (W) × 6.8 (D) cm^3, HPL815 M, Lock & Lock, Hana Cobi Plastic Co, Ltd., Seoul, Korea).

Three PBSA film samples (size: 5.0 × 5.0 cm^2, thickness: 50 µm, initial weight: approximately 0.2 g) were sandwiched between two layers of soil (each layer was 1 cm thick and weighed 200 g, as shown in Figure 5A). Four treatments were conducted, including soil, sterile soil, and soil supplemented with low/high doses of *A. terreus* HC fungal hyphae. In the sterile soil group, the soil used was sterilized by an autoclave (121 °C, 40 min) before the trial. For the low- and high-dose groups, 10 mL (cell dry weight approximately 0.04 g) or 50 mL (cell dry weight approximately 0.16 g) of mycelial suspension were added, respectively. Six holes were drilled in the lid of the respective box for aeration, and the boxes were placed in an incubator at 25 ± 1 °C under moisture-controlled conditions.

For all treatments, water was added until the soil water content reached 50% of the maximum water holding capacity (WHC) according to the EN17033 and OECD Guideline (https://www.iso.org/obp/ui/#iso:std:iso:23517:ed-1:v1:en/, accessed on 25 December 2021). One piece of film was collected from each box weekly and weighed. Each plastic film was brushed softly and washed with distilled water several times to remove the attached soil and microorganisms. The washed plastic samples were placed in a Petri dish and dried overnight in an oven at 60 °C before weighing. After weekly sampling, the whole box was weighed and refilled with distilled water to maintain a 50% WHC water content. The degree of degradation was evaluated by WL as described previously.

2.8. Analysis of Lipolytic Enzyme Activities in Culture Supernatant

The lipolytic enzyme activity analytical procedure was modified based on a previous study [41]. *A. terreus* HC was cultured in basal medium with PBSA plastic film pieces

(2.5 cm × 5.0 cm, 0.1 g) at 30 °C for 60 days. After 30 days of incubation, 30 mL of fresh carbon-free basal medium was added to the culture broth and incubated for an additional 30 days. After 60 days of incubation, all the PBSA film pieces were taken from the broth. The total amount of the culture broth was collected and filtered through Qualitative Filter Paper NO. 1 (TOYO ADVANTEC) and a 0.22 μm filter to remove residual hyphae and the visible suspended matter, and a centrifugal concentrator (molecular weight cutoff (MWCO) 10 kDa, Sartorius) was used at 5000× g at 4 °C to condense the culture fluid. Then, the sample was dialyzed with a cellulose membrane (EIDIA, catalog number: UC27-32-100, 14,000 Dalton MWCO) in 0.1 M Tris-HCl. The dialyzed sample and crude protein suspension were stored at −20 °C for further enzyme activity analysis.

The lipolytic enzyme activity assay was modified based on previous studies [39,42]. p-nitrophenyl esters are usually used as substrates to measure lipase/esterase activity [43]. Thus, 900 μL of crude protein suspension was added to 100 μL of 10 mM 4-nitrophenyl acetate (C2), 4-nitrophenyl butyrate (C4), 4-nitrophenyl caprylate (C8), 4-nitrophenyl decanoate (C10), 4-nitrophenyl dodecanoate (C12), or 4-nitrophenyl palmitate (C16) (Sigma–Aldrich Co.), which were dissolved in DMSO as the substrate. The solution was incubated at 30 °C for 30 min. Fluorescence analysis of the chromogenic pNP products derived from the hydrolyzed p-nitrophenyl substrates was conducted by a microplate reader (Thermo Scientific™ Varioskan™ LUX, VLBL00 D0, Thermo Fisher Scientific Inc., Taipei, Taiwan) at a wavelength of 405 nm. For the preparation of the pNP calibration standard curve, 0, 25, 50, 75, 100, or 125 μg of pNP was dissolved in DMSO buffer and incubated at 30 °C for 30 min. We used a microplate reader to measure the optical density at 405 nm ($O.D_{405}$). The enzyme activity unit (Unit) is expressed as the production of 1 μmole of nitrophenol product per minute per mL of crude protein.

2.9. Analysis of Lipolytic Enzyme Activities in Soil

The soil samples used in this assay were those containing PBSA residue after 30 days of degradation as described in the above section titled "Section 2.7". To analyze the lipolytic enzyme activities of the soil containing decomposed PBSA films, we conducted the following procedures with slight modifications [44]. Individual soil samples (0.05 g) were suspended in 1 mL of 100 mM NaH_2PO_4/NaOH buffer (pH 7.25) and mixed well at 30 °C and 100 rpm for 30 min. Then, 80 μL of 25 mM nitrophenyl esters with different lengths of carbon substrates (i.e., 4-nitrophenyl acetate (C2), 4-nitrophenyl butyrate (C4), 4-nitrophenyl caprylate (C8), 4-nitrophenyl decanoate (C10), 4-nitrophenyl dodecanoate (C12), or 4-nitrophenyl palmitate (C16) (Sigma–Aldrich Co.) were dissolved individually in isopropanol (Sigma–Aldrich Co.) and mixed well at 30 °C and 100 rpm for 30 min. To measure the pNP released from individual substrates, we prepared a buffer control without soil suspension. A sample of soil background without pNP substrates was also prepared as a control. To stop the reaction, the solutions were cooled on ice for 10 min. Afterward, each solution was centrifuged at 3000× g and 4 °C for 10 min. Then, 100 μL of supernatant was collected via pipette, and the OD_{405} was determined by a microplate reader (Thermo Scientific™ Varioskan™ LUX, VLBL00 D0, Taipei, Taiwan) to measure pNP products derived from the hydrolyzed p-nitrophenyl substrates. To prepare the pNP calibration standard curve, 0, 25, 50, 75, 100, or 125 μg of pNP was dissolved in 100 mM NaH_2PO_4/NaOH buffer and incubated at 30 °C for 30 min. The standard contents were cooled on ice for 10 min and centrifuged at 2000× g and 4 °C for 10 min. A microplate reader was used to measure the OD value of each soil sample.

2.10. Phytotoxicity Assessment of the Soil Containing PBSA Residues

This experiment followed a previous method with some modifications [45]. The soil was collected from the experimental farm of National Taiwan University and sieved through a 3 mm stainless steel sifting screen. Four pieces of PBSA films were buried in the device utilized for the soil burial test for phytotoxicity assessment. The total weight of plastic films (i.e., approximately 0.8 g) was a mass percentage of approximately 0.2% of the total soil weight (i.e., 400 g). The total weight of the soil system was determined every week, and distilled water was replenished to maintain the water content as described above. After 30 days of incubation, soil containing PBSA residue was filled into 128 holes of a plug tray ($3 \times 3 \times 3$ cm^3 for each hole). Each group contained 64 seeds of Chinese cabbage (*Brassica rapa* L. ssp. *chinensis* var. Maruha), which were sown on the surface of the soil to observe germination. After the seeds were germinated and cultivated until the seedlings with two true leaves appeared (~8 days of cultivation), the germination rates were calculated, and the fresh weights of the seedlings were determined to evaluate the growth.

To evaluate the long-term impact on plant growth, soil samples containing PBSA residue were filled into 3-inch pots (300 g soil/pot). Each group contained 8–10 seedlings of Chinese cabbage, which were grown in potting soil for 7 days and transplanted into each pot. The rate of fertilizer application followed that of a previous study [46]. During the cultivation period, 0.05 g chemical fertilizer (Sinon Chemical Industry Co., Ltd., Taichung, Taiwan, with an N:P:K ratio of 14:15:10) was applied every week. After 21 days of cultivation, the shoot fresh and dry weights of the plants at five-leaf ages were measured.

3. Results

3.1. Isolation and Identification of Two Elite PBSA-Degrading Fungal Strains

We applied a rapid screening platform (i.e., clear zone method) to screen and isolate PBSA-degrading microorganisms from soil samples in this study. As shown in Figure S1, several transparent circles with different sizes were formed on the PBSA agar plates. We noticed that the numbers of clear zones varied with sampling location. Among the samples, those derived from the tomato field located in Xiushui township (CHT) and the paddy field in Hsinchu (HCP-A) showed better degrade abilities than those collected in other places. It was previously shown that microbial degradation is associated with temperature [47]. We found that the PBSA degradation rates were elevated at relatively high temperatures. This phenomenon was observed among the samples of the deep compost soil at National Taiwan University (NTUCS-D), the paddy fields in Hsinchu (HCP-A, B), and site D of the Xindian riverside (Xindian D). When clear zones were formed on the plates, a single colony of individual PBSA-degrading microorganisms was isolated by the streaking plate method.

According to the morphological characteristics of the isolates with hyphae and spores, we assumed that these isolates were fungi. We isolated dozens of fungal strains from the soil sampling sites, eight of which showed obvious PBSA-degrading activities and were sequenced (Figure S2). Two strains, L30 and HC, were selected based on their growth rates and clear zone sizes. These two strains were derived from a lemon field located in Chaozhou township, Pingtung County, and the B side of a paddy field in Hsinchu. L30 and HC showed high sequence similarity with *Aspergillus fumigatus* (99.74%) and *Aspergillus terreus* (99.78%) by the partial sequences of ITS sequences, respectively. A phylogenetic tree was built from the partial sequences of ITS to elucidate the taxonomic position of these strains (Figure 1A). *A. oryzae* RIB40 (ATCC42149) was used as a positive control for PBSA degradation tests in this study [48]. The clear zones as well as the morphologies of these strains formed on PBSA agar plates are shown in Figure 1B.

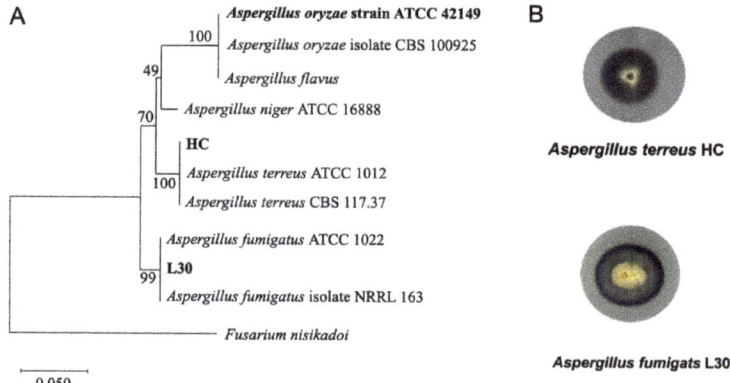

Figure 1. Phylogenetic tree of ITS and clear zone formation on PBSA agar plates. (**A**) A phylogenetic tree of two selected fungal strains (*Aspergillus terreus* HC and *Aspergillus fumigatus* L30) was constructed from a comparison of sequences of approximately 400 bp of the internal transcribed spacer (ITS) region using neighbor-joining analysis of a distance matrix with Kimura's two-parameter model. Bootstrap values (expressed as percentages of 1000 replications) greater than 75% are shown at branch points. A reference fungal strain, *Aspergillus oryzae* RIB40 (ATCC42149), was used as a control in this study, and *Fusarium nisikadoi* was used as an outgroup. The scale bar represents 0.05 substitutions per nucleotide position. (**B**) Morphology of the elite PBSA-degrading fungal strain.

3.2. Two Elite Fungal Strains Showed High PBSA Biodegradation Efficacy

After coincubation with individual elite fungal strains and PBSA film in a carbon-free basal medium for 30 days, hyphae were observed on the surface of the plastic film (Figure 2A, upper panel). After removing the hyphae, several cracks and cavities were observed on the surface of the plastic films treated with *A. terreus* HC or *A. oryzae* ATCC42149 (Figure 2A, bottom panel). On the other hand, the films treated with *A. fumigatus* L30 became relatively thin and transparent in comparison with untreated films (Figure 2A, bottom panel). The degradation rates of individual elite fungal strains were determined by the weight loss (WL) of the films. As shown in Figure 2B, the weights of the plastic films were significantly reduced by biodegradation. The WL of the plastic films treated with *A. terreus* strain HC, *A. fumigatus* strain L30, and *A. oryzae* RIB40 for 30 days were 47.5%, 30.2%, and 27.5%, respectively. Since *A. terreus* HC showed better degrading ability than the other strains, we used this strain for subsequent soil burial tests.

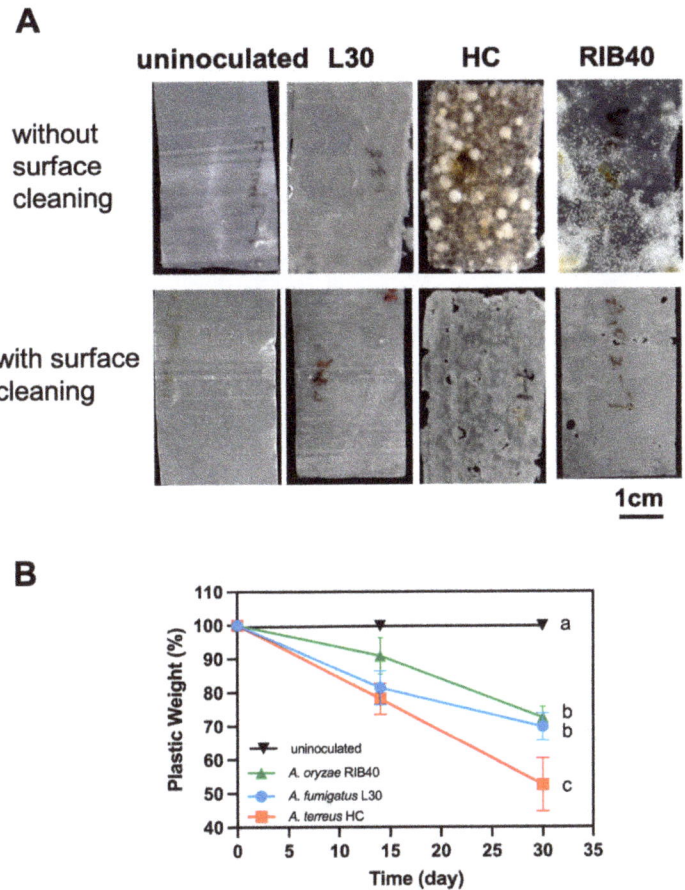

Figure 2. Colonization of fungal strains on PBSA plastic films and their degradation abilities. (**A**) Biodegradation PBSA films after 30 days of incubation. Twelve pieces of PBSA films (size: 2.5 × 5.0 cm^2; thickness: 50 μm) were cocultured with *A. terreus* HC, *A. fumigatus* L30, and *A. oryzae* RIB40 (ATCC42149) at 30 °C in a carbon-free basal medium. The upper panel shows the attached hyphae of individual strains on the surface of PBSA plastic films after 30 days of incubation. The bottom panel shows the surface of plastic films after hyphae removal by rinsing with DDW. (**B**) Weight loss of PBSA plastic film against time. The sampling times were 0, 14, 30 days after incubation. The values are expressed as the mean ± standard deviation of three biological replicates ($p < 0.05$; Tukey's post hoc ANOVA test).

3.3. Effect of Replacement or Addition of Fresh Medium on PBSA Plastic Film Degradation

To evaluate the degradation rates of the elite strains over a relatively long period of time, we conducted a degradation experiment after 60 days of incubation. As shown in Figure 3, the degradation rate of *A. fumigatus* strain L30 or *A. terreus* strain HC decreased after 30 days of incubation, and the latter was obviously changed compared with the former. We deduced that nutrient depletion occurred during this incubation period. Accordingly, we evaluated two strategies to rescue the degradation rates. One strategy was to add fresh medium (30 mL), and the other strategy was to replace all of the old medium (100 mL) after 30 days of incubation. In comparison with those of medium without change, we found that the degradation rate of either *A. fumigatus* strain L30 (Figure 3A) or *A. terreus* strain HC (Figure 3B) was increased when 30 mL of fresh medium was added into the culture fluid after 30 days of incubation, although there was no significant difference in the degradation ability of the former strain. The degradation rates of these strains on the 60th day remained almost the same as those before 30 days of incubation. For the trials to replace the whole old culture broth (100 mL), there were remarkable differences between the two strains. As shown in Figure 3A, the degradation rate of *A. fumigatus* strain L30 was improved significantly in comparison with that of strain in the medium that was unchanged after 30 days of incubation. In contrast, the degradation rate of *A. terreus* strain HC was almost reduced to zero after replacing the old incubation broth with a fresh medium (Figure 3B).

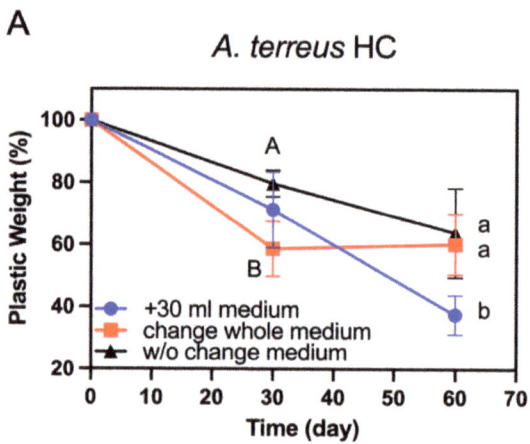

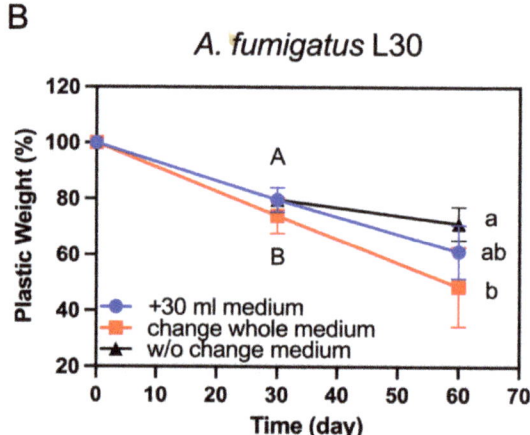

Figure 3. Determination of the effect of replacement or addition of fresh medium on PBSA plastic film degradation after 60 days of incubation. (**A**) Weight loss of PBSA film against time with *A. fumigatus* strain L30. (**B**) Weight loss of PBSA film against time with *A. terreus* HC. w/o medium change: maintaining the original medium during the whole incubation period (60 days); replacement of all culture medium (100 mL) or addition of 30 mL of fresh medium were carried out on the 30th day of incubation. The value of individual treatment is expressed as the row mean ± standard deviation. Statistical analyses were based on the degradation rates (g/day) on the 30th (in uppercase letters A and B) and 60th (in lowercase letters a and b) days of incubation.

3.4. Appearance of PBSA Plastic Films Degraded by Two Elite Aspergillus Strains

To observe erosion or decomposition of the plastic surface by elite strains after 30 days of incubation, SEM analysis was conducted. As shown in Figure 4 and Figure S3, the SEM images showed a PBSA film with an intact surface when not inoculated with *Aspergillus* (Figure S3). When inoculated with *Aspergillus* strains, filamentous hyphae of *A. fumigatus* L30 or *A. terreus* HC densely covered the surface and even intertwined inside and penetrated the plastic film, as shown in Figure 4(Aa,Ba). While removing the hyphae, many cracks and holes were observed on the film surface (Figure 4(Ab,Bb). Moreover, internal erosion also occurred, as shown in Figure 4(Ac,Bc), and irregular cavities were observed inside the PBSA films. The visual results suggest that biodegradation occurred not only on the surface of PBSA films but also inside the films.

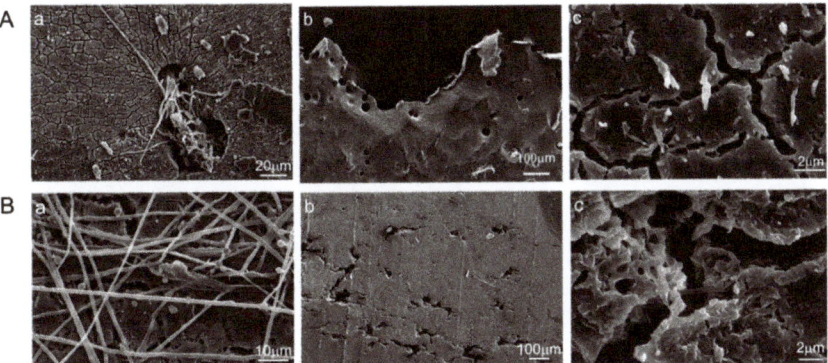

Figure 4. Scanning electron microscopic images of plastic films degraded by two elite fungal strains. PBSA films degraded by *A. terreus fumigatus* L30 (**A**) and HC (**B**) after 30 days of incubation. (**Aa**) The attachment and network of *A. terreus* L30 hyphae on the surface of PBSA film, (**Ab**) cracks and holes on the plastic surface, (**Ac**) irregular cavity in internal of plastic film. (**Ba**) *A. fumigatus* HC hyphae intertwined inside the plastic film and formed holes on the plastic surface, (**Bb**) cracks and holes on the plastic surface, and (**Bc**) internal erosion of the plastic film.

3.5. Changes in the Chemical Composition of PBSA Films during Biodegradation

To confirm the chemical composition of PBSA films before and after biodegradation, we conducted an NMR analysis. The ^{1}H-NMR spectrum of PBSA is shown in Figure 5 with a chemical structure illustration. The peaks marked with 1, 2, 3, 4, and 5 (Figure 5) indicated protons of SA (1), AA (2, 3), and BDO (4, 5). We compared the intensity of terminal H moieties with the five main signals, and the intensity ratios of the PBSA film without degradation were 4.00:620.35 (1), 4.00:218.66 (2), 4.00:1121.22 (3 and 5), and 4.00:837.26 (4) (Figure 5A). After biodegradation with *A. terreus* HC for 30 days, lower intensity ratios were shown as 4.00:335.32 (1), 4.00:117.41 (2), 4.00:564.05 (3 and 5), and 4.00:442.18 (4) (Figure 5B). As shown in Figure 5, the M_n value decreased from 37610.6 to 20079.0. The weakening of the relative intensity ratios and the decrease in the M_n suggested that structural weakening occurred in the PBSA films due to decomposition by *A. terreus* HC.

A

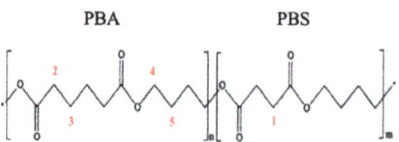

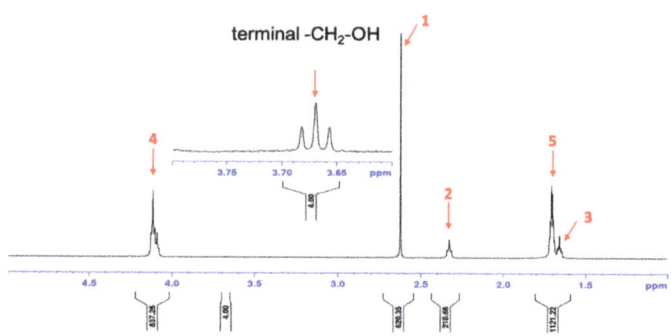

molecular weight (Mn) : 37610.6
mole %: BA:37.0%, AA:13.0%, BDO: 49.9%

B

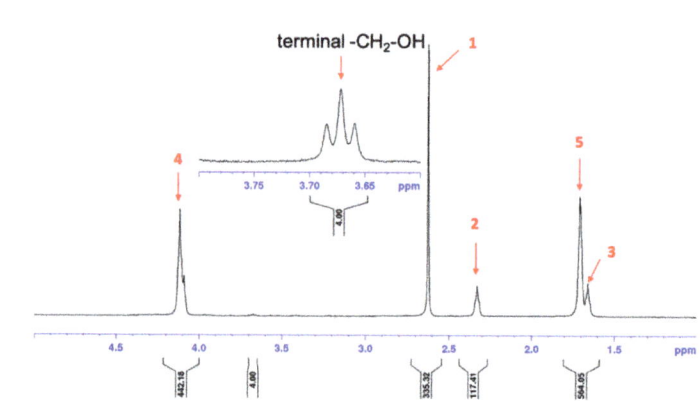

molecular weight (Mn) : 20079.0
mole %: BA:37.5%, AA:13.1%, BDO:49.4%

Figure 5. NMR spectra of PBSA films after degradation by *A. terreus* HC. (**A**) PBSA film without degradation. (**B**) PBSA film degraded by *A. terreus* HC for 30 days. Compared with the original PBSA film, the PBSA film, after 30 days of biodegradation, showed a lower number average molecular weight (Mn) but a similar monomer composition ratio. Peaks 1–5 correspond to the main signals of protons shown in (**A**,**B**). Peak 1 represents succinic acid (SA), peaks 2 and 3 represent adipic acid (AA), and peaks 4 and 5 represent 1,4-butanediol (BDO).

3.6. Biodegradation of PBSA Films in Farmland Soil

To simulate the degradation of PBSA films under general soil environments, we conducted a soil burial test (Figure 6A,B). Soil samples were collected from NTU farmland in September (late summer) and December (winter) of 2020 to evaluate the effect of seasonal conditions on biodegradation. The PBSA films retained approximately 98% of their weight after being buried in sterile soil for 28 days (Figure 6B,C). The PBSA films were degraded by approximately 95% while buried in the nonsterile soil collected in September after 28 days of incubation (Figure 6C), whereas the films were degraded by 61.3% when buried in soil collected in December (Figure 6). We evaluated the effects of supplementation with different doses of *A. terreus* HC fungal mycelia on degradation rates. For the soil collected in September, the weights of PBSA were reduced by 54.9% and 69.3% when the nonsterile soil was supplemented with low and high doses of fungal mycelia, respectively, after 14 days of incubation (Figure 6C). The film weights achieved 92.2% and 93.4% reductions after 28 days of degradation at low and high fungal mycelia doses, respectively (Figure 6C). For the soil collected in December, the weights of PBSA were reduced by 23.7% and 37.4% when the nonsterile soil was supplemented with low and high fungal mycelia doses, respectively, after 14 days of incubation (Figures 6D and S4A). This result indicated that the addition of a high dose of *A. terreus* HC fungal mycelia to winter soil could improve the degradability (+14%) in the early stage of incubation (14 days). After 28 days of incubation, the weight of PBSA films was reduced by 70.2% and 78% with low and high fungal mycelia doses, respectively, and there was no significant difference between the two treatment groups (Figure 6C).

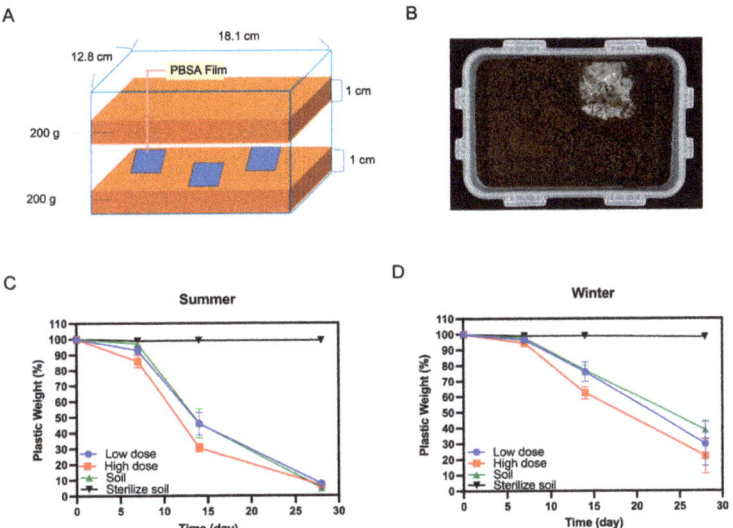

Figure 6. Soil burial test for biodegradation of PBSA films. (**A**) Schematic diagram of the device for the soil burial test. (**B**) A piece of degraded PBSA film before analysis. (**C**) Biodegradation of PBSA films buried in summer soil. (**D**) Biodegradation of PBSA films buried in winter soil. PBSA plastic films were buried in NTU farmland soil that was collected from late summer (September) or winter (December) in 2020. Low (10 mL of mycelial suspension) or high (50 mL of mycelial suspension) doses of *A. terreus* HC fungal mycelia were inoculated into the soil. The values for the respective weights of PBSA film are the mean ± standard deviation of triplicate samples ($p < 0.05$; Tukey's post hoc ANOVA test).

3.7. Lipolytic Enzyme Activities of A. terreus HC and Soil

We extracted crude enzymes from the supernatant of *A. terreus* HC broth after 60 days of incubation in the presence of PBSA films. Lipolytic enzymes included lipases and esterases. We used *p*-nitrophenyl fatty acid esters with different chain lengths (C2, C4, C8, C10, C12, and C16) as substrates to investigate the substrate specificity of the potential lipolytic enzymes of *A. terreus* HC. As shown in Figure 7A, the lipolytic enzymes produced by *A. terreus* HC displayed relatively high activities for C8 to C16 chain fatty acids (i.e., medium to long-chain fatty acids), and the maximal activity was observed for C10 fatty acids. Intriguingly, we noticed that although the *p*-nitrophenyl ester with a short chain (i.e., 4-nitrophenyl acetate, C2) has often been selected as a candidate substrate for lipolytic enzyme activity assays in the literature, it was not detected in the extracellular medium of *A. terreus* HC after PBSA degradation (Figure 7A).

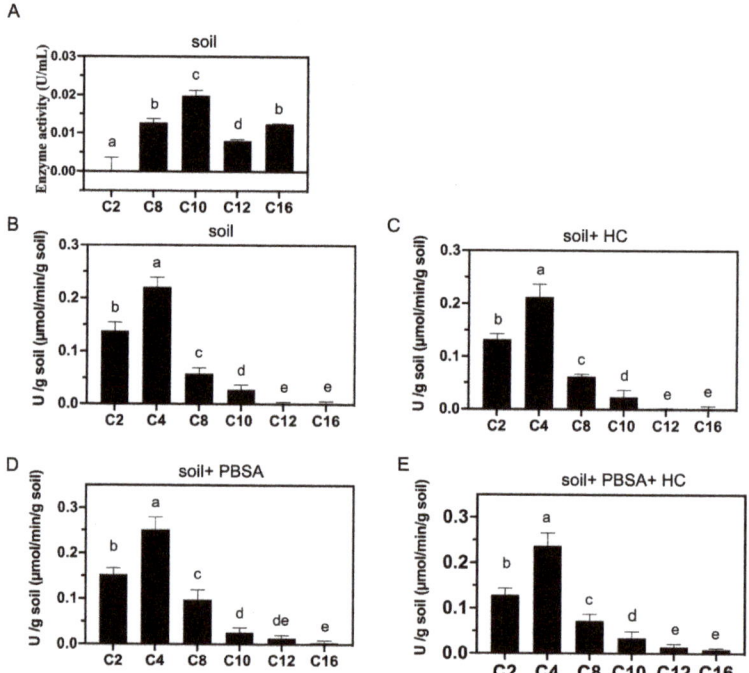

Figure 7. Lipolytic enzyme activities in the *A. terreus* HC culture broth and soil. (**A**) The lipolytic enzyme activities of *A. terreus* HC in the culture broth were determined by chromogenic nitrophenyl esters with different chain lengths as substrates (i.e., *p*-nitrophenyl esters). The supernatant of the *A. terreus* HC culture was collected from the culture fluid incubated with PBSA film and *A. terreus* HC for 60 days. (**B**) The lipolytic enzyme activities of soil. (**C**) The lipolytic enzyme activities of soil supply with *A. terreus* HC. (**D**) The lipolytic enzyme activities of soil buried with PBSA. (**E**) The lipolytic enzyme activities of PBSA buried soil supply with *A. terreus* HC. The assay was conducted after 30 days of PBSA degradation in soil. The substrates used were as follows: C2, 4-nitrophenyl acetate; C4, 4-nitrophenyl butyrate; C8, 4-nitrophenyl caprylate; C10, 4-nitrophenyl decanoate; C12, 4-nitrophenyl dodecanoate; and C16, 4-nitrophenyl palmitate. The results are presented as the mean ± standard deviation ($p < 0.05$; Tukey's post hoc ANOVA test).

After conducting the soil burial test (Figure 7B–E), we further determined the lipolytic enzyme activities of the soil samples (NTU farmland) with the same substrates mentioned above. As shown in Figure S5A, lipolytic enzyme activities of the test soil were detected against all the substrates of C2 to C16 chain fatty acids. We found that the activities were relatively high against those of short-chain fatty acids (C2 to C4) and decreased with increasing carbon chain length (C8 to C16). When PBSA films were buried in the soil for 30 days, we found that the lipolytic enzyme activities against the substrates with C8 and C12 chain lengths were remarkably higher than those in soil without PBSA films (Figure S5B and Figure 7D,E). Furthermore, we noticed that when a suspension of *A. terreus* HC mycelia was added to the test soil at the beginning of the experiment, the lipolytic enzyme activities against the substrates with C10 and C16 chain lengths were dramatically higher than those without fungal addition, which were 1.3- and 5.8-fold increased, respectively (Figure S5B).

3.8. Phytotoxicity Assessment

Phytotoxicity assessment was conducted to evaluate whether there were potential adverse effects derived from the soil containing PBSA residues or *A. terreus* HC on the agricultural environment or plant growth. Chinese cabbage, a popular green leafy vegetable in many Asian countries, was used as the test material. As shown in Figure 8, the germination rates in each treatment group were not significantly different and were 55%, 58.8%, 57.5%, and 60% for the soil, soil + PBSA, soil+ *A. terreus* HC, and soil + PBSA+ *A. terreus* HC treatments, respectively, and the average biomass of the seedlings did not significantly vary, with values of 0.10 g, 0.11 g, 0.12 g, and 0.11 g, respectively (Figure 8A).

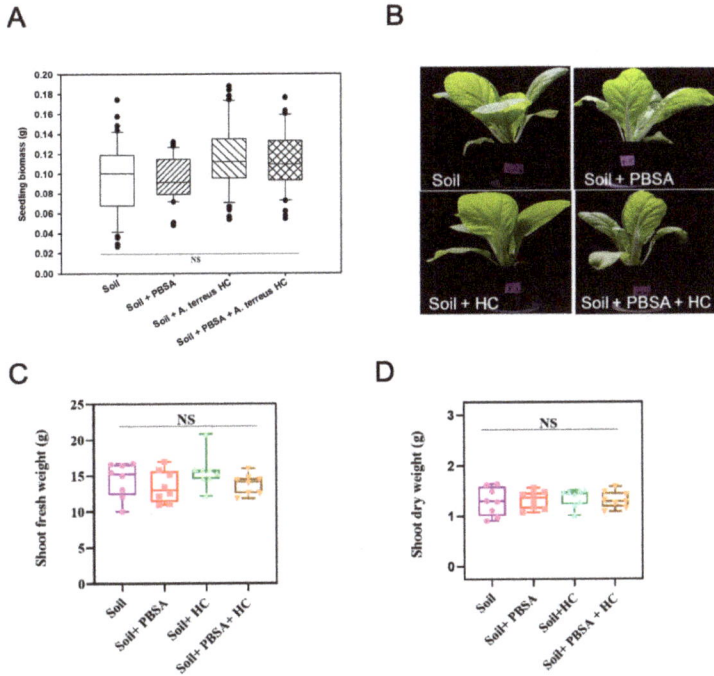

Figure 8. Phytotoxicity assessment of PBSA plastic film residues or *A. terreus* HC on plant growth. Seeds of Chinese cabbage were cultivated in soil containing 30-day degraded PBSA film residues to determine the seedling vigor for assessing phytotoxicity. (**A**) The biomass of Chinese cabbage seedlings was assessed after 8 days of germination. (**B**) The phenotype of Chinese cabbage at 21 days of cultivation. (**C,D**) Fresh and dry weights of shoots after 21 days of cultivation in soil containing PBSA residues. Ns: no significant difference among the treatments. The results are presented as the mean ± standard deviation ($p < 0.05$; Tukey's post hoc ANOVA test).

For the assessment of adult plants, we found no significant difference among the treatments either in morphology or in biomass (shoot fresh and dry weight) of Chinese cabbage after 21 days of cultivation in the PBSA-containing soil (Figure 8B–D). The average fresh weights were 14.44, 13.51, 15.60, and 13.77 (g), and the dry weights were 1.29, 1.32, 1.35, and 1.31 (g) for the soil, soil + PBSA, soil+ *A. terreus* HC, and soil + PBSA+ *A. terreus* HC treatments, respectively. These results indicated that neither PBSA residue nor *A. terreus* HC influenced the germination rate, seedling biomass, and growth of the test green leafy vegetable.

4. Discussion

4.1. Individual Aspergillus Strains Showed Unique PBSA Degradation Characteristics

In this study, two PBSA-degrading *Aspergillus* strains (*A. terreus* HC and *A. fumigatus* L30) were isolated from farmland soil in Taiwan. As shown in Figure 3, the degradation rate of *A. fumigatus* L30 or *A. terreus* HC was increased, while 30 mL of fresh basal medium was added to the incubation broth after 30 days of incubation (i.e., the blue curves). Some literature mentioned that supplementation with other nutrients or substrates can stimulate the degradation ability, which may be due to increased microbial biofilm growth or the provision of an easily metabolized nutrient source for microbes [49–52]. We deduced that this advantageous effect was due to an increase in cell growth and promotion of the synthesis of degrading enzymes by adding fresh basal medium during fed-batch cultivation.

On the other hand, we noticed that while the whole suspension (100 mL) of cell culture was removed and replaced with the equivalent amount of fresh basal medium, the PBSA degradation (30–60 days) of *A. fumigatus* L30 was not affected (i.e., the red curve in Figure 3A), whereas that of *A. terreus* HC was almost terminated (i.e., the red curve in Figure 3B). The case of *A. terreus* HC was reminiscent of the phenomenon previously reported by Bottone and colleagues [53]. They found that some filamentous fungi, such as *Mucor* and *Aspergillus* spp., the inhibitory substances were self-produced during their culture under minimal medium, which precluded the new hyphal backgrowth in the fresh medium. They also observed that even inoculation with fresh spore suspension did not result in new hyphal growth. These inhibitory substances could be bicarbonate, which was identified in the respiratory metabolism of *Aspergillus nidulans* [54,55]. Accordingly, we assumed that *A. terreus* HC produced the still unidentified substance to inhibit its metabolic activity while the whole culture suspension was replaced by a fresh basal medium, although this remains to be elucidated.

4.2. PBSA Degradation Efficiency of the Winter Soil Can Be Improved by Addition of Fungal Mycelia

To evaluate the effect of temperature on PBSA degradation, we sampled the farmland soil in different seasons (September and December) for the soil burial test. As shown in Figure 6, a higher degradation rate was shown in the treatment using late summer (September) soil than that using early winter (December) soil. Hoshino and colleagues proposed PBSA degradation rates under different weather conditions, which were correlated with the accumulation of daily effective temperatures (i.e., effective temperature = $T-10$ °C, T: daily temperature) [56]. We deduced that increasing the temperature increases the rate of effective biodegradation by microbes in the soil sample. As mentioned above, the degradation activity of the indigenous winter soil was lower than that of the summer soil (Figure 6). We found that adding a high dose (50 mL of mycelial suspension) of *A. terreus* HC mycelia to the winter soil sample could improve the degradation rate by 15% at day 14. Such an effect of bioaugmentation with elite microbial inoculant was also reported in previous studies. For example, Ishii and colleagues reported that the addition of 10 mg of wet cells of *A. fumigatus* NKCM1706 into the soil could promote sixfold PBS degradation at 30 °C [57]. On the other hand, such a bioaugmentation effect was not observed in the trial using the summer (September) soil sample (Figure 6C). The degradation rate of the September soil group without fungal mycelia inoculation was relatively high on day 14 of incubation, suggesting that indigenous microbes already possessed high degradation activities. Accordingly, even supplementation with a high dose of *A. terreus* HC mycelia did not exceed the background effect. Taken together, these results suggest that supplementation with proper amounts of elite microbial cultures, such as *A. terreus* HC mycelia, is a practical strategy to enhance PBSA biodegradation activity in environments with poor decomposition.

4.3. PBSA Could Induce the of Lipolytic Enzyme Activities in Farmland Soil

The lipolytic enzyme activities of the NTU farmland soil samples were determined and shown in Figure 7. We found that the soil showed lipolytic activities against all the substrates of C2 to C10 chain fatty acids even in the absence of PBSA film. When PBSA films were buried in the soil for 30 days, the activities were significantly increased against the substrates with C8 and C12 chain lengths compared with the activities without PBSA addition (Figure 7D). This result indicated that PBSA induced the synthesis of lipolytic enzymes of indigenous microbes to degrade substrates with medium and long carbon chains in soil. In a previous study, Yamamoto-Tamura and colleagues observed a similar phenomenon [16]. The authors reported that the activity of soil esterase was remarkably induced upon the addition of PBSA film. Accordingly, we assumed that the presence of PBSA induced esterase production from indigenous microbes in the soil. Moreover, we noticed that when *A. terreus* HC mycelia were added to the soil with PBSA films (i.e., soil + PBSA + HC) (Figure 7E), the lipolytic enzyme activities against the substrates with C10 and C16 chain lengths were further increased in comparison with those without the fungal supplement (i.e., soil + PBSA) (Figure 7D and Figure S5). Taken together, these results, with respect to soil lipolytic enzyme activities, showed that the NTU farmland soil mainly contains short-carbon-chain enzymes, but when the soil contained buried PBSA films, indigenous soil microorganisms were triggered to secrete long carbon chain-degrading enzymes. Moreover, while adding plastic decomposing microbes can further promote the secretion of long carbon chain-degrading enzymes, we deduced that the elite fungal strain *A. terreus* HC can degrade PBSA into shorter-chain fatty acids, which are then catabolized by the other indigenous microorganisms in the soil. This hypothesis remains to be proved by further analysis.

4.4. No Observed Adverse Effect on the Growth of Leafy Vegetable by PBSA Degradation or the Addition of Elite Fungal Culture in Farmland Soil

Plastic pollution has become a global concern for ecosystem health and biodiversity conservation. People are increasingly interested in understanding the impact of biodegradable plastic on ecosystems [4,5,58,59]. The analysis of the toxicity of biodegradable mulch to plants is mainly carried out by measuring plant growth in soil containing biodegradable plastic film fragments [60–64]. Since the sensitivity of plants to a variety of biodegradable materials is species dependent, a few studies have shown that some products may alter plant development, while a few others showed certain mulches to be likely safer for use in agricultural environments [4]. Wang and colleagues reported that soil containing the residue of PBS-based copolymers (P(BS-co-SA)) did not affect the growth of Chinese cabbage [6]. Furthermore, approximately 0.6~2% (w/w) of PLA, PHB, PBAT, or the innovative biodegradable material Mater-Bi® was individually buried in the testing soils for 6–7 months, and plant seeds were sown inside. These studies reported that there was no adverse effect on either the germination rates of cress, barley s, Brassica, and sorghum or the dry weight of seedlings [61,63,64]. On the other hand, Fritz and colleagues found that when the soil contained 2% (w/w) polyester amide mulch film fragments, the plant biomass decreased 20–50% in cress, millet, and rapeseed [60]. In this study, we conducted a phytotoxicity assessment to evaluate whether there were potential adverse effects derived from soil containing 0.2% PBSA films or that supplemented with *A. terreus* HC on plant growth. As shown in Figure 8, we did not observe an obvious adverse influence of PBSA degradation or the addition of fungal culture on the seed germination rate, seedling, and mature plant weight of Chinese cabbage. Whether PBSA or *A. terreus* HC in soil affects other kinds of plants, soil microbial communities, or other organisms requires further research.

5. Conclusions

In this study, we selected two elite PBSA-degrading *Aspergillus* strains, namely, *A. terreus* HC and *A. fumigatus* L30, from soil samples in Taiwan. It is noteworthy that biodegradation of PBSA by *A. terreus* is reported for the first time. The PBSA films placed in the carbon-free basal medium were approximately 42% degraded by *A. terreus* HC within 30 days of incubation, while the films were 26% degraded by *A. fumigatus* L30. In the soil burial test, *A. terreus* HC showed over 90% and 75% degradation rates for summer and winter soil environments, respectively. When adding a high dose of *A. terreus* HC mycelia to the winter soil, the degradation efficacy of PBSA was further improved. This suggests that the application of an elite fungal mycelial suspension to soil with a low degrading ability can aid in its degradability. According to the results of the soil burial test, it can be deduced that PBSA can induce the synthesis of lipolytic enzymes of indigenous microbes to degrade substrates with medium and long carbon chains in the soil. In the phytotoxicity test, the degradation of PBSA films or the addition of fungal mycelia did not cause obvious adverse effects on the growth of seedlings or adult plants of Chinese cabbage. Taken together, the results of this study not only advance our understanding of the biodegradation of PBSA by elite *Aspergillus* strains but also provide insight into improving the efficiency of biodegradation in soil environments with moderate temperature.

Supplementary Materials: The following supporting information can be downloaded at: https://www.mdpi.com/article/10.3390/polym14071320/s1, Figure S1. Clear zones formed by PBSA-degrading microorganisms derived from different sampling sites. Figure S2. Morphology of the respective elite PBSA-degrading fungal strain. Figure S3. Scanning electron microscopy of an uninoculated PBSA plastic film. Figure S4. Weight loss (%) of PBSA plastic films against degradation time in soil burial tests. Figure S5. Lipolytic enzyme activities in the soil.

Author Contributions: Conceptualization, C.-T.L., experiments, H.-L.C. and Y.-T.T.; experimental data analysis, H.-L.C. and Y.-T.T.; writing—original draft preparation, C.-T.L., H.-L.C. and Y.-T.T.; writing—review and editing, C.-T.L., H.-L.C. and Y.-T.T.; soil burial—experiments assistance, W.-S.T.; NMR analysis service, J.-A.W.; research and experimental design recommendations, S.-L.C., S.-J.H. and S.-L.K. C.-T.L. is the corresponding author. All authors have read and agreed to the published version of the manuscript.

Funding: This study was supported by grants from the Ministry of Science and Technology (MOST 108-2313-B-002-058-MY3, MOST 109-2321-B-005-027).

Institutional Review Board Statement: Not applicable.

Informed Consent Statement: Not applicable.

Data Availability Statement: The data presented in this study are available on request from the corresponding author.

Acknowledgments: We would like to thank the Joint Center for Instruments and Researchers, College of Bio-Resources and Agriculture, NTU, for their technical assistance with SEM.

Conflicts of Interest: The authors declare no conflict of interest. All authors have read and approved the final version of the manuscript.

References

1. Bishop, G.; Styles, D.; Lens, P.N. Recycling of European plastic is a pathway for plastic debris in the ocean. *Environ. Int.* **2020**, *142*, 105893. [PubMed]
2. Geyer, R.; Jambeck, J.R.; Law, K.L. Production, use, and fate of all plastics ever made. *Sci. Adv.* **2017**, *3*, e1700782. [PubMed]
3. Horton, A.A.; Walton, A.; Spurgeon, D.J.; Lahive, E.; Svendsen, C. Microplastics in freshwater and terrestrial environments: Evaluating the current understanding to identify the knowledge gaps and future research priorities. *Sci. Total Environ.* **2017**, *586*, 127–141. [PubMed]
4. Serrano-Ruiz, H.; Martin-Closas, L.; Pelacho, A.M. Biodegradable plastic mulches: Impact on the agricultural biotic environment. *Sci. Total Environ.* **2021**, *750*, 141228. [CrossRef] [PubMed]
5. Sintim, H.Y.; Flury, M. Is Biodegradable Plastic Mulch the Solution to Agriculture's Plastic Problem? *Environ. Sci. Technol.* **2017**, *51*, 1068–1069. [CrossRef]

6. Wang, L.; Zhang, F.-S. Characterization of a novel sound absorption material derived from waste agricultural film. *Constr. Build. Mater.* **2017**, *157*, 237–243.
7. Feuilloley, P.; Cesar, G.; Benguigui, L.; Grohens, Y.; Pillin, I.; Bewa, H.; Lefaux, S.; Jamal, M. Degradation of polyethylene designed for agricultural purposes. *J. Polym. Environ.* **2005**, *13*, 349–355. [CrossRef]
8. Zhang, G.S.; Liu, Y.F. The distribution of microplastics in soil aggregate fractions in southwestern China. *Sci. Total Environ.* **2018**, *642*, 12–20. [CrossRef]
9. Brodhagen, M.; Peyron, M.; Miles, C.; Inglis, D.A. Biodegradable plastic agricultural mulches and key features of microbial degradation. *Appl. Microbiol. Biotechnol.* **2015**, *99*, 1039–1056. [CrossRef]
10. Kitamoto, H.K.; Shinozaki, Y.; Cao, X.-H.; Morita, T.; Konishi, M.; Tago, K.; Kajiwara, H.; Koitabashi, M.; Yoshida, S.; Watanabe, T. Phyllosphere yeasts rapidly break down biodegradable plastics. *AMB Express* **2011**, *1*, 44.
11. Zhao, J.-H.; Wang, X.-Q.; Zeng, J.; Yang, G.; Shi, F.-H.; Yan, Q. Biodegradation of poly(butylene succinate-co-butylene adipate) by *Aspergillus versicolor*. *Polym. Degrad. Stab.* **2005**, *90*, 173–179. [CrossRef]
12. Puchalski, M.; Szparaga, G.; Biela, T.; Gutowska, A.; Sztajnowski, S.; Krucińska, I. Molecular and supramolecular changes in polybutylene succinate (PBS) and polybutylene succinate adipate (PBSA) copolymer during degradation in various environmental conditions. *Polymers* **2018**, *10*, 251.
13. Palai, B.; Mohanty, S.; Nayak, S.K. Synergistic effect of polylactic acid (PLA) and Poly (butylene succinate-co-adipate)(PBSA) based sustainable, reactive, super toughened eco-composite blown films for flexible packaging applications. *Polym. Test.* **2020**, *83*, 106130.
14. Lee, S.-M.; Lee, J.-W. Characterization and processing of biodegradable polymer blends of poly (lactic acid) with poly (butylene succinate adipate). *Korea-Aust. Rheol. J.* **2005**, *17*, 71–77.
15. Koitabashi, M.; Sameshima–Yamashita, Y.; Watanabe, T.; Shinozaki, Y.; Kitamoto, H. Phylloplane fungal enzyme accelerate decomposition of biodegradable plastic film in agricultural settings. *Jpn. Agric. Res. Q.* **2016**, *50*, 229–234.
16. Yamamoto-Tamura, K.; Hiradate, S.; Watanabe, T.; Koitabashi, M.; Sameshima-Yamashita, Y.; Yarimizu, T.; Kitamoto, H. Contribution of soil esterase to biodegradation of aliphatic polyester agricultural mulch film in cultivated soils. *AMB Express* **2015**, *5*, 10. [CrossRef]
17. Haider, T.P.; Volker, C.; Kramm, J.; Landfester, K.; Wurm, F.R. Plastics of the Future? The Impact of Biodegradable Polymers on the Environment and on Society. *Angew. Chem. Int. Ed.* **2019**, *58*, 50–62. [CrossRef]
18. Kovacic, F.; Babic, N.; Krauss, U.; Jaeger, K. Classification of lipolytic enzymes from bacteria. *Aerob. Util. Hydrocarb. Oils Lipids* **2019**, *24*, 255–289.
19. Ahmad, A.; Tsutsui, A.; Iijima, S.; Suzuki, T.; Shah, A.A.; Nakajima-Kambe, T. Gene structure and comparative study of two different plastic-degrading esterases from Roseateles depolymerans strain TB-87. *Polym. Degrad. Stab.* **2019**, *164*, 109–117.
20. Tsutsumi, C.; Hayase, N.; Nakagawa, K.; Tanaka, S.; Miyahara, Y. The enzymatic degradation of commercial biodegradable polymers by some lipases and chemical degradation of them. *Macromol. Symp.* **2003**, *197*, 431–442.
21. Hoshino, A.; Isono, Y. Degradation of aliphatic polyester films by commercially available lipases with special reference to rapid and complete degradation of poly(L-lactide) film by lipase PL derived from *Alcaligenes* sp. *Biodegradation* **2002**, *13*, 141–147. [CrossRef] [PubMed]
22. Shinozaki, Y.; Morita, T.; Cao, X.H.; Yoshida, S.; Koitabashi, M.; Watanabe, T.; Suzuki, K.; Sameshima-Yamashita, Y.; Nakajima-Kambe, T.; Fujii, T.; et al. Biodegradable plastic-degrading enzyme from *Pseudozyma antarctica*: Cloning, sequencing, and characterization. *Appl. Microbiol. Biotechnol.* **2013**, *97*, 2951–2959. [CrossRef] [PubMed]
23. Tserki, V.; Matzinos, P.; Pavlidou, E.; Vachliotis, D.; Panayiotou, C. Biodegradable aliphatic polyesters. Part I. Properties and biodegradation of poly(butylene succinate-co-butylene adipate). *Polym. Degrad. Stab.* **2006**, *91*, 367–376. [CrossRef]
24. Hajighasemi, M.; Tchigvintsev, A.; Nocek, B.; Flick, R.; Popovic, A.; Hai, T.; Khusnutdinova, A.N.; Brown, G.; Xu, X.; Cui, H. Screening and characterization of novel polyesterases from environmental metagenomes with high hydrolytic activity against synthetic polyesters. *Environ. Sci. Technol.* **2018**, *52*, 12388–12401.
25. Nakajima-Kambe, T.; Toyoshima, K.; Saito, C.; Takaguchi, H.; Akutsu-Shigeno, Y.; Sato, M.; Miyama, K.; Nomura, N.; Uchiyama, H. Rapid monomerization of poly (butylene succinate)-co-(butylene adipate) by *Leptothrix* sp. *J. Biosci. Bioeng.* **2009**, *108*, 513–516.
26. Waring, S.; Bremner, J. Ammonium production in soil under waterlogged conditions as an index of nitrogen availability. *Nature* **1964**, *201*, 951–952.
27. Sato, S.; Saika, A.; Shinozaki, Y.; Watanabe, T.; Suzuki, K.; Sameshima-Yamashita, Y.; Fukuoka, T.; Habe, H.; Morita, T.; Kitamoto, H. Degradation profiles of biodegradable plastic films by biodegradable plastic-degrading enzymes from the yeast *Pseudozyma antarctica* and the fungus *Paraphoma* sp. B47-9. *Polym. Degrad. Stab.* **2017**, *141*, 26–32.
28. Hu, X.; Gao, Z.; Wang, Z.; Su, T.; Yang, L.; Li, P. Enzymatic degradation of poly (butylene succinate) by cutinase cloned from *Fusarium solani*. *Polym. Degrad. Stab.* **2016**, *134*, 211–219.
29. Merugu, R. Studies on PHB (polyhydroxybutyrate) degradation by some species of Aspergillus. *Int. J. ChemTech Res.* **2012**, *4*, 1111–1113.
30. Vasile, C.; Pamfil, D.; Rapa, M.; Darie-Nita, R.N.; Mitelut, A.C.; Popa, E.E.; Popescu, P.A.; Draghici, M.C.; Popa, M.E. Study of the soil burial degradation of some PLA/CS biocomposites. *Compos. Part B Eng.* **2018**, *142*, 251–262. [CrossRef]

31. Jung, H.-W.; Yang, M.-K.; Su, R.-C. Purification, characterization, and gene cloning of an *Aspergillus fumigatus* polyhydroxybutyrate depolymerase used for degradation of polyhydroxybutyrate, polyethylene succinate, and polybutylene succinate. *Polym. Degrad. Stab.* 2018, *154*, 186–194.
32. Nishida, H.; Tokiwa, Y. Distribution of poly (β-hydroxybutyrate) and poly (ε-caprolactone) aerobic degrading microorganisms in different environments. *J. Environ. Polym. Degrad.* 1993, *1*, 227–233.
33. Yang, J.; Yang, Y.; Wu, W.M.; Zhao, J.; Jiang, L. Evidence of Polyethylene Biodegradation by Bacterial Strains from the Guts of Plastic-Eating Waxworms. *Environ. Sci. Technol.* 2014, *48*, 13776–13784. [CrossRef] [PubMed]
34. Hoang, K.C.; Tseng, M.; Shu, W.J. Degradation of polyethylene succinate (PES) by a new thermophilic Microbispora strain. *Biodegradation* 2007, *18*, 333–342. [CrossRef]
35. Manter, D.K.; Vivanco, J.M. Use of the ITS primers, ITS1F and ITS4, to characterize fungal abundance and diversity in mixed-template samples by qPCR and length heterogeneity analysis. *J. Microbiol. Methods* 2007, *71*, 7–14.
36. Ragonezi, C.; Caldeira, A.T.; Rosário Martins, M.; Salvador, C.; Santos-Silva, C.; Ganhão, E.; Klimaszewska, K.; Zavattieri, A. Molecular approach to characterize ectomycorrhizae fungi from Mediterranean pine stands in Portugal. *Braz. J. Microbiol.* 2013, *44*, 657–665.
37. Hall, T. BioEdit: A user-friendly biological sequence alignment editor and analysis program for Windows 95/98/NT. *Nucleic Acids Symp. Ser.* 1999, *41*, 95–98.
38. La Mantia, F.P.; Ascione, L.; Mistretta, M.C.; Rapisarda, M.; Rizzarelli, P. Comparative investigation on the soil burial degradation behaviour of polymer films for agriculture before and after photo-oxidation. *Polymers* 2020, *12*, 753.
39. Popovic, A.; Hai, T.; Tchigvintsev, A.; Hajighasemi, M.; Nocek, B.; Khusnutdinova, A.N.; Brown, G.; Glinos, J.; Flick, R.; Skarina, T. Activity screening of environmental metagenomic libraries reveals novel carboxylesterase families. *Sci. Rep.* 2017, *7*, 44103.
40. Yoshida, S.; Hiraga, K.; Takehana, T.; Taniguchi, I.; Yamaji, H.; Maeda, Y.; Toyohara, K.; Miyamoto, K.; Kimura, Y.; Oda, K. A bacterium that degrades and assimilates poly (ethylene terephthalate). *Science* 2016, *351*, 1196–1199.
41. Sander, D.; Yu, Y.; Sukul, P.; Schäkermann, S.; Bandow, J.E.; Mukherjee, T.; Mukhopadhyay, S.K.; Leichert, L.I. Metaproteomic Discovery and Characterization of a Novel Lipolytic Enzyme from an Indian Hot Spring. *Front. Microbiol.* 2021, *12*, 1136.
42. Hwang, B.-Y.; Kim, J.-H.; Kim, J.; Kim, B.-G. Screening of Exiguobacterium acetylicum from soil samples showing enantioselective and alkalotolerant esterase activity. *Biotechnol. Bioprocess Eng.* 2005, *10*, 367.
43. Gilham, D.; Lehner, R. Techniques to measure lipase and esterase activity in vitro. *Methods* 2005, *36*, 139–147.
44. Margesin, R.; Feller, G.; Hammerle, M.; Stegner, U.; Schinner, F. A colorimetric method for the determination of lipase activity in soil. *Biotechnol. Lett.* 2002, *24*, 27–33. [CrossRef]
45. Souza, P.M.S.; Sommaggio, L.R.D.; Marin-Morales, M.A.; Morales, A.R. PBAT biodegradable mulch films: Study of ecotoxicological impacts using *Allium cepa*, *Lactuca sativa* and HepG2/C3A cell culture. *Chemosphere* 2020, *256*, 126985. [PubMed]
46. Wong, W.-T.; Tseng, C.-H.; Hsu, S.-H.; Lur, H.-S.; Mo, C.-W.; Huang, C.-N.; Hsu, S.-C.; Lee, K.-T.; Liu, C.-T. Promoting effects of a single *Rhodopseudomonas palustris* inoculant on plant growth by *Brassica rapa* chinensis under low fertilizer input. *Microbes Environ.* 2014, *29*, 303–313.
47. Bano, K.; Kuddus, M.; Zaheer, M.R.; Zia, Q.; Khan, M.F.; Ashraf, G.M.; Gupta, A.; Aliev, G. Microbial Enzymatic Degradation of Biodegradable Plastics. *Curr. Pharm. Biotechnol.* 2017, *18*, 429–440. [CrossRef]
48. Maeda, H.; Yamagata, Y.; Abe, K.; Hasegawa, F.; Machida, M.; Ishioka, R.; Gomi, K.; Nakajima, T. Purification and characterization of a biodegradable plastic-degrading enzyme from *Aspergillus oryzae*. *Appl. Microbiol. Biotechnol.* 2005, *67*, 778–788. [CrossRef]
49. Kliem, S.; Kreutzbruck, M.; Bonten, C. Review on the Biological Degradation of Polymers in Various Environments. *Materials* 2020, *13*, 4586. [CrossRef]
50. Mohanan, N.; Montazer, Z.; Sharma, P.K.; Levin, D.B. Microbial and Enzymatic Degradation of Synthetic Plastics. *Front. Microbiol.* 2020, *11*, 580709. [CrossRef]
51. Singh, B.; Sharma, N. Mechanistic implications of plastic degradation. *Polym. Degrad. Stab.* 2008, *93*, 561–584. [CrossRef]
52. Wilkes, R.A.; Aristilde, L. Degradation and metabolism of synthetic plastics and associated products by *Pseudomonas* sp.: Capabilities and challenges. *J. Appl. Microbiol.* 2017, *123*, 582–593. [CrossRef] [PubMed]
53. Bottone, E.J.; Nagarsheth, N.; Chiu, K. Evidence of self-inhibition by filamentous fungi accounts for unidirectional hyphal growth in colonies. *Can. J. Microbiol.* 1998, *44*, 390–393. [CrossRef] [PubMed]
54. Rodriguez-Urra, A.B.; Jimenez, C.; Duenas, M.; Ugalde, U. Bicarbonate gradients modulate growth and colony morphology in *Aspergillus nidulans*. *FEMS Microbiol. Lett.* 2009, *300*, 216–221. [CrossRef] [PubMed]
55. Ugalde, U. Autoregulatory Signals in Mycelial Fungi. In *Growth, Differentiation and Sexuality*; Springer: Cham, Switzerland, 2016; Volume 1, pp. 185–202.
56. Hoshino, A.; Sawada, H.; Yokota, M.; Tsuji, M.; Fukuda, K.; Kimura, M. Influence of weather conditions and soil properties on degradation of biodegradable plastics in soil. *Soil Sci. Plant Nutr.* 2001, *47*, 35–43.
57. Ishii, N.; Inoue, Y.; Tagaya, N.; Mitomo, H.; Nagai, D.; Kasuya, K.-I. Isolation and characterization of poly (butylene succinate)-degrading fungi. *Polym. Degrad. Stab.* 2008, *93*, 883–888.
58. Bandopadhyay, S.; Sintim, H.Y.; DeBruyn, J.M. Effects of biodegradable plastic film mulching on soil microbial communities in two agroecosystems. *PeerJ* 2020, *8*, e9015. [CrossRef]
59. Li, C.; Moore-Kucera, J.; Lee, J.; Corbin, A.; Brodhagen, M.; Miles, C.; Inglis, D. Effects of biodegradable mulch on soil quality. *Appl. Soil Ecol.* 2014, *79*, 59–69. [CrossRef]

60. Fritz, J.; Sandhofer, M.; Stacher, C.; Braun, R. Strategies for detecting ecotoxicological effects of biodegradable polymers in agricultural applications. *Macromol. Symp.* **2003**, *197*, 397–409. [CrossRef]
61. Muroi, F.; Tachibana, Y.; Kobayashi, Y.; Sakurai, T.; Kasuya, K. Influences of poly(butylene adipate-co-terephthalate) on soil microbiota and plant growth. *Polym. Degrad. Stab.* **2016**, *129*, 338–346. [CrossRef]
62. Qi, Y.L.; Yang, X.M.; Pelaez, A.M.; Lwanga, E.H.; Beriot, N.; Gertsen, H.; Garbeva, P.; Geissen, V. Macro- and micro- plastics in soil-plant system: Effects of plastic mulch film residues on wheat (*Triticum aestivum*) growth. *Sci. Total Environ.* **2018**, *645*, 1048–1056. [CrossRef] [PubMed]
63. Rychter, P.; Biczak, R.; Herman, B.; Smylla, A.; Kurcok, P.; Adamus, G.; Kowalczuk, M. Environmental degradation of polyester blends containing atactic poly(3-hydroxybutyrate). Biodegradation in soil and ecotoxicological impact. *Biomacromolecules* **2006**, *7*, 3125–3131. [CrossRef] [PubMed]
64. Sforzini, S.; Oliveri, L.; Chinaglia, S.; Viarengo, A. Application of Biotests for the Determination of Soil Ecotoxicity after Exposure to Biodegradable Plastics. *Front. Environ. Sci.* **2016**, *4*, 68. [CrossRef]

Article

Influence of Non-Rubber Components on the Properties of Unvulcanized Natural Rubber from Different Clones

Nussana Lehman [1], Akarapong Tuljittraporn [1], Ladawan Songtipya [2,3], Nattapon Uthaipan [4], Karnda Sengloyluan [4], Jobish Johns [5], Yeampon Nakaramontri [6] and Ekwipoo Kalkornsurapranee [1,*]

1. Division of Physical Science, Faculty of Science, Prince of Songkla University, Hat-Yai 90110, Thailand; nussana1708@gmail.com (N.L.); akarapong04@gmail.com (A.T.)
2. Center of Excellence in Bio-Based Materials and Packaging Innovation, Program of Packaging and Materials Technology, Faculty of Agro-Industry, Prince of Songkla University, Hat-Yai 90110, Thailand; ladawan.so@psu.ac.th
3. Faculty of Agro-Industry, Prince of Songkla University, Hat-Yai 90110, Thailand
4. Sino-Thai International Rubber College, Prince of Songkla University, Hat-Yai 90110, Thailand; nuthaipan@gmail.com (N.U.); karnda.seng@gmail.com (K.S.)
5. Department of Physics, Rajarajeswari College of Engineering, Bangalore 560074, India; jobish_johns@rediffmail.com
6. Sustainable Polymer & Innovative Composite Materials Research Group, Department of Chemistry, Faculty of Science, King Mongkut's University of Technology Thonburi, Bangkok 10140, Thailand; yeampon.nak@kmutt.ac.th
* Correspondence: ekwipoo.k@psu.ac.th

Abstract: Natural rubber from different *Hevea braziliensis* clones, namely RRIM600, RRIT251, PB235 and BPM24, exhibit unique properties. The influences of the various fresh natural rubber latex and cream concentrated latex on the non-rubber components related properties were studied. It was found that the fresh natural rubber latex exhibited differences in their particle size, which was attributed to the non-rubber and unique signature of clones which affect various properties. Meanwhile, the cream concentrated latex showed the protein contents, surface tension, and color of creamed latex to be lower than the fresh natural latex. However, TSC, DRC, viscosity, particle size and green strength of concentrated latex were found to be higher than the fresh natural latex. This is attributed to the incorporation of HEC molecules. Also, the rubber particle size distribution in the RRIM600 clone exhibited a large particle size and uniform distribution, showing good mechanical properties when compared to the other clones. Furthermore, the increased green strength in the RRIM600 clone can be attributed to the crystallization of the chain on straining and chain entanglement. These experimental results may provide benefits for manufacturing rubber products, which can be selected from a suitable clone.

Keywords: fresh natural latex; non-rubber components; clones; cream concentrated latex; hydroxyl ethyl cellulose

1. Introduction

Natural rubber latex (NRL) is obtainable from a rubber plant in the form of latex. Latex is the white milk-like fluid which is obtained by wounding the rubber plant. *Hevea braziliensis* is the most common commercial source of latex today. Fresh natural rubber latex consists of two main components. 25–41% of dry rubber content or hydrocarbon (*cis*-1,4-polyisoprene) and the other non-rubber components consisting of mainly carbohydrates, proteins, lipids, minerals, and salt content in an aqueous serum phase (Table 1) [1,2]. Fresh natural rubber latex can be contaminated by micro-organisms because it contains various nutritious substances otherwise known as non-rubber components [3]. Normally, natural rubber latex spontaneously coagulates shortly after it comes out of the tree. Due to the bacterial attack often occurring on the protein constituents which act as colloidal stabilizers to keep the latex water dispersible, coagulation is prevented [2,4]. In order to facilitate preservation, high amounts of ammonia are added to the latex. Ammonia is added to the latex to maintain in the form of liquid, and this is concentrated either by creaming or centrifuging. The resulting concentration can be transported as a liquid. Fresh natural latex can be transformed to concentrated natural latex in order to maintain the constant quality of concentrated natural latex, and to generate the economic value for latex's transportation. The concentrated natural latex can be produced by various processes, including evaporation, electrocantation, centrifuging and creaming processes [3].

Table 1. Composition of fresh natural rubber latex [2].

Constituents	% Composition
Rubber particles (cis-1,4-polyisoprene)	30–40
Protein	2.0–3.0
Lipids	0.1–0.5
Resins	1.5–3.5
Ash	0.5–1.0
Sugars	1.0–2.0
Water	55–65

The creaming process is popular, as it often avoids the use of sophisticated equipment, thus offering a simple and cost-effective route to concentrate the latex. The creaming process is a chemical process involving the addition of creaming agents into the vessels containing field latex to hasten phase separation (upper rubber fraction and lower serum) [5]. The creamed latex generally contains around 50–60% of dry rubber content. It is well-known that concentrated natural rubber latex is used in many applications. Mainly, it has been used in gloves, condoms, toys, balloons, catheters, medical tubing, elastic threads, latex foams, etc. [6–8]. NR has outstanding strength, along with excellent dynamic properties, low hysteresis loss, high tensile strength and resistance to forms of fatigue such as chipping, cutting or tearing [9]. However, fresh natural rubber latex from various *Hevea brasiliensis* clones consists of different non-rubber components. Composition of fresh natural latex depends on factors such as soil condition, fertilizer quality, tapping system, season, and, in particular, the natural rubber clonal variety [10]. A previous study on the properties of various *Hevea brasiliensis* clones (i.e., RRIM600, PB235, and RRTI408) clearly showed the difference in protein and lipid contents. The protein and lipid contents, together with gel content, play essential roles in controlling various properties of unvulcanized NR. It is noted that the non-rubber components, especially proteins and phospholipids, have been found to strongly affect the various properties of raw NR and its vulcanizates. For instance, the proteins present in latex play a major role in deciding the properties of latex products such as elasticity, modulus and barrier functions. This relates to the presence of nitrogenous amino acids, which might act as cure accelerators, antioxidants and thermal stabilizers in NR. In addition, the whole lipid content, especially the phospholipids, was found to be inversely proportional to the tack properties of NR and unsaturated fatty acids act as a plasticizer for rubber by lowering the plasticity of NR [11,12]. Furthermore, fresh natural rubber latex from various *Hevea brasiliensis* clones consists of high molecular weight (MW) components and a wide distribution of molar mass (MMD). The molecular structures of two NR clones (i.e., RRIM600 and PB235) were analyzed by size exclusion chromatography (SEC) and it was found that the RRIM600 clone had bimodal MMD, whereas the PB235 clone had unimodal MMD [12]. Therefore, the non-rubber content, especially proteins and lipids, have been found to strongly affect various properties of raw NR and predominantly influence the green strength of un-vulcanized NR. Furthermore, unsaturated fatty acids can act as plasticizers in natural rubber latex [13]. The properties of fresh natural rubber latex are varied ($P < 0.01$) as a function of clone type, tapping method and climate factors. For example, dry rubber content (DRC) is generally decreased in the beginning of the dry season (May to June), while simultaneously the nitrogen and ash contents (%) are increased in the same period [14].

The present work aimed to study the effect of non-rubber components on the properties of fresh natural latex and creamed concentrated latex. Fresh natural latex from 4 different clones (RRIM600, RRIT251, PB235 and BPM24) were chosen to study their different dry rubber content and molar mass distributions. The yield of fresh natural rubber latex from four different clones is comparatively more upon introducing new genetic varieties of *Hevea Brasiliensis* recommended by the Rubber Research Institute of Thailand (RRIT). The fresh natural latex was collected from plantations in Songkhla province, Thailand, during the late part of tapping season in Thailand (May and June).

The impact of non-rubber components on both biochemical and physicochemical indicators in liquid (latex) and dry states (film) was investigated. Protein contents, molecular weight (MW), DRC, TSC, surface tension, viscosity and morphology were measured for liquid latex, and the mechanical properties of the dry state (unvulcanized natural rubber film) were investigated.

2. Materials and Methods

2.1. Materials

Natural rubber latex samples from four *Hevea brasiliensis* clones, namely RRIM600, RRIT251, PB235 and BPM24, were collected from plantations in Songkhla province, Thailand, during the late part of tapping season in Thailand (i.e., May–June). 28% of ammonia solution was added as a preservative to the fresh field NR latex and it prevents coagulation of latex. Hydroxyethylcellulose (HEC) was used as a creaming agent purchased from Brenntag Ingredients Public Company Limited (Bangkok, Thailand). Potassium laurate was purchased from Lucky Four Co., Ltd., (Bangkok, Thailand) and used as a pH modifier for the creaming process of fresh field NR latex.

2.2. Preparation of Creamed Concentrated Latex

Creamed concentrated latex was prepared using a formulation, as shown in Table 2. The percentages of dry rubber content of four *Hevea brasiliensis* clones, namely RRIM600, RRIT251, PB235 and BPM24, are found to be 41.0%, 40.8%, 42.5% and 24.0%, respectively. 20% (w/w) potassium laurate was added into the treated latex with continuous mechanical stirring for 5 min before incorporation of the creaming agent. The creamed concentrated latex with hydroxyethyl cellulose (HEC) as a creaming agent was dissolved in deionized water at a concentration of 1% (w/w). The mixture was thoroughly stirred at 120 rpm for 30 min. The mixture was then incubated at room temperature for 7 days. After 7 days, the latex was found to be separated into two layers, rubber particles at the top and aqueous serum at the bottom. The aqueous serum phase was removed and the upper rubber fraction—the creamed concentrated latex—was finally extracted from the mixture for further investigation.

Table 2. Formulation is used to prepare creamed concentrated latex.

Chemicals	Quantity, phr
Fresh filed NR latex	100
HEC (1% w/w in water)	0.4
20% w/w potassium laurate	0.3

2.3. Characterization and Measurements

2.3.1. Analysis and Testing of the Fresh Natural Latex and Creamed Concentrated Latex

Analysis and testing of latex including the total solid content (TSC) and dry rubber content (DRC) were performed according to ISO 124 and ISO 126, respectively. The viscosity of latex was also measured using a Brookfield digital viscometer, model LVDV - III Ultra, with spindle no. 2 at a speed of 60 rpm. Furthermore, the surface tension was determined by Du Noüy ring method [15]. The measurement was performed by an instrument known as a Tensiometer.

2.3.2. The Protein Contents

The protein content was estimated from its nitrogen content by the Kjeldahl method according to the Association of Official Analytical Chemists (AOAC) and Official Methods of Analysis of Fertilizers (OMAF) [16]. A 0.1 g rubber sample was mixed with a mixture of catalysts (0.65 g) and then digested in concentrated sulfuric acid (H_2SO_4) until the rubber was completely digested. Then, an alkaline solution (67% w/v NaOH) was added to the mixture of the digested solution. After that, the solution was distilled, and the distillate was collected in a boric acid solution. Finally, the distillate was titrated with 0.01 N H_2SO_4 to determine the ammonia content, allowing for estimating the protein content by multiplying the nitrogen content (mass) with 6.25.

2.3.3. Molecular Weight and Polydispersity Index

Molecular weight (Mw), number-average molecular weight (Mn) and polydispersity index (PDI) of the fresh natural latex were investigated by gel perforation chromatography (GPC) technique (1260 infinity II GPC/SEC MDS, Agilent Technologies, Germany) with a refractive index detector. The solution of fresh natural latex samples with a concentration of 0.001 g/mL was prepared by using tetrahydrofuran (THF) as a solvent before filtrating through a 0.45 μm membrane. The THF was also applied as a mobile phase with a flow rate of 1.0 mL/min under 40 °C.

2.3.4. Morphological Properties

Particles and particle distribution of the lattices were also analyzed using a laser particle size analyzer, Coulter model LS230 particle size analyzer. Furthermore, morphological properties of the latex particles were examined with a transmission electron microscope (TEM), model Jem 2010, Japan with 160 kV with a magnification of 10,000×. In TEM technique, the latex was first diluted with deionized water to a concentration of 0.025 wt%. An aqueous solution of OsO_4 (2 wt%) was then added into the diluted latex and allowed to stain the rubber molecule overnight [17]. The stained samples were then examined by TEM.

2.3.5. Mechanical Properties

The tensile testing was performed using a universal testing machine (model H10KS, Hounsfield, UK). The tests were performed with a crosshead speed of 500 mm/min at room temperature using dumbbell-shaped specimens according to ASTM D412. In the case of hardness, the samples were tested using a Shore A durometer (Frank GmbH, Hamburg, Germany) as per ASTM D2240. Furthermore, the color of NRs from various *Hevea brasiliensis* clones was characterized with a Lovibond colorimeter according to ASTM D3157 and with a HunterLab spectrophotometer.

3. Results and Discussion

3.1. Characterization of Fresh Natural Latex

The physical properties of fresh natural latex from four different *Hevea brasiliensis* clones were measured. Table 3 shows the TSC (total solid content), DRC (dry rubber content), protein contents, surface tension, viscosity and color of different fresh natural latex collected from a variety of clones. TSC, DRC and protein contents showed different results for NRs from various clones. The DRC of RRIM600, RRIT251 and PB235 clones exhibited the same trend (40–42%), whereas the BPM24 clone showed the least value. Moreover, protein contents estimated from Kjeldahl method also showed different values. This is consistent with the observation that clones and the environment affect the metabolism of latex regeneration [18]. The viscosity of the fresh natural latex depends on TSC and DRC due to a high TSC, and DRC may limit the yield by hindering latex flow. In addition, the surface tension is the important parameter that decides the physical properties of fresh natural latex with respect to its intrinsic properties and its applications (wetting and dipping). The results of surface tension of all clones showed the same trend (42–45 mN/m).

Table 3. TSC, DRC, protein content, surface tension, viscosity, lovibond comparator and particle size of the fresh natural latex from four different *Hevea brasiliensis* clones.

Properties	*Hevea Brasiliensis* Clones			
	RRIM600	RRIT251	PB235	BPM24
Total solid content (%)	45.30 ± 0.16	43.47 ± 0.13	45.16 ± 0.05	28.03 ± 0.03
Dry rubber content (%)	40.90 ± 3.45	40.82 ± 0.28	42.47 ± 0.58	23.61 ± 1.90
Protein content (wt%)	3.03 ± 0.009	3.33 ± 0.007	2.64 ± 0.004	3.31 ± 0.002
Viscosity (cps) (Spin no.1/Speed 60 rpm)	12.50	11.50	12.00	5.00
Surface tension (mN/m)	45.50 ± 0.70	45.00 ± 1.41	42.00 ± 0.00	45.00 ± 1.41
Lovibond comparator	2.50–3.00	3.00–4.00	8.00–10.00	12.00–14.00
Particle size (μm) (mean)	1.66	1.59	1.27	1.24

The color of all clones selected were determined with a Lovibond colorimeter and HunterLab spectrophotometer, and they are summarized in Table 3. A photo of the rubber sample is also shown in Figure 1. The yellow color of NR is not only due to the presence of non-rubber constituents but also depends on clonal and seasonal variations, soil types and tapping frequency. The distinctive yellow color in NR has been attributed to the presence of carotenoids. Mostly, the non-rubber components affect the color of products made from it. Therefore, the color change of the samples is likely to arise from the oxidation of non-rubbers such as proteins and lipids [19].

Figure 1. The color of the fresh natural latex from four different *Hevea brasiliensis* clones.

3.2. Rubber Particle Size Distribution of the Fresh Natural Latex

Figure 2 shows the particle size distribution of fresh natural latex from four different *Hevea brasiliensis* clones. Results showed that the diameters of rubber particles in all clones varied between 0.40 μm and 5.00 μm. The typical quasi-unimodal particle size distribution for RRIM600 and RRIT251 clones was observed, whereas PB235 and BPM24 clones were bimodal. This confirms the presence of two populations of chains in rubber, leading to determination of the average molar masses, as well as the size and shape of natural rubber molecules. The main difference between the four clones therefore lies in the relative quantity of short chain. It is anticipated that the high number of short chains in the bimodal distribution for PB235 and BPM24 clone yielded many chain ends or terminals of NR molecules and chain ends together with more non-rubber components than the other clones.

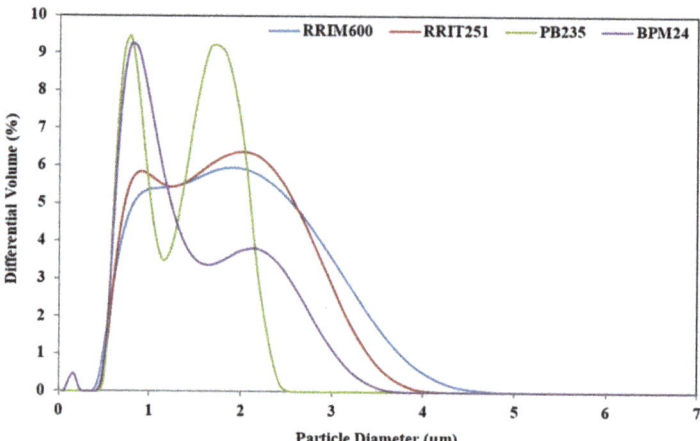

Figure 2. Particle size distribution of the fresh natural latex from four different *Hevea brasiliensis* clones.

Rubber particle size in RRIM600 and RRIT251 clones are found to be larger in the range between 0.50 μm and 4.50 μm. The fine particles with size lower than 0.50 μm are observed in BPM24 clone together with more portions of larger rubber particles in the range between 0.40 μm and 3.50 μm and shows wider particle size distribution than other clones, whereas PB235 clone shows narrower distribution in the range of 0.45–2.52 μm. The difference in the range of particle size might be attributed to the non-rubbers and the unique signature of the clones [20]. These results are in good agreement with the results of transmission electron micrographs as shown in Figure 3. It is clearly observed that the latex of RRIM600 and RRIT251 clones consist of large particles (Figure 3a,b), while the latex from PB235 and BPM24 clones (Figure 3c,d) consist of both large and small particles. Rubber particles are observed as spherically shaped and some were pear-shaped; regardless, their particle size, either they were large or small rubber particles for all clones selected [21]. It was observed that some elongated particles appeared as a result of aggregation of non-rubber in the analyzed sections of BPM24, consisting of many nodules of non-rubber on the particle surfaces.

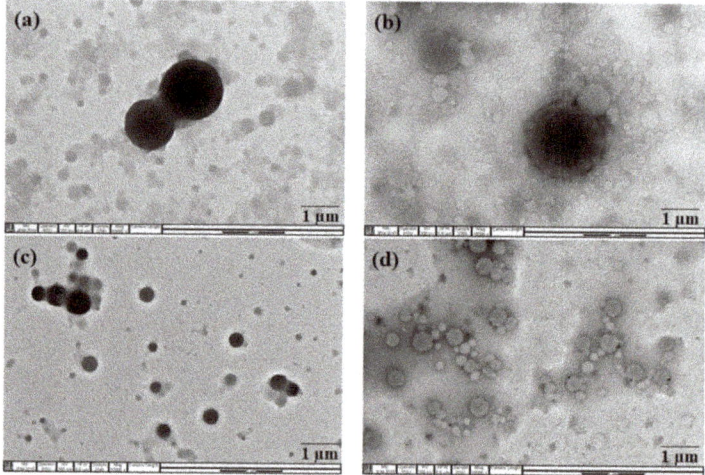

Figure 3. Transmission electron micrographs of (**a**) RRIM600; (**b**) RRIT251; (**c**) PB235; and (**d**) BPM24. (×10,000).

3.3. Molar Mass Distribution of the Fresh Natural Latex

The molar mass and molar mass distributions (MMD) in the natural rubber play an important role in their bulk properties. The molar mass distribution of natural rubber from RRIM600, RRIT251, PB235 and BPM24 clones are shown in Table 4. MMD, in the case of all clones, confirmed that the rubber from RRIM600 and RRIT251 are bimodal MMD, but the rubber from PB235 and BPM24 clones are unimodal MMD. The properties of NRs from the four clones depend on the relative quantity of short chains. It is anticipated that the high number of short chains in this bimodal distribution yielded many chain ends or terminals of NR molecules [10]. The polydispersity index (PDI) of the BPM24 clone exhibited a superior value when compared to the other clones and exhibited an MWD curve with unimodal distribution with almost constant value skewed in case of low molecular weight. However, the size of rubber particles might influence the molecular weight of their rubber particles in the latex.

Table 4. Molecular weight and polydispersity index of fresh natural latex from four different *Hevea brasiliensis* clones.

Hevea Brasiliensis Clones	$M_n \times 10^5$ (g/mol)	$M_w \times 10^6$ (g/mol)	Polydispersity Index (PDI)
RRIM600	3.81	1.96	5.14
RRIT251	3.71	1.68	4.53
PB235	4.34	2.02	4.66
BPM24	1.37	1.09	7.92

3.4. Mechanical Properties of Fresh Natural Latex

Table 5 shows the mechanical properties of NR films collected from four different *Hevea brasiliensis* clones. The green strength of elastomers has been commonly attributed to long-chain branching, interactions between polar groups, the presence of a gel, chain entanglements, and crystallization on stretching [22]. The variations in these factors are responsible for the different green strength in the case of NRs from various rubber clones. It was seen that RRIM600, RRIT251 and BPM24 clones showed higher stress at break compared to PB235. This might be attributed to the lower protein content (Table 3) that corresponds to the levels of short chains *cis*-1,4-isoprene. It was presumed that the non-rubber components in rubber molecules caused the formation of loosely crosslinked structures or a gel in natural rubber. Generally, the area under stress-strain curve indicates the toughness of a material. In Figure 4, it is seen that the PB235 clone showed the least area due to the least toughness among the other clones, and this was related to the results reported previously [10,11]. This finding clearly indicates that the non-rubber components (i.e., protein content) acted as the reinforcing filler and enhanced the rigidity of rubber. They also form macro-gel in rubber molecules by providing stronger rubber networks with high moduli, stiffness, and hardness [11,23–25]. This indicates that the removal of proteins or non-rubber components lead to reduce the moduli, stress at break and hence the stiffness of rubber [25]. In the previous work, [10,22,26] rubber molecular chains comprise of long-chain isoprene units and the chain ends consist of one protein end group and another phospholipid end group. It can be seen that the α-terminal group with mono- or di-phosphate groups associated with phospholipids, whereas the ω-terminal is a dimethylallyl group that interacts with proteins [26]. In Figure 5, it is clearly seen that these molecules could interact with both the functional terminal groups via hydrogen bonding and ionic bonds derived from metal ions. The proteins in natural rubber are considered to originate branch points by hydrogen bonding, as well as the phospholipids are linked to another phospholipid molecule in other chain ends via hydrogen bonding or ionic bonds derived from metal ions. Moreover, proteins at ω-terminal could interact with other protein molecules through hydrogen bonding [10,26]. In addition, the increase in green strength can be attributed to the number of branch points per chain and chain entanglement. The long chain branching in rubber molecule plays an important role in the higher green strength [22]. It was found that a high polydispersity index in case of RRIM600 and BPM24 clones exhibited the higher green strength. These findings might have resulted from the distribution of larger rubber particles and not that of smaller particles. Thus, it is reasonable to assume that the long chain branching in natural rubber affects the green strength. Therefore, it is concluded that the green strength is related to the non-rubber components, branch structure, chain entanglements, and particle size in the natural rubber molecules.

Table 5. Mechanical properties of NR from fresh natural latex collected from four different *Hevea brasiliensis* clones.

Properties	*Hevea Brasiliensis* Clones			
	RRIM600	RRIT251	PB235	BPM24
100% modulus (MPa)	0.36 ± 0.01	0.37 ± 0.04	0.32 ± 0.01	0.40 ± 0.02
300% modulus (MPa)	0.38 ± 0.01	0.39 ± 0.06	0.34 ± 0.04	0.42 ± 0.02
500% modulus (MPa)	0.40 ± 0.02	0.42 ± 0.08	0.36 ± 0.01	0.45 ± 0.04
Green strength (MPa)	0.86 ± 0.07	0.73 ± 0.09	0.47 ± 0.07	0.89 ± 0.06
Elongation at break (%)	969 ± 13	884 ± 4	850 ± 45	877 ± 30
Hardness (Shore A)	15.0 ± 0.8	15.5 ± 0.5	13.0 ± 0.8	14.0 ± 0.5

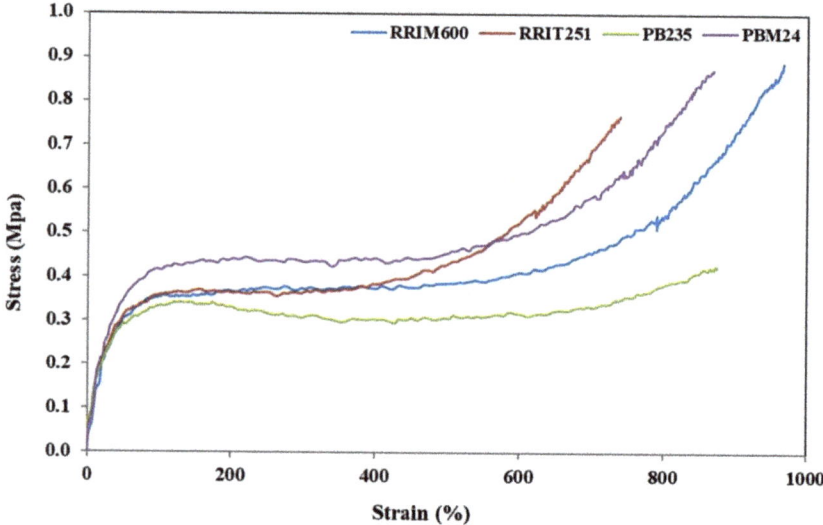

Figure 4. Stress-strain curves of NRs from the four *Hevea brasiliensis* clones.

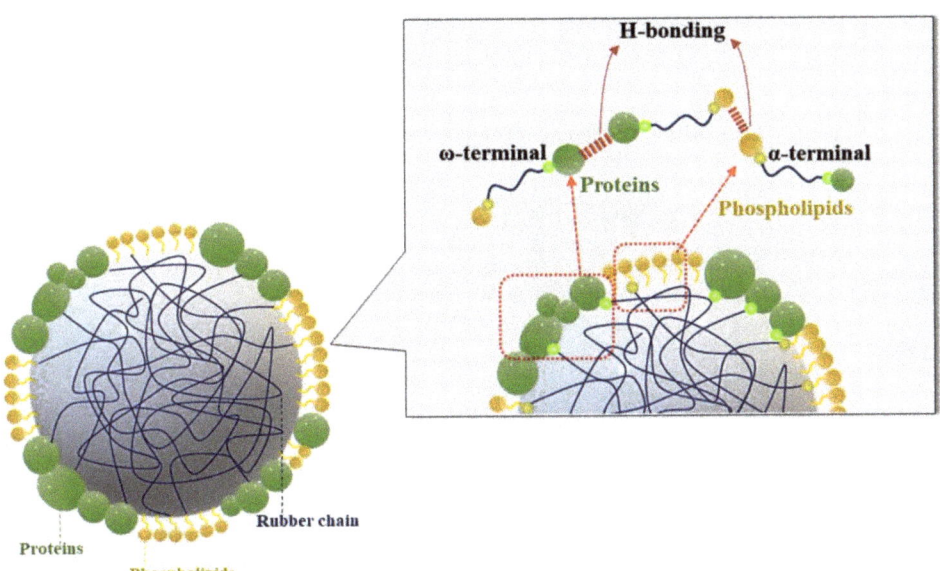

Figure 5. Schematic representation of α- and ω-terminal groups on the rubber latex particle surface.

3.5. Characterization of Cream Concentrated Latex from Four Different Hevea Brasiliensis Clones

Creaming method is a physicochemical process in which the creaming agent is mixed with fresh natural latex and then a phase separation process (upper rubber fraction and lower serum) is applied [4]. This work is aimed at preparing concentrated latex using hydroxyethyl cellulose (HEC) as a creaming agent. Creamed concentrated latex generally contains the creaming agent, as well as ammonia for preservation. It was found that TSC, DRC, viscosity, and particle size values of the cream concentrated latex are clearly higher than those of fresh natural latex, whereas the protein content, surface tension, and the color exhibited lower values. Table 6 shows the TSC and DRC of creamed concentrated latex, which are increased in the case of all clones when compared to the fresh natural latex. This indicates that hydroxyethyl cellulose (HEC) increases the efficiency to separate rubber particles from the serum solution [2,4]. The incorporation of high-molecular weight hydroxyethyl cellulose (HEC) with higher degree of polarity is responsible for the creaming phenomenon. Thus, hydroxyethyl cellulose (HEC) as a hydrophilic colloid dispersed in aqueous medium and covered the surface of rubber particles. Also, the branched segments of the HEC molecules could be well entangled with the neighboring rubber particles (Figure 6), restricting the movement of the rubber particles and causing a larger rubber particle size that led to an increase in viscosity of the latex [2,27]. It is clear that the creamed concentrated latex from all the clones selected exhibit higher TSC, DRC, particle size and viscosity. Furthermore, during the creaming process, ammonia and potassium laurate soap need to be added in order to prevent coagulation and reduce the viscosity in the latex. It is known that potassium laurate soap is a nonionic surfactant used as stabilizing additives for natural rubber latex. The addition of surfactants during storage modifies the surface of the rubber particles or interface between rubber particles and water by reducing the energy difference between rubber hydrocarbon chains and water. It forms a stable colloidal mixture by reducing the surface tension of all clones selected [28]. Moreover, the addition of creaming agent, ammonia, and surfactant leads to chemical transformations of latex particles along the storage time. Therefore, the amount of non-rubber components or protein should be reduced. The protein content of fresh natural latex is decreased after the creaming process, especially in the case of RRIM600 and PB235 clones reduced the values by 27% and 38%, respectively. Proteins associated with rubber particles are reduced by strong alkali hydrolysis known as saponification reaction [26,27,29,30]. Finally, the color of rubber according to Lovibond colorimeter in creamed concentrated latex collected from all clones showed lighter color compared to the fresh natural latex, especially the BPM24 clone shown in Figure 7. The removal of some non-rubber constituents from the latex during storage may be the reason for this light color. It was found that the RRIM600 clone exhibits more transparency than other clones. Therefore, it was concluded that the properties of cream concentrated latex depend on creaming agent, stabilizing agent, and non-rubber components.

Table 6. TSC, DRC, protein contents, surface tension, viscosity, Lovibond comparator and particles size of the cream concentrated latex from four different *Hevea brasiliensis* clones.

Properties	*Hevea Brasiliensis* Clones			
	RRIM600	RRIT251	PB235	BPM24
Total solid content (%)	56.75 ± 0.02	54.46 ± 0.05	66.30 ± 0.05	49.93 ± 0.15
Dry rubber content (%)	54.91 ± 0.06	51.53 ± 1.27	64.80 ± 0.02	48.01 ± 1.81
Protein content (wt%)	2.20 ± 0.002	2.93 ± 0.009	1.64 ± 0.004	2.60 ± 0.001
Viscosity (cps)(Spin no.1/Speed 60 rpm)	46.00	41.00	75.50	7.00
Surface tension (mN/m)	41.50 ± 0.71	41.00 ± 0.00	39.00 ± 0.00	38.50 ± 0.70
Lovibond comparator	2.00–2.50	3.50–5.00	7.00–8.00	5.00–7.00
Particle size (μm) (mean)	1.90 ± 0.00	1.79 ± 0.70	1.52 ± 0.70	1.22 ± 0.64

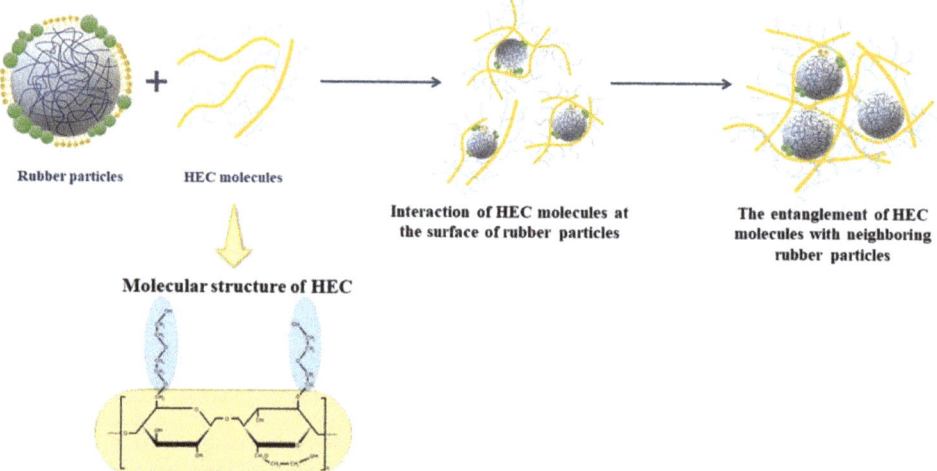

Figure 6. Proposed model for the structures of creamed latex with the incorporation of HEC molecules at the surface of rubber particles.

Figure 7. The color of cream concentrated latex from four different *Hevea brasiliensis* clones.

3.6. Rubber Particle Size Distribution of Fresh Natural Latex and Cream Concentrated Latex from the RRIM600 Clone

Figure 8 shows the particle size distribution of fresh natural latex and cream concentrated latex from the RRIM600 clone. The result shows that the diameter of rubber particles in cream concentrated latex are distributed widely in the range of 0.4–15 µm, whereas the fresh natural latex showed a narrower distribution in the range 0.4–5.0 µm. It is clear that the hydroxyethyl cellulose (HEC) molecules covered the surface of rubber particles and might be attributed to all types of rubber particles that can possibly diffuse from the serum into the creamed layers during the creaming process [27]. Thus, the larger size is observed in the rubber particle in the case of concentrated latex. The results are in good agreement with the TEM measurement, as shown in Figure 9. The cream concentrated latex (Figure 9b) showed a greater number of large particles due to the enclosement of HEC and surfactant layer on the surface of particles.

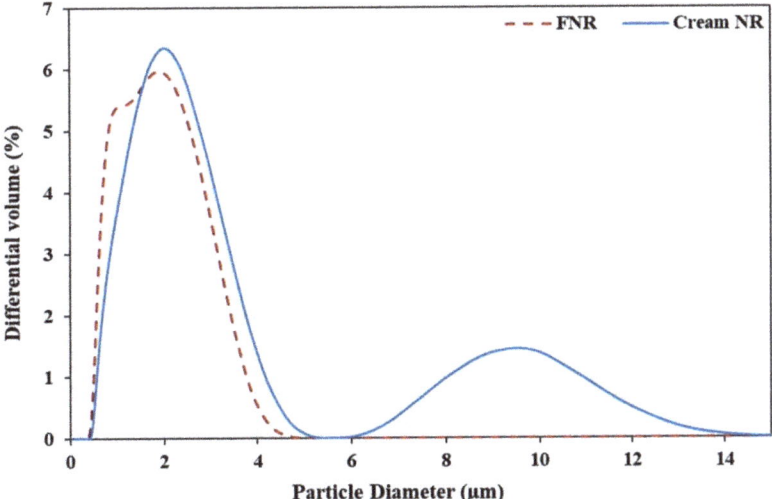

Figure 8. Particle size distribution of the fresh natural latex and cream concentrated latex from the RRIM600 clone.

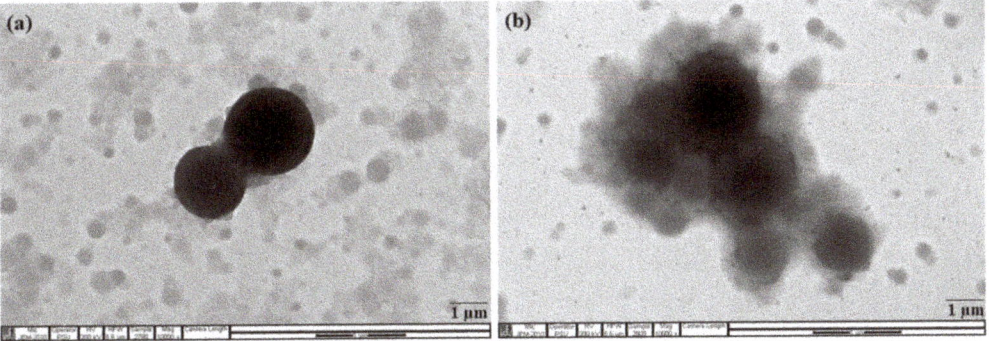

Figure 9. Transmission electron micrographs of (**a**) Fresh natural latex; and (**b**) Cream concentrated latex from the RRIM600 clones. (×10,000).

3.7. Comparison between the Mechanical Properties of Fresh Natural Latex and Cream Concentrated Latex from RRIM600 Clone

Table 7 shows the green strength, moduli at 100, 300, and 500% elongations, elongation at break, and hardness. Also, Figure 10 shows the stress-strain curve of fresh natural latex, and is compared with the cream concentrated latex. The overall properties of cream concentrated latex were shown to be higher than that of the fresh natural latex, except the elongation at break. This can be explained by the long chain branching in HEC molecules and physical entanglement between HEC and rubber chains. This might be attributed to the crystallization on stretching being related to the number of branch points per chain and chain entanglement [26,31]. This physical entanglement could resist the movement of rubber chains. This result led to the increase in modulus, green strength, and hardness by reducing the elongation at break. Furthermore, the toughness observed from the area under stress-strain curve of the fresh latex exhibit slightly higher than creamed latex. This might be related to the non-rubber components due to the higher non-rubber constituents (proteins and lipids) forming strong structures of loose crosslinks in rubber molecules [10]. Therefore, it is concluded that the green strength is related to the incorporation of high-molecular weight HEC molecules as a creaming agent and non-rubber components, along with rubber molecules.

Table 7. Mechanical properties of fresh natural latex and cream concentrated latex from RRIM600 clone.

Properties	Fresh Natural Latex	Cream Concentrated Latex
100% modulus (MPa)	0.36 ± 0.010	0.39 ± 0.005
300% modulus (MPa)	0.38 ± 0.010	0.42 ± 0.005
500% modulus (MPa)	0.40 ± 0.015	0.51 ± 0.012
Green strength (MPa)	0.86 ± 0.070	1.47 ± 0.030
Elongation at break (%)	969 ± 13	834 ± 4
Hardness (Shore A)	13.0 ± 0.8	27.0 ± 0.5

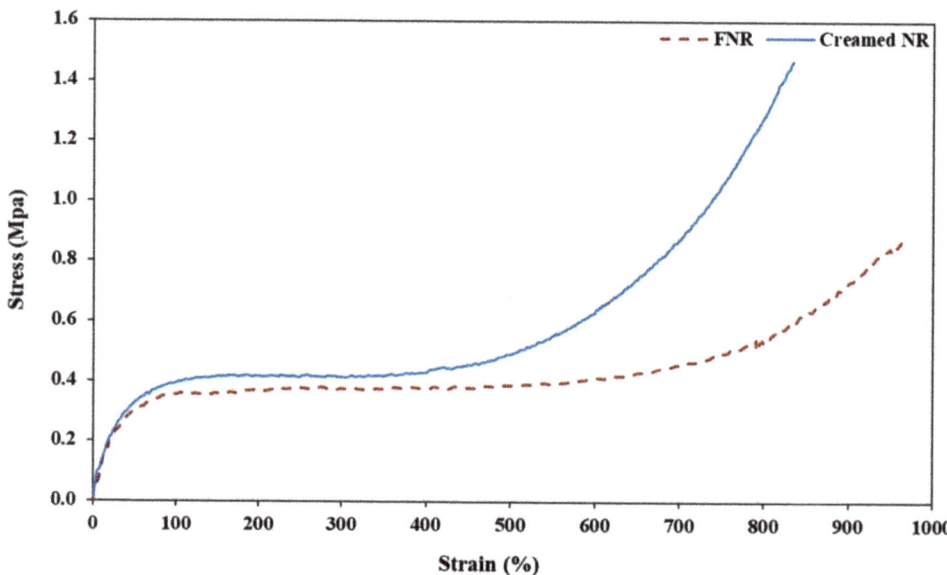

Figure 10. Stress-strain curve of fresh natural latex and cream concentrated latex from RRIM600 clone.

4. Conclusions

Natural rubber was collected from four different *Hevea brasiliensis* clones (i.e., RRIM600, RRIT251, PB235, and BPM24) and the studies showed a variation in their non-rubber components. The amount of non-rubber components in the latex led to better mechanical and physicochemical properties, such as TSC, DRC, viscosity, surface tension, color, particles size, and molar mass distribution. It was found that the RRIM600, RRIT251, and BPM24 clones showed a protein content of about 3.30 wt%, while the PB235 clone exhibited the least protein content. In addition, the rubber particle size distribution in the RRIM600 clone exhibited a large particle size and a uniform distribution. This resulted in better mechanical properties when compared to the other clones.

Furthermore, the cream concentrated latex was successfully prepared using HEC as a creaming agent and potassium laurate as a creaming aid. It was found that the cream concentrated latex showed higher TSC, DRC, viscosity, particle size, and mechanical properties compared to the fresh natural latex. However, except for the surface tension, protein contents and color were found to be lower than the cream concentrated latex. This is attributed to the incorporation of high-molecular weight HEC molecules and surfactant in the cream concentrated latex. Knowledge acquired from this investigation may lead to various applications according to the unique signature of natural rubber from each clone.

Author Contributions: Data curation, writing—original draft preparation, N.L.; Investigation, A.T.; Formal analysis, visualization, L.S.; Methodology, resources, N.U.; Methodology, K.S.; Writing—review and editing, J.J.; Visualization, writing—review and editing, Y.N.; Conceptualization, writing—review and editing, project administration, funding acquisition, E.K. All authors have read and agreed to the published version of the manuscript.

Funding: The Research Fellowship, Faculty of Science, Prince of Songkla University (contract no. 64001).

Institutional Review Board Statement: Not applicable.

Informed Consent Statement: Not applicable.

Data Availability Statement: Not applicable.

Acknowledgments: This research was also supported by the Research Fellowship, Faculty of Science, Prince of Songkla University (contract no. 64001) and Rubber Product and Innovation Development Research (SCIRU63002), Faculty of Science. We would like to thank Sompong Petrat, Rubber Authority of Thailand for supporting the fresh natural rubber latex.

Conflicts of Interest: The authors declare no conflict of interest.

References

1. Ochigbo, S.S.; Lafia-Araga, R.A.; Suleiman, M.A.T. Comparison of two creaming methods for preparation of natural rubber latex concentrates from field latex. *Afr. J. Agric. Res.* **2011**, *12*, 2619–2916. [CrossRef]
2. Suksup, R.; Imkaew, C.; Smitthipong, W. Cream concentrated latex for foam rubber products. In Proceedings of the 4th International Conference on Mechanical, Materials and Manufacturing (ICMMM 2017), Atlanta, GA, USA, 25–27 October 2017; Volume 272, pp. 1–8.
3. Riyajan, S.; Santipanusopon, S. Influence of ammonia concentration and storage period on properties field NR latex and skim coagulation. *J. Elastomers Plast.* **2010**, *63*, 240–245.
4. Suksup, R.; Sun, Y.; Sukatta, U.; Smitthipong, W. Foam rubber from centrifuged and creamed latex. *J. Polym. Eng.* **2019**, *39*, 1–9. [CrossRef]
5. Yip, E.; Cacioli, P. The manufacture of gloves from natural rubber latex. *J. Allergy Clin. Immunol.* **2002**, *110*, S3–S14. [CrossRef] [PubMed]
6. Peng, Z.; Kong, L.X.; Li, S.D.; Chen, Y.; Huang, M.F. Self-assembled natural rubber/silica nanocomposites: Its preparation and characterization. *Compos. Sci. Technol.* **2007**, *67*, 3130–3139. [CrossRef]
7. Krainoi, A.; Poomputsa, K.; Kalkornsurapranee, E.; Johns, J.; Songtipya, L.; Nip, R.L.; Nakaramontri, Y. Disinfectant natural rubber films filled with modified zinc oxide nanoparticles: Synergetic effect of mechanical and antibacterial properties. *Express Polym. Lett.* **2021**, *15*, 1081–1100. [CrossRef]
8. Afreen, S.; Haque, R.K.; Huda, M.K. Troubleshooting for the observed problems in processing latex concentrate from natural resource. In Proceedings of the 4th International Conference on Energy and Environment 2013 (ICEE 2013), Putrajaya, Malaysia, 5–6 March 2013; Volume 16, pp. 1–4. [CrossRef]

9. Rong, J.; Yang, J.; Huang, Y.; Luo, W.; Hu, X. Characteristic tearing energy and fatigue crack propagation of filled natural rubber. *Polymers* **2021**, *13*, 3891. [CrossRef]
10. Nun-anan, P.; Wisunthorn, S.; Pichaiyut, S.; Vennemann, N.; Nakason, C. Novel approach to determine non-rubber content in *Hevea brasiliensis*: Influence of clone variation on properties of un-vulcanized natural rubber. *Ind. Crops Prod.* **2018**, *118*, 38–47. [CrossRef]
11. Nun-anan, P.; Wisunthorn, S.; Pichaiyut, S.; Nathaworn, C.D.; Nakason, C. Influence of nonrubber components on properties of unvulcanized natural rubber. *Polym. Adv. Technol.* **2020**, *31*, 44–59. [CrossRef]
12. Thongpet, C.; Wisunthorn, S.; Liengprayoon, S.; Vaysse, L.; Bonfils, F.; Nakason, C. Effect of rubber clone on fatty acid composition and properties of air-dried sheet. *Adv. Mater. Res.* **2014**, *844*, 194–197. [CrossRef]
13. Kakubo, T.; Matsuura, A.; Kawahara, S.; Tanaka, Y. Origin of characteristic properties of natural rubber—effect of fatty acids on crystallization of *cis*-1,4-polyisoprene. *Rubber Chem. Technol.* **1998**, *71*, 70–75. [CrossRef]
14. Moreno, R.M.B.; Ferreira, M.; Souza, P.D.; Henrique, G.L.; Mattoso, C. Technological properties of latex and natural rubber of *Hevea brasiliensis* clones. *Sci. Agric.* **2005**, *62*, 122–126. [CrossRef]
15. Alam, D.; Yadav, L.; Nath, D.G.; Purushotham, B.B.M.; Kanna, T.R. Determining surface tension of different fluids with the help of tensiometer. *Int. Refereed J. Eng. Sci.* **2017**, *6*, 42–45.
16. Baur, J.F.; Ensminger, G.L. The association of official analytical chemists (AOAC). *J. Am. Oil Chem. Soc.* **1977**, *54*, 171–172. [CrossRef]
17. Thongnuanchan, B.; Ninjan, R.; Kaesaman, A.; Nakason, C. Synthesis of modified natural rubber with grafted poly(acetoacetoxyethyl methacrylate-*co*-methyl methacrylate) and performance of derived adhesives with GTA crosslinker. *Polym. Eng. Sci.* **2018**, *58*, 1610–1618. [CrossRef]
18. Pethin, D.; Nakkanong, K.; Nualsri, C. Performance and genetic assessment of rubber tree clones in Southern Thailand. *J. Agric. Sci.* **2015**, *72*, 306–313. [CrossRef]
19. Rojruthai, P.; Pareseecharoen, C.; Sakdapipanich, J. Physical decoloration in the concentration process of natural rubber. *SPE Polym.* **2021**, *2*, 210–216. [CrossRef]
20. Wei, O.C.; Bahri, A.R.S. Rubber particles: Size, molecular weight and their distributions detected in wild Hevea species. *J. Biol. Agric. Healthc.* **2016**, *6*, 98–103.
21. Shamsul, B.A.R. Evaluation of possible natural latex substitutes from artificially prepared latex using transmission electron microscope. *J. Nat. Sci. Res.* **2014**, *4*, 147–159.
22. Amnuaypornsri, S.; Sakdapipanich, J.; Tanaka, Y. Green strength of natural rubber: The origin of the stress–strain behavior of natural rubber. *J. Appl. Polym. Sci.* **2009**, *111*, 2127–2133. [CrossRef]
23. Junkong, P.; Cornish, K.; Ikeda, Y. Characteristics of mechanical properties of sulphur cross-linked guayule and dandelion natural rubbers. *RSC Adv.* **2017**, *7*, 50739–50752. [CrossRef]
24. Thuong, N.T.; Nghia, P.T.; Kawahara, S. Factors influencing green strength of commercial natural rubber. *Green Process. Synth.* **2018**, *7*, 399–403. [CrossRef]
25. Jong, L. Modulus enhancement of natural rubber through the dispersion size reduction of protein/fiber aggregates. *Ind. Crops Prod.* **2014**, *55*, 25–32. [CrossRef]
26. Zhang, H.; Zhang, L.; Chen, X.; Wang, Y.; Zhao, F.; Luo, F.; Liao, S. The role of non-rubber components on molecular network of natural rubber during accelerated storage. *Polymers* **2020**, *12*, 2880. [CrossRef]
27. Yumae, N.; Kaesaman, A.; Rungvichaniwat, A.; Thepchalerm, C.; Nakason, C. Novel creaming agent for preparation of creamed concentrated natural rubber latex. *J. Elastomers Plast.* **2010**, *42*, 453–470. [CrossRef]
28. Jodar-Reyes, A.B.; Martin-Rodriguez, A.; Ortega-Vinuesa, J.L. Effect of the ionic surfactant concentration on the stabilization/destabilization of polystyrene colloidal particles. *J. Colloid Interface Sci.* **2006**, *298*, 248–257. [CrossRef]
29. Promsung, R.; Nakaramontri, Y.; Uthaipan, N.; Kummerlöwe, C.; Johns, J.; Vennemann, N.; Kalkornsurapranee, E. Effects of protein contents in different natural rubber latex forms on the properties of natural rubber vulcanized with glutaraldehyde. *Express Polym. Lett.* **2021**, *15*, 308–318. [CrossRef]
30. Sansatsadeekul, J.; Sakdapipanich, J.; Rojruthai, P. Characterization of associated proteins and phospholipids in natural rubber latex. *J. Biosci. Bioeng.* **2011**, *111*, 628–634. [CrossRef]
31. Kawahara, S.; Isono, Y.; Kakubo, T.; Tanaka, Y.; Aik-Hwee, E. Crystallization Behavior and Strength of Natural Rubber Isolated from Different Hevea Clone. *Rubber Chem. Technol.* **2000**, *73*, 39–46. [CrossRef]

Article

Amination of Non-Functional Polyvinyl Chloride Polymer Using Polyethyleneimine for Removal of Phosphorus from Aqueous Solution

Sok Kim [1,2], Yun Hwan Park [1] and Yoon-E Choi [1,*]

[1] Division of Environmental Science and Ecological Engineering, Korea University, Seoul 02841, Korea; sokkim81@korea.ac.kr (S.K.); sug4393@korea.ac.kr (Y.H.P.)
[2] OJeong Resilience Institute, Korea University, Seoul 02841, Korea
* Correspondence: yechoi@korea.ac.kr

Abstract: The eutrophication of freshwater environments caused by an excess inflow of phosphorus has become a serious environmental issue because it is a crucial factor for the occurrence of harmful algal blooms (HABs) in essential water resources. The adsorptive removal of phosphorus from discharged phosphorus containing effluents has been recognized as one of the most promising solutions in the prevention of eutrophication. In the present study, a polyvinyl chloride (PVC)-polyethyleneimine (PEI) composite fiber (PEI-PVC) was suggested as a stable and recoverable adsorbent for the removal of phosphorus from aqueous phases. The newly introduced amine groups of the PEI-PVC were confirmed by a comparison between the FT-IR and XPS results of the PVC and PEI-PVC. The phosphorus sorption on the PEI-PVC was pH dependent. At the optimum pH for phosphorus adsorption (pH 5), the maximum adsorption capacity of the PEI-PVC fiber was estimated to be 11.2 times higher (19.66 ± 0.82 mg/g) than that of conventional activated carbon (1.75 ± 0.4 mg/g) using the Langmuir isotherm model. The phosphorus adsorption equilibrium of the PEI-PVC was reached within 30 min at pH 5. From the phosphorus-loaded PEI-PVC, 97.4% of the adsorbed amount of phosphorus on the PEI-PVC could be recovered by employing a desorption process using 1M HCl solution without sorbent destruction. The regenerated PEI-PVC through the desorption process maintained a phosphorus sorption capacity almost equal to that of the first use. In addition, consistently with the PVC fiber, the PEI-PVC fiber did not elute any toxic chlorines into the solution during light irradiation. Based on these results, the PEI-PVC fiber can be suggested as a feasible and stable adsorbent for phosphorus removal.

Keywords: phosphorus; adsorption; adsorbent; polyvinyl chloride; polyethyleneimine (PEI)

Citation: Kim, S.; Park, Y.H.; Choi, Y.-E. Amination of Non-Functional Polyvinyl Chloride Polymer Using Polyethyleneimine for Removal of Phosphorus from Aqueous Solution. *Polymers* 2022, 14, 1645. https://doi.org/10.3390/polym14091645

Academic Editor: Carmine Coluccini

Received: 13 March 2022
Accepted: 18 April 2022
Published: 19 April 2022

Publisher's Note: MDPI stays neutral with regard to jurisdictional claims in published maps and institutional affiliations.

Copyright: © 2022 by the authors. Licensee MDPI, Basel, Switzerland. This article is an open access article distributed under the terms and conditions of the Creative Commons Attribution (CC BY) license (https://creativecommons.org/licenses/by/4.0/).

1. Introduction

Phosphorus (P) is widely applied as an important element in various areas of human activity (e.g., agricultural, living, and bio-industrial areas) [1]. Accordingly, a large number of effluents possessing phosphorus have been discharged into water bodies (e.g., rivers and lakes) from various point and non-point routes, such as mining, industrial, sewage, agricultural soil, and surface runoffs [2–4]. Excessively accumulated phosphorus in water bodies can cause eutrophication, which, in turn, causes serious environmental problems such as harmful algal blooms (HABs) in precious water resources because phosphorus is a limiting nutrient for the proliferation of harmful microalgal and cyanobacterial cells [5]. Therefore, the control and removal of phosphorus from aqueous phases is an immensely important task to maintain the quality of water resources.

Various water treatments based on physiochemical and biological technologies, such as electrodialysis, struvite precipitation, membrane filtration, and biological digestion, are being developed and applied to remove phosphorus from aqueous phases [6–10]. Among these methods, chemical precipitation is used to form ammonium phosphate, is recognized as the most well-established phosphorus removal process, and is reported to have the highest removal efficiency (90–95%); however, it can be economically applied to the treatment of an aqueous solution containing a high phosphate concentration (>50 mg/L). Moreover, in the case of biological and membrane processes, some technical limitations have been claimed, such as low effectiveness and strict control requirements (biological process) and high capital and energy cost requirements (membrane process).

Adsorption technology is considered as a reliable and economical application method for the removal of phosphorus from aqueous solutions, because it does not require any additional separate operation processes, additives, or power sources [11]. Therefore, various types of adsorbents have been designed for application in the adsorption process for the removal of phosphorus (phosphate) from aqueous phases. Synthesized metal hydroxide/oxide-based adsorbents, such as $Mg(OH)_2$ and ZrO_2, have been reported to have considerable phosphate adsorption capacities at 16.3 and 47.4 mg/g, respectively [12]. Clay minerals, such as palygorskite [13] and clinoptilolite [14], have been investigated as adsorbents for the removal of phosphate from aqueous phases. In addition, biochars derived from wood, corncob, rice husk, and saw dust have gained interested as possible adsorbents for the adsorptive removal of phosphate [15]. However, it is difficult to recover these adsorbents from solution after application due to their small size. When considering phosphorus industrial demands and exhausting natural phosphorus sources, and since phosphorus-loaded adsorbents can be another source of phosphorus, the recovery of adsorbent after use may be important from the viewpoint of industrial sustainability [3]. Molded adsorbents using a polymer matrix have been suggested as the solution for the difficult separation of small, formed adsorbents from solution. For example, chitosan is recognized as a potent polymer matrix of adsorbent for the removal of anionic-formed phosphates by electrostatic attraction due to its content of numerous cationic amine groups and possibility to be immobilized in various hydrogel forms (bead and fibers) [16]. However, since it is a biodegradable biopolymer, it is possible to re-discharge adsorbed phosphorus from chitosan-based adsorbents through natural decomposition of the matrix. Indeed, it has been reported that adsorbed cyanotoxins on chitosan can be released again into the aqueous phase through the degradation of the chitosan matrix [17]. Therefore, to prevent secondary pollution by re-discharging target pollutants, the use of a stable matrix should be considered in the development of molded adsorbents.

In the present study, to fabricate an easily recoverable and highly stable adsorbent for the adsorptive removal of phosphorus, the synthetic polymer polyvinyl chloride (PVC), which is a chemically/physically stable, inexpensive, and light-weight commercial thermoplastic [18], was applied as an adsorbent matrix. Since the adsorption target phosphorus generally appears in anionic phosphate forms ($H_2PO_4^-$, HPO_4^{2-}, and PO_4^{3-}) in aqueous phases depending on its pK_a property, it can be electrostatically adsorbed to cationic functional groups such as amine groups. However, the chemical structure of PVC does not possess any functional groups as adsorption sites for anionic phosphates. To aminate the PVC matrix, amine-rich ionic polymer polyethyleneimine (PEI) was applied because PEI modification is a well-known method that can efficiently enhance the adsorption performance of adsorbents for anionic pollutants, including phosphates [2,19,20]. Generally, to fabricate PEI-modified molded adsorbents, at least three separate reaction processes are required, including adsorbent molding, PEI-coating/grafting, washing, and cross-linking processes. However, we expected that the PEI-PVC composite adsorbent (PEI-PVC) would only require polymer dissolution and fabrication processes because the amine groups of PEI molecules can be coupled to the alkyl chloride groups of the PVC backbone via the nucleophilic substitution reaction through chloro-groups during the polymer dissolution step [21]. Consequently, we could fabricate the PEI-PVC adsorbent through the direct injection of the dissolving PEI-PVC composite solution into water. In addition, to obtain an easily separable and surface area-maximized PEI-PVC adsorbent, it was fabricated in fiber form. The enhanced amine groups and decreased chloride of the PEI-PVC fiber compared to those of the pristine PVC could be confirmed using FT-IR and XPS analyses.

Through pH edge, kinetics, and reuse tests using the PEI-PVC fiber for phosphorus, the pH-dependent, rapid, and reusable phosphorus adsorption properties of the PEI-PVC fiber were determined. The estimated phosphorus adsorption capacity of the PEI-PVC through isotherm tests was comparable to that of other reported adsorbents. Moreover, during utilization as an adsorbent, although the pristine PVC matrix could generate toxic chlorines by photo-oxidative degradation, the PEI-PVC fiber did not elute chlorines under light irradiation conditions. Therefore, the suggested PEI-PVC fiber can be feasible and stable adsorbent for phosphorus adsorption. In addition, our results might provide a sustainable way to valorize waste PVC to a capable adsorbent for the treatment of anionic pollutants.

2. Materials and Methods

2.1. Materials

Polyvinyl chloride (PVC, MW of 40 kDa), branched polyethyleneimine (PEI) solution (MW of 750,000, 50% PEI content), and potassium phosphate dibasic (K_2HPO_4, >98%) were purchased from Sigma-Aldrich Korea Ltd. (Seoul, Korea). N,N-dimethylformamide (DMF, 99.8%) was supplied by Duksan Science (Seoul, Korea). HCl (36%) and NaOH (>97%) were purchased from Samchun Chemicals Co., Ltd. (Seoul, Korea) and Daejung Chemicals & Metals Co., Ltd. (Siheung, Korea), respectively.

2.2. Preparation of Adsorbents

To prepare the PEI composite PVC fiber sorbent (PEI-PVC), two grams of PEI solution (50% in solution) was mixed with 30 mL of PVC solution (2 g PVC/30 mL DMF). Next, the mixture was allowed to react at 80 °C for 4 h in a water bath. The reacted mixture was continuously injected into a methanol solution (50% v/v in D.W.) to form a fiber sorbent using a plastic hub needle with air compression. The fabricated PEI-PVC fibers were washed with distilled water several times for the removal of residual DMF and other impurities. The prepared PEI-PVC fiber was then freeze-dried for 24 h. To minimize the effect of humidity and CO_2 on the sorbent, it was stored in a desiccator during period adsorption tests.

2.3. Comparison of Functional Group Characteristics in PVC and PEI-PVC Fibers

The measurement of functional group characteristics on the sorbents was carried out using Fourier transform infrared spectroscopy (FT-IR, Agilent Cary 630 FTIR, Agilent Technology, Santa Clara, CA, USA). FT-IR analyses of adsorbents were conducted within the wavenumber range of 650–4000 cm^{-1} in the attenuated total reflectance (ATR) mode with scanning number of 100 and 0.9 cm^{-1} scanning resolution. Furthermore, to analyze minute changes in the functional groups of the sorbents, the XPS signals for N_{1s} and Cl_{2p} of the PVC and PEI-PVC were analyzed via XPS (X-TOOL, ULVAC-PHI, Kanagawa, Japan).

2.4. Evaluation of Phosphorus Adsorption Performance of Adsorbents

To determine and compare the phosphorus adsorption performances of the PEI-PVC and AC, pH effect (pH edge), kinetics, and isotherm tests were performed in a batch system. In the adsorption experiments, the phosphorus stock solution (initial concentration of phosphorus: 1000 mg/g) was diluted with distilled water to prepare test solutions containing desired initial concentration of phosphorus. For the adsorption test, the adsorbents (adsorbent dosage: 1 g/L) were agitated with the prepared phosphorus solutions in a shaking incubator until reaching the adsorption equilibrium state at 170 rpm and 25 °C. In addition, during the adsorption process, the solution pH was continuously monitored and adjusted to the desired pH range using HCl (0.1 M and 1 M) and NaOH (0.1 M and 1 M) solutions. The detailed conditions of each experiment are summarized in Table 1.

Table 1. Adsorption experimental conditions.

Conditions	Experiments		
	pH Effect	Kinetics *	Isotherm
Working volume (L)	0.03	0.08	0.03
Weight of adsorbents (g)	0.03	0.08	0.03
pH	2–8	5 (4.95–5.05)	5, 6, and 7
Initial phosphorus concentration (mg/L)	98.3	91.7	1–100
Adsorption reaction time (h)	24	8	24

* To determine the adsorption equilibrium time of adsorbents, during the adsorption experiment, solution samples were taken at different reaction times.

After adsorption processes, to measure residual phosphorus concentration in the experimental samples, liquid–solid separation was conducted using high-speed centrifugation at 9000 rpm. The residual phosphorus concentration in the supernatants was detected using inductively coupled plasma-optical emission spectrometry (ICP-OES, Agilent, Santa Clara, CA, USA). The adsorbed amount of phosphorus by the adsorbents could be calculated using Equation (1) as presented below:

$$q = \frac{V_i C_i - V_f C_f}{M} \quad (1)$$

where q (mg/g) is the phosphorus adsorption capacity of the adsorbent. M (g) is the applied adsorbent weight to the adsorption experiment. C_i and C_f are the phosphorus concentrations (mg/L) at the initial and final adsorption states, respectively. V_i (L) is the initial volume of the sample, and V_f (L) is the final volume considering the amount of added acid/alkaline solution for pH adjustment.

2.5. Determination of PEI-PVC Reusability

To determine adsorbent reusability of the used PEI-PVC fiber in phosphorus adsorption, phosphorus-loaded PEI-PVC fiber was prepared through adsorption process (agitating 0.03 g of PEI-PVC fiber with 30 mL of phosphorus solution for 4 h at pH 5 and 25 °C). The initial phosphorus concentration was 98.25 ± 0.30 mg/L. After the sorption process, phosphorus-loaded PEI-PVC was separated from the mixture and briefly rinsed using 30 mL of distilled water one time to remove residual phosphorus solution. Then, the rinsed phosphorus-loaded PEI-PVC was treated in 30 mL of 1M HCl solution for 4 h to dissociate the adsorbed phosphorus from used PEI-PVC. After that, the regenerated PEI-PVC was washed several times using distilled water to remove the remaining HCl solution. The adsorption process was performed again using completely regenerated PEI-PVC under the same conditions as the first adsorption trial. The adsorption and desorption processes using PEI-PVC were repeated in three cycles.

2.6. Measurement of Chlorine Elution from PVC-Based Sorbents

To measure the eluted chlorine levels that were generated from the PVC backbone of the sorbents, PVC and PEI-PVC fibers were agitated in 150 mL of autoclaved distilled water for 24 h under white fluorescent light (50 mol/m^2) at 25 °C. The eluted chlorine concentration in the gathered samples was measured using a residual chlorine meter (HI 96710, Hanna, Woonsocket, RI, USA) following the manufacturer's protocol. All experiments were conducted thrice.

3. Results and Discussion

3.1. Change of Functional Group Properties of PVC Sorbent after PEI Reaction

Since the adsorption target phosphorus usually exhibits anionic phosphate forms in an aqueous solution depending on its pK_a properties, cationic functional groups possessing sorbents can adsorb phosphates by electrostatic attraction. Therefore, the determination of the existing cationic binding sites on the sorbents might be an important factor to confirm the applicability of adsorbents to phosphorus adsorption processes. In the present study, the newly generated cationic functional groups of the PEI-PVC fiber by the PEI and PVC composite were analyzed using FT-IR and XPS instruments. Figure 1 presents the FT-IR spectrum of the PVC and PEI-PVC. As shown in Figure 1, the PEI-PVC fiber showed FT-IR peaks associated with the base material PVC. The peaks in the range of 2921–2852 cm^{-1} correspond to the symmetric and symmetrical stretching bonds of C–H in the PVC molecules [22,23]. It has been reported that peaks at 2965 and 1245 cm^{-1} are related to the C–H stretching of CH–Cl in the PVC molecule [24]. In addition, the peaks at 1430 (C–H bending of –CH_2), 1324 (C–H deforming in plane), 1093 (C–C stretching), and 970 cm^{-1} (CH_2 rocking) indicate the chemical bond properties of the PVC matrix [25]. In the FT-IR spectra of the PEI-PVC, FT-IR adsorption peaks associated with amine groups could be found, whereas those of the PVC did not show. The newly observed broad peak in the range of 3600–3200 cm^{-1} and the strong peak at 1654 cm^{-1} are attributed to the asymmetric N–H stretching of amines and the bending of secondary amines, respectively [26]. In addition, the appearance of the small peak at 1440 cm^{-1} can be attributed to –NH bending or C–N stretching [26,27]. Furthermore, the intensity of the peak at 1245 cm^{-1} assigned to the bending bond of C–H near Cl in the PVC backbone decreased and shifted to 1235 cm^{-1} in the FT-IR spectra of the PEI-PVC. These observed peak changes in the FT-IR result of the PEI-PVC fiber might indicate increased amine groups and decreased Cl amounts in the PEI-PVC fiber due to a nucleophilic substitute reaction between the chloro-groups of the PVC backbone and the amine groups of the PEI molecules [21]. Additional evidence related to these changes in functional groups could be confirmed by XPS analyses for nitrogen (XPS N_{1s}) and chloride (XPS Cl_{2p}) atoms of the PVC-based fibers (Figure 2). When comparing the XPS N_{1s} signals between the PVC and PEI-PVC fibers, the PVC fiber did not show any XPS signals related to the nitrogen-based chemical bonds (i.e., amine groups). However, in the case of the PEI-PVC, the strong N_{1s} signal was newly observed in the binding energy range of 392–404 eV. This signal might be attributed to the amine groups of the coupled PEI molecules with the PVC backbone since N_{1s} peaks of –NH_2 and C–N = C bonds in the PEI molecules could be observed at 398.8 and 399.5 eV [19]. This result is clearly connected to the newly appeared FT-IR peaks of the amine groups in the FT-IR spectra of the PEI-PVC fiber. In the case of XPS Cl_{2p} signals, a higher XPS signal intensity was observed in the XPS result of the PVC than in that of the PEI-PVC. This evidently indicates that the amount of chloride was decreased in the PEI-PVC fiber compared to that of the PVC fiber owing to the substitution reaction between the PEI molecules and the chloride in the PVC.

The decreased amount of chloride in the PEI-PVC compared to the pristine PVC could be additionally supported by the different eluted chlorine concentrations from the PVC and PEI-PVC fibers under light irradiation conditions. It has been reported that PVC can generate toxic chlorines via the photo-degradation of the backbone structure [28]. As shown in the XPS Cl_{2p} results, since the amount of chloride in the PEI-PVC was reduced, it was expected that the elution of chlorines from the PEI-PVC fiber might be reduced compared to that of the PVC. Therefore, the generated chlorines from the PVC and PEI-PVC fibers were compared (Figure 3). As we hypothesized, the pristine PVC fiber released 0.12 ± 0.002 mg/L of chlorine into the aqueous solution during light exposure for 24 h, whereas the PEI-PVC fiber did not. The PEI-PVC-agitated sample showed a chlorine concentration almost the same as that of the control water (0.013 ± 0.001 mg/L). Furthermore, the prevention of toxic chlorine emissions from the PEI-PVC fiber under light conditions might be a meaningful result indicating that the PEI-PVC fiber is directly applicable and a stable adsorbent for the removal of phosphorus in actual environments.

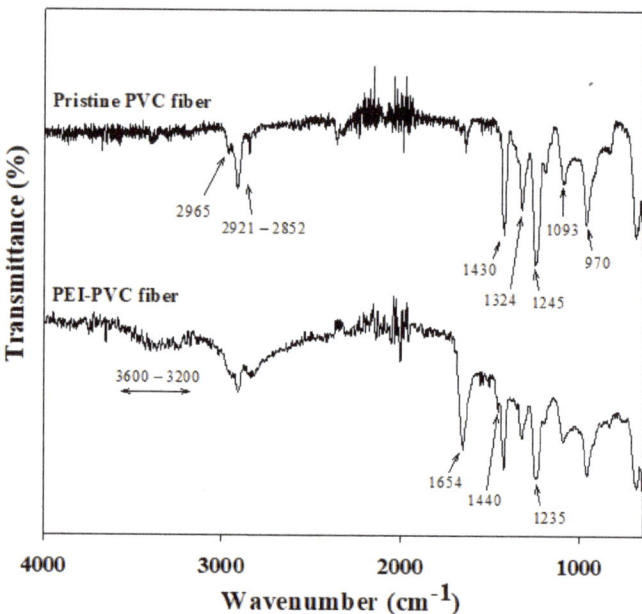

Figure 1. FT-IR spectrum of PVC and PEI-PVC.

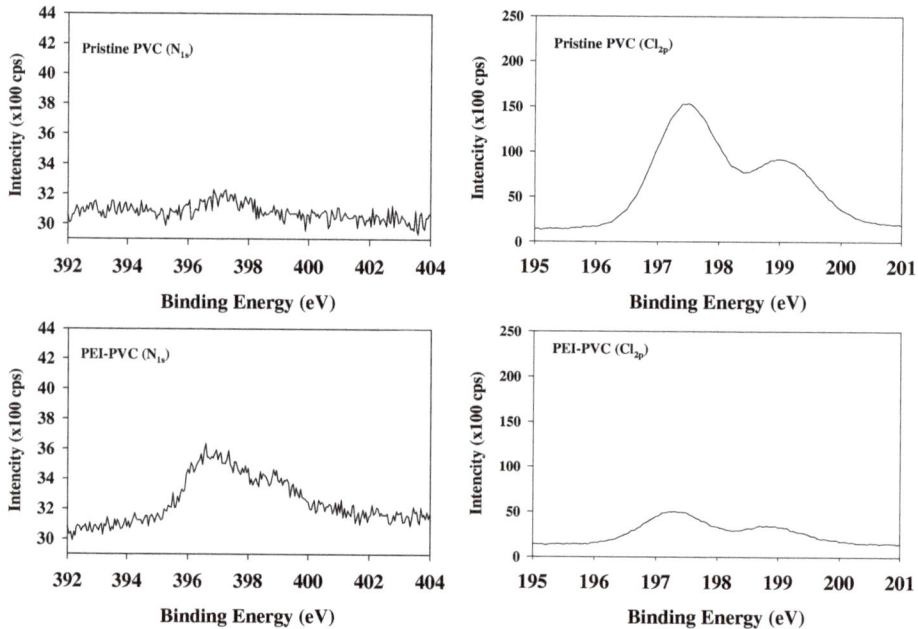

Figure 2. XPS N1s and Cl2p signals of PVC-based sorbent materials.

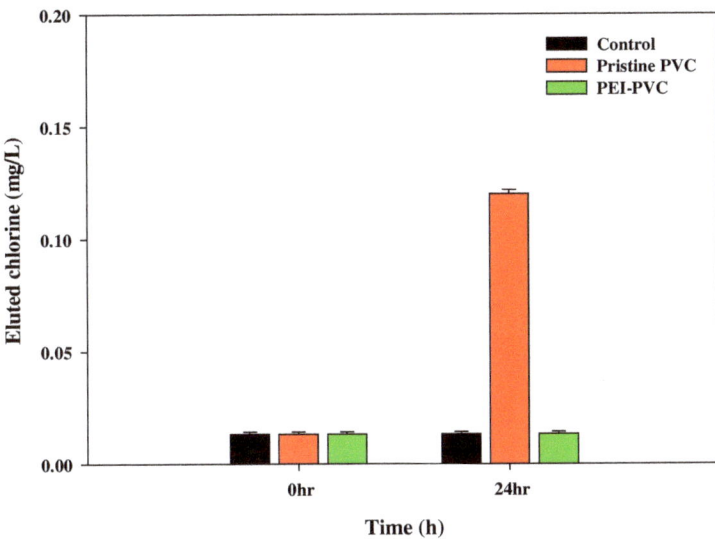

Figure 3. Eluted chlorine levels from PVC and PEI-PVC in an aqueous solution.

3.2. pH Effect on Phosphorus Adsorption

Through the FT-IR and XPS measurements, it was determined that amine groups derived from PEI molecules are the main binding sites of the PEI-PVC fiber, which can electrostatically adsorb phosphates. Since the electrostatic attraction between the PEI-PVC fiber and phosphates can be affected by the solution pH due to the deprotonation of the amine groups of the PEI-PVC and the phosphates in the solution depending on their pK_a properties, the solution pH condition is an important factor in the adsorption process [2]. In the present study, the pH effect on the phosphorus adsorption of the PEI-PVC was evaluated under different solution pH conditions (pH 2–8). In addition, the pH effect on the phosphorus sorption capacity of the conventional adsorbent activated charcoal (AC) was evaluated to make a comparison with that of the PEI-PVC fiber.

Figure 4 illustrates the pH effect on the phosphorus uptakes of the PEI-PVC and AC. As displayed in Figure 4, the PEI-PVC exhibited a pH-dependent phosphorus adsorption property. With an increase in solution pH from 2 to 3.5, the phosphorus sorption capacity of the PEI-PVC increased accordingly. At solution pH 3.5, the phosphorus sorption capacity of the PEI-PVC was recorded as approximately 27 mg/g (26.8 ± 0.48 mg/g), and it showed similar phosphorus sorption capacities until pH 5. Above pH 5, the phosphorus uptake of the PEI-PVC fiber decreased substantially. The pH-dependent phosphorus adsorption of the PEI-PVC fiber can be explained by the electrostatic interaction between the amine groups of the PEI-PVC and phosphate anions, as we assumed previously. It has been reported that the pK_a values of amine groups (e.g., primary, secondary, and tertiary amine groups) in PEI molecules can be observed at approximately pH 4.5, 6.7, and 11.6, respectively [29]. Accordingly, the amine groups on the PEI-PVC derived from PEI molecules can be fully positively charged under acidic conditions below pH 4; therefore, it was expected that the phosphorus sorption uptake of the PEI-PVC might record the maximum value at pH 2. However, the phosphorus sorption uptake of the PEI-PVC at pH 2 was significantly lower than that at pH 3.5. This result might be attributed to the pH dependence of phosphate speciation. Phosphorus can exist as uncharged phosphoric acid (H_3PO_4) and electrostatically adsorbable phosphate anions ($H_2PO_4^-$, HPO_4^{2-}, and PO_4^{3-}) depending on the solution pH condition in the aqueous solution [30]. According to our previous study [2], only 40% of total phosphorus amounts can be presented as the electrostatically adsorbable phosphate anion $H_2PO_4^-$ at pH 2. However, the almost amount of phosphorus can be changed to the phosphate anionic form at pH 3.5. The increasing fraction of anionic phosphate might contribute to the increasing phosphorus uptake of the PEI-PVC in the range of pH 2–pH 3.5 by the enhancement of the electrostatic attraction between the PEI-PVC and phosphate. Above pH 5, although the almost amount of phosphorus can be formed as the adsorbable phosphate anions, since the positivity of the PEI-PVC could be lost by the deprotonation of amine groups depending on their pK_a properties, the phosphorus sorption capacity of the PEI-PVC might be reduced. The conventional adsorbent, AC, did not show pH dependence on phosphorus adsorption capacity. However, the phosphorus sorption capacities of the AC were significantly lower than those of the PEI-PVC fiber under all of the pH conditions.

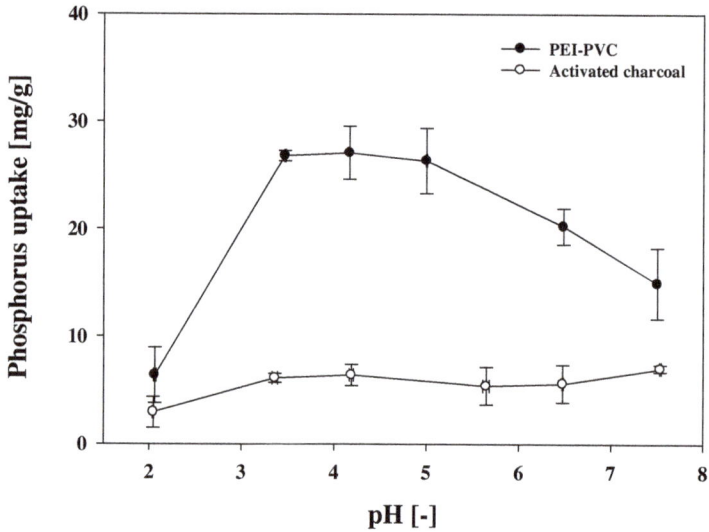

Figure 4. pH effect on the phosphorus sorption uptake of PEI-PVC and activated charcoal.

3.3. Reusability of PEI-PVC Fiber

The pH-dependence property of the PEI-PVC fiber on phosphorus adsorption could be applied to the phosphorus desorption process for the reuse of the PEI-PVC. As discussed above, under highly acidic conditions, since phosphate anions can be changed to uncharged phosphoric acid (H_3PO_4), adsorbed phosphorus might be recovered from phosphate-loaded PEI-PVC. To estimate the desorbed amount of phosphorus, the phosphate-adsorbed PEI-PVC at pH 5 was treated using 1M HCl solution. Figure 5 displays the measured amounts of the adsorbed and desorbed phosphorus through the sorption and desorption processes using the PEI-PVC fiber. As shown in Figure 5, 1.93 ± 0.01 mg of phosphorus could be adsorbed on the PEI-PVC at pH 5 with the first use of the PEI-PVC fiber. In addition, after the treatment of the phosphate-loaded PEI-PVC using the 1M HCl solution, the almost amount of adsorbed phosphorus on the PEI-PVC fiber could be desorbed (1.88 ± 0.05 mg) owing to the loss of electrostatic attraction between the binding site and the sorbate through the speciation change from phosphate anions to uncharged phosphoric acid under highly acidic conditions. From the replicated adsorption and desorption tests, it was found that the regenerated PEI-PVC could adsorb a similar amount of phosphorus to that used in the first trial. In the second and third trials of phosphorus adsorption using regenerated PEI-PVC, 1.78 ± 0.06 and 1.84 ± 0.04 mg of phosphorus were adsorbed on the regenerated PEI-PVC, respectively. When considering the reusability of PEI-PVC, the adsorptive removal process of phosphorus using PEI-PVC might have an economic benefit.

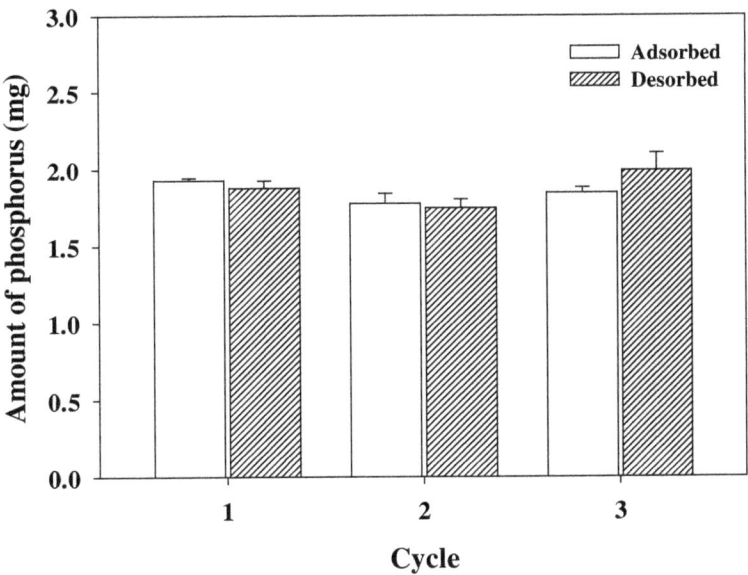

Figure 5. Reusability of PEI-PVC in three cycle applications (adsorbed phosphorus at pH 5 and desorbed in 1 M HCl solution).

3.4. Adsorption Kinetics

Adsorption equilibrium time is an important factor to determine the efficiency of the developed sorbent for the target materials. Therefore, to determine the sorption equilibrium time of the PEI-PVC for phosphorus, a kinetic experiment using the PEI-PVC was performed in a batch system at pH 5. From the kinetic result (Figure 6), the PEI-PVC showed a rapid adsorption equilibrium at approximately 30 min for phosphate anions at pH 5. In the case of the AC, a faster phosphorus adsorption equilibrium was attained than that attained with the PEI-PVC (within 15 min). However, this might be attributable to the significantly low phosphorus sorption capacity of the AC compared to that of the PEI-PVC. To compare the phosphorus sorption kinetic parameters of the PEI-PVC and AC in detail, a pseudo-second-order kinetic model was applied to the experimental kinetic data of the PEI-PVC fiber and AC because it is one of the representative kinetic models used to determine the kinetic properties of adsorbents. The pseudo-second-order kinetic model is represented as Equation (2):

$$q_t = \frac{q_e^2 k_2 t}{1 + q_e k_2 t} \qquad (2)$$

where q_e and q_t are the phosphorus sorption capacity of the sorbents (mg/g) at the sorption equilibrium state and specific sorption reaction time (t), respectively. The parameter k_2 (g/mg min) indicates the estimated adsorption rate constants from the pseudo-second-order model, respectively.

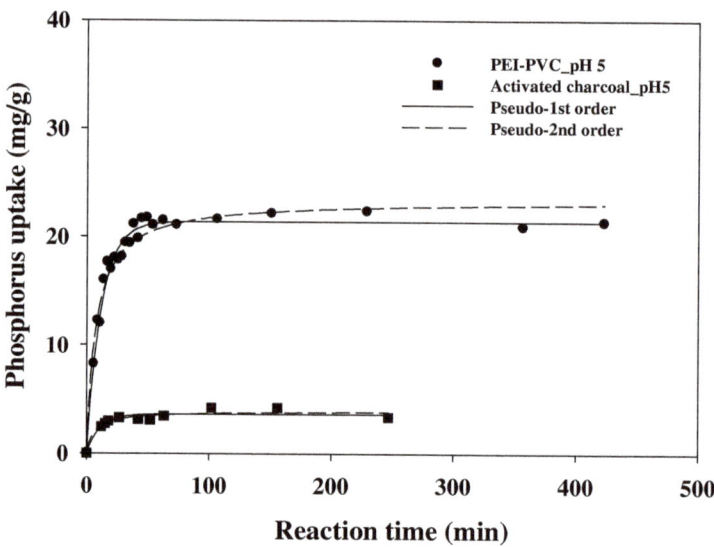

Figure 6. Phosphorus sorption kinetics of PEI-PVC and activated charcoal at pH 5.

The estimated kinetic parameters for the PEI-PVC and AC from the pseudo-second-order kinetic model are summarized in Table 2. According to Table 2, the estimated correlation coefficients (R^2) of the pseudo-second-order model for the phosphorus sorption kinetic data of the PEI-PVC and AC were 0.9585 and 0.9211, respectively. These high R^2 values indicate that the phosphorus sorption kinetic experimental results of the PEI-PVC and AC were predicted well by the pseudo-second-order kinetic model. The sorption reaction rate (k_2) of the AC was estimated as 0.037 ± 0.013 g/mg·min using the kinetic model. This was a value approximately 6.5 times higher than that of the PEI-PVC (0.0057 ± 0.0007 g/mg·min), and it was better connected to the faster sorption equilibrium state of the AC than the PEI-PVC fiber. However, as shown in the experimental kinetic results, the PEI-PVC showed a comparable sorption equilibrium time to that of the AC. Furthermore, the phosphorus sorption capacity (q_e) calculated from the sorption kinetic data of the PEI-PVC was 5.9 times higher (23.39 ± 0.47 mg/g) than that of the AC (3.94 ± 0.21 mg/g) using the pseudo-second-order model. Based on the kinetics result, the PEI-PVC fiber can be suggested as an efficient substitute for conventional adsorbents (i.e., activated carbons) to rapidly counteract the inflow of phosphorus from the effluents of point or non-point pollutant sources.

Table 2. Adsorption kinetic parameters of applied sorbents for phosphorus estimated from pseudo-second-order model at pH 5.

Adsorbents	Kinetic Parameters		
	k_2 (g/mg·min)	q_e (mg/g)	R^2
PEI-PVC	0.0057 ± 0.0007	23.39 ± 0.47	0.9585
AC	0.037 ± 0.013	3.94 ± 0.21	0.9211

3.5. Maximum Phosphorus Sorption Capacity of PEI-PVC Fiber

Together with the sorption kinetic property, the maximum sorption capacity (q_{max}) of the sorbent toward the target pollutant is an essential factor for the evaluation of the adsorption performance of the developed adsorbent. Therefore, to determine the q_{max} of the developed sorbent PEI-PVC fiber, isotherm experiments were performed under various pH conditions (pH 5–7). In addition, the maximum phosphorus (i.e., phosphate anions) sorption capacities of the PEI-PVC fiber were compared with those of the conventional adsorbent (AC). The obtained isotherm results using the PEI-PVC and AC at different pH values are presented in Figure 7. As shown in Figure 7, the phosphorus sorption capacities of the PEI-PVC and AC showed increasing values as the equilibrium phosphorus concentration was increased. In addition, above the specific equilibrium phosphorus concentration, the phosphorus adsorption uptakes of the PEI-PVC and AC reached the adsorption equilibrium state at all pH values. The phosphorus adsorption uptakes of the PEI-PVC fiber at the equilibrium state decreased from approximately 20 mg/g to 13 mg/g as the solution pH was increased from pH 5 to pH 7. The AC showed almost constant phosphorus adsorption capacities (average 1.75 ± 0.4 mg/g) at the adsorption equilibrium state regardless of pH conditions. These phosphorus uptakes of the AC were clearly low compared to those of the PEI-PVC fiber. Therefore, extensively used isotherm models (Freundlich and Langmuir isotherm model equations) were applied to the only experimental isotherm results of the PEI-PVC for the estimation of detailed isotherm parameters. The Freundlich and Langmuir isotherm model equations are presented as Equations (3) and (4), respectively. The estimated isotherm parameters from these model equations are presented in Table 3.

$$q_e = K_f C_f^{\frac{1}{n}} \tag{3}$$

$$q_e = \frac{q_{max,L} b C_f}{1 + b C_f} \tag{4}$$

where q_e is the experimentally calculated phosphorus sorption capacity (mg/g). C_f (mg/L) is the final phosphorus concentration in the solution after the adsorption process. The parameters K_f (L/g) and n are the Freundlich constants, which can be indicators of sorption capacity and sorption intensity, respectively. The parameters in the Langmuir model equation $q_{max,L}$ and b indicate the maximum capacity (mg/g) and sorption affinity (L/mg) of the adsorbent for phosphorus, respectively.

As summarized in Table 3, the Freundlich parameter (K_f) associated with the phosphorus adsorption capacity of the PEI-PVC fiber increased with the decrease in the solution pH from pH 7 (5.12 ± 0.97 L/g) to pH 5 (9.83 ± 1.34 L/g). The dimensionless n value of the PEI-PVC in phosphorus adsorption showed a similar tendency to that of the K_f values. The recorded n values from the isotherm data of the PEI-PVC fiber for phosphorus at pH 5, pH 6, and pH 7 were 5.21 ± 1.05, 5.29 ± 1.10, and 4.14 ± 0.91, respectively. According to Toor and Jin [31], if the n value is larger than 1, it can be determined that the adsorption process is favorable. Therefore, the phosphorus adsorption process with application of the PEI-PVC fiber might be a favorable process. The application of the Langmuir isotherm model to the experimental isotherm data of the PEI-PVC showed better fitting lines, because of its higher determination coefficients (R^2 values: 0.9755, 0.9843, and 0.9844 at pH 5, 6, and 7, respectively), than those of the Freundlich model (R^2 values: 0.9192, 9161, and 0.9046 at pH 5, 6, and 7, respectively) at all solution pH values. The maximum phosphorus sorption capacities ($q_{max,L}$) of the PEI-PVC fiber were estimated from the Langmuir model as 19.66 ± 0.82, 16.54 ± 0.54, and 13.66 ± 0.51 mg/g at pH 5, 6, and 7, respectively. In addition, the isotherm result of the PEI-PVC fiber showed a higher b value at pH 5 (2.08 ± 0.59 L/mg) than at pH 6 (1.65 ± 0.35 L/mg) and pH 7 (0.42 ± 0.08 L/mg) using the Langmuir model. As discussed above section, since the higher amount of amine groups in the PEI-PVC fiber could be positively activated at pH 5 compared to those at pH 6 and pH 7 depending on the pK_a properties of the amine groups, the higher values of the Freundlich and Langmuir isotherm parameters corresponding to the sorption capacity and affinity for phosphorus adsorption could be calculated from the isotherm data.

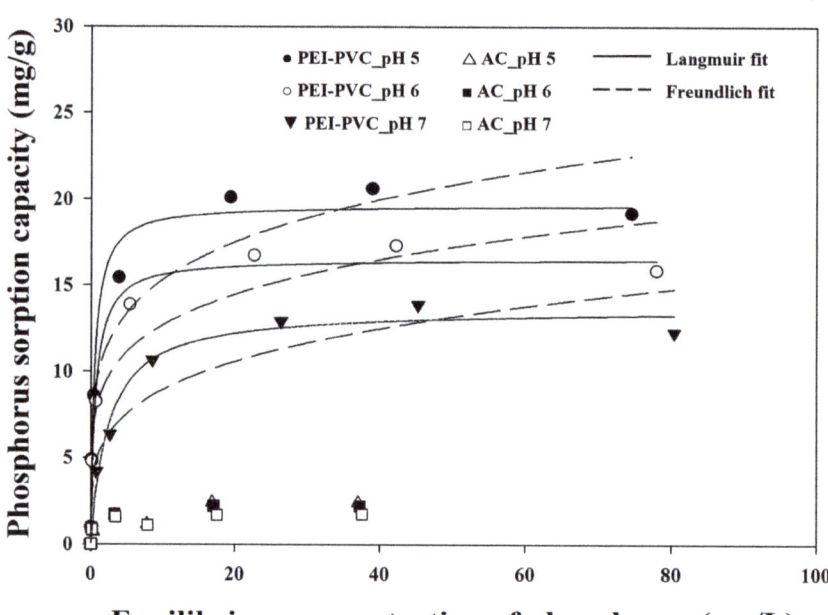

Figure 7. Isotherm results of PEI-PVC and activated charcoal for phosphorus adsorption at pH 5, 6, and 7.

Table 3. Estimated isotherm parameters of PEI-PVC for phosphorus from Freundlich and Langmuir models.

Models	Parameters	pH		
		5	6	7
Freundlich	K_f	9.83 ± 1.34	8.20 ± 1.14	5.12 ± 0.97
	n	5.21 ± 1.05	5.29 ± 1.10	4.14 ± 0.91
	R^2	0.9192	0.9161	0.9046
Langmuir	b	2.08 ± 0.59	1.65 ± 0.35	0.42 ± 0.08
	$q_{max, L}$	19.66 ± 0.82	16.54 ± 0.54	13.66 ± 0.51
	R^2	0.9755	0.9843	0.9844

Furthermore, a comparison of the maximum phosphorus adsorption capacity of the PEI-PVC fiber with that of previously reported adsorbents was performed (Table 4). As shown in Table 4, in general, the adsorbents containing Zr showed higher q_{max} values than those of the PEI-PVC fiber at natural pH. For instance, the q_{max} values for the phosphorus of Zr-loaded orange waste gel, Zr-loaded MUROMAC, and Zr-loaded okara were calculated as 57, 43, and 14.39 mg/g, respectively. Under the condition of pH 3, the covalently cross-linked chitosan by epichlorohydrin showed higher phosphorus adsorption performance (38.22 mg/g) compared to the PEI-PVC fiber (approximately 27 mg/g), whereas the maximum phosphorus uptake of electrostatically cross-linked chitosan by sodium citrate (13.3 mg/g) was smaller than that of the PEI-PVC. Although we could find several adsorbents showing higher q_{max} values for phosphorus than the PEI-PVC fiber, there still remains an opportunity to enhance the phosphorus adsorption performance of the PEI-PVC fiber through the optimization of adsorbent processes and the additional application of adsorption-supporting materials. In addition, as shown in Table 4, the PEI-PVC fiber still displayed superior or comparable phosphorus adsorption performance to other previously reported sorbents. For example, zirconium ferrite [32], Fe-loaded juniper fiber [33], iron–hydroxided eggshell [34], and La(III)-modified fine needle [35] showed maximum phosphorus sorption capacities of 13, 2.31, 14.49, and 6.31 mg/g at pH 7, respectively.

Table 4. Maximum phosphorus adsorption capacities of previously reported adsorbents.

Adsorbent	Maximum Phosphorus Adsorption Capacity (mg/g)	pH Condition	Ref.
$Mg(OH)_2$	5.3	pH 7	[12]
ZrO_2	21.9	pH 7	[12]
Palygorskite	3.7	pH 7	[13]
Clinoptilolite	6.6	pH 5.3	[14]
Zeolite	8.3	pH 5.3	[14]
Fe-loaded juniper fiber	2.31	pH 6.4	[33]
Iron–hydroxide eggshell	14.49	pH 7	[34]
Zirconium ferrite	13	pH 7	[32]
Zr(IV)-loaded apple peels	20.35	pH 2	[36]
Zr(IV)-loaded orange waste gel	57	pH 7	[32]
Zr(IV)-loaded MUROMAC	43	pH 7	[32]
Zn(II)-activated coir pith carbon	5.1	pH 4	[37]
Cross-linked chitosan by epichlorohydrin	38.22	pH 3	[16]
Cross-linked chitosan by sodium citrate	13.3	pH 3	[16]
La(III)-loaded orange waste	13.94	pH 5–7	[38]
La(III)-modified fine needle	6.31	pH 7.1	[35]
Fe(III)-loaded okara	4.78	pH 3	[39]
Zr-loaded okara	14.39	pH 7	[39]

4. Conclusions

In the present study, the amination of a non-functional PVC matrix was performed using a simple chemical reaction process between PEI and PVC to develop an adsorbent for phosphorus removal from aqueous phases. From the FT-IR and XPS-N_{1s} results of the prepared PEI-PVC, the aminated characteristics of the PEI-PVC fiber were well determined. The adsorption of phosphorus on the PEI-PVC was affected by the solution pH. The maximum phosphorus sorption capacity (q_{max}) of the PEI-PVC fiber was estimated as 19.66 ± 0.82 mg/g at pH 5 using the Langmuir equation. At pH 6 and pH 7, the q_{max} values of the PEI-PVC fiber were calculated as 16.54 ± 0.54 and 13.66 ± 0.51, respectively. In the kinetic analysis, the PEI-PVC showed a rapid phosphorus adsorption equilibrium time within 30 min at pH 5. The used PEI-PVC could be regenerated without destructing the adsorbent matrix using the 1M HCl solution by the desorption of the adsorbed phosphorus in the PEI-PVC fiber (desorption rate: >98%). In addition, the regenerated PEI-PVC was reusable for phosphorus adsorption without the loss of sorption capacity. Furthermore, it was found that the PEI-PVC fiber did not discharge toxic chlorines during light irradiation. Based on the results, we can suggest that the stable and reusable PEI-PVC fiber can be a feasible adsorbent for phosphorus removal from aqueous phases. However, to further enhance its phosphorus adsorption performance, additional research on the optimization of adsorbent processes and the additional application of adsorption-supporting materials might be needed.

Author Contributions: Conceptualization, S.K.; methodology, S.K. and Y.H.P.; validation, S.K. and Y.-E.C.; formal analysis, S.K.; investigation, S.K.; resources, S.K. and Y.H.P.; data curation, S.K.; writing—original draft preparation, S.K. and Y.H.P.; review and editing, S.K. and Y.-E.C.; visualization, Y.H.P. and S.K.; supervision, Y.-E.C.; project administration, Y.-E.C. and S.K.; funding acquisition, Y.-E.C. and S.K. All authors have read and agreed to the published version of the manuscript.

Funding: This work was supported by the Korea Environment Industry & Technology Institute (KEITI) through a project to develop eco-friendly new materials and processing technology derived from wildlife, funded by the Korea Ministry of Environment (MOE) (2021003280004). This research was also financially supported by the National Research Foundation of Korea (NRF) (grants NRF-2021R1I1A1A01054658 and NRF-2021R1A6A1A10045235).

Institutional Review Board Statement: Not applicable.

Informed Consent Statement: Not applicable.

Data Availability Statement: The data of manuscript is an original research work and has not been published elsewhere.

Conflicts of Interest: The authors declare no conflict of interest.

References

1. Li, J.; Ianaiev, V.; Huff, A.; Zalusky, J.; Ozersky, T.; Katsev, S. Benthic invaders control the phosphorus cycle in the world's largest freshwater ecosystem. *Proc. Natl. Acad. Sci. USA* **2021**, *118*, e2008223118. [CrossRef]
2. Kim, S.; Park, Y.H.; Lee, J.B.; Kim, H.S.; Choi, Y.-E. Phosphorus adsorption behavior of industrial waste biomass-based adsorbent, esterified polyethylenimine-coated polysulfone-*Escherichia coli* biomass composite fibers in aqueous solution. *J. Hazard. Mater.* **2020**, *400*, 123217. [CrossRef]
3. Bacelo, H.; Pintor, A.M.A.; Santos, S.C.R.; Boaventura, R.A.R.; Botelho, C.M.S. Performance and prospects of different adsorbents for phosphorus uptake and recovery from water. *Chem. Eng. J.* **2020**, *381*, 122566. [CrossRef]
4. Zhao, Y.; Li, Y.; Yang, F. Critical review on soil phosphorus migration and transformation under freezing-thawing cycles and typical regulatory measurements. *Sci. Total Environ.* **2021**, *751*, 141614. [CrossRef]
5. Yindong, T.; Xiwen, X.; Miao, Q.; Jingjing, S.; Yiyan, Z.; Wei, Z.; Mengzhu, W.; Xuejun, W.; Yang, Z. Lake warming intensifies the seasonal pattern of internal nutrient cycling in the eutrophic lake and potential impacts on algal blooms. *Water Res.* **2021**, *188*, 116570. [CrossRef]
6. Xu, C.; Dan, S.F.; Yang, B.; Lu, D.; Kang, Z.; Huang, H.; Zhou, J.; Ning, Z. Biogeochemistry of dissolved and particulate phosphorus speciation in the Maowei Sea, northern Beibu Gulf. *J. Hydrol.* **2021**, *593*, 125822. [CrossRef]
7. Javier, L.; Farhat, N.M.; Vrouwenvelder, J.S. Enhanced hydraulic cleanability of biofilms developed under a low phosphorus concentration in reverse osmosis membrane systems. *Water Res.* **2021**, *10*, 100085. [CrossRef]
8. Li, X.; Shen, S.; Xu, Y.; Guo, T.; Dai, H.; Lu, X. Application of membrane separation processes in phosphorus recovery: A review. *Sci. Total Environ.* **2021**, *767*, 144346. [CrossRef]
9. Guida, S.; Rubertelli, G.; Jefferson, B.; Soares, A. Demonstration of ion exchange technology for phosphorus removal and recovery from municipal wastewater. *Chem. Eng. J.* **2021**, *420*, 129913. [CrossRef]
10. Xia, W.-J.; Guo, L.-X.; Yu, L.-Q.; Zhang, Q.; Xiong, J.-R.; Zhu, X.-Y.; Wang, X.-C.; Huang, B.-C.; Jin, R.-C. Phosphorus removal from diluted wastewaters using a La/C nanocomposite-doped membrane with adsorption-filtration dual functions. *Chem. Eng. J.* **2021**, *405*, 126924. [CrossRef]
11. Liu, Y.; You, Y.; Li, Z.; Yang, X.; Wu, X.; Zhao, C.; Xing, Y.; Yang, R.T. NO_x removal with efficient recycling of NO_2 from iron-ore sintering flue gas: A novel cyclic adsorption process. *J. Hazard. Mater.* **2021**, *407*, 124380. [CrossRef] [PubMed]
12. Lin, J.; He, S.; Wang, X.; Zhang, H.; Zhan, Y. Removal of phosphate from aqueous solution by a novel $Mg(OH)_2/ZrO_2$ composite: Adsorption behavior and mechanism. *Colloid Surf. A Physicochem. Eng. Asp.* **2019**, *561*, 301–314. [CrossRef]
13. Ye, H.; Chen, F.; Sheng, Y.; Sheng, G.; Fu, J. Adsorption of phosphate from aqueous solution onto modified palygorskites. *Sep. Purif. Technol.* **2006**, *50*, 283–290. [CrossRef]
14. Goscianska, J.; Ptaszkowska-Koniarz, M.; Frankowski, M.; Franus, M.; Panek, R.; Franus, W. Removal of phosphate from water by lanthanum-modified zeolites obtained from fly ash. *J. Colloid Interface Sci.* **2018**, *513*, 72–81. [CrossRef] [PubMed]
15. Kizito, S.; Luo, H.; Wu, S.; Ajmal, Z.; Lv, T.; Dong, R. Phosphate recovery from liquid fraction of anaerobic digestate using four slow pyrolyzed biochars: Dynamics of adsorption, desorption and regeneration. *J. Environ. Manag.* **2017**, *201*, 260–267. [CrossRef]
16. Jóźwiak, T.; Filipkowska, U.; Szymczyk, P.; Kuczajowska-Zadrożna, M.; Mielcarek, A. The use of cross-linked chitosan beads for nutrients (nitrate and orthophosphate) removal from a mixture of $P-PO_4$, $N-NO_2$ and $N-NO_3$. *Int. J. Biol. Macromol.* **2017**, *104*, 1280–1293. [CrossRef]
17. Pei, H.-Y.; Ma, C.-X.; Hu, W.-R.; Sun, F. The behaviors of Microcystis aeruginosa cells and extracellular microcystins during chitosan flocculation and flocs storage processes. *Bioresour. Technol.* **2014**, *151*, 314–322. [CrossRef]
18. Abdel-Fattah, E.; Alharthi, A.I.; Fahmy, T. Spectroscopic, optical and thermal characterization of polyvinyl chloride-based plasma-functionalized MWCNTs composite thin films. *Appl. Phys. A* **2019**, *125*, 475. [CrossRef]
19. Kim, H.S.; Park, Y.H.; Nam, K.; Kim, S.; Choi, Y.-E. Amination of cotton fiber using polyethyleneimine and its application as an adsorbent to directly remove a harmful cyanobacterial species, *Microcystis aeruginosa*, from an aqueous medium. *Environ. Res.* **2021**, *197*, 111235. [CrossRef]

20. Kim, S.; Choi, Y.-E.; Yun, Y.-S. Ruthenium recovery from acetic acid industrial effluent using chemically stable and high-performance polyethylenimine-coated polysulfone-*Escherichia coli* biomass composite fibers. *J. Hazard. Mater.* **2016**, *313*, 29–36. [CrossRef]
21. Sneddon, G.; McGlynn, J.C.; Neumann, M.S.; Aydin, H.M.; Yiu, H.H.P.; Ganin, A.Y. Aminated poly(vinyl chloride) solid state adsorbents with hydrophobic function for post-combustion CO_2 capture. *J. Mater. Chem. A* **2017**, *5*, 11864–11872. [CrossRef]
22. Pandey, M.; Joshi, G.M.; Mukherjee, A.; Thomas, P. Electrical properties and thermal degradation of poly(vinyl chloride)/polyvinylidene fluoride/ZnO polymer nanocomposites. *Polym. Int.* **2016**, *65*, 1098–1106. [CrossRef]
23. Slaný, M.; Jankovič, L.; Madejová, J. Near-IR study of the impact of alkyl-ammonium and -phosphonium cations on the hydration of montmorillonite. *J. Mol. Struct.* **2022**, *1256*, 132568. [CrossRef]
24. Consumi, M.; Leone, G.; Bonechi, C.; Tamasi, G.; Lamponi, S.; Donati, A.; Rossi, C.; Magnani, A. Plasticizers free polyvinyl chloride membrane for metal ions sequestering. *Inorg. Chem. Commun.* **2020**, *119*, 108100. [CrossRef]
25. Hezma, A.M.; Elashmawi, I.S.; Rajeh, A.; Kamal, M. Change Spectroscopic, thermal and mechanical studies of PU/PVC blends. *Phys. B Condens. Matter* **2016**, *495*, 4–10. [CrossRef]
26. Kim, M.H.; Hwang, C.-H.; Kang, S.B.; Kim, S.; Park, S.W.; Yun, Y.-S.; Won, S.W. Removal of hydrolyzed Reactive Black 5 from aqueous solution using a polyethyleneimine–polyvinyl chloride composite fiber. *Chem. Eng. J.* **2015**, *280*, 18–25. [CrossRef]
27. Chatterjee, S.; Chatterjee, T.; Woo, S.H. Influence of the polyethyleneimine grafting on the adsorption capacity of chitosan beads for Reactive Black 5 from aqueous solutions. *Chem. Eng. J.* **2011**, *166*, 168–175. [CrossRef]
28. Yousif, E.; Hasan, A. Photostabilization of poly(vinyl chloride)—Still on the run. *J. Taibah Univ. Sci.* **2015**, *9*, 421–448. [CrossRef]
29. Kim, S.; Yun, Y.-S.; Choi, Y.-E. Development of waste biomass based sorbent for removal of cyanotoxin microcystin-LR from aqueous phases. *Bioresour. Technol.* **2018**, *247*, 690–696. [CrossRef]
30. Jang, J.W.; Park, J.J.; Kwan, M.H.; Lim, H. Simultaneous Monitoring Method for Chemical Constituents in Aluminum Etchant. *IEEE Sens. J.* **2012**, *12*, 2180–2185. [CrossRef]
31. Toor, M.; Jin, B. Adsorption characteristics, isotherm, kinetics, and diffusion of modified natural bentonite for removing diazo dye. *Chem. Eng. J.* **2012**, *187*, 79–88. [CrossRef]
32. Biswas, B.K.; Inoue, K.; Ghimire, K.N.; Harada, H.; Ohto, K.; Kawakita, H. Removal and recovery of phosphorus from water by means of adsorption onto orange waste gel loaded with zirconium. *Bioresour. Technol.* **2008**, *99*, 8685–8690. [CrossRef]
33. Han, J.S.; Min, S.-H.; Kim, Y.-K. Removal of phosphorus using AMD-treated lignocellulosic material. *For. Prod. J.* **2005**, *55*, 48–53.
34. Mezenner, N.Y.; Bensmaili, A. Kinetics and thermodynamic study of phosphate adsorption on iron hydroxide-eggshell waste. *Chem. Eng. J.* **2009**, *147*, 87–96. [CrossRef]
35. Wang, X.; Liu, Z.; Liu, J.; Huo, M.; Huo, H.; Yang, W. Removing phosphorus from aqueous solutions using lanthanum modified pine needles. *PLoS ONE* **2015**, *10*, e0142700. [CrossRef]
36. Mallampati, R.; Valiyaveettil, S. Apple Peels—A Versatile Biomass for Water Purification? *ACS Appl. Mater. Interfaces* **2013**, *5*, 4443–4449. [CrossRef]
37. Namasivayam, C.; Sangeetha, D. Equilibrium and kinetic studies of adsorption of phosphate onto $ZnCl_2$ activated coir pith carbon. *J. Colloid Interface Sci.* **2004**, *280*, 359–365. [CrossRef]
38. Biswas, B.K.; Inoue, K.; Ghimire, K.N.; Ohta, S.; Harada, H.; Ohto, K.; Kawakita, H. The adsorption of phosphate from an aquatic environment using metal-loaded orange waste. *J. Colloid Interface Sci.* **2007**, *312*, 214–223. [CrossRef]
39. Nguyen, T.; Ngo, H.; Guo, W.; Zhang, J.; Liang, S.; Tung, K. Feasibility of iron loaded 'okara' for biosorption of phosphorous in aqueous solutions. *Bioresour. Technol.* **2013**, *150*, 42–49. [CrossRef]

Article

Rheological and Thermal Study about the Gelatinization of Different Starches (Potato, Wheat and Waxy) in Blend with Cellulose Nanocrystals

Josefina Chipón [1], Kassandra Ramírez [1], José Morales [2,3] and Paulo Díaz-Calderón [2,3,*]

[1] Escuela de Nutrición y Dietética, Facultad de Medicina, Universidad de los Andes, Chile. Av. Monseñor Alvaro del Portillo N°12.455, Las Condes, Santiago 7620001, Chile; jtchipon@miuandes.cl (J.C.); krramirez@miuandes.cl (K.R.)
[2] Biopolymer Research & Engineering Laboratory (BIOPREL), Escuela de Nutrición y Dietética, Facultad de Medicina, Universidad de los Andes, Chile. Av. Monseñor Alvaro del Portillo N°12.455, Las Condes, Santiago 7620001, Chile; jose.morales@miuandes.cl
[3] Centro de Investigación e Innovación Biomédica (CIIB), Facultad de Medicina, Universidad de los Andes, Chile. Av. Monseñor Alvaro del Portillo N°12.455, Las Condes, Santiago 7620001, Chile
* Correspondence: pdiaz@uandes.cl

Abstract: The goal of this work was to analyze the effect of CNCs on the gelatinization of different starches (potato, wheat and waxy maize) through the characterization of the rheological and thermal properties of starch–CNC blends. CNCs were blended with different starches, adding CNCs at concentrations of 0, 2, 6 and 10% w/w. Starch–CNC blends were processed by rapid visco-analysis (RVA) and cooled to 70 °C. Pasting parameters such as pasting temperature, peak, hold and breakdown viscosity were assessed. After RVA testing, starch–CNC blends were immediately analyzed by rotational and dynamic rheology at 70 °C. Gelatinization temperature and enthalpy were assessed by differential scanning calorimetry. Our results suggest that CNCs modify the starch gelatinization but that this behavior depends on the starch origin. In potato starch, CNCs promoted a less organized structure after gelatinization which would allow a higher interaction amylose–CNC. However, this behavior was not observed in wheat and waxy maize starch. Insights focusing on the role of CNC on gelatinization yielded relevant information for better understanding the structural changes that take place on starch during storage, which are closely related with starch retrogradation. This insight can be used as an input for the tailored design of novel materials oriented towards different technological applications.

Keywords: gelatinization; cellulose nanocrystals; rheology; calorimetry

Citation: Chipón, J.; Ramírez, K.; Morales, J.; Díaz-Calderón, P. Rheological and Thermal Study about the Gelatinization of Different Starches (Potato, Wheat and Waxy) in Blend with Cellulose Nanocrystals. *Polymers* **2022**, *14*, 1560. https://doi.org/10.3390/polym14081560

Academic Editors: Valentina Siracusa, Nadia Lotti, Michelina Soccio and Alexey Iordanskii

Received: 17 March 2022
Accepted: 10 April 2022
Published: 11 April 2022

Publisher's Note: MDPI stays neutral with regard to jurisdictional claims in published maps and institutional affiliations.

Copyright: © 2022 by the authors. Licensee MDPI, Basel, Switzerland. This article is an open access article distributed under the terms and conditions of the Creative Commons Attribution (CC BY) license (https://creativecommons.org/licenses/by/4.0/).

1. Introduction

Starch and cellulose are the most widely distributed polymers in nature. Starch is found in the form of granules which are energy reservoirs for plants [1]. Cellulose, in turn, is part of the structural basis of cell walls in plant tissues, normally forming complexes with hemicellulose and lignin [2]. Although starch and cellulose have been widely used in food science and technology, in recent years, the interest in exploring the novel applications of these polymers and their composites has increased across completely different fields. Indeed, the literature has reported the use of cellulose as a strategy to improve some poor physical properties of starch such as brittleness, low mechanical resistance, high gas permeability and high hygroscopicity [3,4]. This strategy can be useful for the tailored design of starch–cellulose composites destined for food packaging and coating materials [4–7], scaffolds for tissue engineering (e.g., wound healing) [8–10] as well as applications in bioengineering and the pharmaceutical industry [3,8,11].

Likewise, more recently, interest in exploring the use of nanocellulose in the design of starch-based composites has also increased. Among the more interesting choices are cellulose nanocrystals (CNCs). CNCs correspond to the crystalline fractions presented in the elemental fiber of a cellulose microfibril [12,13]. CNCs are normally produced by acid hydrolysis under severe conditions in terms of concentration and temperature, which allows obtaining nanosized crystalline whiskers [14,15]. Depending on hydrolysis conditions and the cellulose source, CNCs have dimensions of approximately 120–200 nm in length and have a negative charge (z-potential lower than −30 mV) [14,15].

Despite the effect of CNCs on the physical properties and functionality of different starches being described in the literature, studies describing the role of CNCs during the gelatinization of starch are scarce. The gelatinization of starch has been well-described in terms of the mechanism of granule swelling, granule disruption, the loss of birefringence and amylose and amylopectin leaching due to the effect of temperature and stirring [16–18]. However, a comprehensive study of the gelatinization of starch in the presence of nanocrystals will help to understand the behavior of starch during the storage when the complex process of self-association and self-assembling will take place. This behavior explains the retrogradation of starch.

Therefore, this work aimed to study the gelatinization of starches from different sources (potato, wheat and waxy maize) in the presence of CNCs produced from cotton cellulose pulp. Pasting parameters closely related to starch gelatinization (such as pasting temperature, peak viscosity, hold viscosity and breakdown viscosity) were assessed in starch–CNC blends by rapid visco-analysis (RVA). The viscoelasticity of these blends were evaluated by rheology test at 70 °C, whereas the gelatinization temperature and enthalpy were assessed by differential scanning calorimetry (DSC).

2. Materials and Methods

2.1. Materials

Native potato (Avebe, Veendam, The Netherlands), wheat (Sigma Aldrich, Darmstadt, Germany) and waxy maize (Sigma Aldrich, Darmstadt, Germany) starches were purchased in powder form. Cellulose nanocrystals (CNCs) produced from cotton fibers were purchased in freeze-dried form from The Process Development Center of University of Maine (Orono, ME, USA). Starch and CNCs were used as received without further purification and stored at room temperature until further use.

2.2. Methods

2.2.1. Sample Preparation

CNC suspensions were prepared for blending with native starch to have a starch suspension with an CNC concentration of 0, 2, 6 and 10% w/w dry cellulose (Equation (1)). For this purpose, well-defined amounts of CNCs were suspended in distilled water (conductivity < 10 mS) only by stirring. Before being mixed with starch, CNCs were sonicated (1500 W, 5 min) at ambient temperature (20 ± 2 °C) to promote the complete dispersion of nanocrystals:

$$\text{CNC concentration } (\%, w/w) = (\text{CNC}_{weight} / (\text{CNC}_{weight} + \text{Starch}_{weight})) \times 100 \quad (1)$$

2.2.2. Gelatinization of Starch–CNC Blends by Rapid Visco-Analysis (RVA)

The gelatinization of starch–CNC blends was carried out by rapid visco-analysis (RVA 4500, Perten Instruments, North Ryde, Australia) in accordance with the methodology proposed by Diaz-Calderon et al. [19] with minor modifications. Two grams of native starch was weighed in aluminum canisters and 25 mL of CNC suspension with different cellulose concentration was transferred using a micropipette. The canister was then inserted into the instrument and viscosity patterns were obtained as a function of temperature: holding at 25 °C during 2 min; heating between 25 and 95 °C at 13.5 °C/min; holding at 95 °C for 5 min; cooling to 25 °C at 13.5 °C/min; and holding at 25 °C for 1 min. The analysis was performed under constant stirring (160 rpm). As the goal of this work was focused on understanding the gelatinization of native starch, the pasting parameters evaluated were pasting temperature (°C), peak viscosity (cP), hold viscosity (cP) and breakdown viscosity (difference between hold viscosity and peak viscosity, cP). All measurements were carried out at least in triplicate.

2.2.3. Rheological Characterization of Gelatinized Starch–CNC Blends

The changes in viscoelasticity of starch–CNC blends were carried out by rheology (Discovery HR-2, TA Instrument, New Castle, DE, USA). Samples analyzed by rheology were prepared using the RVA machine and following the same protocol explained in Section 2.2.2, but with the cooling stage finishing at 70 °C. Once the protocol was finished, the RVA canister was immediately transferred to a controlled bath set at 70 °C. Samples were collected from the canister and transferred to the rheometer.

Apparent viscosity (Pa·s) and shear stress (Pa) were assessed from the rotational flow curve obtained at 70 °C using cone geometry (stainless steel, 40 mm diameter, 0:30:7 angle and 15 μm truncation) in the shear rate range between 1 and 1000 1/s.

Dynamic changes in storage modulus (G', Pa) and loss modulus (G'', Pa) were obtained through a frequency sweep carried out at 70 °C to analyze the behavior of G', G'', and the loss factor (G''/G') as a function of angular frequency from 1 to 648 rad/s at 0.5% strain which was within the linear viscoelastic range (LVR) previously defined by an amplitude sweep (0.1–1000%) at 70 °C. A flat plate geometry (5 cm diameter) was used for analysis and a 300-micron gap was selected for testing. The analysis considered at least five replicates per experimental condition.

2.2.4. Thermal Properties of Gelatinized Starch–CNC Blends

Gelatinization temperature (°C) and enthalpy (J/g starch) were measured by differential scanning calorimetry (DSC 1, Mettler-Toledo, Greinfensee, Switzerland) following the protocol reported by Díaz-Calderón et al. [19]. In order to improve the resolution signal, a higher concentration of the starch weight suspensions was used, keeping the CNC weight previously indicated fractions (0–10% w/w). The starch concentration was 20% w/v. Approximately 60 μL of starch–CNC blend suspensions were loaded into 100 μL aluminum pans and then hermetically sealed. The DSC was calibrated using indium (melting temperature and enthalpy of 156.5 ± 1.56 °C, H = 28.6 ± 1 J/g), and an empty pan was used as a reference. Thermal properties of the suspensions were measured as follows: holding temperature at 5 °C during 3 min, heating from 5 °C to 85 °C at 10 °C/min, and holding at 85 °C during 3 min. The gelatinization temperature (°C) was recorded from the onset of the endothermic peak associated with the starch granule swelling, while gelatinization enthalpy was considered as the area under the endothermic peak. Gelatinization enthalpy was normalized in terms of starch dry mass and was expressed in J/g starch. All measurements were performed at least in triplicate.

2.2.5. Statistical Analysis

Where appropriate, the statistical significance was assessed by a paired t-test (same variances) and ANOVA using the Solver tool in Excel (Office 2016, Microsoft Corp., Redmond, WA, USA).

3. Results

3.1. Gelatinization of Starch–CNC Blends by Rapid Visco-Analysis (RVA)

Viscosity patterns of starch samples resulted from RVA are presented in Figure 1. As expected, the viscosity pattern showed marked differences among the starches tested: potato, wheat and waxy maize. From the viscosity patterns, the pasting parameters that allowed characterizing the gelatinization of starch were assessed, which in our study were pasting temperature, peak viscosity, hold viscosity and breakdown viscosity. The assessments of the pasting parameters for our starch–CNC blends are presented in Table 1.

Significant differences (p-value < 0.05) in the pasting parameters were observed among the pure starch samples (0%CNC). Regarding the pasting parameters of starch–CNC blends, our results show that the presence of CNC did not produce a significant difference (p-value > 0.05) on the pasting temperature. However, CNC produced a significant decrease (p-value < 0.05) in peak viscosity which was proportional to the increase in CNC concentration only in potato starch. In wheat and waxy starch, a decrease in peak viscosity was only detected at the highest CNC concentration tested. On the other hand, hold viscosity (also called through viscosity or hot-paste viscosity [16]) was significant lower (p-value < 0.05) only at the highest CNC concentration tested (10%) in potato and waxy maize starch. This parameter was not significant modified (p-value > 0.05) in the wheat starch blend. Likewise, the breakdown viscosity was significantly reduced (p-value < 0.05) by CNC in potato starch, and only at the highest CNC level tested in wheat and waxy starch. Interestingly, the assessment of the relative decrease only showed a marked decrease in potato starch, whereas in wheat and waxy starch, this ratio was lower only at the highest CNC level tested.

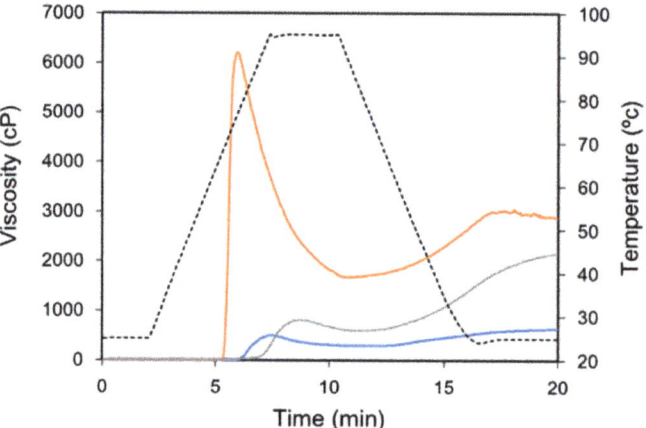

Figure 1. RVA profiles in starch from the different sources: potato (orange line), wheat (grey line) and waxy maize (blue line). Dotted line represents the temperature profile used during analysis.

3.2. Rheological Characterization of Gelatinized Starch–CNC Blends

Rheology analysis by rotational and dynamic tests were carried out at 70 °C on samples completely gelatinized by RVA. Apparent viscosity and shear stress as a function of shear rate recorded in pure starch samples and starch–CNC blends are presented in Figures 2 and 3, respectively. In agreement with the hold viscosity data assessed by RVA (Table 1), the flow curve showed significantly higher viscosity values in potato starch at 70 °C over the complete shear rate range tested, whereas the waxy starch showed the lowest viscosity at 70 °C. Independent of the source, all starches showed non-Newtonian behavior.

Figure 3 shows the effect of different concentrations of CNC on the apparent viscosity and shear stress as a function of shear rate tested at 70 °C in complete gelatinized starches. Our results show that CNC produced only slight changes in apparent viscosity and shear stress. Indeed, well-defined viscosity data recorded at 40 1/s (Table 2) showed that the presence of CNC only significant modified (p-value < 0.05) the apparent viscosity of waxy starch at 6% and 10%, tested at 70 °C. Likewise, the non-Newtonian condition of gelatinized starches tested at 70 °C was not modified by the presence of CNC, independently of the starch source. As expected, and following the same trend observed in Figure 3, gelatinized potato starch–CNC blends showed the highest apparent viscosity values, whereas waxy maize–CNC showed the lowest values of viscosity, independently of the CNC concentration. The same behavior was observed in terms of shear stress as a function of shear rate in the starch–CNC blends.

Table 1. Pasting properties recorded during the gelatinization of starch (potato, wheat and waxy corn) assessed by RVA in starch–CNC blends. Values in brackets correspond to standard deviation. Different upper letters in the same column represent significant differences (p-value < 0.05). The relative decrease was defined as the ratio of peak viscosity over hold viscosity in starch samples containing the same amount of CNC.

CNC (%, db)	Pasting Temperature (°C)			Peak Viscosity (cP)			Hold Viscosity (cP)			Breakdown Viscosity (cP)			Relative Decrease		
	Potato	Wheat	Waxy	Potato	Wheat	Waxy	Potato	Wheat	Waxy	Potato	Wheat	Waxy	Potato	Wheat	Waxy
0	69.2 (0.5) [a]	87.0 (0.1) [a]	79.8 (1.1) [a]	6217.6 (31.1) [a]	810.1 (9.8) [a]	514.2 (11.9) [a]	1638.2 (40.3) [a]	598.6 (8.4) [a]	296.0 (7.0) [a]	4579.4 (172.9) [a]	211.4 (6.7) [a]	218.2 (5.4) [a]	3.80	1.35	1.74
2	69.1 (0.5) [a]	87.0 (0.0) [a]	79.9 (0.4) [a]	3919.4 (98.3) [c]	776.9 (15.8) [b]	498.6 (9.2) [a]	1615.4 (50.7) [a]	562.0 (15.2) [b]	287.4 (5.9) [a]	2304.1 (55.9) [c]	214.5 (3.2) [a]	211.2 (3.6) [a]	2.43	1.38	1.73
6	69.3 (0.5) [a]	87.0 (0.1) [a]	79.8 (0.4) [a]	2667.4 (32.4) [d]	770.8 (13.4) [b]	438.2 (5.8) [b]	1653.0 (15.9) [a]	573.2 (14.9) [a,b]	257.8 (3.8) [b]	1014.4 (22.7) [d]	197.6 (5.9) [b]	180.4 (3.3) [b]	1.61	1.34	1.70
10	68.9 (0.4) [a]	87.1 (0.2) [a]	80.3 (0.9) [a]	1921.0 (14.9) [e]	701.1 (73.8) [c]	395.2 (6.4) [c]	1526.6 (13.9) [b]	578.8 (41.5) [a,b]	245.6 (5.5) [c]	394.4 (11.3) [e]	122.2 (33.4) [c]	149.6 (2.5) [c]	1.26	1.21	1.61

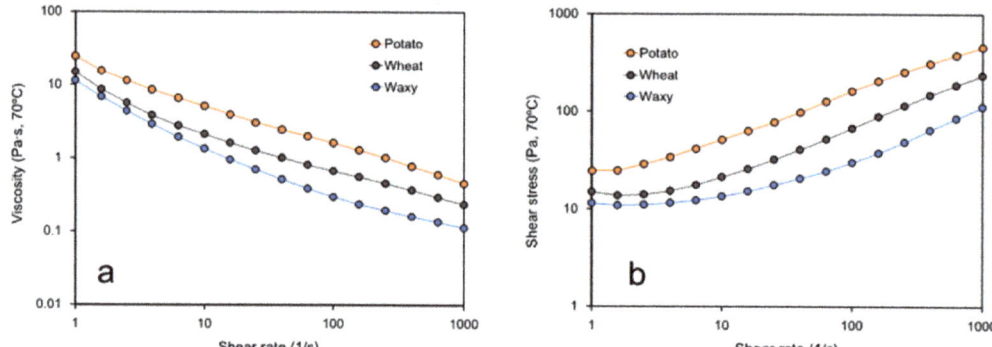

Figure 2. Flow curve in gelatinized pure starch samples (0%CNC) at 70 °C: (**a**) apparent viscosity and (**b**) shear stress as a function of shear rate.

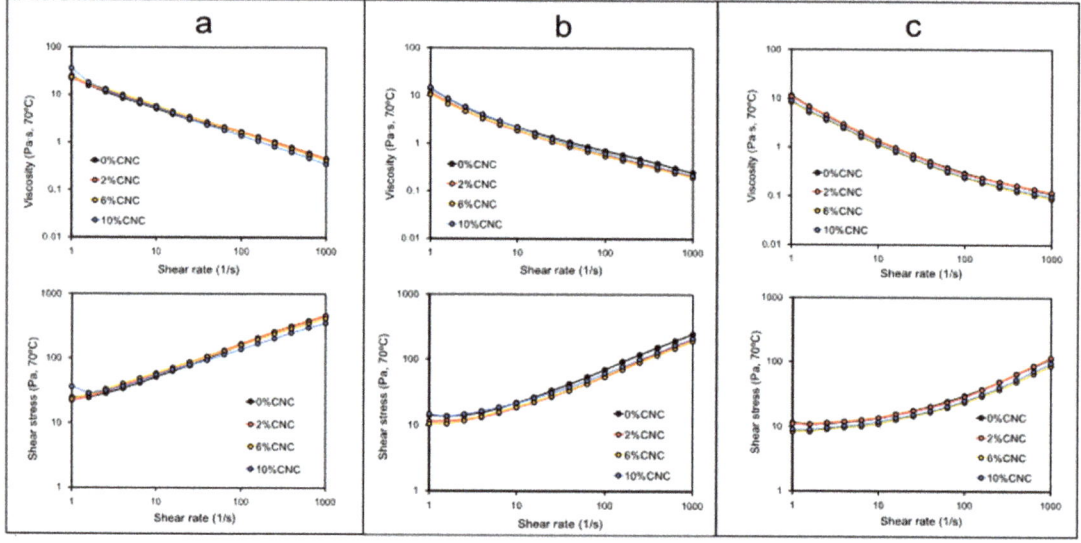

Figure 3. Flow curve in gelatinized starch samples at 70 °C and at different concentrations of CNCs: (**a**) potato; (**b**) wheat; and (**c**) waxy maize. Top plots correspond to apparent viscosity as a function of shear rate, whereas bottom plots correspond to shear stress as a function of shear rate.

The viscoelastic characterization of starch–CNC blends is presented in Figures 4 and 5. Dynamic tests were carried out within the linear viscoelastic range previously defined by an amplitude sweep at 70 °C (Figures S1 and S2, Supplementary File). Wheat and waxy starch did not show significant differences (p-value > 0.05) in G' and both showed a slight slope over the angular frequency range tested at 70 °C (Figure 4a). Potato starch showed a significantly lower G' over the whole angular frequency tested along with a marked slope depicting the strong dependence of G' with the angular frequency. All the gelatinized starch samples showed significantly higher values of G' compared to values of G'', which represent the gel-like condition of starch blends after processing by RVA. The latter is also confirmed from the values of the loss factor at 70 °C which resulted in values lower than 1 in all samples tested (Figure 4b). However, at angular frequencies lower than 100 rad/s, the loss factor in potato starch was constant and with higher values than those assessed in the wheat and waxy starch.

Table 2. Apparent viscosity and shear rate of starch–CNC blends assessed at a shear rate of 40 1/s and 70 °C. Values in brackets correspond to standard deviation. Different upper letters in the same column represent significant differences (p-value < 0.05).

CNC (%, db)	Viscosity (Pa·s, 40 1/s, 70 °C)			Shear Stress (Pa·s, 40 1/s, 70 °C)		
	Potato	Wheat	Waxy	Potato	Wheat	Waxy
0	2.44 (0.08) [a]	1.07 (0.07) [a]	0.51 (0.05) [a]	97.3 (1.5) [a]	42.5 (2.9) [a]	20.4 (1.4) [a]
2	2.64 (0.05) [b]	0.85 (0.02) [b]	0.50 (0.03) [a]	105.2 (2.1) [b]	33.9 (1.0) [b]	19.8 (1.3) [a,c]
6	2.60 (0.03) [b]	0.85 (0.03) [b]	0.40 (0.02) [b]	103.5 (1.3) [b]	33.8 (1.3) [b]	16.5 (1.4) [b]
10	2.33 (0.06) [a]	0.96 (0.09) [a,b]	0.43 (0.01) [b]	92.8 (8.1) [a]	38.2 (4.4) [a,b]	17.2 (1.0) [b,c]

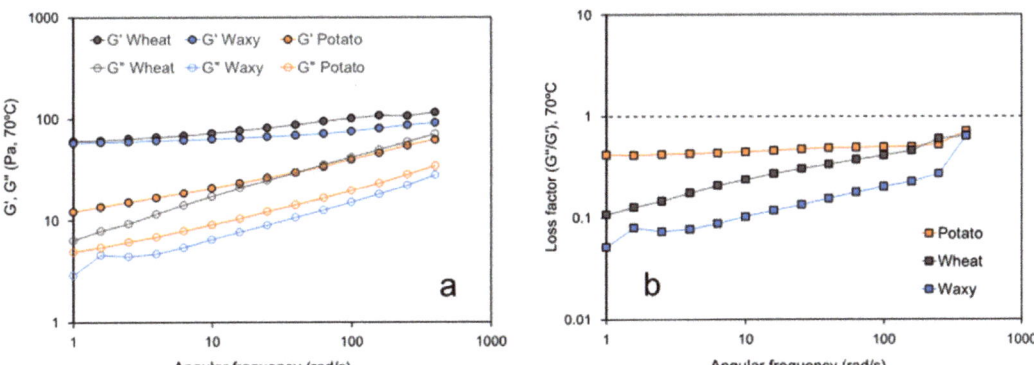

Figure 4. G′ and G″ (a) and loss factor (G″/G′) (b) assessed as a function of angular frequency in gelatinized starch samples at 70 °C.

The effect of CNC on the viscoelasticity of our starch blends is shown in Figure 5, and selected values of G′, G″ and the loss factor assessed at 10 rad/s and 70 °C are presented in Table 3. Our results show that the effect of CNC on the G′ is different depending on the starch source. Thus, CNC produced a significant increase in G′ in potato starch and waxy starch (p-value < 0.05) but a significant decrease in wheat starch. However, this effect was not proportional to the CNC concentration (Table 3). Similar behavior was observed with G″ data. The behavior of G′ and G″ were well reflected by the loss factor data which increased in the case of potato and wheat starch as a function of CNC concentration and decreased in waxy starch in the presence of CNC.

3.3. Thermal Properties of Gelatinized Starch–CNC Blends

Both the gelatinization temperature (°C) and gelatinization enthalpy (J/g starch) of starch–CNC blends are shown in Figure 6. The presence of CNC had little effect on the gelatinization temperature recorded from the onset of the endothermic peak associated with starch granule swelling, which agrees with the behavior of pasting temperature assessed by RVA (Table 1). As far as enthalpy is concerned, significant changes were only observed in potato starch, where the presence of CNC produced a decrease in enthalpy but without significant differences (p-value > 0.05) among CNC concentrations. In both wheat and waxy starch, the presence of CNC did not change the value of energy necessary to trigger the granule swelling and disruption.

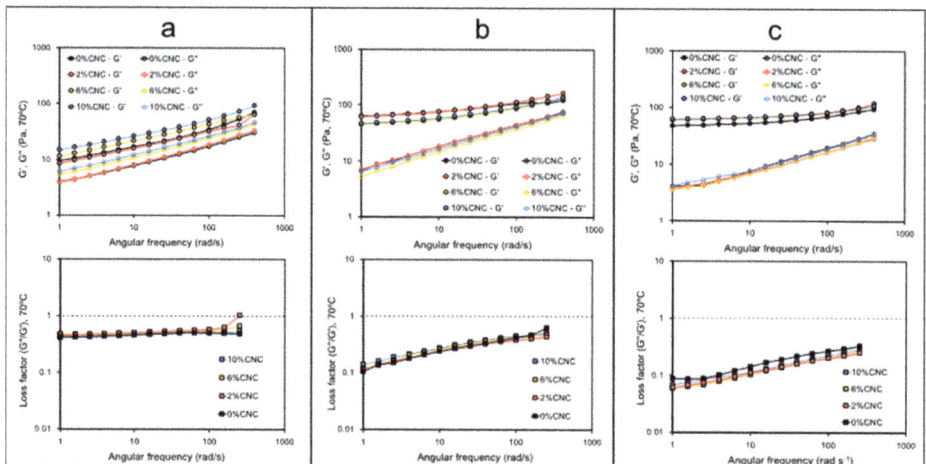

Figure 5. G′, G″ and loss factor (G″/G′) assessed as a function of angular frequency in gelatinized starch samples at 70 °C and at different concentrations of CNCs: (**a**) potato; (**b**) wheat; and (**c**) waxy maize. Top plots correspond to a frequency sweep as a function of shear rate, whereas the bottom plots correspond to the loss factor as a function of shear rate.

Table 3. G′, G″ and loss factor of starch–CNC blends assessed at an angular frequency of 10 rad/s and 70 °C. Values in brackets correspond to standard deviation. Different upper letters in the same column represent significant differences (p-value < 0.05).

CNC (%, db)	G′ (Pa, 10 rad/s, 70 °C)			G″ (Pa, 10 rad/s, 70 °C)			Loss Factor (G″/G′, 10 rad/s, 70 °C)		
	Potato	Wheat	Waxy	Potato	Wheat	Waxy	Potato	Wheat	Waxy
0	16.74 (0.9)[a]	76.43 (3.0)[a]	52.90 (2.4)[a]	7.76 (0.2)[a]	18.32 (1.6)[a]	7.59 (0.9)[a]	0.46 (0.02)[a]	0.24 (0.025)[a]	0.14 (0.072)[a]
2	15.92 (0.8)[a]	76.17 (3.8)[a]	66.28 (4.2)[b]	8.06 (0.3)[a]	18.33 (0.9)[a]	7.30 (0.3)[a]	0.51 (0.01)[b]	0.24 (0.002)[a]	0.11 (0.010)[a]
6	21.65 (1.8)[b]	55.91 (3.7)[b]	65.64 (3.2)[b]	10.92 (0.7)[b]	14.82 (0.9)[b]	6.72 (0.4)[a]	0.51 (0.02)[b]	0.27 (0.06)[a]	0.10 (0.001)[a]
10	26.45 (1.2)[c]	59.26 (2.4)[b]	67.45 (3.7)[b]	12.28 (0.5)[b]	16.63 (0.6)[a,c]	7.69 (0.5)[a]	0.46 (0.01)[a]	0.28 (0.004)[b]	0.11 (0.003)[a]

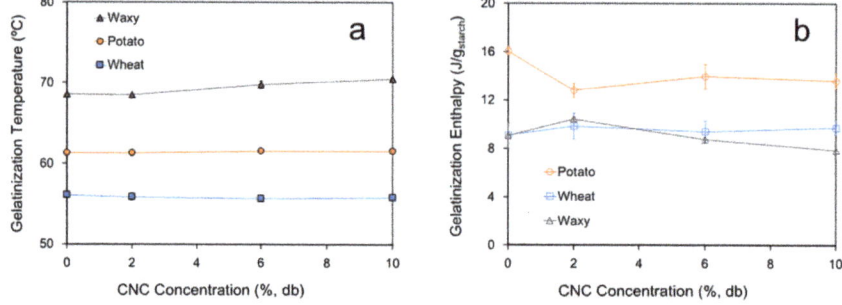

Figure 6. Gelatinization temperature (**a**) and enthalpy (**b**) of starch from different sources (potato, wheat and waxy) as a function of CNC concentration. Continuous lines only correspond to guide the eye.

4. Discussion

Differences in the viscosity patterns among starches showed in Figure 1 can be explained by polymorphism expressed by each starch (e.g., A-type in case of wheat and waxy, B-type for potato), which are dependent on the source, the hierarchical structure and lamellar organization in native granules and the branch-chain length of amylopectin [20]. On the other hand, the assessment of pasting parameters (Table 1) can be used as an approach to understand the effect of CNCs on the gelatinization of starch. Thus, the decrease in peak viscosity suggests that the mechanism of granule swelling and disruption is affected by the presence of CNCs, although this behavior would be dependent of the starch source. The literature has only partially reported the effect of CNCs (or other nanowhiskers) on the pasting properties of gelatinized starch, but with contradictory results. For instance, Cui et al. [21] described that CNCs produced a significant increase in peak viscosity in both waxy maize and sweet potato starch, which was attributed to a cross-linking effect via hydrogen bonding between CNCs and starch granules, resulting in a reduced degree of granule swelling. However, Ji et al. [22] found that chitin nanowhiskers significantly reduced the peak viscosity in potato starch, explaining this behavior by an effect of replacement of starch with chitin nanowhiskers causing a dilution of the concentration of starch, and decreasing the peak viscosity. George et al. [23] recently suggested that a lower hydration property restricts the granule swelling and thereby reduces the peak viscosity. Likewise, the literature hypothesized about the ability of cellulose particles to bind water or compete for free water with leached amylose and ungelatinized granules affecting hydration, mobility and granule swelling [19,24]. This could be another reason why in our study the presence of CNCs reduced the peak viscosity.

On the other hand, the pasting temperature was not changed by CNCs, suggesting that the process by which the starch granules begin to swell is not modified by the presence of CNCs, regardless of the starch origin. These results agree with the study of Cui et al. [21] who reported that the presence of CNCs did not modify the pasting temperature in normal maize, waxy maize and sweet potato starches. However, that study was carried out using samples with marked differences in starch concentration, which resulted in values of the pasting temperature which were identical between waxy and potato. Previously, Ji et al. [22] reported that chitin-based nanowhiskers did not change the pasting temperature of maize starch. Most recently, George et al. [23] reported that the addition of starch nanocrystals in concentrations of up to 15% did not produce significant changes in the pasting temperature of rice and wheat starch.

The hold viscosity was shown to be significantly lower at the highest CNC concentration tested in potato and waxy starch, which agree with the data of apparent viscosity (Table 2). The literature lacks works that describe changes in viscosity in gelatinized starch by the effect of nanoparticles and nanocrystals. BeMiller [25] reviewed the effect of different hydrocolloids on the pasting, paste and gel properties of different starch–hydrocolloids composites, finding that changes in viscosity in gelatinized starches would be in dependent of the hydrocolloid molecular size, and concluding a possible direct correlation between the molecular weight and increase in viscosity. Most recently, Ji et al. [22] reported that the presence of chitin nanowhiskers did not modify the hold viscosity in maize and potato starch pastes, which was explained by a possible effect of the interaction between chitin nanowhiskers and starch molecules via hydrogen bonds among the OH groups of nanowhiskers and starches. Likewise, the literature reported that a significant decrease in breakdown viscosity would suggest a reduced capacity for the swelling of starch granules [26]. This could be the case in the data reported in Table 1; however, the evaluation of the relative decrease in viscosity during gelatinization resulted in potato starch showing a decrease from values of 3.80 to 1.26 by the effect of CNCs (Table 1). In the case of wheat and waxy maize starch, the values of relative decrease remained constants except for the highest CNC concentration, which was lower in both cases. These results support the fact that CNCs mainly modify the RVA pattern and pasting properties in potato starch.

Regarding the apparent viscosity behavior of starch at 70 °C, our results show a significant higher viscosity in potato starch (Figure 2). This behavior was correlated with the substantially longer branch-chains length of the potato starch which once leached out during gelatinization to offer higher resistance to flow [20]. The non-Newtonian behavior of starchy pastes can be explained by the alignment effect of starchy particles in the direction of the shear during the analysis, as has been reported in the literature [20,27–29]. The increase in shear stress as a function of the shear rate (Figure 2b) associated with changes in viscosity has been related with the solid–liquid character of the gelatinized suspensions [30], as well with the increase in interaction between starch particles and fibrils. Agglomerates or network structures able to be broken or aligned during rheological measurements result in higher shear stress values [24,27].

Changes in pasting properties can be better understood through the viscoelastic characterization of starch pastes. The results shown in Figure 4 suggest that once gelatinized, potato starch shows a less ordered structure (lower G', less elastic) compared to wheat and waxy starch. In wheat and waxy, the starchy paste is more elastic at lower angular frequencies, suggesting the time-dependence condition of the structural organization of gelatinized starch tested at 70 °C. This behavior may help to explain the differences previously found in hold viscosity (Table 1) and apparent viscosity (Figures 3 and 4). Since potato starch was shown to be less organized and less elastic once gelatinized, the viscosity in potato starch is higher than that observed in wheat and waxy starch, which is consistent with the higher values of the loss factor observed in the potato starch. These results suggest that the gelatinization mechanism could be different depending on the starch origin or could be influenced by the polymorph organization presented in the native granule. Ai and Jane [20] have reported that the integrity of a swollen granule after gelatinization is essential to understanding the formation of strong gels. The granules of potato starch readily swell and disperse during gelatinization because of the lack of amylose–lipid complex formation which helps to maintain the integrity of swollen starches [20]. Thus, this behavior agrees with results shown in Figure 5 and Table 3, where gelatinized potato starch showed a lower G' and higher loss factor values at 70 °C. Similar behavior was reported by Cui et al. [21] whom despite testing the viscoelasticity of different gelatinized starch–CNC gels at 4 °C, also showed that sweet potato starch has a lower G' and higher loss factor values compared to normal maize and waxy maize starch.

Regarding the effect of CNCs on the viscoelasticity of gelatinized starches assessed at 70 °C, our results showed to be strongly dependent on the starch origin. Only a few studies reporting viscoelastic data in starch–CNC blends are available in the literature. For instance, Cui et al. [21] suggested the occurrence of interactions between CNCs and amylose, allowed by the high specific surface area and hydroxyl groups of CNCs. These interactions would explain why, in their study, CNCs delayed retrogradation in both sweet potato and normal maize starch. Our results show that waxy starch would retain structural order after gelatinization due to an increased G' and decreased loss factor as a function of CNC concentration (Figure 5, Table 3). In turn, in both potato and wheat starch, the presence of CNCs increased the loss factor, suggesting a less organized structure [31]. Interestingly, CNCs decreased the value of G' in wheat starch, whereas in potato starch, CNCs increased G'. However, the potato starch showed higher values of the loss factor compared with the wheat starch despite the fact that CNCs also lead to an increase in G'' in potato starch (Figure 5, Table 3). This behavior could be explained by the higher interaction amylose–CNC promoted by the leaching of amylose during gelatinization, which in the case of potato, is higher than in other starches, as was proposed by Ai and Jane [20]. Thus, the higher amylose–CNC interaction occurring in potato starch could explain the higher values of G' and G'', which was not observed in wheat and waxy starches. Future studies of this point should include using IR-spectroscopy or RAMAN spectroscopy to characterize the type and intensity of these interactions.

Thermal characterization by DSC showed a good correlation with rheology data. The fact that the gelatinization temperature (Figure 6a) and pasting temperature (Table 1) of each starch did not show significant differences ($p > 0.05$) suggests that temperature related to the starting point of starch gelatinization is not modified by CNCs. Similar behavior was reported by Cui et al. [21] and Ji et al. [22]. Differences among the values of gelatinization temperature and pasting temperature could be explained because gelatinization temperature by DSC is detected from the point at which the double helices of amylose and amylopectin begin to unfold, whereas the pasting temperature from RVA is detected from the point at which the granule starts to swell [32]. On the other hand, differences detected between the gelatinization enthalpies of different starches also reflect differences in how these starches gelatinize. Thus, higher values of enthalpy in potato starch agree with the fast swelling and higher dispersion because of the absence of the amylose–lipid complex which helps to maintain the integrity of swollen starches [20]. However, the effect of CNCs reducing the value of the gelatinization enthalpy, which has been reported by other authors in the literature [21,33], can have more than one explanation. For example, Cui et al. [21] explained the reduction in enthalpy by the addition of CNCs due to an effect of the inhibition of gelatinization and conservation of double helices. Likewise, the ability of cellulose particles to bind water or compete for free water with leached amylose affecting hydration and granule swelling was hypothesized in the literature [19,26]. However, a potential effect of amylose–CNC interaction as has been suggested by our rheological characterization should not be neglected. Nonetheless, this behavior could help explain the lower enthalpy observed in wheat and waxy starch which were not modified by the presence of CNCs (Figure 6b). In this sense, as has been proposed in the previous section, a complementary analysis carried out by IR-spectroscopy and NMR could help understand this phenomenon that has not been extensively studied to date and may open up a new area of research.

5. Conclusions

Our results suggest that the presence of CNC produces changes in the mechanism of granule swelling and disruption during gelatinization, along with promoting a certain amylose–CNC interaction, although this behavior would be dependent of the starch origin. Thus, in potato starch, there was a marked decrease in peak viscosity and breakdown viscosity over CNC concentration, both values being consistent with the assessment of relative decrease in viscosity. This behavior was not observed in wheat starch and waxy starch. Hold viscosity measured by RVA agreed with apparent viscosity measured by rotative test, whereas both the peak temperature and gelatinization temperature were not changed by the presence of CNCs in all the starches studied. However, viscoelastic characterization at 70 °C showed higher values of loss factor in potato starch–CNC, suggesting a less organized structure after gelatinization, which in turn showed good correlation with the increase in G″ at different CNC concentration. Our results are presumably influenced by how potato starch gelatinizes and would not be affected by the presence/absence of amylose, as data from wheat and waxy starch suggest. The amylose–CNC interaction in potato starch could explain the higher values of enthalpy detected by calorimetry. However, complementary studies are needed to define the characteristics of these interactions.

These results can be useful in different technological applications based on the use of starch–nanocellulose composites. For instance, for the design of novel composite materials for packaging, bioplastics, bioprinting and food ingredients, to name a few. The potential impact of these biomaterials on sustainable processes opens interesting perspectives in terms of industrial application.

Supplementary Materials: The following supporting information can be downloaded at: https://www.mdpi.com/article/10.3390/polym14081560/s1, Figure S1: G′ (a) and G″ (b) assessed as a function of oscillation strain in gelatinized starch samples at 70 °C. Figure S2: G′ and G″ assessed as a function of oscillation strain in gelatinized starch samples at 70 °C, and at different concentrations of CNC: (a) potato, (b) wheat and (c) waxy maize.

Author Contributions: Conceptualization, P.D.-C.; methodology, P.D.-C.; validation, J.M. and P.D.-C.; formal analysis, J.C., K.R., J.M. and P.D.-C.; investigation, J.C., K.R., J.M. and P.D.-C.; resources, P.D.-C.; data curation, J.C., K.R. and J.M.; writing—original draft preparation, P.D.-C.; writing—review and editing, P.D.-C.; visualization, P.D.-C.; supervision, J.M. and P.D.-C.; project administration, P.D.-C.; funding acquisition, P.D-C. All authors have read and agreed to the published version of the manuscript.

Funding: This research was funded by Fondo Nacional de Desarrollo Científico y Tecnológico FONDECYT grant number 1191375.

Institutional Review Board Statement: Not applicable.

Informed Consent Statement: Not applicable.

Data Availability Statement: Our study do not report supporting results.

Conflicts of Interest: The authors declare no conflict of interest.

References

1. Pfister, B.; Zeeman, S.C. Formation of starch in plant cells. *Cell Mol. Life Sci.* **2016**, *73*, 2781–2807. [CrossRef] [PubMed]
2. Thomas, B.; Raj, M.C.; Athira, K.B.; Rubiyah, M.H.; Joy, J.; Moores, A.; Drisko, G.L.; Sanchez, C. Nanocellulose, a versatile green platform: From bioresources to materials and their applications. *Chem. Rev.* **2018**, *118*, 11575–11625. [CrossRef] [PubMed]
3. Benito-González, I.; López-Rubio, A.; Martínez-Sanz, M. High-performance starch biocomposites with cellulose from waste biomass: Film properties and retrogradation behaviour. *Carbohydr. Polym.* **2019**, *216*, 180–188. [CrossRef]
4. Ilyas, R.A.; Sapuan, S.M.; Ishak, M.R.; Zainudin, E.S. Development and characterization of sugar palm nanocrystalline cellulose reinforced sugar palm starch bionanocomposites. *Carbohydr. Polym.* **2018**, *202*, 186–202. [CrossRef] [PubMed]
5. Becerril, R.; Nerín, C.; Silva, F. Encapsulation systems for antimicrobial food packaging components: An update. *Molecules* **2020**, *25*, 1134. [CrossRef] [PubMed]
6. Fazeli, M.; Keley, M.; Biazar, E. Preparation and characterization of starch-based composite films reinforced by cellulose nanofibers. *Int. J. Biol. Macromol.* **2018**, *116*, 272–280. [CrossRef]
7. Mahardika, M.; Abral, H.; Kasim, A.; Arief, S.; Hafizulhaq, F.; Asrofi, M. Properties of cellulose nanofiber/bengkoang starch bionanocomposites: Effect of fiber loading. *LWT* **2019**, *116*, 108554. [CrossRef]
8. Eichhorn, S.J.; Etale, A.; Wang, J.; Berglund, L.A.; Li, Y.; Cai, Y.; Chen, C.; Cranston, E.D.; Johns, M.A.; Fang, Z.; et al. Current international research into cellulose as a functional nanomaterial for advanced applications. *J. Mater. Sci.* **2022**, *57*, 5697–5767. [CrossRef]
9. Waghmare, V.S.; Wadke, P.R.; Dyawanapelly, S.; Deshpande, A.; Jain, R.; Dandekar, P. Starch based nanofibrous scaffolds for wound healing applications. *Bioact. Mater.* **2018**, *3*, 255–266. [CrossRef]
10. Velásquez, D.; Pavon-Djavid, G.; Chaunier, L.; Meddahi-Pellé, A.; Lourdin, D. Effect of crystallinity on mechanical properties and tissue integration of starch-based materials from two botanical origins. *Carbohydr. Polym.* **2015**, *124*, 180–187. [CrossRef]
11. Bertsh, P.; Arcari, M.; Geue, T.; Mezzenga, R.; Nyström, G.; Fischer, P. Designing cellulose nanofibrils for stabilization of fluid interfaces. *Biomacromolecules* **2019**, *20*, 4574–4580. [CrossRef] [PubMed]
12. Mu, R.; Hong, X.; Ni, Y.; Li, Y.; Pang, J.; Wang, Q.; Xiao, J.; Zheng, Y. Recent trends and applications of cellulose nanocrystals in food industry. *Trends Food Sci. Technol.* **2019**, *93*, 136–144. [CrossRef]
13. Titton Dias, O.A.; Konar, S.; Lopes Leao, A.; Yang, W.; Tjong, J.; Sain, M. Current state of applications of nanocellulose in flexible energy and electronic devices. *Front. Chem.* **2020**, *8*, 420. [CrossRef] [PubMed]
14. Reid, M.S.; Villalobos, M.; Cranston, E.D. Benchmarking cellulose nanocrystals: From the laboratory to industrial production. *Langmuir* **2017**, *33*, 1583–1598. [CrossRef]
15. Vasconcelos, N.F.; Feitosa, J.P.A.; da Gama, F.M.P.; Morais, J.P.S.; Andrade, F.K.; de Souza, M.D.S.M.; de Freitas Rosa, M. Bacterial cellulose nanocrystals produced under different hydrolysis conditions: Properties and morphological features. *Carbohydr. Polym.* **2017**, *155*, 425–431. [CrossRef]
16. Balet, S.; Guelpa, A.; Fox, G.; Manley, M. Rapid Visco Analyser (RVA) as a tool for measuring starch-related physiochemical properties in cereals: A review. *Food Anal. Method* **2019**, *12*, 2344–2360. [CrossRef]
17. Belitz, H.D.; Grosch, W.; Schieberle, P. *Food Chemistry*, 4th ed.; Springer: Berlin, Germany, 2009; pp. 248–339.
18. BeMiller, J.N.; Huber, K.H. Carbohydrates. In *Fenemma's Food Chemistry*, 4th ed.; Damodaran, S., Parkin, K.L., Fennema, O.R., Eds.; CRC Press: New York, NY, USA, 2008; pp. 83–154.
19. Díaz-Calderón, P.; MacNaughtan, B.; Hill, S.; Foster, T.; Enrione, J.; Mitchell, J. Changes in gelatinisation and pasting properties of various starches (wheat, maize and waxy maize) by the addition of bacterial cellulose fibrils. *Food Hydrocoll.* **2018**, *80*, 274–280. [CrossRef]
20. Ai, Y.; Jane, J.-l. Gelatinization and rheological properties of starch. *Starch/Stärke* **2015**, *67*, 213–224. [CrossRef]
21. Cui, S.; Li, M.; Zhang, S.; Liu, J.; Sun, Q.; Xiong, L. Physicochemical properties of maize and sweet potato starches in the presence of cellulose nanocrystals. *Food Hydrocoll.* **2018**, *77*, 220–227. [CrossRef]

22. Ji, N.; Liu, C.; Zhang, S.; Yu, J.; Xiong, L.; Sun, Q. Effects of chitin nano-whiskers on the gelatinization and retrogradation of maize and potato starches. *Food Chem.* **2017**, *214*, 543–549. [CrossRef]
23. George, J.; Nair, S.G.; Kumar, R.; Ssemwal, A.D.; Sudheesh, C.; Basheer, A.; Sunooj, K.V. A new insight into the effect of starch nanocrystals in the retrogradation properties of starch. *Food Hydrocoll. Health* **2021**, *1*, 100009. [CrossRef]
24. Cao, Y.; Zavattieri, P.; Youngblood, J.; Moon, R.; Weiss, J. The relationship between cellulose nanocrystal dispersion and strength. *Constr. Build. Mater.* **2016**, *119*, 71–79. [CrossRef]
25. BeMiller, J.N. Pasting, paste and gel properties of starch-hydrocolloid combinations. *Carbohydr. Polym.* **2011**, *86*, 386–423. [CrossRef]
26. Qiu, S.; Yadav, M.P.; Liu, Y.; Chen, H.; Tatsumi, E.; Yin, L. Effects of corn fiber gum with different molecular weights on the gelatinization behaviors of corn and wheat starch. *Food Hydrocoll.* **2016**, *53*, 180–186. [CrossRef]
27. Ayala Valencia, G.; Freitas Moraes, I.C.; Gilles Hilliou, L.; Lourenco, R.V.; Sobral, P.J.A. Nanocomposite-forming solutions based on cassava starch and laponite: Viscoelastic and rheological characterization. *J. Food Eng.* **2015**, *166*, 174–181. [CrossRef]
28. Fang, F.; Tuncil, Y.E.; Luo, X.; Tong, X.; Hamaker, B.R.; Campanella, O.H. Shear-thickening behavior of gelatinized waxy starch dispersions promoted by the starch molecular characteristics. *Int. J. Biol. Macromol.* **2019**, *121*, 120–126. [CrossRef] [PubMed]
29. Ma, S.; Zhu, P.; Wang, M.; Wang, F.; Wang, N. Effect of konjac glucomannan with different molecular weights on physicochemical properties of corn starch. *Food Hydrocoll.* **2019**, *96*, 663–670. [CrossRef]
30. Agoda-Tandjawa, G.; Berot, D.S.; Blassel, C.; Gaillard, C.; Garnier, C.; Doublier, J.-L. Rheological characterization of microfibrillated cellulose suspensions after freezing. *Carbohydr. Polym.* **2010**, *80*, 677–686. [CrossRef]
31. Ottenhoff, M.-A. A Multi-Technique Study of the Retrogradation of Concentrated Starch Systems. Ph.D. Thesis, The University of Nottingham, Loughborough, UK, 2003.
32. Phimolsiripol, Y.; Siripatrawan, U.; Henry, C.J.K. Pasting behaviour, textural properties and freeze-thaw stability of wheat flour-crude malva nut (*Scaphium scaphigerum*) gut system. 2011). *J. Food Eng.* **2011**, *105*, 557–562. [CrossRef]
33. Cano, A.; Fortunati, E.; Cháfer, M.; González-Martínez, C.; Chiralt, A.; Kenny, J.M. Effect of cellulose nanocrystals on the properties of pea starch-poly(vinyl alcohol) blends films. *J. Mater. Sci.* **2015**, *50*, 6979–6992. [CrossRef]

Communication

Super-Hydrophobic Magnetic Fly Ash Coated Polydimethylsiloxane (MFA@PDMS) Sponge as an Absorbent for Rapid and Efficient Oil/Water Separation

Mengqi Zhao [1,2,*], Xiaoqing Ma [1], Yuxi Chao [1], Dejun Chen [1] and Yinnian Liao [1,*]

1. School of Chemical Engineering and Technology, Xinjiang University, Urumqi 830017, China
2. State Key Laboratory of Chemistry and Utilization of Carbon Based Energy Resources, Urumqi 830017, China
* Correspondence: xjzmq1205@xju.edu.cn (M.Z.); liaoyinnian@zcst.edu.cn (Y.L.)

Abstract: In this study, magnetic fly ash was prepared with fly ash and nano-magnetic Fe_3O_4, obtained by co-precipitation. Then, a magnetic fly ash/polydimethylsiloxane (MFA@PDMS) sponge was prepared via simple dip-coating PDMS containing ethanol in magnetic fly ash aqueous suspension and solidifying, whereby Fe_3O_4 played a vital role in achieving the uniformity of the FA particle coating on the skeletons of the sponge. The presence of the PDMS matrix made the sponge super-hydrophobic with significant lubricating oil absorption capacity; notably, it took only 10 min for the material to adsorb six times its own weight of n-hexane (oil phase). Moreover, the MFA@PDMS sponge demonstrated outstanding recyclability and stability, since no decline in absorption efficiency was observed after more than eight cycles. Furthermore, the stress–strain curves of 20 compression cycles presented good overlap, i.e., the maximum stress was basically unchanged, and the sponge was restored to its original shape, indicating that it had good mechanical properties, elasticity, and fatigue resistance.

Keywords: super hydrophobic; magnetic fly ash; polydimethylsiloxane; oil/water separation

1. Introduction

With global industrialization, oil spills remain a major source of water pollution and are among the most difficult challenges facing the world today. They are not only harmful to the natural ecosystem, but also have long-term adverse effects on human health and the economy [1,2]. However, treatment methods, such as flotation [3], combustion and linseed oil, have the disadvantages of poor selectivity and low efficiency [4].

The contact angle of water droplets on a lotus leaf is about 156°. As such, water droplets can easily slide off the surface of a leaf, taking surface pollutants with them and thereby achieving a self-cleaning effect. This phenomenon known as the "lotus leaf effect". Inspired by this natural phenomenon, biomimetic superhydrophobic membrane materials based on polymer materials have been developed through polymerization, vapor/liquid phase deposition, membrane transfer, micro-contact printing, electrostatic spinning and other processes. Superhydrophobic materials have a wide range of applications [5–7]. In recent years, the surface structures of lotus leaves have been mimicked, and interfaces similar to those of hydrophobic biomaterials with completely different wettability properties for oil and water have been designed and synthesized on the basis of surface chemical bionics. Such structures can effectively achieve efficient separation of oil and water. Superhydrophobic materials prepared by mimicking the surface characteristics of natural superhydrophobic materials are collectively known as biomimetic superhydrophobic materials. With the aim of adsorbing organic pollutants and purifying oily wastewater, research on such structures is rapidly growing in scope [8–10].

Superhydrophobicity means that water droplets are spherical on the surface with a contact angle greater than 150 degrees. True intrinsic superhydrophobicity does not exist, and the maximum water contact angle for flat materials is only 119 degrees. However, metals, ceramics and polymers can be made superhydrophobic by certain treatments, i.e., creating a suitable surface roughness or through the modification of a low surface energy material. For example, metals do not have superhydrophobic properties, but if their surface is roughened by corrosion etching and the surface energy is reduced by fluorination, a contact angle of more than 150 degrees can be obtained, thus turning them into superhydrophobic materials. In contrast, the surface energy of polymers is usually very low, making it easier to convert them into superhydrophobic materials [11].

For example, non-stick pans are made of polytetrafluoroethylene, which is superhydrophobic as long as the surface is rough. The lower the surface energy of the material (i.e., the more energy the molecules on the surface of the material have compared to the molecules inside), the better the hydrophobicity. When a low surface energy material has a microscopic rough structure, an air film will be formed between the water droplets and the material, preventing water from wetting the surface and thus forming a superhydrophobic state [12].

The surface energy of synthetic polydimethylsiloxane (PDMS) is only 22 mN/m. The compound is fluorine-free, environmentally-friendly and cheap, and it can be made to be highly soluble in organic solvents with curing agents [13,14]. PDMS has Si-O-Si as its main chain and has both organic groups and inorganic structures. Its Si-O bond length and bond angle are large, and it is easy to deform. As such, it can grafted/coated in the preparation of low surface energy biomimetic materials with special wettability properties, making it possible to achieve selective adsorption or filtration of oil–water mixtures [15,16]. Wang [17] prepared a super-hydrophobic PDMS@Fe_3O_4/MS sponge by an impregnation method. The sponge has the advantages of low cost, simple operation and high oil–water separation efficiency. The same author [18] synthesized a hydrophobic PDMS/reduced graphene oxide composite with an adsorption capacity for n-hexane that was 18.5 times its own weight. Zhu [19] connected cement gel particles with PDMS by covalent bonds. This not only changed the morphology of the particles, but also increased their contact angle, providing a research basis for PDMS organic/inorganic hybrid biomimetic materials. Finally, Yu et al. prepared Fe_3O_4 nanoparticles by the solvothermal method. They then used ammonium persulfate (APS) as an initiator to synthesize magnetic polystyrene oil-absorbing materials with different coating rates by emulsion polymerization of styrene and divinylbenzene (DVB) at different doses on the nanoparticle surfaces [20].

Fly ash is a kind of solid waste, whose main components are silica and alumina. Fly ash itself has adsorption capacity. Compared with other materials, fly ash has a large specific surface area and a porous structure, and thus, has good adsorption characteristics. The use of fly ash as a raw material has obvious advantages in terms of the required source materials.

The innovation of this study is the preparation of a new composite functional polymeric bionanomaterial, MFA@PDMS, for oily wastewater treatment. Based on the concept of "waste to waste", the natural porous structure of solid waste fly ash was used to construct a PDMS porous skeleton to make it rough and improve its oil–water separation efficiency [21,22]. Magnetic Fe_3O_4 nanoparticles were used to modify the natural porous structure of the fly ash in order to prepare the magnetic porous skeleton. Then, the superhydrophobic bionanomaterials were combined with the magnetic porous skeleton material coating/resin to prepare fluorine-free oil–water separation polymer bionanomaterials.

The preparation process is simple and the resulting magnetic sponges are super hydrophobic and super lipophilic. Additionally, the magnetic properties of the material make it possible to separate oil–water mixtures by an external magnetic field [23]. Magnetic nanomolecular sieve composites have the advantages of high magnetization performance and surface area. In addition, Fe_3O_4 nanoparticles were shown to improve the mechanical properties of the sponge, which remained superhydrophobic when it was under tension or compression. The tested sponge also showed good mechanical stability, oil stability and reusability in terms of its superhydrophobicity and oil absorption [21].

2. Materials and Methods

2.1. Materials

The fly ash (FA) used in this study was obtained from the Kanas Power Plant in Xinjiang Autonomous Region, China. Its chemical composition was as follows: SiO_2 (45.9 wt%), Al_2O_3 (19.03 wt%), CaO (16.39 wt%), SO_3 (6.01 wt%), Fe_2O_3 (5.65 wt%), K_2O (2.07 wt%), MgO (2.02 wt%) and Na_2O (0.954 wt%). The PDMS prepolymer and its curing agent were obtained from American Dow Corning. $FeCl_2 \cdot 4H_2O$, $FeCl_3 \cdot 6H_2O$ and sodium citrate all came from Macklin Biochemical (Shanghai, China) Co., Ltd. Ammonia water was obtained from Sigma-Aldrich (Shanghai, China) Trading Co., Ltd.

2.2. Preparation of Magnetic Fly Ash/Polydimethylsiloxane (MFA@PDMS) Sponge

(1) Preparation of magnetic fly ash: First, we poured $FeCl_2 \cdot 4H_2O$ (1.72 g) and $FeCl_3 \cdot 6H_2O$ (4.72 g) into a three-port flask, added 80 mL distilled water, and stirred for 10 min at 80 °C. Then, we added ammonia water dropwise (25 %) until a pH of 10 was reached. Black material could be seen in the process of dropping, and the reaction continued for 30 min after the solution had turned black. Next, 20 mL 0.3 m sodium citrate solution was added and the mixture heated to 90 °C for 30 min. We then added about 8 g of fly ash and continued stirring for 2 h. At the end of the reaction, we applied solid–liquid separation with an external magnetic field and repeated washing with distilled water. Then, we dried the compound at 80 °C for 12 h and, finally, used mechanical grinding to obtain the magnetic fly ash nanomaterials ($FA@Fe_3O_4$).

(2) Preparation of magnetic fly ash @PDMS bionic material: First, we placed 0.20 g magnetic fly ash, 4.00 g PDMS prepolymer and 0.40 g curing agent (10:1) into a beaker. The mixture was stirred, and 2.00 g anhydrous ethanol was added and mixed evenly. We then slowly added 4 mL water, stirred and allowed the mixture to emulsify for 15 min. It was poured into a sand core funnel mold for forming, and then placed in a drying oven at 120 °C for curing for 1.5 h to obtain dark brown spongy magnetic material (MFA@PDMS).

2.3. Adsorption Capacity, Separation Efficiency and Reusability Test

(1) To evaluate the adsorption capacity, a certain mass of the material was first immersed in an organic solvent and weighed after reaching mass absorption equilibrium. The measurement was repeated three times and the average value was taken. The adsorption capacity Q of the material was calculated according to Formula (1).

$$Q = \frac{(M_1 - M_0)}{M_0} \quad (1)$$

where Q (g/g) is the mass-based adsorption capacity, M_1 (g) is the mass of the sponge after oil absorption and M_0 (g) is the mass of the sponge before oil absorption.

(2) Next, we built an oil–water separation device and calculated the oil–water separation efficiency, R, of the sponge according to Formula (2).

$$R = \frac{V_C}{V_0} \times 100\% \quad (2)$$

where R (%) is the oil–water separation efficiency, V_c (mL) is the volume of the original solvent or oil collected after separation and V_0 (mL) is the volume of the original solvent or oil before separation.

(3) The repeatability of the material was initially evaluated using a simple electronic universal testing machine. First, the material was immersed in an organic solvent to achieve maximum absorption and its mass was weighed and recorded as M_2. Afterwards, the absorbed oil was recovered by mechanical compression and the mass of the material was determined again and recorded as M_3. The reusability of the sponge was assessed by absorption-compression cycles, each time weighing the material to calculate the adsorption

multiplier (i.e., the weight of the material after the adsorption of an organic solvent/the weight of the material itself).

2.4. Wenzel Model

In Equation (3), r represents the roughness factor of the solid surface ($r > 1$), which is actually the magnitude of the ratio of the actual area on the surface to the projected area; θ_W represents the apparent contact angle with the rough surface; and θ_Y is the contact angle in an ideal state [12].

$$\cos\theta_W = \frac{r(\gamma_{SV} - \gamma_{SL})}{\gamma_{LV}} = r \cdot \cos\theta_Y \qquad (3)$$

Under the condition that the solid surface itself is hydrophilic, if the surface roughness increases, the hydrophobicity of the surface will not increase. In contrast, under the condition that the solid surface itself has a certain degree of hydrophobicity, if the surface roughness increases, it will also have a certain effect on the hydrophobicity of the solid surface, and the level of hydrophobicity will naturally increase.

2.5. Determination and Characterization

The composite samples were analyzed using an X-ray fluorescence spectrometer (XRF, ARL PERFORM'X, Thermo Fisher Scientific, Waltham, MA, USA) and Fourier transformation infrared spectrometer (FTIR, VERTEX 70 RAMI, Bruker Corporation, Billerica, MA, USA). The morphologies of samples were comparatively observed by scanning electronic microscopy (SEM, SU8010, Hitachi, Ltd., Tokyo, Japan). The morphologies of the samples were observed using a transmission electron microscope (TEM, JEM-2100, JEOL Ltd. Akishima City, Tokyo, Japan). Magnetic analyses of the samples were conducted using a vibrating sample magnetometer (VSM, Lake Shore Corporation, Columbus, OH, USA). The stress–strain analyses of the samples were conducted using an electronic universal testing machine (CMT6103, MTS Systems Corporation, Eden Prairie, MN, USA).

3. Results and Discussion

3.1. Materials Characterization

3.1.1. TEM Analysis of Nanometer Fe_3O_4

The sample powder was dried using a 200 nm copper mesh. The electron acceleration voltage of the transmission electron microscope (TEM) was 100 kV.

A TEM image of the magnetic nano-Fe_3O_4 is shown in Figure 1. As shown, the average particle size of the magnetic nanospheres was 30 nm, and the particle size was uniform.

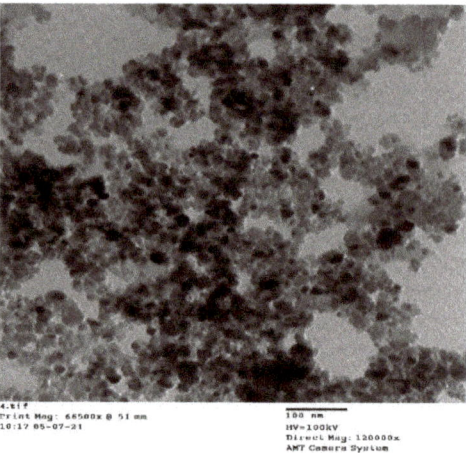

Figure 1. Transmission Electron Microscope Image of Fe_3O_4.

3.1.2. SEM Analysis

As Figure 2a,b shows, the surface of PDMS was extremely uniform and the folds were extremely smooth, which is the reason why the sponge material had such good mechanical properties and elasticity. Additionally, its internal porous structure increased its oil absorption performance to a certain extent.

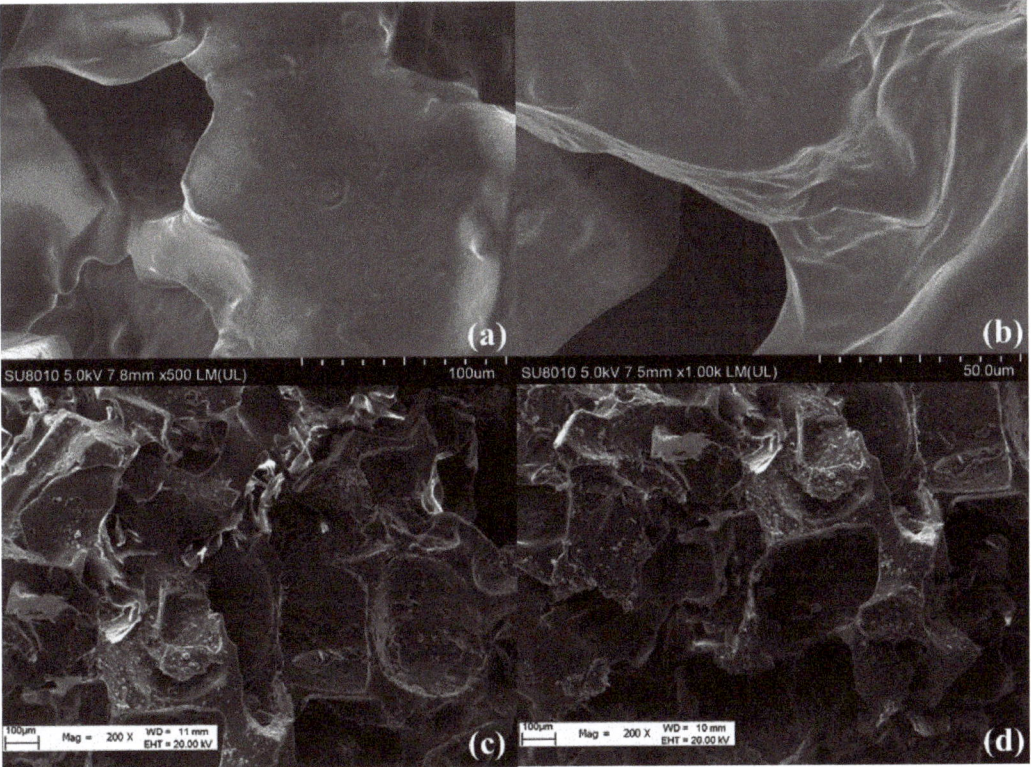

Figure 2. SEM of (**a**,**b**) PDMS and (**c**,**d**) MFA@PDMS.

As Figure 2c,d shows, the morphology changed significantly with modification with MFA; the smooth surface of the PDMS became rough, which was favorable for hydrophobicity. Additionally, because of the natural porous structure of FA, the MFA@PDMS had a rough, honeycomb porous structure, which undoubtedly increased the hydrophobicity of the sponge surface and the oil absorption ability of the sponge.

3.1.3. FT-IR Analysis

The Fourier transform infrared spectrum (FT-IR) of the MFA@PDMS is shown in Figure 3. The absorption peak at 2964 cm^{-1} corresponds to -CH$_3$ stretching vibrations; that at 1262 cm^{-1} was attributed to the sharp symmetric deformation vibration peak of Si-CH$_3$; that at 793 cm^{-1} corresponded to that Si-C tensile vibration peak [23]; that near 1083 cm^{-1} was attributed to the asymmetric stretching vibrations of Si-O-Si bond and Si-O-Al bonds, which was caused by the polymerization of the fly ash aluminosilicate or silicon-oxygen tetrahedron structure. Finally, a crystal lattice absorption peak of Fe$_3$O$_4$ appeared at 599 cm^{-1}, indicating that the MFA@PDMS material had been prepared successfully.

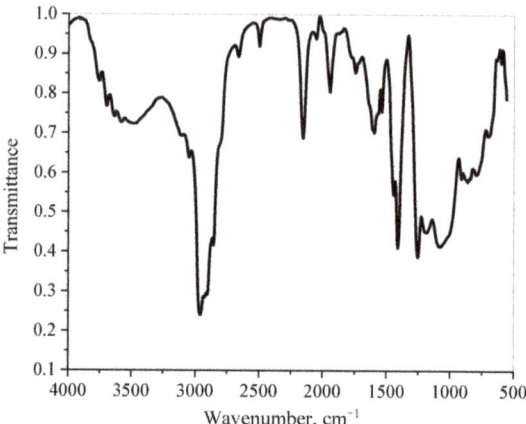

Figure 3. FT-IR Diagram of MFA@PDMS.

3.1.4. Characterization of Magnetic Properties on MFA@PDMS Sponge

The vibrating sample magnetometer (VSM) test results for nano-Fe_3O_4 and MFA@PDMS sponge are shown in Figure 4. As shown, their coercivity was low, and they demonstrated super-paramagnetism. Figure 5 is a graph comparing the natural sedimentation separation and magnetic field separation of the FA@Fe_3O_4 suspension.

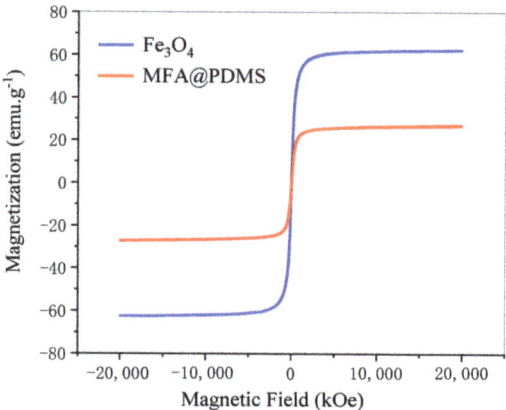

Figure 4. Hysteresis loop of Fe_3O_4 and MFA@PDMS.

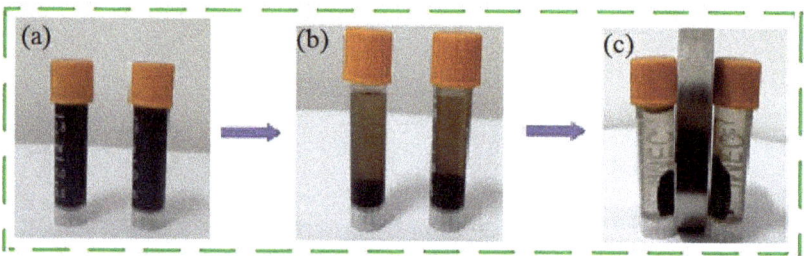

Figure 5. Solid–liquid separation process of FA@Fe_3O_4. (**a**) Suspension; (**b**) Natural sedimentation separation; (**c**) Magnetic field separation.

3.2. Hydrophobicity of MFA@PDMS Sponge

Deionized water dyed with blue ink (aqueous phase) and hexane dyed with oil-soluble black (oil phase) were dropped on the surface of the MFA@PDMS sponge (the volume ratio of hexane to water was 1:1). The experimental results are shown in Figure 6. The deionized water with blue ink formed a typical spherical shape, revealing the superhydrophobicity of the surface of the material. The surface contact angle was calculated to be greater than 150°, while the black hexane penetrated the interior of the parent body. A material with hydrophobic characteristics cannot be wetted by water, but can be wetted by oil. It can be seen that the sponge material had good hydrophobicity to aqueous solution, reflecting its good waterproof performance; this is also key its selectivity for oil–water separation.

Figure 6. Hydrophobic Properties of MFA@PDMS sponge.

3.3. Oil–Water Separation Performance of MFA@PDMS Sponge

3.3.1. Separation Performance for Immiscible Oil/Water

In order to verify the oil–water separation capability of the MFA@PDMS sponge, an experiment was designed (see Figure 7). The MFA@PDMS sponge was placed in a normal hexane aqueous solution containing Sudan red II dye. The hexane solution was then moved to an oily area using a magnet. It was observed that the red normal hexane was quickly absorbed by the sponge (the red circles correspond to the locations of the red hexane), with the oil-bearing water quickly becoming transparent, indicating that that material has excellent oil–water separation capability (the separation efficiency can reach more than 95%), can effectively and quickly remove n-hexane (oil phase) from water, and can be remotely controlled via an external magnetic field, thereby facilitating the separation, recovery and transportation of oils.

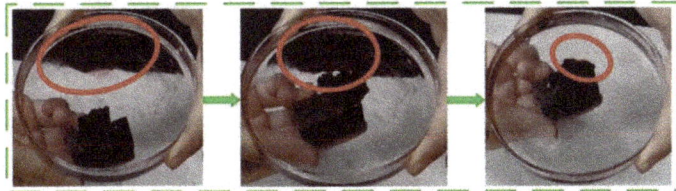

Figure 7. Oil–water separation using the MFA @ PDMS sponge.

According to the experimental process shown in Figure 8, the adsorption times of pure PDMS, FA@PDMS, Fe_3O_4@PDMS and MFA@PDMS sponges for n-hexane were tested; the results are shown in Figure 8. As indicated, the adsorption multiple of the MFA@PDMS sponge was the highest, reaching six times of its own weight, indicating that the addition of MFA significantly improved the oil–water separation effect of the PDMS.

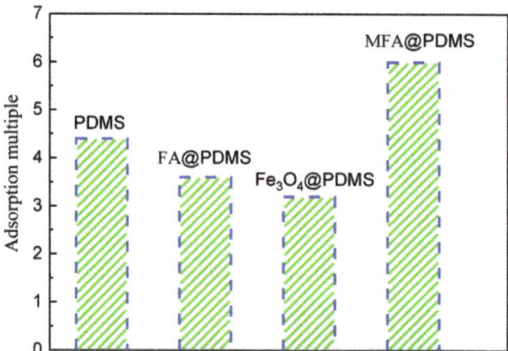

Figure 8. Comparison of adsorption times.

3.3.2. Separation Performance for Oil–Water Emulsion

In general, the particle size of the dispersed phase of oil–water emulsions is much smaller than 20 μm, and a stable protective layer is formed on the surface of droplet molecules. As such, it is difficult to break emulsions, making oil–water separations extremely difficult. In this study, the separation efficiency of MFA@PDMS for different types of emulsified oils (n-hexane, phenol and canola oil) was investigated via the experimental method shown in Figure 9. The results are shown in Figure 10.

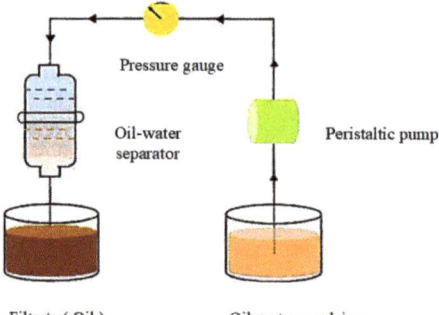

Figure 9. Schematic diagram of oil–water separation process of emulsion mixture.

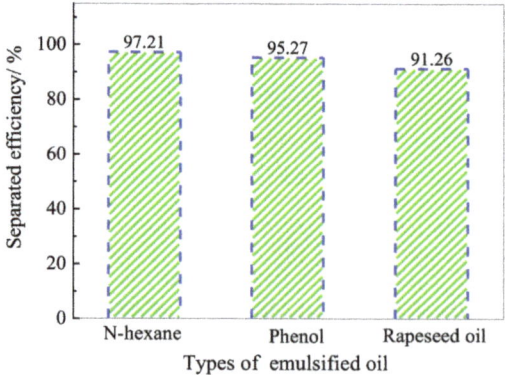

Figure 10. Schematic diagram of emulsion oil–water separation efficiency changes over time.

This oil–water separation device was based on the hydrophobicity and lipophilicity of the MFA@PDMS based material. The MFA@PDMS was clamped in the middle of the left tube and the oil–water mixture was poured into the wide-mouth collection container on the right side. The red oil quickly passed through the MFA@PDMS and flowed into the collection container on the lower left side, while the blue deionized water was blocked on the top side of the MFA@PDMS. Finally, the oil–water mixture was successfully separated. When the oil–water mixture came into contact with the MFA@PDMS surface, its superhydrophobic and super lipophilic properties caused the oil contaminants to penetrate and pass through the material instantly, while the aqueous solution was blocked by its hydrophobic nature. In this study, after each separation, the sample was washed and dried before being reused for the next oil–water separation.

As shown in Figure 10, the separation efficiency of magnetic MFA@PDMS sponge for n-hexane emulsified oil was the highest, reaching 97.12%, followed by phenol, which reached 95.27%. This was due to the unique structure of fly ash, i.e., its large specific surface area, porosity and abundant aluminum-silicon components, as well as the superhydrophobic performance of PDMS, which can attract organic molecules and chain-like structures to the inner surface of sponge materials through π-π dispersion interactions and the donor receptor effect [6].

3.4. Reuse of MFA@PDMS Sponge

The experiment shown in Figure 11 was designed to characterize the reutilization performance of the sponge bionic material. The sponge material was put into n-hexane dyed with Sudan red II, and it was observed that its thickness increased from 0.6 cm to 1.2 cm after only 10 min. We then removed the sponge bionic material and weighed it to calculate its adsorption multiple (i.e., the weight of the material after absorbing the normal hexane/the weight of the material). The oil phase could be extruded by simple compression, and the material could be reused within 30 min after extrusion. Over eight adsorption-extrusion-drying tests, the oil absorption capacity did not decrease significantly, showing good reusability.

Figure 11. Flexible deformation of materials during oil absorption-extrusion-drying process on MFA@PDMS sponge.

An H5KT-type static mechanical tester was used for our stress–strain performance analysis. The compressive stress–strain relationships under single and 20 cyclic quasi-static compressive loadings are shown in Figures 12 and 13, respectively (the compression rate was 50 mm/min).

As shown in Figures 12 and 13, the maximum stress of the material was 446.9 kPa before adsorption on the sponge and 410 kPa afterwards, indicating that the material has good recyclability. The stress–strain curves of 20 compression cycles showed good overlap, and the maximum stress was basically unchanged, indicating that the sponge has good mechanical properties, elasticity, and fatigue resistance.

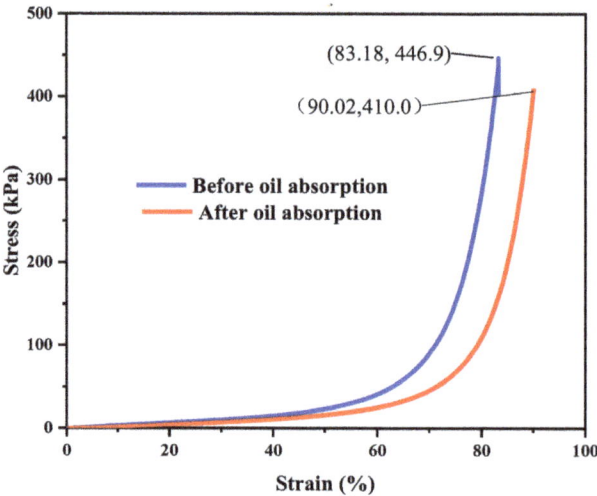

Figure 12. Stress–strain diagram of MFA@PDMS sponge material during a single cycle.

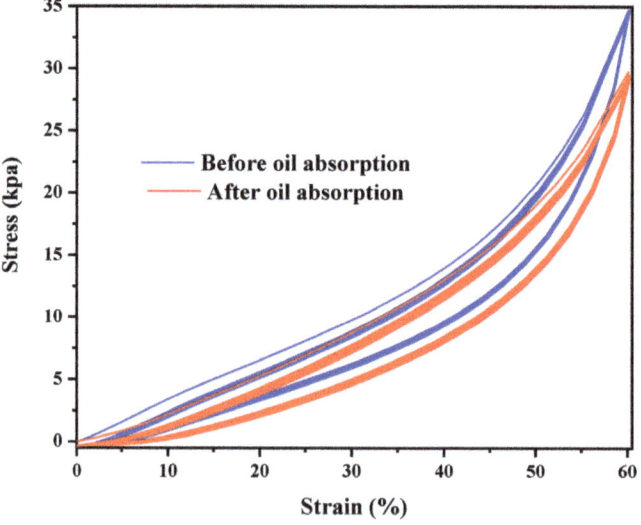

Figure 13. Stress–strain diagram of MFA@PDMS sponge over 20 cycles.

4. Conclusions

In this study, magnetic FA was used to construct a flexible PDMS porous structure, and a MFA@PDMS sponge for oil–water separation was successfully prepared. The sponge was shown to be able to adsorb six times its own weight of n-hexane in only 10 min. It could remotely process oil-bearing waters under the action of an external magnetic field, making it an effective, fast and convenient approach. The MFA@PDMS can be reused by applying a simple oil absorption-extrusion-drying operation which takes around 30 min, after which the oil–water separation performance is not significantly reduced. The stress–strain curves of 20 compression cycles for the sponge showed good overlap, and the maximum stress was basically unchanged, i.e., the sponge was restored to its original shape, indicating that it has good mechanical properties, elasticity, and fatigue resistance.

Author Contributions: Conceptualization and methodology, M.Z. and Y.L.; methodology, M.Z.; formal analysis, M.Z. and X.M.; investigation, M.Z. and X.M.; resources, Y.C. and D.C.; data curation, M.Z. and Y.L.; writing—original draft preparation, M.Z.; writing—review and editing, M.Z. and X.M.; visualization, X.M. and D.C.; supervision, Y.C. and D.C. All authors have read and agreed to the published version of the manuscript.

Funding: This research were funded by the National Natural Science Foundation of China (21968035, U2003123) and the Provincial Natural Science Foundation of Xinjiang Uygur Autonomous Region (2020D01C021).

Institutional Review Board Statement: Not applicable.

Informed Consent Statement: Not applicable.

Data Availability Statement: The data presented in this study are available on request from the corresponding author.

Acknowledgments: The authors would like to thank Meiling Song from Shiyanjia Lab (www.shiyanjia.com (accessed on 23 Mar 2022)) for the VSM analysis.

Conflicts of Interest: The authors declare no conflict of interest.

References

1. Saleh, T.A.; Ali, I. Synthesis of polyamide grafted carbon microspheres for removal of rhodamine B dye and heavy metals. *J. Environ. Chem. Eng.* **2018**, *6*, 5361–5368. [CrossRef]
2. Hoang, A.T.; Nguyen, X.P.; Duong, X.Q.; Huynh, T.T. Sorbent-based devices for the removal of spilled oil from water: A review. *Environ. Sci. Pollut. Res.* **2021**, *28*, 28876–28910. [CrossRef] [PubMed]
3. Talimi, V.; Thodi, P.; Abdi, M.A.; Burton, R.; Liu, L.; Bruce, J. Enhancing harsh environment oil spill recovery using air floatation system. *Saf. Extreme Environ.* **2020**, *2*, 57–67. [CrossRef]
4. Doshi, B.; Sillanpää, M.; Kalliola, S. A review of bio-based materials for oil spill treatment. *Water Res.* **2018**, *135*, 262–277. [CrossRef]
5. Feng, L.; Li, S.H.; Li, Y.S.; Li, H.; Zhang, L.; Zhai, J.; Song, Y.; Liu, B.; Jiang, L.; Zhu, D. Super-hydrophobic surfaces: From natural to artificial. *Adv. Mater.* **2002**, *14*, 1857–1860. [CrossRef]
6. Kesong, L.; Moyuan, C.; Akira, F.; Jiang, L. Bio-Inspired Titanium Dioxide Materials with Special Wettability and Their Applications. *Chem. Rev.* **2014**, *114*, 10044–10094.
7. Shuto, W.; Kesong, L.; Xi, Y.; Jiang, L. Bioinspired surfaces with superwettability: New insight on theory, design, and applications. *Chem. Rev.* **2015**, *115*, 8230–8293.
8. Hoang, A.T.; Nižetić, S.; Duong, X.Q.; Rowinski, L.; Nguyen, X.P. Advanced super-hydrophobic polymer-based porous absorbents for the treatment of oil-polluted water. *Chemosphere* **2021**, *277*, 130274. [CrossRef]
9. Shami, Z.; Amininasab, S.M.; Katoorani, S.A.; Gharloghi, A.; Delbina, S. NaOH-Induced Fabrication of a Superhydrophilic and Underwater Superoleophobic Styrene-Acrylate Copolymer Filtration Membrane for Effective Separation of Emulsified Light Oil-Polluted Water Mixtures. *Langmuir* **2021**, *37*, 12304–12312. [CrossRef]
10. Li, Z.; Wang, B.; Qin, X.; Wang, Y.; Liu, C.; Shao, Q.; Guo, Z. Superhydrophobic/superoleophilic polycarbonate/carbon nanotubes porous monolith for selective oil adsorption from water. *ACS Sustain. Chem. Eng.* **2018**, *6*, 13747–13755. [CrossRef]
11. Li, X.M.; Reinhoudt, D.; Crego-Calama, M. What do we need for a superhydrophobic surface? A review on the recent progress in the preparation of superhydrophobic surfaces. *Chem. Soc. Rev.* **2007**, *36*, 1350–1368. [CrossRef] [PubMed]
12. Wenzel, R.N. Surface roughness and contact angle. *J. Phys. Chem.* **1949**, *53*, 1466–1467. [CrossRef]
13. Baier, R.E. Surface behaviour of biomaterials: The theta surface for biocompatibility. *J. Mater. Sci. Mater. Med.* **2006**, *17*, 1057–1062. [CrossRef]
14. Martin, S.; Bhushan, B. Transparent, wear-resistant, superhydrophobic and superoleophobic poly (dimethylsiloxane)(PDMS) surfaces. *J. Colloid Interface Sci.* **2017**, *488*, 118–126. [CrossRef] [PubMed]
15. Yu, C.; Yu, C.; Cui, L.; Song, Z.; Zhao, X.; Ma, Y.; Jiang, L. Facile preparation of the porous PDMS oil-absorbent for oil/water separation. *Adv. Mater. Interfaces* **2017**, *4*, 1600862. [CrossRef]
16. Chen, X.; Weibel, J.A.; Garimella, S.V. Continuous oil–water separation using polydimethylsiloxane-functionalized melamine sponge. *Ind. Eng. Chem. Res.* **2016**, *55*, 3596–3602. [CrossRef]
17. Wang, J.C.; Li, Y.; Li, H.; Cui, Z.H.; Hou, Y.; Shi, W.; Jiang, K.; Qu, L.; Zhang, Y.P. A novel synthesis of oleophylic Fe_2O_3/polystyrene fibers by gamma-Ray irradiation for the enhanced photocatalysis of 4-chlorophenol and 4-nitrophenol degradation. *J. Hazard. Mater.* **2019**, *379*, 120806. [CrossRef]
18. Wang, B.; Liu, Q.; Fan, Z.; Liang, T.; Tong, Q.; Fu, Y. Fabrication of PDMS/GA Composite Materials by Pickering Emulsion Method and Its Application for Oil-Water Separation. *Energies* **2021**, *14*, 5283. [CrossRef]
19. Yang, R.L.; Zhu, Y.J.; Chen, F.F.; Qin, D.D.; Xiong, Z.C. Synthesis and structure of calcium silicate hydrate (CSH) modified by hydroxyl-terminated polydimethylsiloxane (PDMS). *Constr. Build. Mater.* **2021**, *267*, 120731.

20. Yu, L.; Hao, G.; Gu, J.; Zhou, S.; Zhang, N.; Jiang, W. Fe_3O_4/PS magnetic nanoparticles: Synthesis, characterization and their application as sorbents of oil from waste water. *J. Magn. Magn. Mater.* **2015**, *394*, 14–21. [CrossRef]
21. Javadian, S. Magnetic superhydrophobic polyurethane sponge loaded with Fe3O4@ oleic acid@ graphene oxide as high performance adsorbent oil from water. *Chem. Eng. J.* **2021**, *408*, 127369.
22. Yang, R.L.; Zhu, Y.J.; Chen, F.F.; Qin, D.D.; Xiong, Z.C. Recyclable, fire-resistant, superhydrophobic, and magnetic paper based on ultralong hydroxyapatite nanowires for continuous oil/water separation and oil collection. *ACS Sustain. Chem. Eng.* **2018**, *6*, 10140–10150. [CrossRef]
23. Zhang, Y.; Ren, F.; Liu, Y. A superhydrophobic EP/PDMS nanocomposite coating with high gamma radiation stability. *Appl. Surf. Sci.* **2018**, *436*, 405–410. [CrossRef]

MDPI
St. Alban-Anlage 66
4052 Basel
Switzerland
www.mdpi.com

MDPI Books Editorial Office
E-mail: books@mdpi.com
www.mdpi.com/books

Disclaimer/Publisher's Note: The statements, opinions and data contained in all publications are solely those of the individual author(s) and contributor(s) and not of MDPI and/or the editor(s). MDPI and/or the editor(s) disclaim responsibility for any injury to people or property resulting from any ideas, methods, instructions or products referred to in the content.

www.ingramcontent.com/pod-product-compliance
Lightning Source LLC
LaVergne TN
LVHW070218100526
838202LV00015B/2057